S0-CBG-578

AN INTRODUCTION TO POPULATION GENETICS THEORY

AN INTRODUCTION TO POPULATION GENETICS THEORY

JAMES F. CROW
UNIVERSITY OF WISCONSIN

MOTOO KIMURA
NATIONAL INSTITUTE OF GENETICS
Japan

HARPER & ROW, PUBLISHERS
New York, Evanston, and London

AN INTRODUCTION TO POPULATION GENETICS THEORY

LIBRARY OF CONGRESS CATALOG CARD NUMBER: 78-103913

To Sewall Wright

CONTENTS

5. SELECTION 173

6. POPULATIONS IN APPROXIMATE EQUILIBRIUM 255

PREFACE

This book is intended primarily for graduate students and advanced undergraduates in genetics and population biology. We hope that it will be of value and interest to others also. It is an attempt to present the field of population genetics, starting with elementary concepts and leading the reader well into the field. At first we intended to include experimental work; but this has been largely omitted, partly in the interest of coherence and partly because the book is already long.

The first two-thirds of the book do not require advanced mathematical background. An ordinary knowledge of the calculus will suffice. For the reader who is not familiar with the mathematical and statistical procedures employed, we have added an appendix. The latter parts of the book, which deal with populations stochastically, use more advanced methods. We have made no attempt to explain all of these, either in the text or in the appendix. The reader with only elementary knowledge will have to accept some of the conclusions on faith. We have tried, however, to present the model and the conclusion, leaving it to the reader to follow as much of the intermediate mathematical manipulation as he wishes. We have also added a number of tables and graphs in these chapters so that the major conclusions are available in this form.

There are problems at the end of the first seven chapters. These are more numerous in the early chapters where the reader is more likely to wish for a means to test his knowledge.

The bibliography is longer than is customary. There are many more articles listed than are referred to. We hope in this manner to provide a list which is a useful guide to the literature and which shows the richness and diversity of research in this field.

Without any implication that they share any responsibility for the choice of content or for errors (although they are specifically responsible for the removal of a number of errors), we should like to thank the many people who have helped in various ways. We have benefited greatly from critical comments by Joseph Felsenstein. Others who have helped by suggesting alternative ways of presentation and by pointing out errors are Takeo Maruyama, Laurence Resseguie, Daniel Hartl, Carter Denniston, Thomas Wolfe, Etan Markowitz, and Tomoko Ohta. We should also like to thank the many students who have used in class the notes that were the forerunner of this book; they have been especially helpful in pointing out ambiguities.

We should like to thank both our institutions, the University of Wisconsin and the National Institute of Genetics in Japan, for leaves that permitted us to work together on several occasions either in the United States or in Japan. The Rockefeller Foundation generously provided the initial support to get us started.

Finally, our indebtedness to Professor Sewall Wright will be apparent in every chapter. In many, many instances it is his pioneering work that gave us something to write about. A number of topics that we have treated only lightly, or not at all, are included in his three-volume treatise, "Evolution and the Genetics of Populations."

J.F.C.
M.K.

AN INTRODUCTION TO POPULATION GENETICS THEORY

INTRODUCTION

We are concerned in this book mainly with population genetics in a strict sense. We deal primarily with natural populations and less fully with the rather similar problems that arise in breeding livestock and cultivated plants. The latter subject, sometimes called quantitative or biometrical genetics, emphasizes economically important measurements where the breeding system is under human control. Although this is not neglected, we emphasize more the behavior of genes and population attributes under natural selection where the most important measure is Darwinian fitness.

As do most sciences, population genetics includes both observations and theory. The observations come both from studies on natural populations and from laboratory experiments. Sometimes the observations are verifications of existing theory, sometimes they are tests to distinguish among alternative theories, or they may lead to totally new ideas. Our emphasis is on the theory, but we shall occasionally make use of experimental or observational data, usually for illustration.

The theory of population genetics is largely mathematical. By a biologist's standards it is highly developed, although a theoretical physicist might well regard it as rather primitive. A mathematical theory that could take into account all the relevant phenomena of even the simplest population would be impossibly complex. Therefore it is absolutely necessary to make simplifying assumptions. To a large extent the success or failure of a theory is determined by the choice of assumptions—by the extent to which the model accounts for important facts, ignores trivia, and suggests new basic concepts. The general pattern of this book will be to start with the simplest models and then to extend these to more complicated, but more realistic, formulations.

We have tried to steer a middle course between completely verbal biological arguments and the rigor of the mathematician. We have not hesitated to appeal to the reader's biological intuition and we often have used derivations and proofs that fail to take into account all the mathematical possibilities when these can fairly clearly be ruled out on biological grounds. Furthermore, we have sometimes used models which are somewhat vague, but which seem to us to have considerable biological interest and generality. We frequently use approximations rather than exact expressions, since we are more interested in finding an approximate solution to a model that seems to be biologically interesting than an exact solution to one that is less interesting or realistic. When a choice is necessary we prefer generality and realism to precision and rigor.

We continue the tradition of Sewall Wright and R. A. Fisher in using heuristic arguments that are not rigorously proven, especially in the use of continuous approximations and diffusion models. This involves a risk of later being shown to be wrong, but the history of the physical sciences is on the side of such a strategy. The view is well expressed by Richard Feynman in his Nobel Prize lecture (*Science* 153: 699, 1966):

> In the face of the lack of direct mathematical demonstration one must be careful and thorough to make sure of the point, and one should make a perpetual attempt to demonstrate as much of the formula as possible. Nevertheless, a very great deal more truth can become known than can be proven.

1 MODELS OF POPULATION GROWTH

Theoretical population genetics is concerned with model building. Any model of nature is an oversimplification, as is any verbal description of a natural process. The model is an attempt to abstract from nature some significant aspect of the true situation.

The models employed in population genetics are mathematical. The model is always unsatisfactory in some respects. Inevitably, it is unable to reflect all the complexities of the true situation. On the other hand, it is usually true that the more closely the model is made to conform to nature the more unmanageable it becomes from the mathematical standpoint. If it is as complex as the true situation, it is not a model. We have to choose some sort of compromise between a model that is so crude as to be unrealistic or misleading and one that is incomprehensible or too complex to handle. Those men who have laid the mathematical foundation for the theory of population genetics—J. B. S. Haldane, R. A. Fisher, and Sewall Wright—have had the capacity of inventing mathematical models that extracted the essence of the situation in a formulation that could be handled mathematically. With increased mathematical sophistication, a more comprehensive and rigorous theory can be developed and much of the current research in theoretical population genetics is concerned with such developments.

Just as the economist considers the broad consequences of individual transactions in a more or less free market, the population geneticist is interested in the overall consequences of a large number of events—births, deaths, choice of mates, and the host of individual circumstances, habits, decisions, and accidents that determine these. As the physicist or chemist works with the statistical averages of molecular behavior and does not try to describe the behavior of each individual molecule, the population geneticist tries to describe the overall effect of a large number of individual events.

It is convenient to divide mathematical models of population structure into two kinds, deterministic and stochastic. With a deterministic model, the population is assumed to be large enough and the factors determining individual birth rates and death rates constant enough that the consequences of random fluctuations can be ignored. This would be true only for an infinite population under highly idealized conditions; but actually many populations are large enough that the "noise" introduced into the system by random processes is small enough in relation to systematic factors that, for the degree of approximation needed, it may be ignored. Deterministic models are much easier to handle mathematically, as will be abundantly clear throughout the book.

Stochastic models take account of the effects of the finiteness of the population and other random elements. Some populations are small enough or the conditions are variable enough that random fluctuations are appreciable. The difficulty is with the mathematical complexity. Only the simplest situations have exact solutions. However, it was the genius of R. A. Fisher (1930, 1958) and Sewall Wright (1931, 1945, 1960) to devise procedures that provided very accurate approximations which have led to deep biological insights. Recently, more sophisticated mathematical techniques and electronic computers have been used to give more exact and extensive results.

In this chapter we describe briefly some of the deterministic models that have been used in population genetics. Genetics is ordinarily concerned more with the relative frequencies of different genes and genotypes in a population than with the size of the total population. Nevertheless, in this chapter we shall consider the population as a whole. The purpose is to introduce the models rather than make use of them in genetic analysis; they will be used later.

We introduce four deterministic models:

1. Discrete, nonoverlapping generations,

2. Continuous random births and deaths,
3. Overlapping generations, discrete time intervals,
4. Overlapping generations, continuous change.

In this book we shall make use of only the first two for genetic problems. Thus far the more realistic models 3 and 4 have been used mainly by ecologists and demographers. But the increasing mutuality of interest of population geneticists, demographers, and ecologists forecasts a greater emphasis on these models in population genetics.

1.1 Model 1: Discrete, Nonoverlapping Generations

This is in many ways the simplest description of population growth. We assume that the parent generation reproduces and that, before the offspring reach reproductive age, the parents have all died (or at least are no longer counted). Time is measured most conveniently in units of generations.

This model is a realistic description of some populations, such as annual plants. For others it may still be a useful first approximation. It is widely used because of its mathematical simplicity.

Let N_t be the number of individuals in the population at time t, measured in generations. If the average number of progeny per individual is w, the population number in generation t can be expressed in terms of the number in the previous generation, $t - 1$, by

$$N_t = wN_{t-1}. \qquad 1.1.1$$

We regard w as a measure of both survival and reproduction; individuals who do not survive to the reproductive age are counted as leaving 0 progeny. We call w the Darwinian fitness, or simply the fitness. Each generation must be counted at the same age. It is often convenient to count the population as zygotes, so that the survival and reproduction of an individual occur within the same generation.

The relation between N_{t-1} and N_{t-2} is the same as that between N_t and N_{t-1}. Therefore, if w remains constant

$$N_t = w(wN_{t-2}) = w^2 N_{t-2}.$$

Continuing this process, $N_t = w^3 N_{t-3}$, and finally

$$N_t = N_0 w^t, \qquad 1.1.2$$

where N_0 is the number in the population in generation 0.

The change in population size is analogous to money invested at compound interest. If $w = 1 + s$, then s is equivalent to the interest rate. If $w > 1$, or $s > 0$, the population is increasing; alternatively, it is decreasing if $w < 1$ or $s < 0$.

Equation 1.1.1 may be written

$$\Delta N_t = sN_t, \tag{1.1.3}$$

where ΔN_t is $N_{t+1} - N_t$. This shows that the proportion by which the population changes in one generation, $\Delta N_t / N_t$, is given by s.

If the population is composed of several kinds of individuals (perhaps several different genotypes) with different fitnesses, then the whole population is increasing or decreasing at the same rate as if all the individuals were of average fitness, the appropriate average being the weighted arithmetic mean. This is easily demonstrated as follows:

Let $n_1, n_2, \ldots, n_k$ be the numbers of each of k types in the population in generation $t-1$, and let the fitnesses of these types be $w_1, w_2, \ldots, w_k$. Assume that $n_1 + n_2 + \cdots + n_k = N$ and let,

$$\overline{w} = \frac{n_1 w_1 + n_2 w_2 + \cdots + n_k w_k}{n_1 + n_2 + \cdots + n_k} = \frac{\Sigma n_i w_i}{N_{t-1}}.$$

Next generation there will be $w_1 n_1$ of type 1, $w_2 n_2$ of type 2, and so on. Thus, letting $\overline{w} = \Sigma n_i w_i / \Sigma n_i$,

$$\begin{aligned} N_t &= w_1 n_1 + w_2 n_2 + \cdots + w_k n_k = \Sigma w_i n_i \\ &= \overline{w} N_{t-1}. \end{aligned} \tag{1.1.4}$$

If the fitnesses of the different types are inherited, those with greater fitness will be represented in greater proportion in the next generation. Hence $\overline{w}$ may be expected to change from generation to generation. In Chapter 5 the amount of this change will be discussed.

A related question is this: What happens when $\overline{w}$ is not constant from generation to generation? In this case, let $\overline{w}_1, \overline{w}_2, \ldots, \overline{w}_t$ be the value of $\overline{w}$ in the different generations. Then

$$\begin{aligned} N_t &= \overline{w}_{t-1} N_{t-1} = \overline{w}_{t-1} \overline{w}_{t-2} N_{t-2} = \overline{w}_{t-1} \overline{w}_{t-2} \cdots \overline{w}_0 N_0 \\ &= N_0 \prod_{0}^{t-1} \overline{w}_i \\ &= N_0 G_{\overline{w}}^t, \end{aligned} \tag{1.1.5}$$

where $G_{\overline{w}}$ is the geometric mean of the w's (see A.1.2).

To summarize: With a discrete generation model and variable fitnesses, the weighted arithmetic mean is appropriate for contemporary differences, while the geometric mean is appropriate for averaging over different generations.

1.2 Model 2: Continuous Random Births and Deaths

In this model we regard the population number N_t at time t as being very large. In an infinitesimal time interval Δt a fraction $b\Delta t$ of the population produce an offspring and a fraction $d\Delta t$ die. Thus the change in population number during this interval is

$$\begin{aligned} \Delta N_t &= (b - d)N_t \, \Delta t \\ &= mN_t \, \Delta t, \end{aligned} \qquad \text{1.2.1}$$

where $m = b - d$.

As $\Delta t \to 0$, this becomes

$$\frac{dN_t}{dt} = mN_t. \qquad \text{1.2.2}$$

In integrated form, $mt = \log_e N_t - \log_e N_0$, or

$$N_t = N_0 e^{mt}. \qquad \text{1.2.3}$$

Time can be measured in any convenient units—hours, days, weeks, or years; in any case m is measured in the reciprocals of the same units (per hour, per day, etc.).

In its strictest form this model applies only to such situations as bacterial growth in an unrestricted environment. Each individual is regarded as equally likely to die or reproduce in any instant. On the other hand, most populations are neither decreasing nor increasing very rapidly. The age distribution is often very near to equilibrium so that average birth and death rates are nearly constant. Under these circumstances the model may be a very good approximation. This is especially true if we are interested in changes in proportions of different types and such changes are slow compared with the average life span of an individual.

Comparison of 1.1.2 with 1.2.3 shows the correspondence between w and m. If $w = e^m$, or $m = \log_e w$, the formulae are equivalent. Therefore if the time unit is chosen so that $m = \log_e w$, the continuous population changes such that, if counted at time intervals corresponding to these units, the numbers correspond to the discrete population. Thus measuring time in such units makes t in this sense equivalent to the number of generations in a discrete population.

If w is nearly 1, and therefore s is very small, then $\log_e (1 + s)$ is very nearly s. Thus for very slow change s and m are very similar.

The comparison of the two systems may be made clearer, perhaps, with a concrete example. Consider first a discrete population that reproduces once each year, each parent giving rise to two progeny and dying immediately after (or at least not being counted in the ensuing generation). Thus $\bar{w} = 2$. The population doubles each year; $N_t = 2N_{t-1}$ and $N_t = N_0 2^t$,

A second population is growing continuously at such a rate as to double in a year's time. $N_t = N_0 2^t$. From 1.2.3, $e^m = 2$, or $m = \log_e 2 = .693$. Thus, $N_t = N_0 e^{.693t}$.

A third population is also growing continuously, but has a growth rate equivalent to 100% interest compounded continuously. That is to say, $m = 1$ and $dN_t/dt = N_t$. Integrating, $N_t = N_0 e^t$; each year the population increases by a factor $e = 2.71$.

More generally, the relations of the w's, s's, and m's of the three populations are:

Population 1: $w = 1 + s$,
Population 2: $m = \log_e w = \log_e(1 + s)$,
Population 3: $m = s$.

Figure 1.2.1 shows these relationships graphically.

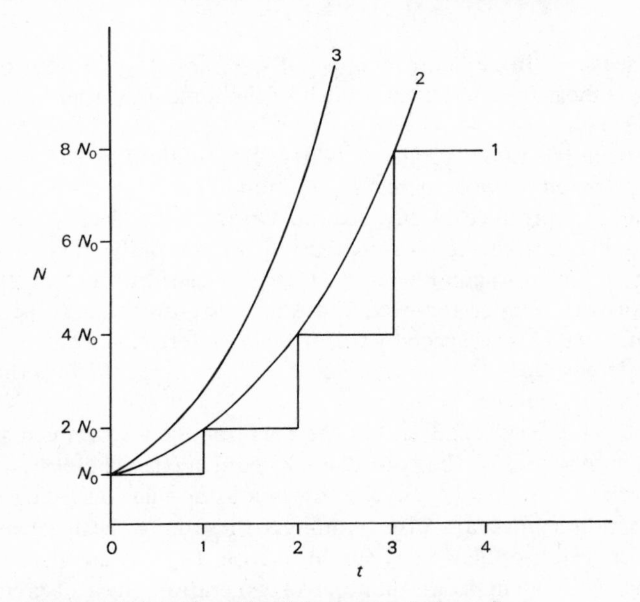

Figure 1.2.1. Growth of three populations. In population 1 there are discrete generations with growth rate $w = 1 + s$, such that $N_t = N_0 w^t$. In population 2 growth is continuous such that $N_t = N_0 e^{mt}$ and $m = \log_e w$. In population 3 growth is continuous such that $N_t = N_0 e^{mt}$ and $m = s$. The graphs are drawn for $s = 1$.

We shall call w the *fitness*, or the *Wrightian fitness*, after Sewall Wright, R. A. Fisher (1930, 1958) designated m as the *Malthusian parameter*. Although used here as measures of the absolute rate of increase, w and m are also used for the relative rates of different types (see Chapter 5).

Notice that if there are no births we can write 1.2.1 as

$$dN_t = \hat{d} N_t \, dt,$$

which, on integration, is

$$N_t = N_0 e^{-\hat{d}t}.$$

(The death rate is written with a caret over the d to distinguish it from the derivative sign.) Hence the life expectancy curve is exponential with each survivor having a constant probability of surviving for another time unit.

If there are different fitnesses of the different types in the population, 1.2.2 becomes

$$\begin{aligned} \frac{dN}{dt} &= m_1 n_1 + m_2 n_2 + \cdots + m_k n_k \\ &= \bar{m} N, \end{aligned} \tag{1.2.4}$$

where

$$\bar{m} = \frac{\Sigma n_i m_i}{\Sigma n_i} = \frac{\Sigma n_i m_i}{N}.$$

Hence, as with w in the discrete model, it is appropriate to replace m with the weighted average of the m's in the population.

When $\bar{m}$ varies in time we can think of time as broken into k intervals of length Δt during each of which $\bar{m}$ is different. Then 1.2.3 becomes

$$\begin{aligned} N_t &= N_0 (e^{\bar{m}_1 \Delta t_1})(e^{\bar{m}_2 \Delta t_2}) \cdots (e^{\bar{m}_k \Delta t_k}) \\ &= N_0 e^{\bar{m}_1 \Delta t_1 + \bar{m}_2 \Delta t_2 + \cdots + \bar{m}_k \Delta t_k} \\ &= N_0 e^{\bar{\bar{m}} t}, \end{aligned} \tag{1.2.5}$$

where

$$\bar{\bar{m}} = \frac{\Sigma \Delta t_i \bar{m}_i}{\Sigma \Delta t_i} = \frac{\Sigma \Delta t_i \bar{m}_i}{t}.$$

Thus, in contrast to w in the discrete model, with a continuous model the appropriate average is the arithmetic mean of the m's. This is reasonable, of course, since if $m = \log w$, the arithmetic mean of m is the log of the geometric mean of w.

As mentioned earlier, if there are different types in the population with different m's and if the fitnesses are heritable, then the average value of m

will change from generation to generation as the fitter types make a disproportionate contribution to future generations. This is another way of saying that natural selection will generally lead to an increase in the average fitness.

Our intuition tells us that the amount by which the average fitness changes per unit time will be related to the variability in fitness among the different types in the population. We shall now show that the rate of increase is given by the mean squared deviation or the variance.

Consider, as before, that there are k types with frequencies $n_1, n_2, \ldots, n_k$ and fitnesses $m_1, m_2, \ldots, m_k$ measured in Malthusian parameters. Assume that the fitness in each type is perfectly heritable and that the populations do not interbreed. This would be appropriate to separate species, or to asexually propagating clones. The formulae would also apply to a single locus in a haploid population, to a plasmon, or to competition between self-fertilized strains

We wish to know how $\overline{m}$ changes with time. From the definition of $\overline{m}$, given by 1.2.4,

$$\frac{d\overline{m}}{dt} = \frac{d}{dt}\left[\frac{\Sigma m_i n_i}{N}\right]. \tag{1.2.6}$$

The m_i's are constant, but the n_i's are variable; therefore the weighted average $\overline{m}$ and $N = \Sigma n_i$ are also variable. From the rules for differentiation,

$$\frac{d\overline{m}}{dt} = \frac{N\Sigma m_i \dfrac{dn_i}{dt} - (\Sigma m_i n_i)\dfrac{dN}{dt}}{N^2}.$$

But, from 1.2.4,

$$\frac{dn_i}{dt} = m_i n_i$$

and

$$\frac{dN}{dt} = \overline{m}N.$$

Substituting these and recalling that $\Sigma m_i n_i = N\overline{m}$, we obtain

$$\frac{d\overline{m}}{dt} = \frac{\Sigma n_i m_i^2 - N\overline{m}^2}{N} = V_m, \tag{1.2.7}$$

where V_m is the variance of the m_i's (see A.2.4).

We have just demonstrated a rather special example of R. A. Fisher's "Fundamental Theorem of Natural Selection," which he presented in 1930. It says that the rate of change in fitness at any instant, measured in Malthusian parameters, is equal to the variance in fitness at that time. The variance in this case is the variance in fitness among the types. In a Mendelian population with different genotypes there will be crossing among the types. As expected, the rate of change in fitness is then determined not by the total variance, but by the part of the variance that is associated with transmissible gene differences. The way in which this variance may be computed will be shown in Chapter 5.

1.3 Model 3: Overlapping Generations, Discrete Time Intervals

We now consider a more complicated model. The time is divided into discrete units, which are usually short relative to the life span. The model is especially appropriate for species, such as many birds and mammals, which have a specific breeding season, but which may survive for several such seasons. It is also a good approximation for organisms, such as man, where the population, although changing continuously, is censused at discrete intervals. In man, time would ordinarily be measured in years or in 5-year intervals. We follow closely the procedures of Leslie (1945, 1948) as given by Moran (1962, p. 7).

Let n_{xt} be the number of individuals of age x at time t. More precisely, "age x" means "in the age interval x to $x + 1$" as in ordinary discourse. In bisexual forms it is convenient to adopt some convention in regard to the sexes; one such is to count only the females. Let p_x be the probability of surviving from age x to age $x + 1$. Likewise, let b_x be the average number of progeny produced by an individual of age x, counting only those progeny that survive long enough to be counted in the next time interval. To be concrete, we shall discuss the model as if time were measured in years.

For convenience we shall assume that no individual lives more than 5 years, although the procedure is obviously capable of being extended for any number of time intervals. Starting from $t = 0$, the number of individuals of age 0 at time 1 will be the number born to parents of age 0, $n_{00} b_0$, plus those born to parents of age 1, $n_{10} b_1$, and so on. In turn, the number of age 1 will be the number who were age 0 at time 0 multiplied by the probability of surviving from age 0 to age 1, $n_{00} p_0$, and so on for the other ages. Writing these as recurrence relations, we obtain

$$
\begin{aligned}
n_{01} &= n_{00} b_0 + n_{10} b_1 + n_{20} b_2 + n_{30} b_3 + n_{40} b_4, \\
n_{11} &= n_{00} p_0, \\
n_{21} &= n_{10} p_1, \\
n_{31} &= n_{20} p_2, \\
n_{41} &= n_{30} p_3,
\end{aligned}
$$

and

$$n_5 = 0. \tag{1.3.1}$$

From these equations the composition of the population a year later can be worked out. For example,

$$n_{02} = n_{01}b_0 + n_{11}b_1 + n_{21}b_2 + n_{31}b_3 + n_{41}b_4,$$
$$n_{12} = n_{01}p_0,$$
$$n_{22} = n_{11}p_1,$$
$$n_{32} = n_{21}p_2,$$
$$n_{42} = n_{31}p_3,$$

and so on for later years.

Table 1.3.1 gives data on the composition of a population in successive years, starting with the highly artificial situation of 10,000 individuals of

Table 1.3.1. The age composition of a population starting with 10,000 new born. The survival probabilities at age 0, 1, 2, 3, and 4 are .8, .9, .9, .7, and 0; the birth rates for the same ages are 0, .3, .5, .5, and .2. A stable age distribution and a constant rate of population growth are attained in about 20–25 units of time.

TIME	PROPORTION OF AGE					POPULATION	N_t
t	0–1	1–2	2–3	3–4	4–5	NUMBER, N_t	$\overline{N_{t-1}}$
0	1.000	0	0	0	0	10,000	
1	.000	1.000	.000	.000	.000	8,000	.800
2	.250	.000	.750	.000	.000	9,600	1.200
3	.300	.160	.000	.540	.000	12,000	1.250
4	.294	.222	.133	.000	.350	12,960	1.080
5	.268	.310	.264	.158	.000	9,835	.759
6	.265	.187	.244	.207	.097	11,267	1.146
7	.288	.203	.161	.210	.139	11,785	1.046
8	.280	.235	.187	.148	.151	11,532	.979
9	.275	.230	.217	.172	.106	11,247	.975
10	.277	.214	.201	.190	.117	11,553	1.027
12	.278	.225	.198	.172	.126	11,627	.995
14	.278	.220	.198	.180	.124	11,763	1.0092
16	.278	.222	.199	.177	.124	11,847	1.0023
18	.278	.221	.198	.178	.124	11,967	1.0054
20	.278	.222	.199	.178	.124	12,067	1.0041
25	.2782	.2222	.1985	.1779	.1239	12,343	1.0045
Limit	.2782	.2222	.1985	.1778	.1239	∞	1.0045

age 0. The survival probabilities are taken to be $p_0 = .8$, $p_1 = .9$, $p_2 = .9$, $p_3 = .7$, $p_4 = 0$. The birth rates are $b_0 = 0$, $b_1 = .3$, $b_2 = .5$, $b_3 = .5$, and $b_4 = .2$. The information is shown graphically in Figure 1.3.1. Note that the population age distribution fluctuates for several generations, and then reaches a constant proportion for each age. At this time the population is increasing at a constant rate of 0.45% per year.

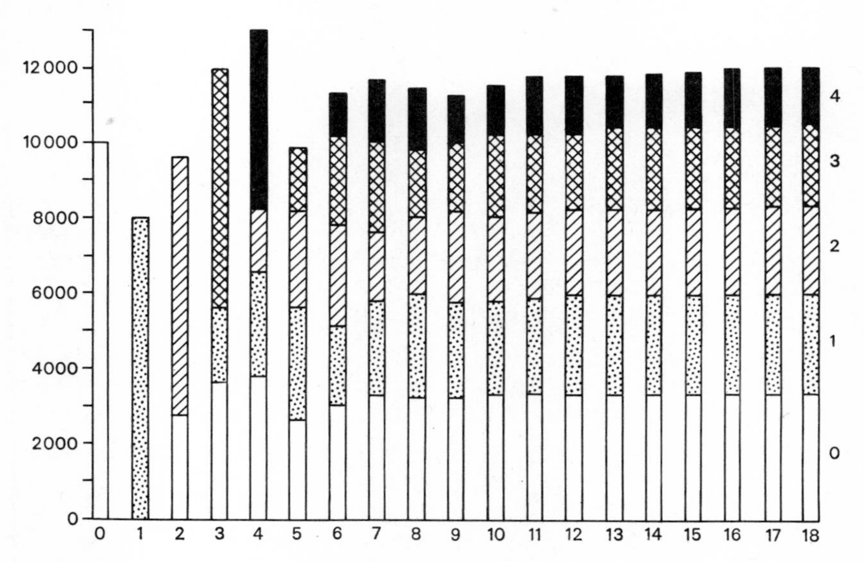

Figure 1.3.1. The total number and age distribution of a population starting with 10,000 individuals of age 0. The survival probabilities at the 5 ages are $p_0 = .8$, $p_1 = .9$, $p_2 = .9$, $p_3 = .7$, $p_4 = 0$. The birth rates at the corresponding ages are $b_0 = 0$, $b_1 = .3$, $b_2 = .5$, $b_3 = .5$ and $b_4 = .2$. After wild fluctuations in the first few generations the population settles to a constant age distribution and a rate of increase of 0.45% per unit time. The total population scale is on the left ordinate, the age at the right. The number of time units is given along the bottom.

The example illustrates the general principle that any population with a fixed schedule of age-specific birth and death rates eventually reaches a characteristic age distribution which remains stable. When this stage is reached the growth rate is constant and equations 1.1.1 and 1.1.2 are applicable.

We can carry out a more complete analysis, including a demonstration that stability of age distribution is eventually attained, by using matrix notation. The necessary techniques are given in Sections 7 and 8 of the Appendix.

In matrix form, the relations 1.3.1 become

$$\begin{pmatrix} n_{01} \\ n_{11} \\ n_{21} \\ n_{31} \\ n_{41} \end{pmatrix} = \begin{pmatrix} b_0 & b_1 & b_2 & b_3 & b_4 \\ p_0 & 0 & 0 & 0 & 0 \\ 0 & p_1 & 0 & 0 & 0 \\ 0 & 0 & p_2 & 0 & 0 \\ 0 & 0 & 0 & p_3 & 0 \end{pmatrix} \begin{pmatrix} n_{00} \\ n_{10} \\ n_{20} \\ n_{30} \\ n_{40} \end{pmatrix} \qquad 1.3.2$$

or, in abbreviated notation,

$$\boldsymbol{N_1} = \boldsymbol{MN_0},$$

from which

$$\boldsymbol{N_t} = \boldsymbol{M^t N_0}. \qquad 1.3.3$$

Thus the composition of any future year can be obtained by successive multiplication of matrices.

The characteristic equation (see A.8) is

$$|\boldsymbol{M} - \lambda \boldsymbol{I}| = 0, \qquad 1.3.4$$

which on expansion gives

$$\lambda^5 - b_0 \lambda^4 - p_0 b_1 \lambda^3 - p_0 p_1 b_2 \lambda^2 - p_0 p_1 p_2 b_3 \lambda - p_0 p_1 p_2 p_3 b_4 = 0. \qquad 1.3.5$$

We now let λ_1 stand for the positive real root of this equation. From the fact that all coefficients after the first are negative we know that there is one and only one positive real root. It is also the largest in absolute value (for a demonstration of this see Moran, 1962, p. 8). It can be found from 1.3.5 by trial and error.

Consider now the equation

$$\boldsymbol{Ml} = \lambda_1 \boldsymbol{l}, \qquad 1.3.6$$

where $\boldsymbol{M}$ is the original matrix and $\boldsymbol{l}$ is a column vector (corresponding to l, m, n in equation A.8.3). The solutions for the l's can be gotten, after λ_1 is determined, from 1.3.6. We assume that the roots are unequal.

When t is large λ_1^t is much larger than $\lambda_2^t, \lambda_3^t, \ldots,$ so that we can write (see A.8.9)

$$\left.\begin{aligned} n_{0t} &= Cl_0 \lambda_1^t \\ n_{1t} &= Cl_1 \lambda_1^t \\ n_{4t} &= Cl_4 \lambda_1^t \end{aligned}\right\} \text{ approximately,} \qquad 1.3.7$$

where C is a constant determined by the initial conditions.

The population thus has a stationary age distribution with individuals of age 0, 1, 2, 3, and 4 in the ratio $l_0 : l_1 : l_2 : l_3 : l_4$. Notice that these ratios do not depend on the initial age distribution.

To solve for the l's we write 1.3.6 in the form

$$\begin{aligned} l_0 \lambda_1 &= l_0 b_0 + l_1 b_1 + l_2 b_2 + l_3 b_3 + l_4 b_4, \\ l_1 \lambda_1 &= l_0 p_0, \\ l_2 \lambda_1 &= l_1 p_1, \\ l_3 \lambda_1 &= l_2 p_2, \\ l_4 \lambda_1 &= l_3 p_3. \end{aligned}$$

From the second equation, $l_1 = l_0 p_0 / \lambda_1$. From the third, $l_2 = l_1 p_1 / \lambda_1 = l_0 p_0 p_1 / \lambda_1^2$. Continuing this pattern, we obtain the solutions

$$\begin{aligned} l_0 &= l_0, \\ l_1 &= l_0 p_0 \lambda_1^{-1}, \\ l_2 &= l_0 p_0 p_1 \lambda_1^{-2}, \\ l_3 &= l_0 p_0 p_1 p_2 \lambda_1^{-3}, \\ l_4 &= l_0 p_0 p_1 p_2 p_3 \lambda_1^{-4} \end{aligned} \tag{1.3.8}$$

The l's give the ratios of the numbers in the different age groups. These can be converted into proportions by dividing by the total number.

Notice that when $\lambda_1 = 1$, the stable age distribution is determined entirely by the survival rates. For example, the number of age 4 is proportional to $p_0 p_1 p_2 p_3$, which is the probability of survival to this age.

We now inquire into the biological meaning of λ_1, the largest root. When the age distribution is stabilized, then all age groups change at the same rate. By summing 1.3.7 we obtain

$$\Sigma n_{it} = C(\Sigma l_i) \lambda_1^t.$$

But, when $t = 0$, $\Sigma n_{i0} = N_0$, the initial population number. Therefore $C \Sigma l_i = N_0$, and we can write

$$N_t = N_0 \lambda_1^t, \tag{1.3.9}$$

for a population with a stable age distribution.

Comparison with equation 1.1.2 reveals the meaning of λ_1. This corresponds to w, the population fitness. Thus, when a stable age distribution is attained, the whole population (and each age group within it) changes by a factor λ_1 in each time unit.

We have considered only five age classes. The generalization to any number of classes is obvious. A more general and rigorous discussion of this model is given by Moran (1962, p. 7ff).

This section brings out the circumstances under which the simple equations of Section 1 are applicable. They clearly apply when the generations are discrete and nonoverlapping. But we see also that if the population has had a given schedule of birth and death rates for a time long enough to reach age-distribution stability, then the equations of Section 1 are appropriate. Many populations in nature change in total size at a rate that is slow relative to the life duration; under these circumstances there would be an approximate age-distribution stability.

In summary, a population with a fixed schedule of age-specific birth and death rates eventually attains a state when the numbers of individuals in the various age groups are in a fixed ratio. The ratio of numbers in the different age groups at this stage is given by the l's of equation 1.3.8. At this stage the population is changing in total size at a rate determined by λ_1, the largest eigenvalue. When λ_1 is greater than 1 the population is increasing; when it is less than 1, the population is decreasing; when $\lambda_1 = 1$, the population is stable in size.

The hypothetical population in Table 1.3.1 and Figure 1.3.1 has a value of $\lambda_1 = 1.00449$, which corresponds to an eventual rate of increase of about 0.45% per year.

Table 1.3.2 gives the transition matrix for the female population of the United States in 1964, from Keyfitz and Murphy (1967). The survival and

Table 1.3.2. The transition matrix for the female population in the United States for birth and death rates of 1964 (Keyfitz and Murphy, 1967). The values are for 5-year periods. The principal diagonal gives the probability of surviving from one period to the next and the first row gives the number of female births per woman during this age period.

				AGE					
0–4	5–9	10–14	15–19	20–24	25–29	30–34	35–39	40–44	45–49
.0000	.0010	.0878	.3487	.4761	.3377	.1833	.0761	.0174	.0010
.9966	0	0	0	0	0	0	0	0	0
0	.9983	0	0	0	0	0	0	0	0
0	0	.9979	0	0	0	0	0	0	0
0	0	0	.9968	0	0	0	0	0	0
0	0	0	0	.9961	0	0	0	0	0
0	0	0	0	0	.9947	0	0	0	0
0	0	0	0	0	0	.9923	0	0	0
0	0	0	0	0	0	0	.9887	0	0
0	0	0	0	0	0	0	0	.9831	0

fertility values are for 5-year periods. The dominant root, λ_1, is 1.082. This tells us that when a population with this structure of age-specific birth and death rates reaches a stable age distribution it will be increasing at a rate of 8.2%. This is the rate per 5-year time unit, so the annual rate of increase would be about one-fifth of this, or 1.6%.

An ordinary mortality or "life" table gives the proportion remaining alive at various ages with a given schedule of births and deaths. From equations 1.3.8 we know that the table gives the relative numbers in each age group in an age-stable population that is neither increasing nor decreasing ($\lambda_1 = 1.000$). A growing population will have more in the younger age groups; a decreasing population will have more in the older age groups.

1.4 Model 4: Overlapping Generations, Continuous Change

The methods of the previous section are specifically adapted to an organism, such as many annually reproducing plants, where reproduction is seasonal but where the typical individual may live through several such seasons. In many other organisms, ranging from short-lived insects to man, births as well as deaths occur at all times. If the population is large the discreteness introduced by individual births and deaths is lost in the large total and the process of population change can be regarded as essentially continuous. Mathematically we think of the process as the limit of the previous method as the age intervals get smaller and smaller.

The equations for dealing with such a population were developed by Lotka (1925, 1956) and Fisher (1930, 1958). Our procedure follows Fisher (1958, pp. 25–27). To be concrete, we shall think of the human population where time is measured in years.

We let $l(x)$ be the probability of survival from birth to age x. (We do not intend to imply that x can take only integral values; at some instant a person may be 27.4658 years old.) Let the probability of reproducing during the infinitesimal age interval from x to $x + dx$ be $b(x)\,dx$. Then the probability of living to age x and reproducing during the next time interval dx is $l(x)b(x)\,dx$. The expected number of offspring per individual for his whole lifetime is this quantity summed over all ages, or

$$\int_0^A l(x)b(x)\,dx \qquad 1.4.1$$

where A is the highest age at which reproduction is possible. For convenience mathematically we replace A with ∞, which doesn't change anything since $b(x)$ is 0 for all ages beyond A. If the integral is greater than 1 the population will eventually increase, although there may be temporary decreases because of changes in the age distribution; if the quantity is less than 1 the population will eventually decrease.

In the previous section we saw that a population with a constant set of age-specific birth and death rates attains a stable age distribution. In this state the population will increase or decrease at a constant rate. To correspond to λ or w in the discrete case we use e^m in the continuous case, where m is the Malthusian parameter (see Section 1.2). We now wish to see how m can be determined from the birth and death rates.

Of the population now alive (at time t, say), those of age x were born x years ago at time $t - x$. Let the number of births occurring in the interval dt now be $B(t)dt$. $B(t)$ is the instantaneous birth rate at time t; if this rate continued for a year, there would be $B(t)$ births. The birth rate x years ago was $B(t - x)$. Of those born at that time a fraction $l(x)$ will still be alive and of these a fraction $b(x)\,dx$ will give birth during the interval dx. Thus the current birth rate of persons of age x will be $B(t - x)l(x)b(x)$, and this, summed over all ages, is the total birth rate at time t. This gives us

$$B(t) = \int_0^\infty B(t - x)l(x)b(x)\,dx. \qquad \textbf{1.4.2}$$

If the population has achieved age-distribution stability, its size (and therefore its birth rate) is increasing at a rate m. In x years the birth rate will have increased by a factor e^{mx}. Therefore, the rate x years ago was fraction e^{-mx} of the current rate. That is to say,

$$B(t - x) = B(t)e^{-mx}.$$

Substituting this into 1.4.2 gives, after cancelling $B(t)$ on both sides of the equation,

$$1 = \int_0^\infty e^{-mx}l(x)b(x)\,dx. \qquad \textbf{1.4.3}$$

This provides a means for calculating m if $l(x)$ and $b(x)$ are known for all ages.

This equation is analogous to 1.3.5. Perhaps the analogy is more apparent if we divide by λ^5 and rewrite 1.3.5 as

$$1 = \lambda^{-1}b_0 + \lambda^{-2}p_0 b_1 + \lambda^{-3}p_0 p_1 b_2 + \cdots. \qquad \textbf{1.4.4}$$

Just as 1.4.4 has only one real root, so does 1.4.3. In these equations λ^{-x} corresponds to e^{-mx}, $l(x)$ to $p_0 p_1 p_2 \cdots p_{x-1}$, and $b(x)$ to b_x. In actual practice, m is usually computed from discrete data, since birth and death rates are usually given at discrete (e.g., yearly) intervals.

When a population has reached age-distribution stability the formulae of Section 1.2 are applicable, in particular 1.2.2 and 1.2.3, so

$$\frac{dN}{dt} = mN \qquad \text{and} \qquad N_t = N_0 e^{mt}. \tag{1.4.5}$$

The Malthusian parameters corresponding to the birth and death rates in the United States in recent years are given in Table 1.4.1. As mentioned

Table 1.4.1. The Malthusian parameters of the United States population corresponding to birth and death rates in particular years. This is the rate at which the population would increase if the population came to age-distribution equilibrium with these rates of births and deaths at each age. Source is U.S. Census data.

DATE	MALTHUSIAN PARAMETER
1940	0.0010
1945	0.0045
1950	0.0137
1955	0.0198
1960	0.0208
1961	0.0205
1962	0.0188
1963	0.0171
1964	0.0156
1965	0.0121
1966	0.0097
1967	0.0074

earlier, these give the rate at which the population would increase if it were in equilibrium for the age distribution. The value 0.016 obtained earlier for 1964 data is seen to be in close agreement despite being based on analysis of 5-year units. The changes in the Malthusian parameter as a result of changing death and, more importantly, birth patterns in recent years is striking in the data. At the 1940 rates the population would have been nearly stable in size. Then there were substantial increases in fertility rates, followed by a decline in recent years.

The way in which mortality acts at various ages in an insect species is illustrated by the data in Table 1.4.2. These come from an exceptionally thorough study by Varley (1947), also discussed by Haldane (1949). The total mortality in these gall flies is about 99%, compensated of course by a correspondingly high fertility. The table shows the comparison between the death rate and the Malthusian parameter m, given by the negative of the log-survival.

Table 1.4.2. An illustration of the death rates at different ages in the life cycle of a gall insect, *Urophora jaceana*. The data are given as the death rate and as the Malthusian parameter, $m = \log_e$ (survival rate). Data from Varley (1947).

MONTH	DENSITY/METER^2	CAUSE OF DEATH	DEATH RATE	m
July	203.0			
		Infertile eggs	.090	−.094
	184.7			
		Failure to form gall	.201	−.224
	147.6			
		Died in gall	.020	−.021
	144.6			
		Eurytoma curta	.455	−.607
August	78.8			
		Parasites	.365	−.455
	50.0			
		Probably mice	.616	−.957
Winter	19.2			
		Mice	.635	−1.009
	7.0			
		Unknown	.257	−.297
	5.2			
		Birds and parasites	.308	−.368
May	3.6			
		Floods	.436	−.573
July	2.03			
Total			.990	−4.605

Note that the Malthusian parameters are additive while the survivals are multiplicative. The total Malthusian parameter, -4.605, is $\log_e 0.01$ as it should be .

1.5 Fisher's Measure of Reproductive Value

Fisher (1958, p. 27) also asks a genetically relevant question: To what extent does an individual of age x contribute to the ancestry of future generations?

In order to answer this, he defines the quantity $v(x)$, the *reproductive value* at age x. (We have also used the letter V to stand for the variance but we shall make clear from the context which meaning is intended).

Obviously the reproductive value is 0 for an individual who is past the reproductive age. It clearly is lower at birth than a few years later, since a person of age 10, say, has a better chance of surviving to reproduce than has a child at birth. Furthermore, the reproduction would begin sooner and, in a growing population, this would increase the total contribution to future years by an earlier start. The value might be expected to be maximum somewhere near the beginning of the reproductive period.

To define the reproductive value we note first that the number of births from parents of age x is proportional to $e^{-mx}l(x)b(x)\,dx$, as shown in the previous section. If we now think of this cohort of persons followed through the rest of their lifetimes, their total contribution is proportional to

$$\int_x^\infty e^{-my}l(y)b(y)\,dy.$$

The exponential term serves to diminish the value of children born a long time in the future. This is analogous to the situation where the present value of a loan or investment is greater if it is to be paid soon rather than later. This is reversed, of course, if m is negative (as would be the present value of a loan or investment if interest rates were negative or the investment were decreasing in value). The reproductive value is proportional to the total contribution per individual of this age, so we divide the contribution by the number of persons of that age. This leads to the definition of reproductive value at age x,

$$v(x) = \frac{\int_x^\infty e^{-my}l(y)b(y)\,dy}{e^{-mx}l(x)}. \qquad 1.5.1$$

If $x = 0$, the denominator is equal to 1. Likewise, if $x = 0$, the numerator is equal to 1 from equation 1.4.3. Therefore, the reproductive value at birth is 1 , and $v(x)$ is a measure of the reproductive value of an individual of age x relative to that of a newborn child.

In developing 1.5.1 we discussed the situation as if the population were in age-distribution equilibrium. On the other hand, we can accept the definition as given and apply it to populations in general. From this a remarkable property emerges: Irrespective of the age distribution, the total reproductive value of a population increases at a rate given by m.

This can be shown as follows. First we rewrite 1.5.1 as

$$e^{-mx}l(x)v(x) = \int_x^\infty e^{-my}l(y)b(y)\,dy.$$

Differentiating both sides with respect to x leads to

$$e^{-mx}\left[v(x)\frac{dl(x)}{dx} + l(x)\frac{dv(x)}{dx} - v(x)l(x)m\right] = -e^{-mx}l(x)b(x).$$

Cancelling e^{-mx} and dividing both sides by $v(x)l(x)$ gives

$$\frac{1}{l(x)}\frac{dl(x)}{dx} + \frac{1}{v(x)}\frac{dv(x)}{dx} - m = -\frac{b(x)}{v(x)}.$$

The leftmost term, with sign changed, is simply the death rate $d(x)$, for it is the rate of decrease in the number of age x expressed as a fraction of the proportion alive at that age. Making this substitution and rearranging, we obtain

$$\frac{dv(x)}{dx} - v(x)\,d(x) + b(x) = mv(x). \qquad 1.5.2$$

The first term is the rate of change in the reproductive value of an individual as his age increases. The second is the rate of decrease in reproductive value per individual caused by deaths of individuals of age x. The third is the rate of increase in reproductive value from new births; this is simply the instantaneous birth rate, since the value of each newborn child, $v(0)$, is equal to 1. The left side of the equation, then, is the net change in reproductive value of the population contributed by an individual of age x, either by growing older, by dying, or by giving birth. The $n(x)$ individuals of age x then contribute $mn(x)v(x)$ to the increase in reproductive value of the population.

Thus the rate of change in reproductive value for each age group is given by m. Adding up all ages, we have

$$\frac{dv}{dt} = mv$$

and

$$v_t = v_0\, e^{mt}, \qquad 1.5.3$$

where v is the total reproductive value of the population.

This demonstrates Fisher's principle: The rate of increase in total reproductive value is equal to the Malthusian parameter times the total reproductive value, regardless of the age distribution. This means that the equations of Section 2 become applicable for populations not at age equilibrium if each individual in the population is weighted by the reproductive value appropriate to his age.

1.6 Regulation of Population Number

We have said nothing so far about population regulation. It is obvious that a population cannot grow exponentially forever. It must eventually reach a state where m becomes 0 or negative, or where in a discrete model w becomes 1 or less. The growth rate is eventually limited by all the factors that collectively make up the carrying capacity of the environment.

In population genetics we are mainly concerned with the changes in proportions of different types of individuals, rather than total numbers. We shall consider some examples of this under various types of population regulation. However, we shall ignore until later in the book the complications introduced by Mendelian inheritance.

We shall be dealing in this section with continuous models of the type introduced in Section 1.2. We are assuming the kind of model described in that section, or if the population has a more complicated structure we assume that it has reached age-distribution stability. Alternatively, in principle we could deal with reproductive values rather than actual numbers by weighting each individual by the reproductive value appropriate to its age.

Equation 1.2.2 or 1.4.5 can be modified to take regulation into account by writing

$$\frac{dN}{dt} = rN[1 - f(N)]. \tag{1.6.1}$$

The quantity r is the intrinsic rate of increase—the rate at which the population would grow if it had unlimited food supply and room for expansion. The function $f(N)$ implies some change in the rate of increase with the size of the population. The regulation may be, for example, by limitation of food supply, by the space available, by the accumulation of toxic products, or by territorial behavior patterns.

A particularly simple model is provided by letting $f(N)$ be a linear function of N, say N/K, where K is a constant sometimes called the carrying capacity of the environment. Such a population will grow approximately exponentially as long as N is much smaller than K, but as N approaches K the rate will decrease until size stability is reached at $N = K$.

Substituting N/K for $f(N)$ in 1.6.1 and rearranging, we have

$$\frac{dN}{dt} = \frac{rN(K - N)}{K}. \tag{1.6.2}$$

The equation may be rewritten as

$$\frac{dN}{N} + \frac{dN}{K - N} = r\,dt,$$

which is readily integrated to give

$$t = \frac{1}{r} \ln \frac{N_t(K - N_0)}{(K - N_t)N_0}. \tag{1.6.3}$$

Here and throughout the book ln means $\log_e$.

For example, if the intrinsic rate of increase of a population is 1% per year ($r = .01$) and the carrying capacity, K, is 5000, the time, t, required to change the number from $N_0 = 1000$ to $N_t = 2000$ is

$$t = \frac{1}{.01} \ln \frac{2000 \times 4000}{3000 \times 1000} = 98 \text{ years.}$$

If there were no regulation, the time required (see 1.2.3) would be

$$t = \frac{1}{r} \ln \frac{N_t}{N_0} = 69 \text{ years.}$$

Notice that, whether there is regulation or not, the time required for a certain change is proportional to $1/r$.

We can also write 1.6.3 in the inverse form, giving the number at time t as a function of t and the initial number N_0. This is

$$N_t = \frac{K}{1 + C_0 e^{-rt}}, \qquad 1.6.4$$

where

$$C_0 = \frac{K - N_0}{N_0}.$$

This function is shown graphically in Figure 1.6.1.

The curve is often called the "logistic" curve of population increase and has been widely used in ecology (Pearl, 1940; Nair, 1954; Slobodkin,

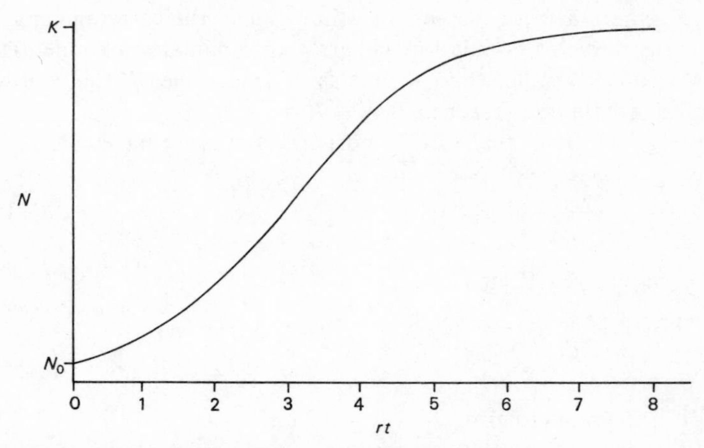

Figure 1.6.1. The logistic curve of population increase. The ordinate is the population number; the abscissa is rt, the product of the intrinsic rate of increase and the time. N_0, the initial population, is taken as 1/20 the value of K, the final number.

1962). Of course it is merely the simplest of a number of equations that could be derived and many populations, natural and experimental, depart widely from the model.

As stated earlier, we are mainly concerned in population genetics with the proportion of different genes and genotypes rather than the total number. We shall see that many of the equations for the proportions of different types are the same, despite quite different mechanisms for regulation of the population number.

The intrinsic rate of increase, r, is closely related to the Malthusian parameter m. We shall use the latter for the actual rate of change in numbers of the population, or of a part of the population, and r for the value this would take in a situation where the growth rate is not regulated.

In Chapter 5 we shall consider the effects of natural and artificial selection on the composition of a population. Now we are considering a simplified system in which the complications of Mendelian genetics are omitted. The essential points are brought out more clearly this way. Furthermore, the general principles that we are trying to show can be illustrated with only two classes; the extension to an arbitrary number presents no difficulties. So we are treating two competing species or asexual clones. Actually the same equations can be adapted to a single locus in a haploid population, to a cytoplasmic particle, to a gene on the Y chromosome or to competition between two self-fertilized strains.

The methods in the following sections come largely from Egbert Leigh.

1. Unregulated Growth Consider two strains, 1 and 2, with numbers n_1 and n_2 and intrinsic growth rates r_1 and r_2. These are the same as the Malthusian parameters when there is no restriction on continuous exponential growth. Let N be the total population number; i.e., $N = n_1 + n_2$. We shall designate by $p_1 = n_1/N$ and $p_2 = 1 - p_1 = n_2/N$ the proportions of the two strains. If there is no regulation of the growth of either strain the rates of increase are

$$\frac{dn_1}{dt} = r_1 n_1$$

and

$$\frac{dn_2}{dt} = r_2 n_2 . \tag{1.6.5}$$

The rate of increase of the total population is

$$\frac{dN}{dt} = r_1 n_1 + r_2 n_2 = \bar{r}N, \tag{1.6.6}$$

where $\bar{r}$ is the mean of the r's, weighted by the numbers in each population.

To obtain the rate of change in the proportions of the two types, we write

$$\begin{aligned}\frac{d\ln(p_1/p_2)}{dt} &= \frac{d\ln(n_1/n_2)}{dt}\\ &= \frac{d\ln n_1}{dt} - \frac{d\ln n_2}{dt} \\ &= \frac{dn_1}{n_1\,dt} - \frac{dn_2}{n_2\,dt}\\ &= r_1 - r_2.\end{aligned} \tag{1.6.7}$$

Notice also that

$$\begin{aligned}\frac{d\ln(p_1/p_2)}{dt} &= \frac{d\ln p_1}{dt} - \frac{d\ln(1-p_1)}{dt}\\ &= \frac{dp_1}{p_1\,dt} + \frac{dp_1}{(1-p_1)\,dt}\\ &= \frac{dp_1}{p_1(1-p_1)\,dt}.\end{aligned}$$

Putting these two equations together gives

$$\frac{dp_1}{dt} = (r_1 - r_2)p_1(1-p_1). \tag{1.6.8}$$

Notice that if we let $p_1 = N/K$ and $r_1 - r_2 = r$ we obtain equation 1.6.2. So, 1.6.8. is the equation of a logistic curve. Despite the fact that both strains are growing exponentially, the proportion of one type (the faster growing one) is increasing according to the logistic equation.

Equation 1.6.8 can be written in another form by noting that $\bar{r} = p_1r_1 + p_2r_2$. Substituting for r_2 in 1.6.8 gives

$$\frac{dp_1}{dt} = p_1(r_1 - \bar{r}). \tag{1.6.8a}$$

This form of the equation suggests the extension to more than two strains. When three or more strains are present the same equation is correct for the rate of change of a particular strain and $\bar{r}$ is the weighted average of the rates of increase of all the strains.

In this model the total population is increasing exponentially at any instant, although the rate of increase, $\bar{r}$, is changing continuously as the faster growing strain replaces the slower. This model is obviously unrealistic for any long period of time. We discuss it here to illustrate the point that equations of the type 1.6.8. can accurately describe the rate of change of the proportion of one type in a mixed population even when the numbers are changing according to quite a different rule. This is also true with various forms of regulation as the following examples show.

2. Logistic Regulation of Total Number A simple model for this situation is given by the equations,

$$\frac{dn_1}{dt} = n_1(r_1 - \bar{r}N/K)$$

and

$$\frac{dn_2}{dt} = n_2(r_2 - \bar{r}N/K). \qquad \textbf{1.6.9}$$

The total population increases logistically until it reaches an equilibrium at $N = K$, as can be seen by writing

$$\frac{dN}{dt} = \frac{d(n_1 + n_2)}{dt} = r_1 n_1 + r_2 n_2 - (n_1 + n_2)\bar{r}N/K.$$

But, since $N = n_1 + n_2$ and

$$N\bar{r} = r_1 n_1 + r_2 n_2,$$

$$\frac{dN}{dt} = \bar{r}N(1 - N/K). \qquad \textbf{1.6.10}$$

This is the equation for logistic growth (see 1.6.2).

At the same time that N is changing according to this rule, we can see what is happening to the proportions of the two types by writing

$$\frac{d\ln(p_1/p_2)}{dt} = \frac{dn_1}{n_1\,dt} - \frac{dn_2}{n_2\,dt} = r_1 - r_2,$$

or following the pattern of 1.6.7 and 1.6.8,

$$\frac{dp_1}{dt} = p_1(r_1 - \bar{r}). \qquad \textbf{1.6.11}$$

Again, the proportion of strain 1 is changing logistically.

Equation 1.6.8 may be written in integrated form as

$$\ln[p/(1 - p)] = C + rt, \qquad \textbf{1.6.12}$$

where $r = r_1 - r_2$ and C is $\ln[p_0/(1 - p_0)]$, a constant determined by the initial composition. This suggests a convenient way of plotting data from selection experiments; by plotting $\ln[p/(1-p)]$ against time one can easily see whether the trend is linear and thus see if the logistic equation is appropriate.

Alternatively, if we wish to know how much time is needed to change the proportion from p_0 to p_t, we can write 1.6.12 as

$$t = \frac{1}{r} \ln \frac{p_t(1 - p_0)}{(1 - p_t)p_0}, \tag{1.6.12a}$$

which gives the time as a function of the frequency of the type of interest. Note that the time required to accomplish a certain change in proportion is strictly proportional to the reciprocal of r.

3. Weaker Population Control The previous example assumed that the total population has an absolute upper limit, K. We now consider a population that is limited, but the limit is proportional to r so that as one type replaces the other the total population increases. A simple model is

$$\frac{dn_1}{dt} = n_1(r_1 - cN)$$

and

$$\frac{dn_2}{dt} = n_2(r_2 - cN). \tag{1.6.13}$$

The total number changes according to

$$\frac{dN}{dt} = N(\bar{r} - cN). \tag{1.6.14}$$

The population reaches a limit when $N = \bar{r}/c$; for this value $dN/dt = 0$.

The same procedure as before leads to the equation for change in proportion of type 1,

$$\frac{dp_1}{dt} = p_1(r_1 - \bar{r}). \tag{1.6.15}$$

This situation is probably quite unusual in nature. The size of the population is usually determined mainly by factors other than the r's. A replacement of the original strain by one with a higher r will cause only a slight increase in the final population number, if indeed there is any change at all.

The point of these three examples is to show that, despite great differences in the way in which the total population changes, the changes in proportion follow the same general rule given by 1.6.15.

4. Regulation by Efficiency of Space or Food Utilization If populations are regulated by the available space, food, or some other limiting factor, the type that wins in the competition may not be the one with the higher intrinsic rate of increase, but rather the one that can maintain the largest numbers in this environment. A simple model illustrating this possibility is given by

$$\frac{dn_1}{dt} = r_1 n_1 (K_1 - N)/K_1$$

and

$$\frac{dn_2}{dt} = r_2 n_2 (K_2 - N)/K_2 . \qquad \textbf{1.6.16}$$

One interpretation of K_1 is that this is the maximum population size that strain 1 can maintain when it is the only species; K_2 has the same meaning for strain 2. Suppose, to bring out the point of interest, that $r_1 = r_2$, but $K_1 \neq K_2$; the two strains differ, not in their intrinsic growth rates, but in the maximum number that this environment can support.

The change in total number is given by

$$\frac{dN}{dt} = rN\left[1 - \frac{n_1}{K_1} - \frac{n_2}{K_2}\right], \qquad \textbf{1.6.17}$$

where $r = r_1 = r_2$. The change in the proportion of type 1 is

$$\frac{dp_1}{dt} = Rp_1p_2 = Rp_1(1 - p_1), \qquad \textbf{1.6.18}$$

where

$$R = Nr\left[\frac{K_1 - K_2}{K_1K_2}\right].$$

Equation 1.6.18 is in the general form of the logistic equation (see 1.6.8), but it is not the same since R is not a constant. However, in many cases R is changing slowly and an equation of the form of 1.6.8 describes the rate of change at any particular time.

In an uncrowded environment the success of a population is determined mainly by its intrinsic rate of increase, r. In a crowded environment the carrying capacity, K, for the species may be more important. MacArthur and Wilson (1967) refer to "r selection" and "K selection." In an uncrowded environment (r selection) types which harvest the most food, even if they are wasteful, have the largest rate of increase. On the other hand, in a crowded environment (K selection) there is a great value on efficiency of utilization rather than simple productivity.

Ecologists have considered in more detail equations of these types. For many purposes it is more meaningful to measure biomass rather than simply count numbers, so that one isn't in the position of equating one mouse with one elephant. But we shall do no more with the subject here and get on to problems that are more strictly genetic.

In Chapter 5 we shall be dealing with the changes in gene and genotype frequencies under selection. The equations will be basically similar to 1.6.15, complicated by the Mendelian mechanism. The purpose of this section, as stated before, is to show that equations in the same form are widely applicable, even though the populations may differ greatly in their states. The total population may be growing, or static, or decreasing and the different genotypes may differ in intrinsic birth rates, death rates, or their response to the environment; yet the equations for changes in proportions may be basically similar. For these reasons, population genetics has usually ignored the total numbers and concentrated on the proportions.

1.7 Problems

1. In a population with discrete generations and with fitness w, how many generations are required to double the population number?
2. How long is required for the population to double with model 2?
3. A population under model 3 has reached age stability. How long, in units of λ, will be required for the population to double? What is the effective generation length, defined as the unit that will give the same answer as problem 1?
4. Suppose you know the age-specific death rates (the probability that an individual of age x will die during the next time unit). What is the life expectancy, that is, the mean length of life? What is the median length of life?
5. Show that equation 1.6.8*a* is correct for any number of strains.
6. What are the median and mean length of life under model 2, expressed in terms of the death rate, d?
7. Show that the time required to change the number from N_0 to N_t in a logistically growing population exceeds that in an unregulated population with the same intrinsic rate of increase by $\ln[(K - N_0)/(K - N_t)]/r$.
8. Again considering a logistic population with carrying capacity K, what is the time required to go from a fraction x to a fraction y of this capacity?
9. One bacterium which reproduces by fission and follows a logistic growth pattern is introduced into each of several ponds. Show that the time required to fill a pond to half its capacity is proportional to the log of the carrying capacity.

2
RANDOMLY MATING POPULATIONS

We are now ready to consider populations with Mendelian inheritance and begin by inquiring into the frequencies of the different genotypes that comprise the population. In the last chapter we were interested in the total number of individuals in the population and in different subpopulations. In genetic studies there is usually greater interest in the relative numbers of different genotypes, so it is convenient to express the numbers of different types as proportions of the total. As long as we use deterministic models the total number in the population is not important and we can deal as well with proportions. With stochastic models the number in the population becomes important in determining the extent of random fluctuations and therefore must be taken into consideration.

Only deterministic models will be considered in this chapter, and for the most part we shall be concerned with a model of discrete generations.

2.1 Gene Frequency and Genotype Frequency

The number of possible genotypes in a population greatly exceeds the number of genes and soon becomes enormous. A diploid population with only two alleles at each of 100 loci would have 3^{100} possible genotypes, a number far larger than the number of individuals in any population. Therefore we effect a great simplification by writing formulae in terms of gene frequencies rather than genotype frequencies. This usually entails some loss of information, for knowledge of the gene frequencies is not sufficient to specify the genotype frequencies; but it is usually possible to do this to a satisfactory approximation by introducing other information, such as the mating system and linkage relations.

Furthermore, in a sexually reproducing population the genes are reassorted by the Mendelian shuffle that takes place every generation. The effects of such reassortment are largely transitory, being undone as fast as they are done. For long-range trends we look to the changes in the frequencies of the genes themselves. As R. A. Fisher (1953) said:

> The frequencies with which the different genotypes occur define the gene ratios characteristic of the population, so that it is often convenient to consider a natural population not so much as an aggregate of living individuals as an aggregate of gene ratios. Such a change of viewpoint is similar to that familiar in the theory of gases, where the specification of the population of velocities is often more useful than that of a population of particles.

At the outset we need formulae to specify the relation between gene frequencies and genotype frequencies. This will also serve to introduce the kind of notation that will be used throughout the book. We start with a single locus with only two alleles.

Consider a population of N diploid individuals, of which

N_{11} are of genotype A_1A_1,
$2N_{12}$ are of genotype A_1A_2, and
N_{22} are of genotype A_2A_2,

where $N_{11} + 2N_{12} + N_{22} = N$. It is sometimes convenient to distinguish, among A_1A_2 heterozygotes, those that received A_1 from the mother and those that received it from the father. We can do this by designating the two numbers as N_{12} and N_{21}. However, in most populations $N_{12} = N_{21}$ and it is not necessary to make any distinction between them.

We designate the frequency or proportion of the three genotypes by P_{11}, $2P_{12}$, and P_{22} as follows:

$$A_1A_1: \quad P_{11} = N_{11}/N,$$
$$A_1A_2: \quad 2P_{12} = 2N_{12}/N,$$
$$A_2A_2: \quad P_{22} = N_{22}/N.$$

From these genotype frequencies, we can write the frequencies of alleles A_1 and A_2, which are designated p_1 and p_2, as follows:

$$p_1 = \frac{N_{11} + N_{12}}{N} = P_{11} + P_{12},$$
$$p_2 = \frac{N_{22} + N_{12}}{N} = P_{12} + P_{22}. \quad \text{2.1.1}$$

The blood groups provide convenient examples, and Table 2.1.1 shows some data on the frequency of MN types in Britain.

Table 2.1.1. Numbers of persons of blood types M, MN, and N in a sample from a British population. Data from Race and Sanger (1962).

PHENOTYPE GENOTYPE	M MM	MN MN	N NN	TOTAL
NUMBER	363	634	282	1,279
FREQUENCY	0.284	0.496	0.220	1.000

p_1 or $p_M = .284 + \frac{1}{2}(.496) = .532$

p_2 or $p_N = .220 + \frac{1}{2}(.496) = .468$

Extension of this principle to multiple alleles causes no difficulty, though the symbolism becomes a little more abstract. Equations of the type of 2.1.1 suggest the nature of the extension. Consider a locus with n alleles. As before, we designate allele frequencies with small letters and genotype frequencies with capital letters.

Alleles:	A_1	A_2	A_i	A_j	A_n
Frequencies:	p_1	p_2	p_i	p_j	p_n

The subscripts i and j are used to designate any two different alleles.

With these symbols, and letting P_{ii} stand for the frequency of genotype A_iA_i and $2P_{ij}$ stand for the heterozygous genotype A_iA_j, we obtain

$$p_i = P_{i1} + P_{i2} + \cdots + P_{ii} + \cdots + P_{ij} + \cdots + P_{in}$$
$$= \sum_{j=1}^{n} P_{ij}. \quad \text{2.1.2}$$

This procedure is applicable in any situation where each genotype is identifiable. The effects of other loci may be ignored if they do not obscure the distinction between the genotypes of the locus under consideration.

Because of dominance, it is often not possible to distinguish all genotypes. For example, in most blood group studies no distinction is made between *AA* and *AO* persons, both being classified simply as belonging to blood group *A*. Under such circumstances the allele frequency can be measured only if there is some knowledge about the way in which the genes are combined into genotypes in the population. The simplest assumption, and fortunately one that is often very closely approximated in many actual populations, is random mating—the subject of the next section.

2.2 The Hardy–Weinberg Principle

With random mating, the relation between gene frequency and genotype frequency is greatly simplified. By random mating we mean that the matings occur without regard to the genotypes in question. In other words, the probability of choosing a particular genotype for a mate is equal to the relative frequency of that genotype in the population.

Notice that it is possible for a population to be at the same time in random mating proportions for some genes and not for others. For example, it is quite reasonable for mating to be random with respect to a blood group or serum protein factor but be nonrandom with respect to genes for skin color or intelligence.

In many respects, random mating among the different genotypes in the population is equivalent to random combination of the gametes produced

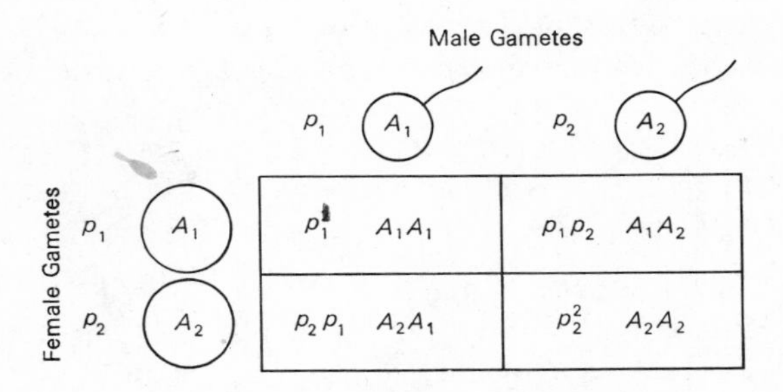

Figure 2.2.1. A demonstration that random combination of gametes leads to the Hardy–Weinberg frequencies of zygotes. The gametes are along the margins, with the zygotes being in the body of the table.

by these individuals. We consider a single locus with two alleles, A_1 and A_2, with frequencies p_1 and p_2 $(p_1 + p_2 = 1)$. The frequencies are taken to be the same in both sexes. The results of random combination of gametes are shown in Figure 2.2.1, where the gamete types are on the margins and the resulting zygotes are in the body of the table.

Hence, the first generation after random mating begins, the proportions of the three genotypes A_1A_1, A_1A_2, and A_2A_2 will be p_1^2, $2p_1p_2$, and p_2^2. Later we shall show that these proportions, gotten by combining gametes at random, are in fact equivalent to what is obtained by random mating.

This principle was first reported for the special case $p_1 = p_2 = .5$ by Yule (1902) and Pearson (1904), and for other values by Castle (1903). Later it was shown for any value of p_1 and p_2 by Hardy (1908) and Weinberg (1908), and has since come to be known as the Hardy–Weinberg principle.

The Hardy–Weinberg principle is thought of by population geneticists in two somewhat different ways. It is sometimes spoken of as the Hardy–Weinberg equilibrium, with the implication that in the absence of factors that change the gene frequencies (mutation, selection, migration, random drift) and with random mating the population will immediately arrive at and remain in the same proportions. Note also that the population variability remains the same from generation to generation. Less inclusively, the principle says simply that, given the gene frequencies and assuming random mating, the zygotic frequencies can be predicted. In this book we shall usually regard the principle in the second way; we shall consider the effects of various influences on the gene frequencies and then use the Hardy–Weinberg principle to relate gene frequencies and genotype frequencies.

It is easy to verify that random mating of genotypes leads to the same expected zygote frequencies as random combination of gametes. Using the same symbols as before, we can enumerate the various mating combinations and their relative frequencies as shown in Table 2.2.1.

From the bottom row in the table, noting from 2.1.1 that $p_1 = P_{11} + P_{12}$ and $p_2 = P_{12} + P_{22}$, we see that in one generation the three genotypes attain the same proportions as given by random combination of gametes.

The fact that the equilibrium is attained immediately after random mating begins, rather than gradually, is of great importance. This means that it is not necessary to inquire into the past history of a population; if mating was at random during the preceding generation, the principle holds. The only exception is the rather unusual case where the gene frequencies are different in the two sexes, a topic that we shall return to later.

Notice that $p_1^2 + 2p_1p_2 + p_2^2 = (p_1 + p_2)^2$. Therefore the random-mating principle can be stated this way: The array of genotype frequencies in the zygotes is given by the square of the array of gene frequencies in the gametes.

Table 2.2.1. Derivation of the Hardy–Weinberg principle by combining parental genotypes in random proportions.

MATING	FREQUENCY OF THIS MATING	PROGENY		
		A_1A_1	A_1A_2	A_2A_2
$A_1A_1 \times A_1A_1$	$(P_{11})^2$	P_{11}^2		
$A_1A_1 \times A_1A_2$	$2(P_{11})(2P_{12})$	$2P_{11}P_{12}$	$2P_{11}P_{12}$	
$A_1A_1 \times A_2A_2$	$2(P_{11})(P_{22})$		$2P_{11}P_{22}$	
$A_1A_2 \times A_1A_2$	$(2P_{12})^2$	P_{12}^2	$2P_{12}^2$	P_{12}^2
$A_1A_2 \times A_2A_2$	$2(2P_{12})(P_{22})$		$2P_{12}P_{22}$	$2P_{12}P_{22}$
$A_2A_2 \times A_2A_2$	$(P_{22})^2$			P_{22}^2
Total	$(P_{11}+2P_{12}+P_{22})^2 = 1$	$(P_{11}+P_{12})^2 = p_1^2$	$2(P_{11}+P_{12})(P_{12}+P_{22}) = 2p_1p_2$	$(P_{12}+P_{22})^2 = p_2^2$

As a numerical example, return to the MN blood group data discussed earlier (Table 2.1.1). From these data the frequency of the M allele, p_M, was computed to be .532 and the frequency of the N allele, p_N, was .468. Thus, by the Hardy–Weinberg principle the three genotypes MM, MN, and NN would be expected to be in the ratio $(.532)^2$, $2(.532)(.468)$, and $(.468)^2$. As shown in Table 2.2.2 these expectations are in remarkably good agreement with the observed numbers.

Table 2.2.2. Comparison of the observed proportion of MN blood types with the proportions expected with random mating. The data are from Table 2.1.1; $p_M =$ .532 and $p_N = .468$.

GENOTYPE	MM	MN	NN	TOTAL
Expected proportion	.283	.498	.219	1.000
Expected number	362.0	636.9	280.1	1279.0
Observed number	363	634	282	1279

$\chi_1^2 = 0.029$, Prob $= .87$

The χ^2 method is described in the appendix. In this instance there is only one degree of freedom, rather than the two that might have been expected. This is because the gene frequency has been estimated from the data, thereby reducing the number of degrees of freedom by one.

A second example: The incidence of the recessive disease, phenylketonuria, which results in mental deficiency is approximately 1 in 10,000. What proportion of the population are heterozygous carriers of this condition?

We let A stand for the normal allele and a for the recessive allele for phenylketonuria, and designate the gene frequencies by p_A and p_a. The Hardy–Weinberg principle shows that p_a^2 is .0001, so p_a is the square root of this, or .01. Since p_A and p_a must add up to 1, p_A is .99. Therefore, the frequency of heterozygotes in the population $(2p_A p_a)$ is $2(.99)(.01) = .0198$.

This example illustrates the important fact that, when the recessive gene is rare, the number of heterozygous carriers of the gene is enormously larger than the number of individuals who carry the gene in the homozygous state. In this example the carriers outnumber the affected in the ratio of 198 to 1.

Still considering a recessive factor, we can inquire as to the parentage of individuals that are homozygous for a recessive factor. If all genotypes are equally viable and fertile, and mating is at random, the matings that could produce homozygous recessive progeny are as shown in Table 2.2.3.

Table 2.2.3. Computation of the relative proportions of different genotypes among the parents of homozygous recessives, *aa*.

MATING	FREQUENCY OF THIS MATING	FREQUENCY OF HOMOZYGOUS-RECESSIVE PROGENY	RELATIVE FREQUENCY OF RECESSIVE PROGENY
$Aa \times Aa$	$(2p_A p_a)^2$	$p_A^2 p_a^2$	p_A^2
$Aa \times aa$	$2(2p_A p_a)(p_a^2)$	$2p_A p_a^3$	$2p_A p_a$
$aa \times aa$	$(p_a^2)^2$	p_a^4	p_a^2
Total		p_a^2	1

Thus, even if persons with phenylketonuria were normally viable and fertile and married at random with the rest of the population, the great majority of phenylketonuric children would still come from normal but heterozygous parents. The ratio of feeble-minded children with normal parents, those with one normal parent, and with neither parent normal would be 9801 : 198 : 1.

One might question the validity of the Hardy–Weinberg assumption for a condition such as phenylketonuria, where affected persons are likely not to have children at all. Nevertheless, the error involved in making this assumption is very small, for the same reason that has already been emphasized.

Homozygous recessives make up such a small fraction of the population that no serious error is made by assuming that all classes marry at random, which is a reasonable assumption for the homozygous-normal and heterozygous classes.

The Hardy–Weinberg principle is *exactly* true only in an infinitely large population in which mating is completely at random; but it is approximately correct for the great majority of genes in most cross-fertilizing species.

The principal departures are of two types: (1) inbreeding and (2) assortative mating. Inbreeding occurs when mates are more closely related than if they were chosen at random. The most extreme examples are self-fertilizing species, but even in species with obligatory cross-fertilization there may be some inbreeding mainly because of the frequent geographical propinquity of close relatives. Ways in which inbreeding can be measured and its effects taken into account are given in Chapter 3.

Assortative mating occurs when mates are phenotypically more alike than with random mating. For example, tall men tend to marry tall women and two people are more likely to marry if they have similar intelligence. There are also numerous examples in experimental animals of mating preferences that lead to assortative mating in heterogeneous populations. Assortative mating is discussed in Chapter 4.

A population that is in Hardy–Weinberg frequencies when the zygotes are formed may show significant departures at adult stages due to differential mortality. Usually the younger the stage counted, the closer is the realization to random-mating proportions. On the other hand, differential fertility does not affect the Hardy–Weinberg relation as long as there is no correlation between the fertility of an individual and its choice of mate.

Despite these numerous discrepancies of detail, the random-mating assumption fits very well a great many real situations. The most important reason is that the genotype proportions are attained after only a single generation. This means that a perturbation for any reason is rectified immediately rather than slowly over a number of generations.

It might be of interest to examine some circumstances that are exceptions to the above statement in that the equilibrium frequencies are approached gradually rather than suddenly. One of these is with sex-linked genes where the gene frequencies are different in the two sexes; this will be discussed later in the chapter (Section 2.5). Another is a population in which each individual can reproduce both asexually and sexually, as can many plants.

Let C be the proportion of progeny that are produced clonally or asexually and S the proportion sexually by random mating. Let P_t be the proportion of genotype AA at generation t and p be the frequency of gene A. We assume that C is independent of the genotype of the individual. Notice that the genotype frequency P_t changes from generation to generation whereas the gene frequency p remains constant. In the following generation the AA plants can be produced in two ways: (1) AA plants, with frequency P_t, reproducing clonally with probability C and yielding a fraction CP_t of the total progeny, and (2) plants that are homozygous or heterozygous for allele A mating at random with probability S and producing Sp^2AA progeny. Thus, in the next generation the frequency of AA genotypes will be

$$P_{t+1} = CP_t + Sp^2.$$

Rearranging, and recalling that $C + S = 1$,

$$\begin{aligned} P_{t+1} - p^2 &= CP_t + Sp^2 - p^2 \\ &= CP_t - p^2(1 - S) \qquad \textbf{2.2.1} \\ &= C(P_t - p^2). \end{aligned}$$

There is the same relation between P_t and P_{t-1} as between P_{t+1} and P_t. Thus

$$\begin{aligned} P_t - p^2 &= C(P_{t-1} - p^2) \\ &= C^2(P_{t-2} - p^2) \qquad \textbf{2.2.2} \\ &= C^t(P_0 - p^2) \end{aligned}$$

where P_0 is the initial frequency of the genotype AA. Unless $C = 1$ (exclusively clonal reproduction), the limit of C^t as t increases is 0, and P_t becomes closer and closer to p^2. The same procedure can be applied to any other genotype. Each genotype frequency approaches the Hardy–Weinberg value, but only gradually at a rate determined by C. The smaller the fraction of asexual progeny, the more rapidly the random-mating proportions are approached.

Another situation that leads to a gradual rather than a sudden attainment of Hardy–Weinberg proportions is overlapping generations. We can consider random mating under population model 2 (continuous random deaths and births) by supposing that in any small time interval of length dt, a fraction of the population, chosen at random, dies and a new fraction is produced by random mating of the existing population. The details of this process are given by Moran (1962, pp. 23–25). It is clear that the Hardy–Weinberg proportions would be approached asymptotically as the proportion of the population derived from mating increases and the old survivors constitute a smaller and smaller fraction of the population.

In the simple Moran model the life expectancy is exponential. Thus the final equilibrium would be reached strictly only after an infinite period of time. However, most organisms have a finite life expectancy, and random-mating proportions would be attained as soon as the last survivor from the parent generation had died.

2.3 Multiple Alleles

Additional alleles do not introduce any serious difficulty. The principles of the preceding section are easily extended to include any number of alleles. For multiple alleles the Hardy–Weinberg principle can be stated this way:

If p_i is the frequency of allele A_i and p_j is the frequency of some other allele A_j, the expected frequencies of the genotypes with random mating are

$$A_i A_i : P_{ii} = p_i^2, \tag{2.3.1}$$

$$A_i A_j : 2P_{ij} = 2p_i p_j, (P_{ij} = P_{ji}). \tag{2.3.2}$$

As an example consider the blood group alleles O, A, and B. The allele frequencies are p_O, p_A and p_B and the blood groups are as shown below.

BLOOD GROUP	GENOTYPES	FREQUENCY
O	OO	p_O^2
AB	AB	$2p_A p_B$
A	AA or OA	$p_A^2 + 2p_O p_A$
B	BB or OB	$p_B^2 + 2p_O p_B$

The equations 2.3.1 and 2.3.2 enable one to determine the expected genotype frequencies when the gene frequencies are given. For the reverse problem, to determine the gene frequencies when the genotype frequencies are given, the relevant formula is 2.1.2.

The problem is more difficult when not all genotypes are distinguishable. It is necessary to make some assumption about the mating system; usually one assumes random mating unless there is evidence to the contrary. Consider, for example, a 3-allele system in which A_1 is dominant to both A_2 and A_3 and A_2 is dominant to A_3. Such a system is frequently found: for example, blood groups A_1, A_2, and O when B is disregarded; B^{26}, B^2, and B blood groups in chickens; agouti, chinchilla and white in rabbits.

Suppose P is the observed proportion of A_1-, Q the proportion of A_2-, and R of $A_3 A_3$. Equating the observed and the expected proportions gives

$$P = p_1^2 + 2p_1 p_2 + 2p_1 p_3,$$
$$Q = p_2^2 + 2p_2 p_3,$$
$$R = p_3^2,$$

which on solving gives

$$p_3 = \sqrt{R},$$
$$p_2 = \sqrt{Q + R} - \sqrt{R},$$
$$p_1 = 1 - \sqrt{Q + R}.$$

When there are more phenotypes than genes there is no unique solution to the equations such as those just given. In this case the usual method of estimating the gene frequencies from the phenotype frequencies is the maximum-likelihood method of Fisher (see A.9). As seen in the appendix, the solution in the previous paragraph is in fact the maximum-likelihood solution.

The problem of classifying and enumerating the different possible genotypes with a large number of multiple alleles is a difficult and intriguing problem. For a discussion, see Cotterman (1953, 1954), Bennett (1956), Hartl and Maruyama (1968), and Cotterman (1969).

2.4 *X*-linked Loci

For alleles on the X chromosome the genotype frequencies in the male are the same as the gene frequencies, since the male is haploid for this chromosome. The female frequencies, on the other hand, are the same as for diploid autosomal loci, since there are two X chromosomes. (We are discussing organisms in which the male is heterogametic; when the female is heterogametic the following statements should be reversed accordingly.)

This means that there will usually be a large difference in the frequency of an X-linked character in the two sexes. For example, a trait caused by a recessive gene, a, with frequency p would have a frequency p in males, but p^2 in females. If the gene were rare, the frequency in males would be much greater than in females. A classical example of a human X-linked trait is color blindness. One of the earliest and largest studies was by Waaler (1927). He found that in a group of 18,121 school children in Oslo, about 8% of the boys were color-blind, but only 0.4% of the girls. The data are in Table 2.4.1.

If we take the proportion of color-blind boys as p, we can estimate the proportion of girls as p^2. As can be seen from Table 2.4.1, the observed

Table 2.4.1. Frequencies of color blindness among school children in Oslo. Data from Waaler (1927).

	MALES		FEMALES		
			OBSERVED		EXPECTED ON BASIS OF MALE FREQUENCY
	NUMBER	PROPORTION	NUMBER	PROPORTION	
Color-blind	725	$.0801 = p$	40	.0044	$p^2 = .0064$
Normal	8324	$.9199 = 1 - p$	9032	.9956	

number does not agree very well with this expectation. We can use the χ^2 test to see if the disagreement is too great to be reasonably attributed to chance.

To obtain the expected numbers by the maximum-likelihood method, we let p be the relative frequency of the allele for color blindness and A, B, C, and D be the observed numbers of color-blind males (725), normal males (8324), color-blind females (40) and normal females (9032), respectively. Then, if m is the proportion of males and N the total observed number ($N = A + B + C + D = 18{,}121$), the probability of the observed results is

$$\text{Prob} = K(mp)^A[m(1-p)]^B[(1-m)p^2]^C[(1-m)(1-p^2)]^D,$$

where K is a constant determined by A, B, C, and D. Letting $L = \log \text{Prob}$,

$$L = (A+B)\log m + (C+D)\log(1-m) + (A+2C)\log p$$
$$+ (B+D)\log(1-p) + D\log(1+p)$$

plus a constant. Differentiating with respect to m and p, and equating to zero, we obtain:

$$\frac{\partial L}{\partial m} = \frac{A+B}{m} - \frac{C+D}{1-m} = 0,$$

$$\frac{\partial L}{\partial p} = \frac{A+2C}{p} - \frac{B+D}{1-p} + \frac{D}{1+p} = 0.$$

Solving these equations,

$$m = \frac{A+B}{N} = \frac{9049}{18121},$$

as expected, and

$$p = \frac{-B + \sqrt{B^2 + 4(A+2C)[A+B+2(C+D)]}}{2[A+B+2(C+D)]} = .0772.$$

The expected numbers in the four classes are Nmp, $Nm(1-p)$, $N(1-m)p^2$ and $N(1-m)(1-p^2)$.

The expected numbers are given along with the observed numbers in each of the four classes in Table 2.4.2. The number of degrees of freedom is $3 - 2 = 1$, since 2 degrees of freedom were lost by estimating m and p from the data. The probability of obtaining a disagreement as great as this or greater by chance is about .03, so we look for an explanation.

A plausible basis for the discrepancy is revealed by a finer analysis. There are four kinds of color blindness: (1) protanopia, or red blindness; (2) protanomaly, or partial red blindness; (3) deuteranopia, or green blindness;

Table 2.4.2. Statistical analysis by the χ^2 method of the data in Table 2.4.1. The expected numbers are computed on the assumption of random mating by the method of maximum likelihood.

	OBSERVED NUMBER	EXPECTED NUMBER	DEVIATION	$\frac{\text{DEVIATION}^2}{\text{EXPECTED}}$
Color-blind males	725	698.6	26.4	1.00
Normal males	8,324	8,350.4	−26.4	0.08
Color-blind females	40	54.1	−14.1	3.67
Normal females	9,032	9,017.9	14.1	0.02
Total	18,121	18,121.0		4.77

$\chi_1^2 = 4.77 \qquad P = .03$

and (4) deuteranomaly, or partial green blindness. The numbers of the four types are given in Table 2.4.3.

If we let p_1, p_2, p_3, and p_4 stand for the frequencies of the four kinds of abnormal X chromosomes, these proportions can be estimated directly from the proportions in males. From these proportions we can estimate the expected proportions in females. The expected proportions in females in Table 2.4.3 are based on the assumption that the capacities for red and green vision

Table 2.4.3. A finer resolution of the data on color blindness by taking into account different types of color blindness.

	MALES		FEMALES		
	NUMBER	PROPORTION	NUMBER	PROP.	EXPECTED PROPORTION BASED ON MALE FREQUENCIES
Protanopia	80	$.0088 = p_1$	0	.0000	$p_1^2 = .0001$
Protanomaly	94	$.0104 = p_2$	3	.0003	$p_2^2 + 2p_1p_2 = .0003$
Deuteranopia	93	$.0103 = p_3$	1	.0001	$p_3^2 = .0001$
Deuteranomaly	458	$.0506 = p_4$	36	.0040	$p_4^2 + 2p_3p_4 = .0036$
Normal	8324	$.9199 = p_5$	9032	.9956	
Total	9049	1.0000	9072	1.0000	

are complementary; that is, the heterozygote for a red deficiency and a green deficiency has normal vision. On the other hand, protanomaly and protanopia are regarded as noncomplementary, with the heterozygote being protanomalous. Likewise deuteranomaly acts as if dominant to deuteranopia. There is supporting evidence for this hypothesis from other sources; for example, there is an instance where a woman with normal vision produced two kinds of sons—red deficient and green deficient.

Notice in Table 2.4.3 that the female frequencies fit the expectation very well indeed. Ideally we could test for agreement by a χ^2 test, using maximum-likelihood estimates based on the frequencies in both sexes, but the procedure hardly seems necessary in this case.

The close fit of the data to the expectation supports the hypothesis that the red and green components are complementary. Notice that this analysis does not distinguish between a single locus and two loci. If it is assumed that there are two separate loci, the quantities p_1, p_2, p_3, p_4, and p_5 are interpreted as chromosome frequencies rather than allele frequencies. Evidence since these data were collected has shown that there are two loci and recombination has been reported.

2.5 Different Initial Gene Frequencies in the Two Sexes

Usually the gene frequencies are the same in the two sexes, but under some unusual circumstances they may be different. One possibility is in animal or plant breeding where the original hybrids are made up of males of one strain and females of another. Another is in a natural population where most of the migrants are of one sex. For example, in human populations there are examples where most of the migrants to a newly colonized area were male, and many of the marriages were between local women and immigrant men.

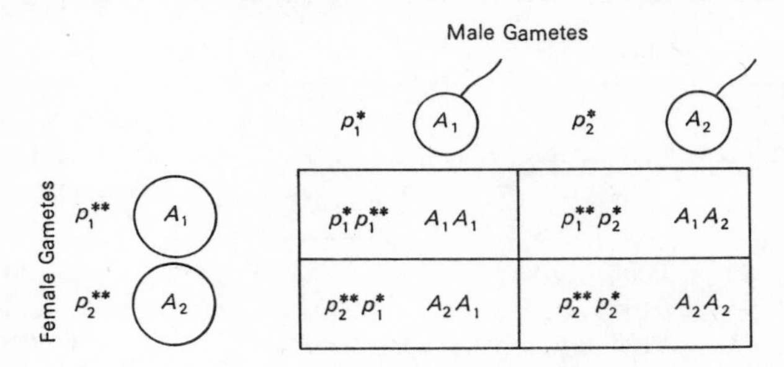

Figure 2.5.1. Random combination of gametes when the allele frequencies are different in the two sexes. The single and double asterisks refer to male and female frequencies, respectively.

For an autosomal locus, the Hardy–Weinberg frequencies are attained, but after a delay of one generation. Consider a population in which the frequencies of the alleles A_1 and A_2 are p_1^* and p_2^* in males and p_1^{**} and p_2^{**} in females. If the gametes are combined at random, we obtain the results in Figure 2.5.1.

In the next generation the frequency of gene A_1 in both sexes is

$$\begin{aligned} p_1 &= p_1^* p_1^{**} + \tfrac{1}{2} p_1^* p_2^{**} + \tfrac{1}{2} p_1^{**} p_2^* \\ &= \tfrac{1}{2}(p_1^* + p_1^{**}), \end{aligned} \tag{2.5.1}$$

and in all following generations the genotypes are in the proportions p_1^2, $2p_1p_2$, and p_2^2. As expected, the gene frequency is the unweighted average of what it was originally in the two sexes.

The demonstration that random mating of the different genotypes gives the same results as random combination of the gametes follows the same general method as was used in Table 2.2.1. Likewise, extension to multiple alleles is straightforward and will not be discussed here.

With an X-linked locus, starting out with different allele frequencies in the two sexes, the situation is quite different. The equilibrium, instead of being attained in two generations as for an autosomal locus, is reached only gradually.

Consider a particular X-linked allele, A, in a multiple-allelic series. Let the frequency of this allele in generation t be p_t^* in males and p_t^{**} in females.

Since a male always gets his X chromosome from his mother, (1) the allele frequency in males will always be what it was in females a generation earlier. Likewise, (2) the frequency in females will be the average of the two sexes in the preceding generation, since each sex contributes one X chromosome. Finally, (3) the mean frequency of the gene will be the weighted average of that in the two sexes, attaching twice as much weight to the female frequency as the male because the female has twice as many X chromosomes. This quantity must be a constant, since the mean frequency of the gene does not change. These statements may be stated mathematically as follows, where $t - 1$ designates the previous generation.

$$\begin{aligned} &(1)\ p_t^* = p_{t-1}^{**}, \\ &(2)\ p_t^{**} = \tfrac{1}{2}p_{t-1}^* + \tfrac{1}{2}p_{t-1}^{**}, \\ &(3)\ \bar{p}_t = \tfrac{1}{3}p_t^* + \tfrac{2}{3}p_t^{**} = \tfrac{1}{3}p_{t-1}^* + \tfrac{2}{3}p_{t-1}^{**} = \bar{p}_{t-1} = \bar{p}, \end{aligned} \tag{2.5.2}$$

where $\bar{p}$ is a constant throughout the process.

From the third equation we have $p_{t-1}^* = 3\bar{p} - 2p_{t-1}^{**}$. Substituting this into the second equation, after some algebraic rearrangement we obtain a relation between the gene frequency in the female sex from generation to generation.

$$p_t^{**} - \bar{p} = -\tfrac{1}{2}(p_{t-1}^{**} - \bar{p}). \qquad \textbf{2.5.3}$$

Since the same relationship holds for p_{t-1}^{**} and p_{t-2}^{**},

$$\begin{aligned} p_t^{**} - \bar{p} &= (-\tfrac{1}{2})^2(p_{t-2}^{**} - \bar{p}) \\ &= (-\tfrac{1}{2})^3(p_{t-3}^{**} - \bar{p}) \end{aligned}$$

and, finally,

$$p_t^{**} - \bar{p} = (-\tfrac{1}{2})^t(p_0^{**} - \bar{p}), \qquad \textbf{2.5.4}$$

where p_0^{**} is the allele frequency in females in the initial generation. Notice that these formulae do not depend on equal numbers of males and females in the population.

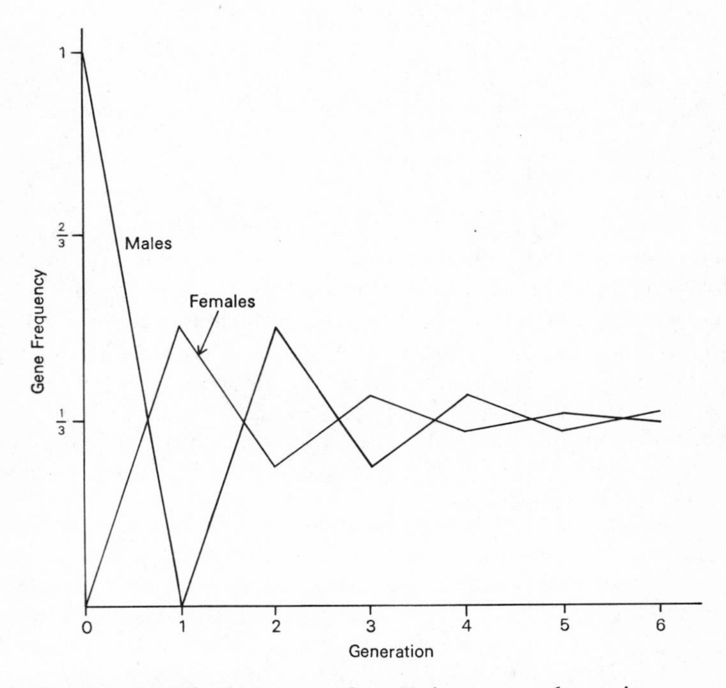

Figure 2.5.2. The frequency of an *X*-chromosomal gene in the two sexes. In generation 0 the gene has a frequency of 1 in males and 0 in females. The weighted average frequency in the two sexes remains $\frac{1}{3}$ throughout.

Thus the approach to equilibrium is gradual. Since $(-1/2)^t$ approaches 0 as t becomes large, the gene frequency must approach $\bar{p}$ as a limit. In each successive generation the gene frequency in females is only half as far from the final value, $\bar{p}$, as it was in the previous generation, but in the opposite direction. The population moves toward equilibrium in a zig-zag manner, like a series of damped vibrations.

From equation 2.5.2, it can be seen that the gene frequency in males follows that of females but is delayed by one generation. This oscillatory approach to equilibrium for an X-linked gene with different initial frequencies in the two sexes was first shown by Jennings (1916). It is shown graphically in Figure 2.5.2.

2.6 Two Loci

In Section 2.3 it was stated that, with random mating, the Hardy–Weinberg relation was attained in a single generation, irrespective of the number of alleles. But it is possible for each of two loci to be in random-mating frequencies, yet for them not to be in equilibrium with each other. We shall see that, contrary to the situation with a single locus, the equilibrium relation between two or more loci is not attained immediately, but gradually over a number of generations. We still assume that there is no selection.

Consider two linked loci, with the recombination frequency between them equal to c. Unlinked loci can be regarded as a special case where c is 1/2. Assume that there are n alleles at locus A and m at locus B.

A alleles:	A_1	A_2	A_i	A_j	A_n
Frequency:	p_1	p_2	p_i	p_j	p_n
B alleles:	B_1	B_2	B_k	B_l	B_m
Frequency:	q_1	q_2	q_k	q_l	q_m

Let $P_t(A_i B_k)$ be the frequency of the chromosome $A_i B_k$ among the gametes produced in generation t. We want to inquire into the frequency of this chromosome in successive generations. For brevity, we shall designate $P_t(A_i B_k)$ simply as P_t.

The gametes produced in generation t may be of two kinds; they may be the product of a recombination, with probability c, or they may not have been involved in a recombination at the preceding meiosis, with probability $1 - c$. If no recombination has occurred, the probability of the $A_i B_k$ chromosome is the same as it was in the preceding generation, which will be designated as P_{t-1}. If a crossover has occurred, the A and B alleles in a gamete produced by the tth generation will have come from different gametes in the $t - 1$th generation; that is, one from the egg and one from the sperm that produced the individual whose gamete we are discussing. Since we are assuming random mating, the egg and sperm are independent, and therefore the A locus and the B locus are independent. Thus, in the gamete produced by the tth generation, the probability of the A allele being A_i and the B allele being B_k is simply the product of the two gene frequencies, $p_i q_k$.

Putting these together, we have

$$P_t = (1 - c)P_{t-1} + cp_i q_k, \tag{2.6.1}$$

and after subtracting $p_i q_k$ from both sides, we obtain

$$P_t - p_i q_k = (1 - c)(P_{t-1} - p_i q_k). \tag{2.6.2}$$

There is the same relation between P_{t-1} and P_{t-2} as there is between P_t and P_{t-1}. Therefore we can write

$$\begin{aligned} P_t - p_i q_k &= (1 - c)^2(P_{t-2} - p_i q_k) \\ &= (1 - c)^3(P_{t-3} - p_i q_k), \end{aligned}$$

until finally,

$$P_t - p_i q_k = (1 - c)^t(P_0 - p_i q_k), \tag{2.6.3}$$

where P_0 is the initial value of the frequency of chromosome $A_i B_k$.

Thus the frequency of the $A_i B_k$ chromosome gets closer and closer to the value $p_i q_k$; each generation the departure from the final value is reduced by a fraction equal to the recombination value. For unlinked loci, the chromosome frequency goes half way to the equilibrium value each generation. See Figure 2.6.1.

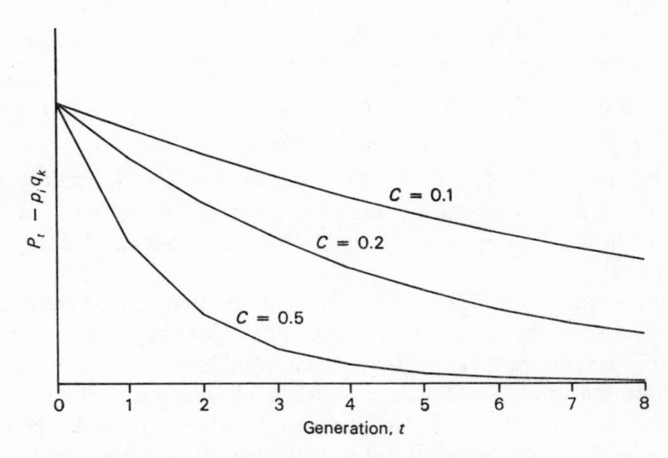

Figure 2.6.1. The rate of approach to gametic phase or linkage equilibrium. The ordinate is the difference between the frequency of a gamete type and its equilibrium value, $P_t - p_i q_k$.

The equilibrium value is approached only gradually, and theoretically is never attained. For comparison it is convenient to speak of the time required to go halfway to the final value, in the same way that one speaks of the halflife or median life of a radioactive element.

The median equilibrium time is given by solving for t the equation

$$(1-c)^t = \tfrac{1}{2},$$

which leads to

$$t = \frac{\log \frac{1}{2}}{\log(1-c)}. \qquad \textbf{2.6.4}$$

When c is small, $\log_e(1-c)$ is very close to $-c$. So, using logs to the base e, $\log \frac{1}{2} = -.693$, and $t = .693/c$ approximately for small values of c. For example, with independent gene loci ($c = \frac{1}{2}$), the median time is 1 generation by equation 2.6.4. When $c = .1$, the approximation is satisfactory and t is $.693/.1$, or about 7 generations. When $c = .01$, the time is about 69 generations; when $c = .001$, about 693 generations.

We can summarize the main results of this section in two statements:

1. The approach to equilibrium between coupling and repulsion phases of the gametes (usually called gametic phase equilibrium or "linkage" equilibrium) is gradual. The rate of approach depends on the rate of recombination between the two loci, and for independent loci is 50% per generation.

2. We can broaden the meaning of the Hardy–Weinberg principle by regarding chromosomes or gametes rather than single genes. With random mating the frequency of any diploid genotype is given by the appropriate term in the expansion of the square of the array of gamete frequencies. These frequencies are attained immediately if the gamete frequencies are the same in both sexes.

These principles are illustrated by the following example.

GENOTYPE	IN TERMS OF GAMETE FREQUENCIES	FINAL EQUILIBRIUM IN TERMS OF GENE FREQUENCIES
$A_i B_k/A_i B_k$	P_{ik}^2	$p_i^2 q_k^2$
$A_i B_k/A_j B_l$	$2P_{ik}P_{jl}$	$2p_i p_j q_k q_l$
$A_i B_l/A_j B_k$	$2P_{il}P_{jk}$	$2p_i p_j q_k q_l$

Therefore, at equilibrium, the two linkage phases (coupling and repulsion) in the double heterozygote are equally frequent.

Notice also that the frequency of any composite genotype can be obtained by simple multiplication. The frequency of the genotype $A_i A_j B_k B_k$, for example, is $(2p_i p_j)(q_k^2)$; the frequency of $A_i A_i B_k B_k$ is $p_i^2 q_k^2$; and $A_i A_j B_k B_l$ is $(2p_i p_j)(2q_k q_l)$.

Crossing over in the two sexes is not necessarily the same. The formula given earlier can easily be modified to take this into consideration. A (successful) gamete has equal chances of having come from a male or a female, since the total contribution of the two sexes is necessarily equal. Hence, letting c_m and c_f stand for the recombination fraction in males and females, the formula 2.6.1 becomes

$$\begin{aligned} P_t &= \tfrac{1}{2}[(1 - c_m)P_{t-1} + c_m p_i q_k] + \tfrac{1}{2}[(1 - c_f)P_{t-1} + c_f p_i q_k] \\ &= (1 - \bar{c})P_{t-1} + \bar{c} p_i q_k, \end{aligned} \qquad 2.6.5$$

where $\bar{c} = \tfrac{1}{2}(c_m + c_f)$.

Therefore, when crossing over differs in males and females it is sufficient to replace c in equations 2.6.2 and 2.6.3 by the mean value in the two sexes. In organisms like Drosophila and Bombyx, where meiotic crossing over is restricted to one sex (female in Drosophila, male in Bombyx), the approach to equilibrium is just half as fast as it would be if it occurred at the homogametic rate in both sexes.

The results embodied in equations 2.6.2 and 2.6.3 were first given by Robbins (1918) who obtained them by considering all possible genotypes. The simpler and more elegant procedure used here is due to Malécot (1948).

The results of this section show that, in a population that has been mating at random for many generations, a strong association between traits is not due to linkage unless the amount of recombination between the responsible loci is very small. Such persisting associations as are found in natural populations are probably due to pleiotropy or to recent amalgamation of populations with incomplete mixing, that is to say nonrandom mating.

With selection and epistasis, especially with strong interaction and close linkage, there can be permanent gametic phase disequilibrium. We discuss this in Chapter 5.

2.7 More Than Two Loci

It is possible for each group of two among a set of three loci to be in random combination, yet for the whole set not to be random. We should expect, however, that with random mating all gametic combinations would eventually reach equilibrium proportions. That this is in fact correct has been shown, first for three loci by Geiringer (1945), and more generally by Bennett (1954*a*).

The reader who trusts his intuition on this point is invited to skip the rest of this section. For the skeptic we give the derivation for three loci.

Let $P_t(ABC)$ be the probability that a gamete from generation t carries alleles A_i, B_j, and C_k at the three loci under consideration. Let p_i, q_j, and r_k be the frequencies of these three alleles. No restriction is placed on the number of alleles at any locus; but we are discussing a particular one from each. For simplicity of writing we drop the subscripts since we refer to only one allele at each locus.

We designate by a the frequency of the recombinational event of the type that separates A from B and C, b that of the type that separates B, and c that for C. For example, if the gene order is ABC, a is the probability of a crossover between the A and B loci, and b is that of a double crossover. In the special case of independent loci, $a = b = c = 1/4$. We let $K = 1 - a - b - c$ be the probability of a nonrecombinant gamete.

We can obtain an ABC gamete in four ways:

(1) From an ABC chromosome that has not undergone any crossover. The probability is K times the frequency of this chromosome in the gametes produced by the preceding generation.

(2) From a recombination that combines B and C from one parent with A from the other. The probability of this is $ap_i P(BC) = apP(BC)$, where $P(BC)$ is the probability of a gamete carrying B_j and C_k in the previous generation irrespective of whether it carries A or not.

(3) Recombination that combines B_j with A_i and C_k, with probability $bqP(AC)$.

(4) A recombination that combines C with A and B, with probability $crP(AB)$.

Putting these together, we obtain

$$P_t(ABC) = KP_{t-1}(ABC) + apP_{t-1}(BC) + bqP_{t-1}(AC) + crP_{t-1}(AB).$$

We now subtract from the left side of the equation the quantity

$$p[P_t(BC) - qr] + q[P_t(AC) - pr] + r[P_t(AB) - pq]$$

and from the right side the equivalent quantity

$$p(1 - b - c)[P_{t-1}(BC) - qr] + q(1 - a - c)[P_{t-1}(AC) - pr] \\ + r(1 - a - b)[P_{t-1}(AB) - pq].$$

That these two expressions are equivalent can be seen by reference to equation 2.6.2. After making these subtractions we note that, since $K = 1 - a - b - c$, $1 - b - c = K + a$, $1 - a - c = K + b$, and $1 - a - b = K + c$. Making these substitutions in the right side, subtracting pqr from both sides, and rearranging, we arrive at the following equation:

$$\begin{aligned}[P_t(ABC) - pqr] - p[P_t(BC) - qr] - q[P_t(AC) - pr] - r[P_t(AB) - pq] \\ = K\{[P_{t-1}(ABC) - pqr] - p[P_{t-1}(BC) - qr] - q[P_{t-1}(AC) - pr] \\ - r[P_{t-1}(AB) - pq]\}.\end{aligned} \qquad 2.7.1$$

If we represent the left side of 2.7.1 by the function $L_t(ABC)$, we can write

$$\begin{aligned}L_t(ABC) &= KL_{t-1}(ABC) \\ &= K^t L_0(ABC).\end{aligned} \qquad 2.7.2$$

The intuitive conclusion that linkage equilibrium is approached is proven by noting first that, since $K < 1$, $L_t(ABC)$ approaches 0. Furthermore, from 2.6.1, each of the terms such as $P_t(AB) - pq$ approaches 0. Thus $P_t(ABC)$ must approach pqr.

The extension to more than three alleles can be done in a similar way. Bennett (1954*a*) has shown how to define a function such as $L_t(ABC)$ for four or more loci which satisfies an equation like 2.7.2.

2.8 Polyploidy

In the main this book will be restricted to diploid populations (or, sometimes for simplicity, haploids). However, we shall occasionally make brief comments about polyploid populations, usually without proof, but with references to the pertinent literature.

Polyploid inheritance is complicated by the large number of essentially different modes of gamete formation, determined not only by the frequency of recombination between two loci but also by the recombination between each of these and its centromere. These modes have been classified in a systematic way by Fisher (1947*a*).

The principle of Sections 2.6 and 2.7, that with random mating the chromosome frequencies approach the product of the component allele frequencies, is also true for polyploids. Bennett (1954) has shown this for two loci in tetraploids and hexaploids and has given explicit formulae for the frequencies at any time in terms of the initial frequencies and the probabilities of the different modes of gamete formation. Bennett's methods are rather sophisticated and an elementary derivation for tetraploids has been given by Crow (1954).

A peculiarity of polyploid inheritance is that the simple Hardy–Weinberg principle is not always true. The single-locus equilibrium is not attained in a single generation, as with diploids, but is approached gradually. Furthermore, except for a locus that is completely linked to its centromere, the equilibrium zygote frequencies are not the products of the gene frequencies. We shall illustrate this for a tetraploid, following the methods of Bennett (1968).

Assume that there are n alleles at this tetraploid locus, with frequencies $p_1, p_2, \ldots, p_n$. Let $P_{ii,t}$ be the frequency of the $A_i A_i$ (diploid) gamete in generation t. Likewise let $2P_{ij,t}$ be the frequency of an $A_i A_j$ gamete. We shall designate the frequency of double reduction by the traditional symbol, α. By this, we mean the probability of a homoallelic gamete from a plant of genotype $A_1 A_2 A_3 A_4$; it is a function of the amount of recombination between the locus and the centromere.

If there has been double reduction, the gamete is necessarily homoallelic and the probability of its being $A_i A_i$ is simply the frequency of this gene in the population. In the $1-\alpha$ cases where double reduction does not occur, two of the four chromosomes are chosen. In 1/3 of the cases, the two genes will have come from the same gamete in the previous generation, in which case the probability of their both being A_i is the same as in the gametes of the previous generation. In the other 2/3, the two genes will come from different gametes in the previous generation, in which case they are independent and the probability of their both being A_i is p_i^2. Putting all this together,

$$P_{ii,t} = (1-\alpha)(\tfrac{1}{3}P_{ii,t-1} + \tfrac{2}{3}p_i^2) + \alpha p_i. \qquad 2.8.1$$

Similarly,

$$P_{ij,t} = (1-\alpha)(\tfrac{1}{3}P_{ij,t-1} + \tfrac{2}{3}p_i p_j). \qquad 2.8.2$$

At equilibrium, $P_{ij,t} = P_{ij,t-1} = \hat{P}_{ij}$. Making this substitution in 2.8.2 leads to

$$\hat{P}_{ij} = (1-f)p_i p_j, \qquad f = 3\alpha/(2+\alpha). \qquad 2.8.3$$

Likewise,

$$\hat{P}_{ii} = (1-f)p_i^2 + fp_i. \qquad 2.8.4$$

Therefore, at equilibrium the proportion of heteroallelic gametes is reduced by a fraction f, and the homozygotes are correspondingly increased. We have used the letter f intentionally, because equations like 2.8.3 and 2.8.4 will occur in Chapter 3 when we consider the reduction in heterozygosity caused by inbreeding in diploids, and f corresponds to the coefficient of inbreeding.

When the gene is near the centromere, α approaches 0. When the gene is more and more distant from the centromere, α approaches 1/7 as a limit. When $\alpha = 1/7$, $f = 1/5$; so we can say that the limiting decrease in heteroallelic gametes is 1/5.

The Hardy–Weinberg principle holds true in the sense that the zygote frequencies are given by the products of the appropriate gamete frequencies, just as in diploids. Thus the frequency of a zygote of type $A_iA_iA_iA_j$ would be $2(P_{ii})(2P_{ij})$.

Finally, by subtracting $\hat{P}_{ij}$ from both sides of 2.8.2 and simplifying we find

$$\begin{aligned} P_{ij,t} - \hat{P}_{ij} &= \frac{1-\alpha}{3}(P_{ij,t-1} - \hat{P}_{ij}) \\ &= \left(\frac{1-\alpha}{3}\right)^t (P_{ij,0} - \hat{P}_{ij}), \end{aligned} \qquad 2.8.5$$

which shows the rate at which the equilibrium is approached.

2.9 Subdivision of a Population: Wahlund's Principle

In nature or in domestic animals and plants the population is often structured. One possibility is that it is divided into subpopulations or isolates, between which there is partial or complete isolation. In such strains the gene frequencies may diverge, either because of different environments that favor different genotypes, or simply by chance if the subpopulations are small. We inquire into the effect of amalgamation of previously isolated subpopulations.

The key relationship is expressible in terms of the variance in gene frequencies among the subpopulations. The formula was discussed by Wahlund (1928) and is often called Wahlund's principle.

Imagine that there are k subpopulations, completely isolated from each other and of sizes $n_1, n_2, \ldots$ and n_k. Mating within each subpopulation is assumed to be at random. Let the frequency of allele A be $p_1, p_2, \ldots$ and p_k in these subpopulations. Then the mean proportion of AA homozygotes in the whole population is

$$\frac{n_1 p_1^2 + n_2 p_2^2 + \cdots + n_k p_k^2}{n_1 + n_2 + \cdots + n_k} = \overline{p^2}. \qquad 2.9.1$$

Now suppose that these populations are pooled into a single panmictic unit. The average frequency of the A allele is now (as before) $\bar{p}$, the weighted average of the frequencies in the different populations. Then the proportion of AA homozygotes in the pooled population after one generation of random mating is $\bar{p}^2$.

Recall that the variance is $V_p = \overline{p^2} - \bar{p}^2$ (see A.2.2 and A.2.4). Hence

$$\bar{p}^2 = \overline{p^2} - V_p, \qquad 2.9.2$$

where V_p is the variance in the frequency of the gene A among the k subpopulations.

This explains why the proportion of individuals with recessive traits is reduced by migration between previously isolated communities. Since the variance is always positive, there will always be a decrease unless the gene frequency is identical in the subpopulations. The magnitude of the decrease will depend on the diversity of frequencies among the populations, as measured by the variance.

The previous discussion has referred to a situation where two or more populations are pooled, and then matings occur without regard to the origin of the individuals. The situation is somewhat different if the first matings are all between individuals from different populations. We shall consider only two populations.

If p_1 and p_2 represent the frequency of the A gene in the two populations, the proportion of AA homozygotes in the F_1 hybrids is $p_1 p_2$. The gene frequency in the F_1 is the mean of the two parent population frequencies, or $\bar{p} = (p_1 + p_2)/2$. The variance in the two original populations (equally weighted) is

$$V_p = \tfrac{1}{2}p_1^2 + \tfrac{1}{2}p_2^2 - \bar{p}^2 = \tfrac{1}{4}(p_1 - p_2)^2. \quad \textbf{2.9.3}$$

Note that $\bar{p}^2 - V_p = p_1 p_2$, which is the frequency of AA homozygotes in the F_1 population.

For comparison, the proportion of AA homozygotes in the three populations is:

(1)	Separate populations	$\bar{p}^2 + V_p$,
(2)	F_1 population	$\bar{p}^2 - V_p$,
(3)	F_2 and later	$\bar{p}^2$.

Hybridization between two populations causes an initial decrease in homozygosity, followed by a rise to a point halfway between. This argument does not consider linkage, the effect of which is to slow the approach to the final value.

2.10 Random-mating Proportions in a Finite Population

The Hardy–Weinberg proportions are realized exactly only in an infinite population. For one thing, a finite population is subject to chance deviations from the expected proportions. There is also a systematic bias because of the discreteness of the possible numbers of different genotypes. The bias can become important if there are a number of individually very rare alleles. For example, one might determine the allele frequencies from a natural population and then wish to inquire if these are in random-mating proportions. The problem has been considered in detail by Hogben (1946) and Levene (1949).

Consider a population of size N. Since we are considering diploid populations, there are $2N$ genes per locus. Let p_i be the proportion of allele A_i in this population; hence there are $2Np_i$ representatives of the A_i gene. Then we regard the zygotes as made up by combining these $2N$ genes at random in pairs. The probability of drawing an A_i allele is $2Np_i/2N$; after this is done, the probability of drawing another A_i allele from the remaining genes is $(2Np_i - 1)/(2N - 1)$. Thus the expected proportion of $A_i A_i$ individuals, given that there are exactly $2Np_i$ A_i alleles, is

$$P(A_i A_i) = \frac{2Np_i}{2N} \cdot \frac{2Np_i - 1}{2N - 1} = p_i^2 - p_i(1 - p_i)f, \quad \textbf{2.10.1}$$

where $f = 1/(2N - 1)$. Likewise the expected proportion of $A_i A_j$ heterozygotes is

$$P(A_i A_j) = \frac{2Np_i}{2N} \cdot \frac{2Np_j}{2N - 1} + \frac{2Np_j}{2N} \cdot \frac{2Np_i}{2N - 1} = 2p_i p_j(1 + f). \qquad 2.10.2$$

Thus the heterozygotes are increased by a fraction $f = 1/(2N - 1)$ and the homozygotes are correspondingly decreased, in comparison with the proportions in an infinite population with the same allele frequencies.

As a simple example, consider a population with only two alleles at the A locus, A_1 and A_2. Assume further that the A_1 allele is represented only once. Thus $p_1 = 1/2N$ and $p_2 = 1 - p_1$. Substituting these values into 2.10.1 we obtain 0 for the frequency of A_1A_1, as we should; for if there is only one A_1 gene there can be no homozygotes. Furthermore, substituting into 2.10.2 leads to a frequency of A_1A_2 heterozygotes of $1/N$; this is also correct, since only one heterozygote exists in the population of N individuals.

In Chapter 3 we shall see that within a finite population there is an opposite effect, a decrease in heterozygosity. However, this decrease is strictly due to changes in the gene frequencies due to random gains and losses in a small population. Within the population the relation between the gene and genotype frequencies is given by the Hardy–Weinberg principle, with the slight correction given here.

2.11 Problems

In all problems, unless the contrary is stated, assume random mating.

1. In a population there are 8 times as many heterozygotes as homozygous recessives. What is the frequency of the recessive gene?
2. Show that, for a very rare recessive gene, the proportion of heterozygous carriers is approximately twice the frequency of the recessive gene.
3. If 16% of the population are Rh− (*dd*), what fraction of the Rh+ population (*DD* and *Dd*) are homozygous?
4. From the data in problem 3, what fraction of the children from a large group of families where both parents were Rh+ would be expected to be Rh+?
5. Show that if the *A* and *B* antigens of the *ABO* blood group system were caused by two dominant genes, independently inherited, the product of the frequency of *A* and *B* should equal the product of *O* and *AB*.
6. What is the maximum proportion of heterozygotes with two alleles? With three alleles? With *n* alleles? (See A.10)
7. From the data given on color blindness, what fraction of women would be of normal vision, but carrying two different color-blind factors?

8. Show that in a randomly-mating population with two alleles half the heterozygotes have heterozygous mothers.
9. Here are some hypothetical data on the frequencies of the *ABO* blood groups:

Genotype:	*OO*	*OA*	*AA*	*OB*	*BB*	*AB*
Frequency:	.40	.30	.08	.12	.04	.06

What would the frequencies be next generation if mating were at random?
10. Letting p, q, and r stand for the frequencies of the *A*, *B*, and *O* blood group alleles, what is the probability that two persons chosen at random have the same blood group?
11. An outrageously careless hospital gives transfusions at random. What proportion would be mismatches? (To refresh your memory, group *O* can give to anyone, *A* to *A* or *AB*, *B* to *B* or *AB*, and *AB* only to *AB*.)
12. Show that if p is the frequency of a recessive allele, the average proportion of recessive children when one parent is of the dominant phenotype and the other of the recessive phenotype is $p/(1 + p)$, and when both parents are of the dominant phenotype is $[p/(1 + p)]^2$.
13. The two ratios, $p/(1 + p)$ and $[p/(1 + p)]^2$, are sometimes called Snyder's ratios and the fact that one is the square of the other is sometimes used as a test for recessive inheritance. Does this discriminate between a trait caused by a single pair of recessive genes and one caused by simultaneous homozygosity for several recessive genes?
14. Does the answer to problem 13 depend on whether the genes are independent or not? Must they be in gamtiec phase equilibrium?
15. A. G. Searle (*J. Genet.* 56: 1–17, 1959) reports the following frequencies of coat colors of cats in Singapore. The observed numbers were as follows:

FEMALES			MALES	
DARK	CALICO	YELLOW	DARK	YELLOW
$+/+$	$+/y$	y/y	$+$	y
63	55	12	74	38

Use the maximum-likelihood method to compute the gene frequency and test by Chi-square the agreement with the hypothesis of random-mating proportions.
16. Give an example of a set of gamete frequencies for three loci such that any two are in linkage equilibrium, but the set of three are not in equilibrium.

17. Genes A and B are linked with 20% recombination between them. An initial population is composed of AB/AB, AB/ab, and ab/ab plants in the ratio of 1 : 2 : 1. The population is allowed to pollinate at random.
    a. What would be the frequencies of the four kinds of chromosomes in the next generation?
    b. What would be the frequency of the AB/aB genotype in the next generation?
    c. What would be the chromosome frequencies when equilibrium is reached?
    d. What would be the frequency of the AB/aB genotype at equilibrium?
    e. How many generations would be required for the population to go halfway to equilibrium?
18. Two homozygous strains *aa bb* and *AA BB* are crossed. The A and B loci are on separate chromosomes. Show that these loci are in linkage equilibrium in the F_2 generation. Why doesn't equation 2.6.3 apply?
19. Show that in an autotetraploid the value of α is 1/7 as the locus becomes far enough from the centromere to be independent of it.
20. If p_1, p_2, p_3, and p_4 represent the frequencies of alleles A_1, A_2, A_3, and A_4 in a randomly mating tetraploid population that has reached equilibrium, and the relevant locus is very far from the centromere, what will be the frequency of $A_1A_1A_1A_1$ plants? $A_1A_1A_2A_2$? $A_1A_2A_3A_4$?
21. The equations $p_O = \sqrt{O}$, $p_A = 1 - \sqrt{B + O}$, and $p_B = 1 - \sqrt{A + O}$, where O, A, and B represent the frequencies of these three blood groups, are often used to estimate the gene frequencies. Show that these are not the solutions to the maximum-likelihood equations. Derive these equations from the relations below equation 2.3.2.
22. Two plausible hypotheses that explain the much greater incidence of early baldness in males than in females are (1) an autosomal dominant that is normally expressed only in males and (2) an X-linked recessive. If the first hypothesis is correct, and q is the frequency of the gene for baldness, what proportion of the sons of bald fathers are expected to be bald? What proportion from nonbald fathers? What are the corresponding expectations on the X-linked recessive hypothesis?
23. Harris (*Ann. Eugen.* 13: 172–181, 1946) found that 13.3% of males in a British sample were prematurely bald. He also found that of 100 bald men, 56 had bald fathers. Show that this is consistent with the sex-limited dominant hypothesis but not the sex-linked recessive. (You may want to satisfy yourself that the expected fraction of bald sons when the father is bald is the same as the expected fraction of bald fathers when the son is bald. It is easier to get data by selecting a group of bald men and inquiring about their fathers than it is to wait for their sons to grow up.)

24. Show that if a group of previously isolated populations are pooled the proportion of heterozygotes for alleles A_i and A_j is equal to the average proportion of heterozygotes before pooling minus twice the covariance of the 2-allele frequencies. Show also that when there are only two alleles the covariance is minus the variance.
25. Prove the statements in the legend of Figure 2.11.1. Assume that the base

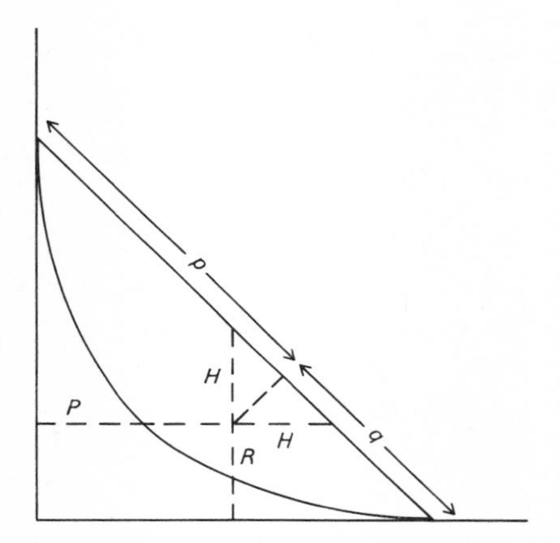

Figure 2.11.1. Representation of a population as a point in a 2-dimensional diagram. *P*, *H*, and *R* represent the frequencies of the genotypes *AA*, *Aa*, and *aa*; *p* and *q* are the frequencies of the *A* and *a* alleles. *P* is given by the distance from the vertical axis, *R* by that from the horizontal axis. *H* is given by either the horizontal or vertical distance to the hypotenuse of the triangle. All possible populations lie within the triangle; populations in Hardy–Weinberg ratios lie along the parabola.

and altitude are each equal to 1. In particular, show that $P + H + R = 1$, and that the perpendicular line from the point to the hypotenuse divides it in the ratio $p : q$. Show also that the equation of the Hardy–Weinberg parabola is $P^2 - 2PR + R^2 - 2P - 2R + 1 = 0$. [You might find it useful to note that, with random-mating proportions, $H^2 = 4PR$.]

26. Another way of representing a population was used by De Finetti (1926). Prove that, if the altitude of the triangle is 1, $P + H + R = 1$. Show also that the perpendicular from the point to the base divides it in the ratio $p : q$. See Figure 2.11.2.

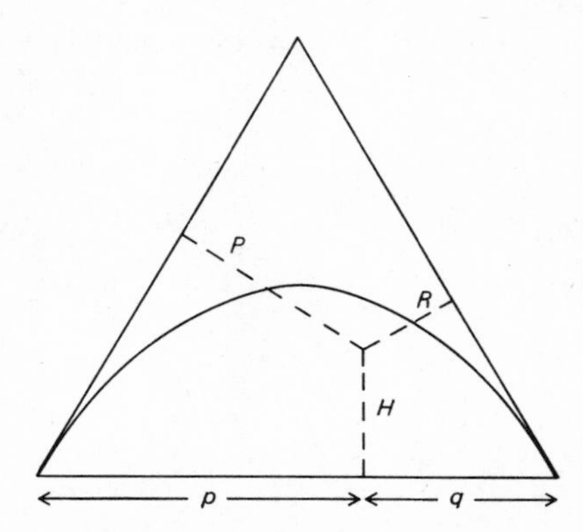

Figure 2.11.2. De Finetti diagram of a population in triangular coordinates. *P*, *H*, and *R* are given by the perpendicular distances to the three sides. The vertical line divides the base in the ratio of the gene frequencies, *p*, and *q*. Points in Hardy–Weinberg ratios lie along the curve.

27. Use Wahlund's principle or the definition of the variance to show that with *n* alleles the minimum homozygosity with random mating occurs when all alleles are equally frequent.
28. Assume two segregating loci, each with two alleles. Let P_{AB}, P_{Ab}, P_{aB}, and P_{ab} be the frequency of the four chromosomes and *c* the amount of recombination between the two loci. A conventional measure of linkage disequilibrium is $D = P_{AB}P_{ab} - P_{Ab}P_{aB}$. What is the equilibrium value of *D*? How fast is the equilibrium approached?

3 INBREEDING

Inbreeding occurs when mates are more closely related than they would be if they had been chosen at random from the population. Related individuals have one or more ancestors in common, so the extent of inbreeding is related to the amount of ancestry that is shared by the parents of the inbred individuals. Alternatively stated, the degree of inbreeding of an individual is determined by the proportion of genes that his parents have in common.

An immediate consequence of this sharing of parental genes is that the inbred individual will frequently inherit the same gene from each parent. Thus inbreeding increases the amount of homozygosity. So one observable effect of inbreeding is that recessive genes, previously hidden by heterozygosity with dominant alleles, will be expressed. Since most such genes are harmful in one way or another, inbreeding usually leads to a decrease in size, fertility, vigor, yield, and fitness. There are also likely to be loci segregating in a population where a heterozygote is fitter than either of the two corresponding homozygotes. In this case, too, inbreeding leads to a decreased fitness.

Another consequence of consanguineous mating within the population is greater genetic variability, since similar genes tend to be concentrated in the same individuals. Usually, because of the correlation between genotype and phenotype, this leads to an increase in phenotypic variability.

Inbreeding may follow either of two patterns. There may be a certain amount of consanguineous mating within a population, with the consequences just mentioned. On the other hand, the inbreeding may be such as to break the population into subgroups. An extreme example is continued self-fertilization in which the population (if it is of constant size and each parent contributes equally to the next generation) is divided into a set of subpopulations of one individual each. Likewise, a pattern of repeated sib mating could lead to a series of isolated populations of size 2. As a third example, there may be a natural population which is divided into isolated subpopulations, within each of which mating is random or nearly so. The effect will be that each subpopulation becomes more homozygous, and therefore the whole population does. The individual subpopulations become more uniform genetically; but, since they become homozygous for different genes, the population as a whole becomes more variable. Of course there may be only partial isolation, with intermediate consequences.

A point that at first seems paradoxical is that within a subpopulation there is an increase in homozygosity despite the fact that mating within this group is random. The reason, as will be discussed in Section 3.11, is that there are random changes in the frequencies of the individual alleles and these, on the average, lead to a decrease in heterozygosity. As an extreme example, self-fertilization can be regarded as random mating (i.e., random combination of gametes) within a population of one. The gene frequencies at different, previously heterozygous loci change from 1/2 to 0 or 1.

Whether inbreeding leads to subdivision or not, it can be measured in the same way—by Wright's (1922) coefficient of inbreeding, f, which measures the proportion by which the heterozygosity has been decreased. As we shall show later, other population properties can also be related to f. However, before discussing f, we shall illustrate with two simple examples the effect of continued inbreeding.

3.1 Decrease in Heterozygosity with Inbreeding

The qualitative effect of continued inbreeding can be seen by examining the most extreme form, self-fertilization. In a self-fertilized population the progeny of homozygotes are like their parents, whereas the progeny of heterozygotes are 1/2 heterozygotes and 1/4 each of the two homozygous types.

Thus, in each generation the proportion of heterozygous loci is reduced by half and the homozygous types are correspondingly increased. This is illustrated in Table 3.1.1.

Table 3.1.1. The changes in the probabilities of different genotypes with continued self-fertilization. D, H, and R stand for the initial proportions of dominant, heterozygous, and recessive types.

GENERATION	FREQUENCY OF GENOTYPE		
	AA	Aa	aa
0	D	H	R
1	$D+H/4$	$H/2$	$R+H/4$
2	$D+3H/8$	$H/4$	$R+3H/8$
3	$D+7H/16$	$H/8$	$R+7H/16$
4	$D+15H/32$	$H/16$	$R+15H/32$
Limit	$D+H/2$	0	$R+H/2$

If H_0 is the initial proportion of heterozygotes, the proportion after t generations of self-fertilization is $H_0/2^t$. If the original population were panmictic, with AA, Aa, and aa genotypes in the proportions p^2, $2pq$, and q^2 $(p+q=1)$, the individual lines eventually become homozygous. The probability of being AA is $D+H/2=p^2+pq=p(p+q)=p$; likewise the probability of being aa is q. Thus, the population becomes broken into separate lines, each homozygous for one or the other of the genes in the ratio of their original frequencies in the population. Notice one other fact: There has been no change in the gene frequency. Inbreeding per se does not change the proportions of the various genes, only the way they are combined into homozygous and heterozygous genotypes.

With less extreme forms of inbreeding the results are similar, though the change in heterozygosity is less rapid. The results for continued brother-sister mating are shown in Table 3.1.2. Again there is a decrease in the proportion of heterozygotes, with the amount deducted being divided equally and added on to the two homozygous types.

These results may be obtained by writing out all the possible matings generation after generation, as was done by the early investigators (Fish, 1914; Jennings, 1916). This and several other systems of recurrent inbreeding were worked out by these authors. The papers are now mainly of historical interest since more general methods are available. We shall discuss them

Table 3.1.2. The decrease in heterozygosity with successive generations of brother-sister mating.

GENERATION	RELATIVE HETEROZYGOSITY	DECREASE IN HETEROZYGOSITY	RATES OF CHANGE IN HETEROZYGOSITY	
t	$\frac{H_t}{H_0}=P$	$\frac{H_0-H_t}{H_0}=f$	$\frac{H_t}{H_{t-1}}$	$\frac{H_{t-1}-H_t}{H_{t-1}}$
0	1	0	1	0
1	2/2	0	1	0
2	3/4	1/4	3/4 = .750	.250
3	5/8	3/8	5/6 = .833	.167
4	8/16	8/16	8/10 = .800	.200
5	13/32	19/32	13/16 = .812	.188
6	21/64	43/64	21/26 = .808	.192
Limit	0	1	λ = .809	.191

in Sections 3.4 and 3.8, where the results of this table will appear as a special case.

In this example the heterozygosity follows a simple rule. The numerator in successive generations is given by the Fibonacci series in which each term is the sum of the two preceding terms, while the denominator doubles each generation. The number 1 in the second row is written as 2/2 to make the sequence more obvious. The reduction in heterozygosity, expressed as a fraction of the initial heterozygosity, is the same regardless of the initial gene frequencies and, as we shall show later, the number of alleles.

The relative heterozygosity, H_t/H_0, has been called by Wright (1951) the *panmictic index*, for which he used the letter P. $1-P$ is the inbreeding coefficient, for which Wright has used the letter F. (We shall use the lower case f in order to reserve F for multiple-locus inbreeding effects.)

The last two columns give the rate of change in heterozygosity. Notice that the ratio H_t/H_{t-1} after a few oscillations rapidly approaches a constant value. The limiting value of the ratio of heterozygosity to that in the previous generation is usually designated by λ (Fisher, 1949).

3.2 Wright's Inbreeding Coefficient, *f*

Wright's (1922) original derivation of the inbreeding coefficient, f, was through correlation analysis. An alternative approach using only probability rules has been developed by Haldane and Moshinsky (1939), Cotterman (1940), and

Malécot (1948). They distinguish between two ways in which an individual can be homozygous for a given locus. The two homologous genes may be: (1) alike in state, that is to say, indistinguishable by any effect they produce (or perhaps, when molecular genetics has become sufficiently precise, alike in their nucleotide sequence), and (2) identical by descent, in that both are derived from the same gene in a common ancestor.

We follow the notation of Cotterman in designating an individual whose two homologous genes are identical by descent as *autozygous*. If the two alleles are of independent origin (as far as known from our pedigree information), the individual is *allozygous*. The effect of inbreeding is to increase that part of the homozygosity that is due to autozygosity. (Notice that an individual can be homozygous without being autozygous, if the two homologous genes are alike in state but not identical by descent. Conversely, an autozygous individual can be heterozygous for this locus if one of the two alleles has mutated since their common origin, although this is negligibly rare if only a small number of generations is being considered.)

The inbreeding coefficient, f, is defined as the probability that the individual is autozygous for the locus in question. Alternatively stated, it is the probability that a pair of alleles in the two gametes that unite to form the individual are identical by descent.

An individual with inbreeding coefficient f has a probability f that the two genes at a particular locus are identical and a probability $1 - f$ that they are not identical, and therefore independent. If they are independent the frequencies of the genotypes will be given by the binomial formula. If they are identical, the frequencies of the gene pairs will be simply the frequencies of the alleles in the population. Thus, for two alleles, A_1 and A_2, with frequencies p_1 and p_2 $(p_1 + p_2 = 1)$, the genotype frequencies are:

	ALLOZYGOUS		AUTOZYGOUS	
Homozygous, A_1A_1:	$p_1^2(1-f)$	+	$p_1 f$	
Heterozygous, A_1A_2:	$2p_1p_2(1-f)$			3.2.1
Homozygous, A_2A_2:	$p_2^2(1-f)$	+	$p_2 f$	
Total	$1-f$		f	

Notice that when f is 0 these formulae reduce to the usual Hardy–Weinberg proportions. When $f = 1$ the population is completely homozygous. Thus f ranges from 0 in a randomly mating population to 1 with complete homozygosity. How to compute f from a pedigree will be shown later.

Multiple alleles introduce no difficulty. The genotype frequencies are a natural extension of the results for two alleles. The frequencies are

$$A_iA_i : p_i^2(1-f) + p_i f \qquad 3.2.2$$

for homozygous genotypes, and

$$A_iA_j : 2p_ip_j(1-f) \qquad 3.2.3$$

for heterozygous genotypes.

The expected proportion of heterozygous genotypes with inbreeding coefficient f, H_f, is given by

$$H_f = \sum_{i \neq j} p_i p_j(1-f) = H_0(1-f); \quad f = \frac{H_0 - H_f}{H_0}, \qquad 3.2.4$$

where H_0 is a constant equal to the proportion of heterozygotes expected with random mating ($f = 0$). The summation is over all combinations of values of i and j except when these are equal.

This proves the assertion made earlier that the inbreeding coefficient measures the fraction by which the heterozygosity has been reduced. We have written the formula as if, when $f = 0$, the population is in Hardy–Weinberg proportions. However, for any measured f (as determined, for example, from a pedigree), the heterozygosity, H, is $H_0(1-f)$, where H_0 is whatever the heterozygosity would have been in the absence of the observed inbreeding. To be concrete, the inbreeding coefficient for the child of a cousin marriage is 1/16 (as we shall show later); therefore the child of such a marriage is 15/16 as heterozygous as if his parents had the same relationship as a random pair in this population.

There is a simple relationship between the correlation coefficient, r, and the inbreeding coefficient, f. If we assign numerical values to each allele, then the inbreeding coefficient, f, is the correlation between these values in a pair of uniting gametes. In fact, Wright's original derivation of the inbreeding coefficient was through correlation methods.

The relationship between r and f can be shown in the following way. For convenience we assign the value 1 to allele A_1 and 0 to allele A_2, though we would get the same result with any values. The calculations are shown in Table 3.2.1.

Since the sum of the genotype frequencies is equal to 1, the weighted sum and the mean of any value are the same. For example, the sum (and mean) of the egg value, X, is $[p_2^2(1-f) + p_2 f](0) + [p_1p_2(1-f)](1) + [p_2p_1(1-f)](0) + [p_1^2(1-f) + p_1 f](1)$, which after some algebraic simplification reduces to p_1. The other calculations are given in the table, using the standard formula for calculation of r given in A.4.3.

The calculations in Table 3.2.1. are made by assuming that there are only two alleles and letting them have the values 0 and 1. The correlation interpretation of f, however, is completely general. Table 3.2.2 gives the same

Table 3.2.1. Demonstration of the equivalence of the inbreeding coefficient, f, and the coefficient of correlation, r_{xy}, between the genetic values of the uniting gametes.

EGG	SPERM	FREQUENCY OF THIS COMBINATION	VALUE OF EGG X	VALUE OF SPERM Y	X^2	Y^2	XY
A_2	A_2	$p_2^2(1-f)+p_2 f$	0	0	0	0	0
A_1	A_2	$p_1 p_2(1-f)$	1	0	1	0	0
A_2	A_1	$p_2 p_1(1-f)$	0	1	0	1	0
A_1	A_1	$p_1^2(1-f)+p_1 f$	1	1	1	1	1
Sum or Mean		1	p_1	p_1	p_1	p_1	$p_1^2(1-f)+p_1 f$

$$\bar{X} = p_1 p_2(1-f) + p_1^2(1-f) + p_1 f = p_1.$$

Likewise, $\bar{Y} = \overline{X^2} = \overline{Y^2} = p_1.$

$$r_{xy} = \frac{\overline{XY} - \bar{X}\bar{Y}}{\sqrt{(\overline{X^2} - \bar{X}^2)(\overline{Y^2} - \bar{Y}^2)}} = \frac{p_1^2(1-f) + p_1 f - p_1^2}{p_1 - p_1^2} = f.$$

demonstration without restriction as to number of alleles and letting the contribution of the alleles differ. Furthermore, if the genic values are summed over k loci the covariance will be

$$f \sum_k \sum_i p_{ik} a_{ik}^2,$$

where p_{ik} and a_{ik} are the frequency and value of the ith allele at the kth locus. This is f times the variance. Hence f is the expected value of the correlation between the genetic values of two uniting gametes, regardless of the number of loci and number of alleles under consideration.

The equivalence of r and f suggests an interpretation of the correlation coefficient. If a measurement can be thought of as being the sum of a number of elements, then the correlation coefficient is the measure of the fraction of these elements that are common to the two measurements, the other elements being chosen at random. This interpretation is useful in many branches of science. In quantitative genetics the elements can obviously be interpreted as cumulatively acting genes.

The computation of f will be discussed in Section 3.4.

Table 3.2.2. Demonstration of the equivalence of the inbreeding coefficient and the correlation between the genetic value of the uniting gametes regardless of the contribution of the individual genes and the number of alleles. The contribution, or value, of allele A_i is assumed to be a_i, measured as a deviation from the mean value.

EGG	SPERM	FREQUENCY OF THIS COMBINATION	VALUE OF EGG X	VALUE OF SPERM Y	X^2	Y^2	XY
A_i	A_i	$p_i^2(1-f)+p_i f$	a_i	a_i	a_i^2	a_i^2	a_i^2
A_i	A_j	$p_i p_j(1-f)$	a_i	a_j	a_i^2	a_j^2	$a_i a_j$

$$V_X = \sum_i p_i a_i^2,$$

since the variance of the egg value is the sum of the squares of the allele values, each weighted by its frequency. V_Y is the same.

$$\text{Cov}_{XY} = (1-f)\Big[\sum_i p_i^2 a_i^2 + \sum_{i \neq j} p_i p_j a_i a_j\Big] + f \sum_i p_i a_i^2.$$

But the quantity in brackets is equal to $[\sum p_i a_i]^2$ which is equal to 0, because the sum of the deviations from the mean is 0. Therefore,

$$\text{Cov}_{XY} = f \sum p_i a_i^2.$$

The correlation coefficient, being the ratio of the covariance to the geometric mean of the two variances (which in this case are the same), is f, as was to be shown.

$$r_{XY} = \text{Cov}_{XY}/V_X = f.$$

3.3 Coefficients of Consanguinity and Relationship

We have used the inbreeding coefficient of an individual I, f_I, to give the probability that two homologous genes in that individual are identical by descent. Or, as just shown, this is the correlation between the genetic value of the two gametes that united to produce the individual. Since inbreeding of the progeny depends on the consanguinity of the parents we can use the inbreeding coefficient as a measure of this.

We define the *coefficient of consanguinity*, f_{IJ}, of two individuals I and J as the probability that two homologous genes drawn at random, one from each of the two individuals, will be identical. The answer to this is clearly the same as the inbreeding coefficient of a progeny produced by these two individuals. Hence the inbreeding coefficient of an individual is the same as the coefficient of consanguinity of its parents (Malécot, 1948).

There is a bewildering plethora of alternative names for this coefficient. Malécot, who introduced the idea, called it the *coefficient de parenté*. Falconer (1960) calls it the *coancestry*. Kempthorne (1957) translated *parenté* into *parentage*. Malécot himself has, on at least one occasion, translated it into *kinship*. We shall use either consanguinity or kinship.

A different measure of relatedness, introduced much earlier and still widely used, is Wright's (1922) *coefficient of relationship*, r_{IJ}, defined as:

$$r_{IJ} = \frac{2f_{IJ}}{\sqrt{(1 + f_I)(1 + f_J)}}. \qquad 3.3.1$$

For two individuals that are not inbred, the coefficient of relationship is exactly twice the coefficient of consanguinity.

As we shall show later, the coefficient of relationship is the correlation between the genic, or genetic, values of the two individuals. If the genes act without dominance or epistasis, and there is no effect of the environment on the trait being measured, this is the expected correlation. We shall also show later the effect of dominance on the correlation between relatives (Section 4.3).

3.4 Computation of *f* from Pedigrees

The procedure for computing the inbreeding or consanguinity coefficient from a pedigree follows directly from the definition of f. Consider the pedigree in Figure 3.4.1.

In this pedigree individual I is inbred because both his parents are descended from a single common ancestor, A. All unrelated ancestors, which are irrelevant to the inbreeding of I, are omitted from the pedigree. We ask for the probability that I is autozygous; i.e., that the homologous genes contributed to I by gametes b and e are both descended from the same gene in ancestor A. We shall use the notation Prob($c = b$) to mean the probability that c and b carry identical genes for the locus under consideration.

Prob($c = b$) = 1/2, since the gene in b has an equal chance of having come from C or from B's other parent. Likewise, Prob($c = a$) = 1/2. The probability that a and a' carry identical genes may be obtained as follows:

Let the two alleles in A be called W and Z. Then there are four equally likely possibilities for gametes a and a': (1) W and W, (2) Z and Z, (3) W and Z, and (4) Z and W. In the first two cases they are identical, so the probability is 1/2 that a and a' get the same gene from A. However, there is an additional possibility if ancestor A is inbred, for in this case the two alleles W and Z

may both be descended from some more remote ancestor not shown in the figure. The probability that A is autozygous, is, by definition, the inbreeding coefficient of A, f_A. Altogether, if A is inbred, $\text{Prob}(a = a') = \frac{1}{2} + \frac{1}{2}f_A = \frac{1}{2}(1 + f_A)$; if A is not inbred $\text{Prob}(a = a') = 1/2$.

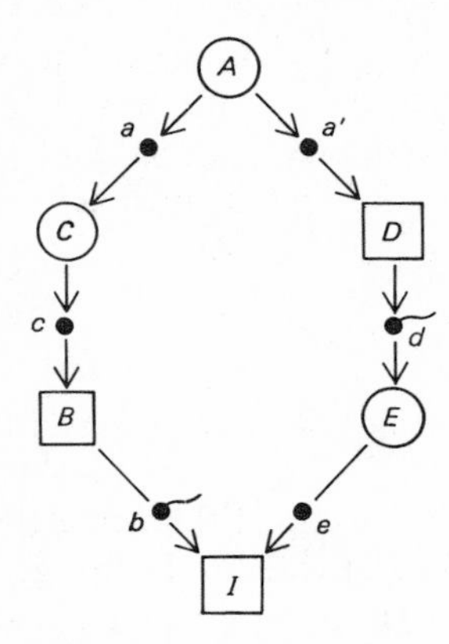

Figure 3.4.1. A simple pedigree with inbreeding. Circles and squares denote females and males respectively. Eggs and sperms are designated by small letters. Ancestors that do not contribute to the inbreeding of I are omitted.

Continuing around the path $BCADE$, $\text{Prob}(a' = d) = \text{Prob}(d = e) = 1/2$. Summarizing, b and e will carry identical genes only if b, c, a, a', d, and e do so. Therefore, since all these probabilities are independent

$$f_I = f_{BE} = \text{Prob}(b = e) = \underset{b=c}{\frac{1}{2}} \times \underset{c=a}{\frac{1}{2}} \times \underset{a=a'}{\frac{1}{2}(1 + f_A)} \times \underset{a'=d}{\frac{1}{2}} \times \underset{d=e}{\frac{1}{2}}$$

$$= (\tfrac{1}{2})^5(1 + f_A).$$

If A is not inbred (and according to information given in this pedigree she is not) the inbreeding coefficient of I is simply $(1/2)^5$. Notice that whether B, C, D, and E is inbred is irrelevant, since, for example, the probability that c and b are identical is independent of the gene contributed by B's other parent. The general rule is that the contribution of a path of relationship through a common ancestor is $(1/2)^n(1+f_A)$ where n is the number of individuals in the path from one parent to the ancestor and back through the other parent.

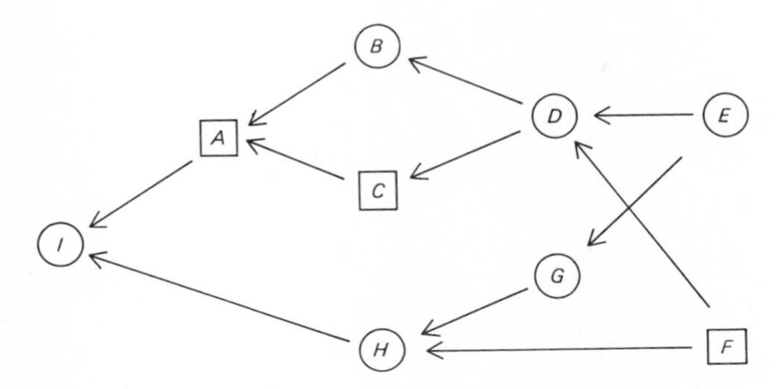

Figure 3.4.2. A more complicated pedigree; $f_I = 3/32$.

In more complicated pedigrees there may be multiple paths through an ancestor or more than one common ancestor. Consider the pedigree in Figure 3.4.2. The contributions to the inbreeding coefficient of I from the various paths are as follows. The common ancestor in a path is underlined.

PATH	CONTRIBUTION TO f
$ABD\underline{E}GH$	$(1/2)^6 = 1/64$
$ACD\underline{E}GH$	$(1/2)^6 = 1/64$
$ABD\underline{F}H$	$(1/2)^5 = 1/32$
$ACD\underline{F}H$	$(1/2)^5 = 1/32$

As we are considering only a single locus, the paths are all mutually exclusive; if I is autozygous for a pair of genes inherited through one path it cannot at the same time be autozygous for a pair inherited through another. Therefore the total probability for autozygosity is the sum of the probabilities for the separate paths, in this case 3/32.

This pedigree was not complicated by the common ancestor of any path being inbred. Individual A is inbred, but this is irrelevant since A is not a common ancestor. Only inbreeding of E or F would matter. This complication arises in the pedigree in Figure 3.4.3, where there are several inbred individuals in the pedigree and two of these, B and D, are common ancestors of one or more paths.

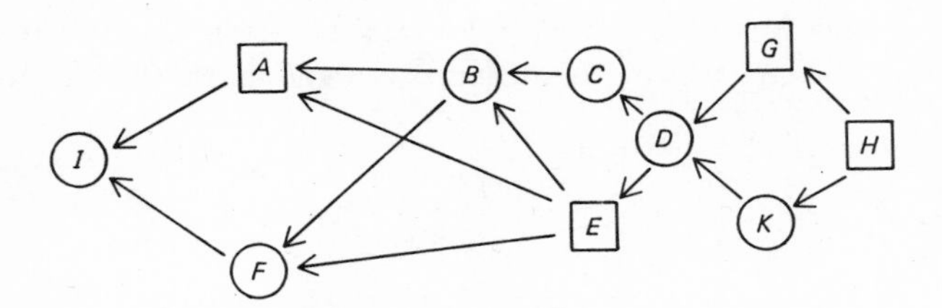

Figure 3.4.3. A still more complicated pedigree; $f_I = f_{AF} = .428$.

We begin by noting the inbreeding coefficients of D and B. The inbreeding coefficient of D, f_D, is $(1/2)^3$; $f_B = (1/2)^3(1 + f_D) = 9/64$. The components of f_I through the various paths are:

$$
\begin{aligned}
&A\underline{E}F: && (1/2)^3 && = .1250\\
&A\underline{B}F: && (1/2)^3(1 + f_B) && = .1426\\
&AB\underline{E}F: && (1/2)^4 && = .0625\\
&A\underline{E}BF: && (1/2)^4 && = .0625\\
&ABC\underline{D}EF: && (1/2)^6(1 + f_D) && = .0176\\
&AE\underline{D}CBF: && (1/2)^6(1 + f_D) && = .0176\\
& && f_I = f_{AF} && = .4278
\end{aligned}
$$

Notice that the inbreeding of B is taken into consideration in path ABF where B is a common ancestor, but ignored in the other paths where B is not the common ancestor. A path such as $ABCDEBF$ is not included because B enters twice; the contribution to this path is included in path ABF by the term $(1 + f_B)$.

To summarize: The inbreeding coefficient of an individual I, or the coefficient of consanguinity of his parents, J and K, is the sum of a series of terms, one for each path leading from a parent to a common ancestor and back through the other parent. The general formula is

$$f_I = f_{JK} = \Sigma[(1/2)^n(1 + f_A)], \qquad 3.4.1$$

where the summation is over all possible paths, n is the number of individuals in the path (counting J and K, but not I) and f_A is the inbreeding coefficient of the common ancestor at the apex of this path.

A path cannot pass through the same individual twice. No reversal of direction is permitted except at the common ancestor; always go against the arrows in going from one parent to the ancestor, and with them coming back through the other. It is helpful in avoiding counting the same path twice to adopt the convention of starting all paths with the same parent (the male, say) and ending with the other.

In the earlier literature the procedure given for computing f was to count the number of steps between individuals in a path rather than the number of individuals. The results are of course the same either way. Formula 3.4.1 was first given in the present form by Wright (1951). We use it because it follows more naturally from our derivation than the earlier form, and because it is easily adapted to X-linked genes.

An X-chromosome gene that is in a gamete produced by a male must be the same as was in the egg from which this male came. Therefore the probability of identity by descent in these two gametes is 1, rather than 1/2 as it would be for a female or an autosomal locus. Hence each male in a path multiplies the probability of identity through this path by 1 rather than 1/2, and the effect is as if the males were not counted at all. Furthermore, a male does not receive an X-chromosome from his father, so a path involving two successive males makes no contribution to the probability of identity of X-chromosomal loci.

Therefore the rule for obtaining the inbreeding coefficient for a sex-linked locus in females is: Proceed as usual except that only females in a path are counted and any path with two successive males is omitted entirely.

As examples, consider again the three pedigrees in Figure 3.4.1, 3.4.2, and 3.4.3. There is no meaning to the inbreeding coefficient of a male since he has only one X chromosome; so the pedigree in 3.4.1 is not of interest. In Figure 3.4.2, f_I is 1/32 (path $ABD\underline{E}GH$) + 1/8($ABD\underline{F}H$) = 5/32. Paths $ACD\underline{F}H$ and $ACD\underline{E}GH$ have successive males and are omitted. In the same manner, in Figure 3.4.3, $f_D = 0$, $f_B = 1/4$, and $f_I = 5/8$.

3.5 Phenotypic Effects of Consanguineous Matings

In Section 2.2 the frequency of the recessive gene causing phenylketonuric feeble-mindedness was given as approximately 1/100. Therefore with random mating the frequency of persons homozygous for the gene is the square of this, or 1/10,000. We now inquire how much this is enhanced with consanguineous marriage.

The probability of an affected child as given by 3.2.1 is $p^2(1-f)+pf$, where p is the frequency of the recessive allele and f is the inbreeding coefficient of the child. If the parents are cousins their coefficient of consanguinity or the inbreeding coefficient of their child is 1/16, as computed from Figure 3.5.1. With $p = 1/100$ and $f = 1/16$ the expected frequency of homozygous recessives is 115/160,000, or approximately 7/10,000, a 7-fold increase compared with the risk when the parents are unrelated.

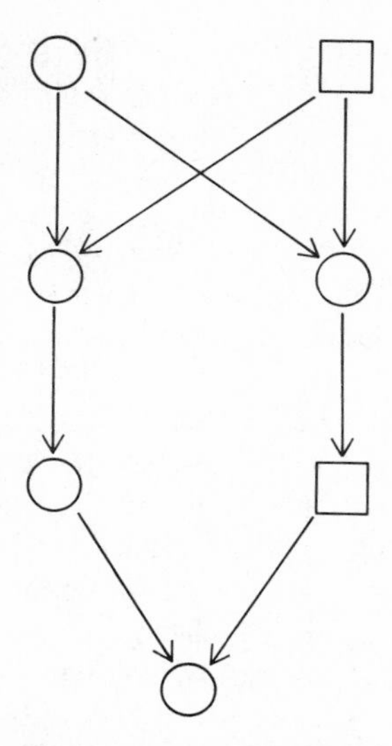

Figure 3.5.1. Pedigree of an individual whose parents are cousins. $f = \frac{1}{2}^5 + \frac{1}{2}^5 = 1/16$.

Table 3.5.1 illustrates the way in which the relative frequency of a disease in the children of cousin marriages increases with the rarity of the gene.

We can also ask the question the other way around: What proportion of the persons affected with recessive traits come from consanguineous marriages?

This proportion, K, may be obtained by dividing the number of affected from consanguineous marriages by the total number of affected. If c is the

Table 3.5.1. The proportion of homozygous-recessive individuals in a population with random mating and when the parents are first cousins for various frequencies of the recessive allele.

GENE FREQUENCY	FREQUENCY OF AFFECTED WHEN		RATIO
	$f = 0$	$f = 1/16$	
0.1	0.01	0.016	1.6
0.01	0.0001	0.00072	7.2
0.005	0.000025	0.000335	13.4
0.001	0.000001	0.000063	63

proportion of consanguineous marriages in the population of size N, the number of affected from consanguineous marriages is $Nc[p^2(1-f)+pf]$, where f is the coefficient of consanguinity of the parents and p is the recessive gene frequency. The total number of affected in the population is $N[p^2(1-\bar{f})+p\bar{f}]$, where $\bar{f}$ is the mean inbreeding coefficient in the population. Therefore

$$K = \frac{c[p^2(1-f)+pf]}{p^2(1-\bar{f})+p\bar{f}} = \frac{c[p+(1-p)f]}{p-p\bar{f}+\bar{f}}, \qquad 3.5.1$$

or, since $p\bar{f}$ is usually very small

$$K = \frac{c[p+(1-p)f]}{p+\bar{f}} \qquad 3.5.2$$

approximately.

The most common consanguineous marriage is between first cousins. For $f = 1/16$, 3.5.1 becomes

$$K = \frac{c(1+15p)}{16[p+\bar{f}(1-p)]},$$

or approximately

$$\frac{c(1+15p)}{16(p+\bar{f})}. \qquad 3.5.3$$

Some representative values of K are given in Table 3.5.2.

This shows that, even though consanguineous marriages are very rare, a substantial fraction of diseases caused by recessive genes comes from such marriages if p is .01 or less. Consanguinity of the parents is one of the strongest kinds of evidence of recessive inheritance.

On the other hand, if the recessive gene is common, the increased incidence with consanguinity is very slight. Cystic fibrosis of the pancreas appears to be due to a simple recessive factor, yet there is no appreciable rise in incidence from consanguineous marriages, because the allele frequency is so high. The parental-consanguinity rate is much higher for recessive traits where the gene frequency is low.

Most recessive genes are carried concealed in the heterozygous condition. We can get some idea of the total number of such genes carried by normal persons through a study of consanguineous marriages. A study by Jan A. Book showed that about 16% of the children of first cousin marriages in Sweden had a genetic disease, and if diseases of more doubtful etiology were included the number rose to 28%. The corresponding figures for the control population with unrelated parents were 4% and 6%. Thus cousin marriage, by these

Table 3.5.2. The proportion, K, of cases of recessive conditions expected from first-cousin marriage for various values of gene frequency (p), frequency of cousin marriage (c), and population inbreeding coefficient ($\bar{f}$).

c	$\bar{f}$	p	K
0.005	0.0005	0.1	0.008
0.005	0.0005	0.03	0.015
0.005	0.0005	0.01	0.034
0.005	0.0005	0.003	0.093
0.005	0.0005	0.001	0.212
0.005	0.0005	0.0003	0.392
0.01	0.001	0.1	0.015
0.01	0.001	0.03	0.029
0.01	0.001	0.01	0.065
0.01	0.001	0.003	0.163
0.01	0.001	0.001	0.317
0.01	0.001	0.0003	0.483
0.02	0.002	0.1	0.031
0.02	0.002	0.03	0.057
0.02	0.002	0.01	0.120
0.02	0.002	0.003	0.262
0.02	0.002	0.001	0.423
0.02	0.002	0.0001	0.546

data, entails an increased risk of 12% to 22% of having a child with a detectable genetic defect. Since the child of a cousin marriage has an inbreeding coefficient of 1/16, we reason that a completely homozygous individual would have 16 times as many diseases, or approximately 2.0 to 3.5. This is the number of recessive factors per gamete (since a homozygous individual may be regarded as a doubled gamete), so the number per zygote is between 4 and 7. These figures are based on rather limited data, but they furnish a rough idea of the magnitude. The conclusion is that the average human carries hidden the equivalent of some half a dozen deleterious recessive genes that, if made homozygous, would cause a detectable disease.

We can also estimate the amount of genetic weakness that is carried hidden in a heterozygous individual, but which would be expressed as inviability if he were made homozygous. Sutter and Tabah (1958 and earlier) found from a demographic study in two rural provinces in France that children of cousin marriages died before adulthood about 25% of the time, whereas the death rate from unrelated parents was about 12%. Thus, in this environment, cousin marriage increased the risk of death by about 0.13. Making the same calculations as above (i.e., multiplying by 16×2) we estimate that the average individual in this population carries $32 \times .13$ or about 4 hidden "lethal equivalents." We say "lethal equivalents" because one cannot distinguish between 4 full lethals and 8 genes with 50% probability of causing death, or any system where the product of the number of genes and the average effect of each is 4. For a more sophisticated treatment of this subject, making use of all degrees of relationship rather than just cousins, see Morton, Crow, and Muller (1956).

The data on human inbreeding effects have not been very reproducible. The large body of data from Japan show significant heterogeneity effects from city to city. There is danger of confounding inbreeding effects with the effects of social concomitants of consanguineous marriages. For all these reasons, we cannot place too much reliance on the numerical values of the previous paragraph.

It is also to be expected that what is lethal in one environment may be only detrimental in a better one. In much of the world there has been a substantial rise in the standard of living and a decrease in the death rate. This means that the number of lethal equivalents is decreasing.

In Drosophila, where the measures are precise and reproducible, there are about two lethal equivalents per fly. About 2/3 of the viability depression from inbreeding is attributable to monogenic lethals; the rest is the cumulative effect of a much larger number of genes with individually small effects.

3.6 The Effect of Inbreeding on Quantitative Characters

We consider first a theoretical model that is applicable to any measurable trait, such as height, weight, yield, survival, or fertility. For initial simplicity, a single locus with only two alleles is assumed. The model is summarized as follows:

GENOTYPE	A_1A_1	A_1A_2	A_2A_2
FREQUENCY	$p_1^2(1-f)+p_1 f$	$2p_1p_2(1-f)$	$p_2^2(1-f)+p_2 f$
PHENOTYPE	$Y-A$	$Y+D$	$Y+A$

In this model Y is the residual phenotype when the A locus is not considered. Genotype A_2A_2 adds an amount A to the phenotype, and A_1A_1 subtracts an equal amount. (We could just as well assume that both genotypes add to the residual, or that both subtract; this model is chosen arbitrarily and for algebraic simplicity. The same result would be obtained in any case.)

If there were no dominance, the phenotype A_1A_2 would be Y. Under this circumstance, the amount by which the phenotype is changed by substituting an A_2 for an A_1 is always A, which we can call the additive effect of the A locus. D is a measure of dominance. When $D = 0$, there is no dominance; when $D = A$, A_2 is completely dominant; when $D = -A$, A_2 is completely recessive, or A_1 dominant; when $D > A$, there is overdominance, the heterozygote having a higher phenotypic value than either homozygote.

We now obtain an expression for the mean phenotype $\bar{Y}$. This will be given by summing the products of each phenotype and its frequency. (Since the frequencies add up to one, the sum is the same as the mean.) Therefore,

$$\bar{Y} = (Y - A)[p_1^2(1 - f) + p_1 f] + (Y + D)2p_1p_2(1 - f) + (Y + A)[p_2^2(1 - f) + p_2 f],$$

which leads after some algebraic rearrangement to

$$\begin{aligned} \bar{Y} &= Y + A(p_2 - p_1) + 2p_1p_2 D - 2p_1p_2 Df \\ &= G - Hf, \end{aligned} \qquad \textbf{3.6.1}$$

where $G = Y + A(p_2 - p_1) + 2p_1p_2 D$ and $H = 2p_1p_2 D$. G is the average phenotype with random mating and $G - H$ is the average with complete homozygosity. Notice that H is positive if D is positive.

This equation brings out two important facts about the phenotypic consequences of inbreeding. The first is that in the absence of dominance ($D = 0$), there is no mean change with inbreeding. The second fact is that the equation is a linear function of f. This means that whatever the level of dominance, as measured by D, the change with inbreeding is proportional to the inbreeding coefficient. As long as D is positive (i.e., the heterozygote has a larger phenotype than the mean of the two homozygotes), inbreeding will produce a decline.

There are two possible causes of the inbreeding decline that is so universally observed: (1) Favorable genes tend to be dominant or partially dominant ($0 < D < A$), and (2) the heterozygote has a higher phenotype than either homozygote, ($0 < A < D$). Notice that the observation of a linear decline in a quantitative trait cannot discriminate between these possibilities, for it would be expected with either type of gene action, or any mixture of the two. To discriminate between them will require other kinds of evidence.

The extension to more than two alleles is straightforward and will not be given here. We shall consider the extension to more than one locus. Consider a model with two loci, each with two alleles. We shall let A stand for the additive effect of the A locus and B for that of the B locus. If there is no interaction between loci A and B, the model is as follows:

GENOTYPE		A_1A_1	A_1A_2	A_2A_2
	FREQUENCY	$p_1^2(1-f)+p_1f$	$2p_1p_2(1-f)$	$p_2^2(1-f)+p_2f$
B_1B_1	$r_1^2(1-f)+r_1f$	$Y-A-B$	$Y+D_A-B$	$Y+A-B$
B_1B_2	$2r_1r_2(1-f)$	$Y-A+D_B$	$Y+D_A+D_B$	$Y+A+D_B$
B_2B_2	$r_2^2(1-f)+r_2f$	$Y-A+B$	$Y+D_A+B$	$Y+A+B$

If the A and B loci are independent and in gametic phase equilibrium, the frequency of any of the nine classes in the table is given by the product of the frequencies at the borders. We get the mean phenotype as before, by multiplying each phenotype value within the table by its frequency and summing the nine products. After simplification this leads to

$$\bar{Y} = Y + A(p_2 - p_1) + 2p_1p_2 D_A + B(r_2 - r_1) + 2r_1r_2 D_B - 2(p_1p_2 D_A + r_1r_2 D_B)f. \quad 3.6.2$$

As before, there is a linear relation to the inbreeding coefficient (unless $D_A = D_B = 0$), as might be expected from knowledge that this is true for either locus by itself. We now complicate the model by assuming an interaction between the two loci. In population genetics the word *epistasis* is given a meaning broader than its classical meaning so as to include all levels of nonadditive effects between loci. Any circumstance where a substitution at the A locus has a different effect depending on the genotype at the B locus is an example of epistasis. A simple way to construct such a model is to add interaction terms to each of the values already given so that the phenotypes are now:

	A_1A_1	A_1A_2	A_2A_2
B_1B_1	$Y-A-B+I$	$Y+D_A-B-L$	$Y+A-B-I$
B_1B_2	$Y-A+D_B-K$	$Y+D_A+D_B+J$	$Y+A+D_B+K$
B_2B_2	$Y-A+B-I$	$Y+D_A+B+L$	$Y+A+B+I$

Notice that there are nine parameters, Y, A, B, D_A, D_B, I, J, K, and L, which correspond to the nine phenotypes so there is a complete specification of the phenotypes when the parameters are given, and vice versa.

The formula for $\bar{Y}$ may be written as

$$\bar{Y} = G - Hf + Mf^2, \tag{3.6.3}$$

where

$$G = Y + A(p_2 - p_1) + B(r_2 - r_1) + 2D_A p_1 p_2 + 2D_B r_1 r_2$$
$$+ I(r_1 - r_2)(p_1 - p_2) + 2Kr_1 r_2(p_2 - p_1) + 2Lp_1 p_2(r_2 - r_1)$$
$$+ 4Jp_1 p_2 r_1 r_2,$$
$$M = 4p_1 p_2 r_1 r_2 J,$$

and

$$H = 2[p_1 p_2 D_A + r_1 r_2 D_B + p_1 p_2(r_2 - r_1)L$$
$$+ r_1 r_2(p_2 - p_1)K + 4p_1 p_2 r_1 r_2 J].$$

In this model, A and B represent the additive effects of the two loci, and D_A and D_B the dominance effects, as before. I is the effect of pure epistasis without dominance; in other words, the interaction of the additive effect of A with the additive effect of B. K is a measure of interaction and dominance; it is the effect of the A locus on the dominance of the B locus. Likewise, L is the effect of the B locus on the dominance of the A locus. Another way of saying this is that this is the interaction of the additive effect of the B locus with the dominance effect of the A. Finally, J is the epistatic effect of the two dominances; it is the interaction of dominance at the A locus with dominance at the B locus.

From equation 3.6.3 we see that if all the terms involving dominance effects—D_A, D_B, J, K, and L—are 0, there is no inbreeding effect since the coefficients of f and f^2 are 0. Thus epistasis alone, without dominance, does not produce an inbreeding decline.

If $J = 0$ the inbreeding change is linear in f. In order for the coefficient of f^2 to be other than 0, there must be interaction between the two dominance effects.

If A, B, D_A, D_B, K, and L are all positive, then the genes with subscript 2 are associated with increased performance (or yield, or fitness, or whatever is being measured). This also generally means that the alleles with subscript 2 will be more frequent than those with subscript 1. Then if J is positive there will be *diminishing* epistasis during inbreeding. By this, we mean that the curve is concave upward and that homozygosity for two loci reduces performance by less than the sum of the individual effects. Contrariwise if J is negative (assuming $H > 0$), the epistasis is *reinforcing*. That is, the deleterious effect of two loci is more than cumulative. This is sometimes called synergistic.

These general types of epistasis are illustrated in Figure 3.6.1.

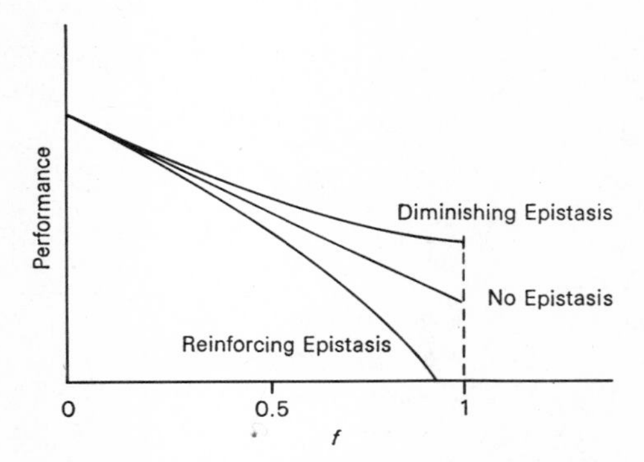

Figure 3.6.1. Decline in performance (or any trait of interest) with inbreeding under no epistasis, diminishing epistasis, and reinforcing epistasis.

3.7 Some Examples of Inbreeding Effects

Table 3.7.1 gives some numerical examples and illustrates the way in which specific levels of dominance and epistasis may be constructed by choosing appropriate values for the parameters. In these examples the gene frequencies are all 1/2. The values are contrived so that fully homozygous individuals ($f = 1$) have an average phenotype of 10.

Models 1 and 2 show that, whether or not there is epistasis, there is no inbreeding effect without dominance; the phenotype is 10, independent of f.

Models 3 and 4 show that the rate of decline with inbreeding by itself cannot distinguish among different levels of dominance. Model 5 is an extreme form of reinforcing epistasis; model 6 is an extreme form of diminishing epistasis. Models 6 and 9 show that quite different systems of gene action can give the same quadratic equation.

Models 7 and 8 show that epistasis, though it is a necessary condition for a nonlinear inbreeding effect, is not sufficient. In both there is epistasis, yet the equations are the same; moreover they are the same as models 3 and 4. Only when J is not 0 is there a nonlinear effect, as 5, 6, and 9 show.

The examples in Table 3.7.1 all assume equal frequency for the two alleles at each locus. Usually this is not the case. Let us assume that the higher phenotypic value is desirable, which means that for most of the models

Table 3.7.1. Some numerical illustrations of the phenotypic values under various models of dominance and epistasis. Also given is the average phenotype, $\bar{Y}$, when $p_1 = p_2 = r_1 = r_2 = .5$.

MODEL	DESCRIPTION	PHENOTYPIC VALUE			AVERAGE PHENOTYPE
1.	No dominance, no epistasis	8	9	10	
	$Y = 10, A = B = 1$	9	10	11	$\bar{Y} = 10$
	$D_A = D_B = I = J = K = L = 0$	10	11	12	
2.	Epistasis, but no dominance	7	8	9	
	$Y = 10, A = B = 2, I = 1$	8	10	12	$\bar{Y} = 10$
	$D_A = D_B = J = K = L = 0$	9	12	15	
3.	Complete dominance, no epistasis	8	10	10	
	$Y = 10, A = B = D_A = D_B = 1$	10	12	12	$\bar{Y} = 11 - f$
	$I = J = K = L = 0$	10	12	12	
4.	Overdominance, no epistasis	10	11	10	
	$Y = 10, A = B = 0, D_A = D_B = 1$	11	12	11	$\bar{Y} = 11 - f$
	$I = J = K = L = 0$	10	11	10	
5.	Complete dominance, complementary recessive genes	7	11	11	
	$Y = 10, A = B = D_A = D_B = 1$	11	11	11	$\bar{Y} = \frac{43}{4} - \frac{1}{2}f - \frac{1}{4}f^2$
	$I = J = K = L = -1$	11	11	11	
6.	Complete dominance, complementary dominant genes	9	9	9	
	$Y = 10, A = B = D_A = D_B = 1$	9	13	13	$\bar{Y} = \frac{45}{4} - \frac{3}{2}f + \frac{1}{4}f^2$
	$I = J = K = L = 1$	9	13	13	
7.	Dominance and epistasis	9	9	9	
	$Y = 10, A = B = D_A = D_B = 1$	9	12	13	$\bar{Y} = 11 - f$
	$I = K = L = 1, J = 0$	9	13	13	
8.	Dominance and epistasis	9	10	9	
	$Y = 10, A = B = D_A = D_B = I = 1$	10	12	12	$\bar{Y} = 11 - f$
	$J = K = L = 0$	9	12	13	
9.	Overdominance and epistasis	10	11	10	
	$Y = 10, A = B = 0, D_A = D_B = 1,$	11	13	11	$\bar{Y} = \frac{45}{4} - \frac{3}{2}f + \frac{1}{4}f^2$
	$I = K = L = 0, J = 1$	10	11	10	

A_2 and B_2 will tend to be more frequent than their alleles. Consider as an example that $p_1 = r_1 = .1$. Then the equations for models 5 and 6 become:

Model 5, complementary recessive genes: $\bar{Y} = G - .007f - .032f^2$,
Model 6, complementary dominant genes: $\bar{Y} = G - .356f + .032f^2$.

Model 5 shows that with a trait depending on simultaneous homozygosity for two rare recessives, there will be very little inbreeding effect when f is small—because the coefficient of f is small, and f^2 is very small. Only when the inbreeding is sufficient for f^2 to be appreciable will its larger coefficient become important. Thus, to the extent that there are detrimental traits depending on multiple homozygosity, inbreeding effects will tend to be nonlinear, with very little effect of slight inbreeding, but with an accelerating effect at very high levels of inbreeding.

With model 6, on the other hand, the quadratic term never becomes important and the linear term dominates for all values of f. Thus for rare genes with duplicate effects the inbreeding effect is linear for all practical purposes.

When the trait considered is survival it is often more natural to measure epistasis as deviations from independence rather than from additivity. Survival probabilities are multiplicative if the genes act independently. It is often advantageous to transform to logs, or to measure fitness in Malthusian parameters.

We summarize this section by stating three conclusions, all of which are apparent from 3.6.3 and are illustrated by the numerical examples in Table 3.7.1.

1. If there is no dominance, there is no mean change in phenotype with inbreeding regardless of the amount of epistasis (models 1 and 2.). If D_A, D_B, J, K, and L are 0, $\bar{Y}$ is not a function of f.

2. If there is dominance, but no epistasis, the effect of inbreeding on the phenotype is linear in f. Usually inbreeding leads to a change in quantitative measures and if there is no epistasis this change is proportional to f (models 3 and 4.) If $I = J = K = L = 0$, the term in f^2 drops out.

3. If there is both dominance and epistasis the inbreeding effect may be quadratic in f (or higher order if more than two loci are involved). With reinforcing type epistasis the inbreeding effect is greater than if the loci were additive; with diminishing epistasis the effect is less (models 5 and 6). However, the change in average phenotype with inbreeding is not necessarily quadratic (models 7 and 8); as long as $J = 0$, the inbreeding effect is linear.

Reliable data on the results of inbreeding uncomplicated by the effects of selection are rare. There is also a difficulty in choosing an appropriate scale of measurement if the linearity of the inbreeding effect is to be tested. Some of the best data come from maize and Table 3.7.2 shows an example.

The yield was measured for inbred lines, crosses, and randomly mated progeny from fields of crosses.

Table 3.7.2. The average yield in bushels per acre of randomly pollinated maize derived from hybrids between inbred lines. Expected yields are based on assumption of no epistasis. Data from Neal (1935).

	HYBRID AVERAGE $f=0$ (G)	INBRED AVERAGE $f=1$ $(G-H)$	RANDOMLY POLLINATED		
			f	EXPECTED YIELD $(G-Hf)$	OBSERVED YIELD
10 two-way hybrids	62.8	23.7	.500	43.3	44.2
4 three-way hybrids	64.2	23.8	.375	49.1	49.3
10 four-way hybrids	64.1	25.0	.250	54.3	54.0

Three kinds of crosses were tested, two-way, three-way, and four way (or double-cross) hybrids. Let A, B, C, and D stand for four lines that have been self-fertilized long enough to be regarded as completely homozygous. Two-way crosses are first generation hybrids between two lines, e.g., $A \times B$; three-way crosses are between a hybrid and a different inbred, e.g., $(A \times B) \times C$; four-way crosses are between two different hybrids, e.g., $(A \times B) \times (C \times D)$.

If a field of two-way hybrids is allowed to pollinate at random the probability that two alleles have come from the same parental inbred line is 1/2. Assuming the parent line to be autozygous, two alleles from the same line are identical; hence the progeny from random pollination have an inbreeding coefficient of 1/2. For a four-way cross the probability of two alleles from the same line is 1/4. For the three-way cross, $(A \times B) \times C$, the probability of two alleles both coming from C is 1/4, from A is 1/16, and from B is 1/16; f is the sum, or 3/8.

The data in Table 3.7.2. show the close agreement with the expected values. Since the inbreeding effect is so nearly linear with the inbreeding coefficient, this implies that epistasis is not very important in corn yield. Either the genes at different loci act additively on yield or opposite interactive effects cancel each other.

Figure 3.7.1 shows data from Drosophila gathered in a different way. In these experiments the cultures were maintained with as little natural selection as possible and the chromosome number 2 was kept heterozygous.

Then after several generations the chromosome, with its accumulated recessive mutations, was made homozygous. That there is an appreciable synergistic effect is clear from the graph. The mutants presumably accumulate linearly with time, but the homozygous viability decreases somewhat more than linearly. The graph shows only the influence of mutations with small effects; chromosomes with lethal mutations are not included.

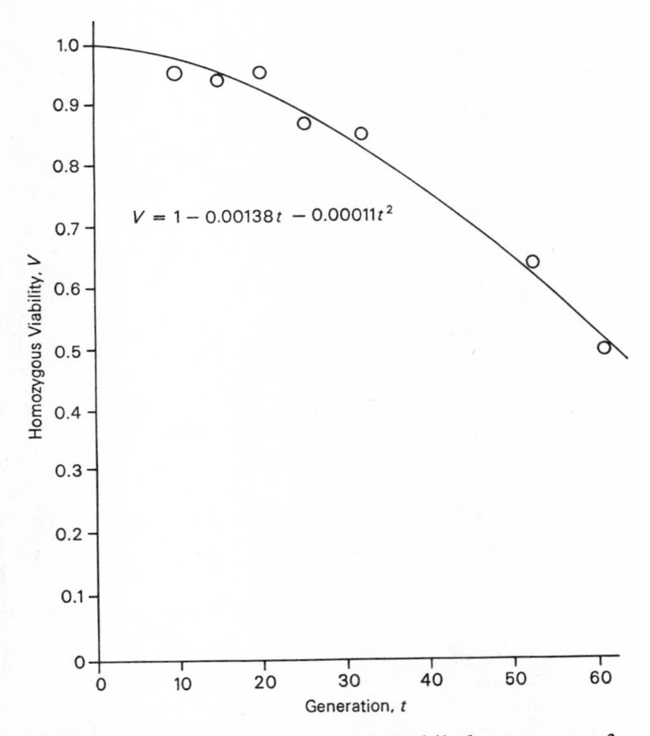

Figure 3.7.1. The viability of Drosophila homozygous for the second chromosome as a function of the number of generations during which mutants were permitted to accumulate. All lethal chromosomes are omitted. There is clearly some reinforcing epistasis for the deleterious effects of the accumulated mutations. Data from T. Mukai.

3.8 Regular Systems of Inbreeding

In Section 3.4 we derived methods for computing the coefficients of inbreeding and consanguinity from pedigrees. The same methods can be used to derive recurrence relations for f in successive generations with regular systems of inbreeding. A detailed treatement of this subject was first given by Wright (1921) and recent summaries are available (Wright, 1951; Li, 1955*a*). We shall illustrate only some of the simpler, but most important cases.

1. Self-fertilization From Figure 3.8.1 and the principles already

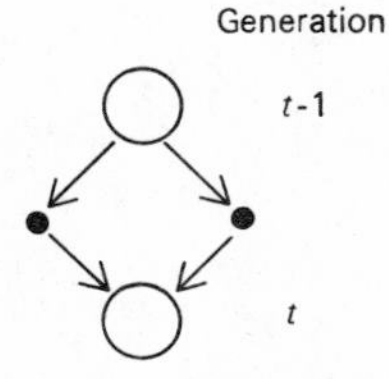

Figure 3.8.1. A diagram of self-fertilization. Zygotes are shown in large circles; gametes by dots.

discussed in Section 3.4,

$$f_t = \tfrac{1}{2}(1 + f_{t-1}), \tag{3.8.1}$$

where f_t is the inbreeding coefficient in generation t and f_{t-1} is the coefficient one generation earlier. To obtain the change in heterozygosity we utilize the relation $f_t = (H_0 - H_t)/H_0$ from equation 3.2.4, where H_t and H_0 are the proportions of heterozygosity in generation t and initially ($t = 0$) .This leads to

$$\begin{aligned} H_t &= \tfrac{1}{2}H_{t-1} \\ &= (\tfrac{1}{2})^2 H_{t-2} = (\tfrac{1}{2})^t H_0 . \end{aligned} \tag{3.8.2}$$

This confirms the result stated in Section 3.1; with self-fertilization the amount of heterozygosity is reduced by one-half each generation. After 10 generations, only 1/1024 of the loci that previously were heterozygous remain heterozygous.

For a matrix treatment of this, see A.7.

2. Sib Mating Recalling previous definitions, the inbreeding coefficient f_I is the probability that two homologous genes in an individual I are identical by descent; the coefficient of consanguinity f_{IJ} is the probability that two homologous genes, one chosen at random from individual I and the other from individual J are identical.

We let f_t be the inbreeding coefficient of an individual in generation t and g_t the coefficient of consanguinity of two individuals (necessarily sibs after the first generation). The inbreeding coefficient of an individual is the same as the coefficient of consanguinity of his parents; that is, $f_t = g_{t-1}$.

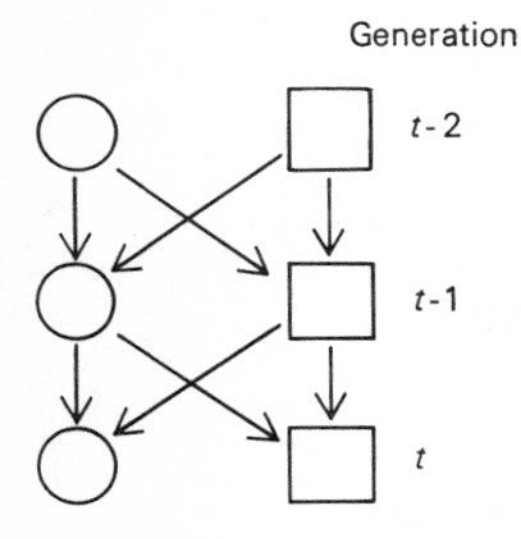

Figure 3.8.2. Continued sib mating.

Two genes in different individuals in generation t came from the same individual in generation $t-1$ with probability 1/2, in which case their probability of identity is $(1 + f_{t-1})/2$. They came from different individuals also with probability 1/2, in which case the probability of identity is g_{t-1}. Putting these together, we have

$$f_t = g_{t-1},$$
$$g_t = \tfrac{1}{4}(1 + f_{t-1}) + \tfrac{1}{2}g_{t-1}. \qquad \textbf{3.8.3}$$

We can, if we wish, eliminate the g's from these two equations and write a relation involving only the inbreeding coefficients.

$$f_t = \tfrac{1}{4}(1 + 2f_{t-1} + f_{t-2}). \qquad \textbf{3.8.4}$$

Letting $f_t = (H_0 - H_t)/H_0$, we can obtain a recurrence relationship for the heterozygosity

$$H_t = \tfrac{1}{2}H_{t-1} + \tfrac{1}{4}H_{t-2}. \qquad \textbf{3.8.5}$$

If $H_0 = H_1 = 1/2$, we obtain the sequence 3/8, 5/16, 8/32, 13/64, etc., as given in Table 3.1.2.

In order to get a more general expression for the heterozygote frequency, H_t, in terms of the initial frequency, H_0, we let $f_t = 1 - h_t$ and $g_t = 1 - k_t$. Then equations 3.8.3 become

$$h_t = \qquad k_{t-1},$$
$$k_t = \tfrac{1}{4}h_{t-1} + \tfrac{1}{2}k_{t-1}, \qquad \textbf{3.8.6}$$

or, in matrix form,

$$\begin{pmatrix} h_t \\ k_t \end{pmatrix} = \begin{pmatrix} 0 & 1 \\ \frac{1}{4} & \frac{1}{2} \end{pmatrix} \begin{pmatrix} h_{t-1} \\ k_{t-1} \end{pmatrix} = \begin{pmatrix} 0 & 1 \\ \frac{1}{4} & \frac{1}{2} \end{pmatrix}^t \begin{pmatrix} h_0 \\ k_0 \end{pmatrix}. \qquad \text{3.8.6a}$$

The rules for matrix multiplication are given in A.7. However, there is a standard procedure for obtaining h_t directly without having to proceed generation by generation. For an explanation using this same example see A.8.

The characteristic equation for this matrix is

$$\begin{vmatrix} 0 - \lambda & 1 \\ \frac{1}{4} & \frac{1}{2} - \lambda \end{vmatrix} = 0,$$

which leads to

$$\lambda^2 - \tfrac{1}{2}\lambda - \tfrac{1}{4} = 0, \qquad \text{3.8.7}$$

of which the largest (and only positive) root is

$$\lambda = \frac{1 + \sqrt{5}}{4} = 0.809. \qquad \text{3.8.8}$$

Using both roots we can write an exact expression for h_t for any generation, t. This is shown with this same example in the appendix starting with equation A.8.13.

We are especially interested in the limiting rate of change in heterozygosity, which is given by the largest root, 3.8.8. Thus, after t is sufficiently large, h_t is approximately λh_{t-1}, which is equivalent to

$$H_t \approx \lambda H_{t-1} \approx 0.809 H_{t-1}. \qquad \text{3.8.9}$$

Comparison with the actual values in Table 3.1.2 shows that this equation becomes exceedingly accurate after about five to six generations.

We can also obtain λ directly and simply from 3.8.5. Dividing both sides of the equation by H_{t-1}, and letting $H_t/H_{t-1} = H_{t-1}/H_{t-2} = \lambda$ (which of course implies a constant rate of decrease in heterozygosity), we obtain equation 3.8.7. directly. This procedure has been regularly used by Wright (1933*a*, 1951).

If we desire a system for comparison of two mating systems in regard to their ultimate progress toward homozygosity, we can ask how many generations are required by the two systems to achieve the same reduction in heterozygosity. We can do this by equating $(\lambda_1)^{t_1}$ to $(\lambda_2)^{t_2}$ where t is the number of generations to achieve a given level of change in heterozygosity and the subscripts refer to the two mating systems. Writing

$$\lambda_1^{t_1} = \lambda_2^{t_2} \qquad \textbf{3.8.10}$$

and taking logarithms of both sides, we obtain

$$\frac{t_1}{t_2} = \frac{\log \lambda_2}{\log \lambda_1}. \qquad \textbf{3.8.11}$$

For example, with self-fertilization $\lambda = .5$ while with sib mating $\lambda = .809$. This leads to

$$\frac{t_1}{t_2} = \frac{\log .500}{\log .809} = \frac{-.3010}{-.0921} = 3.27.$$

So we can say that asymptotically one generation of self-fertilization is equal to a little more than three generations of sib mating.

This method of comparing mating systems was first used by Fisher (1949). There are of course many properties of a mating system other than the rate of change in heterozygosity. These can be studied by writing out all possible matings and noting how their frequencies change in successive generations. This quickly becomes unmanageable unless matrix methods are used. The matrix procedure first introduced by Haldane (1937*a*) and further developed by Fisher (1949) can be used to obtain a general solution.

3. More Complicated Systems Figure 3.8.3 shows a pedigree in which, starting with the third generation, all matings are between double first cousins.

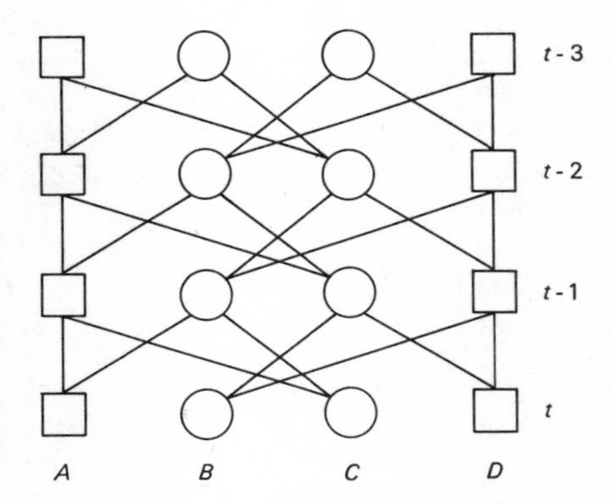

Figure 3.8.3. Repeated mating of double first cousins. This is the maximum avoidance of inbreeding in a population of size 4. $\lambda = .920$.

To write equations for this system we let f_t be the coefficient of inbreeding of an individual in generation t, as before. Notice that there are two kinds of relationship in generation t. We let g_t be the coefficient of consanguinity of nonsibs such as A and B, and j_t be that of sibs such as A and C. Thus

$$g_t = f_{AB} = f_{BC} = f_{CD} = f_{AD},$$
$$j_t = f_{AC} = f_{BD}.$$

From these definitions it follows that:

$$\begin{aligned} f_t &= g_{t-1}, \\ g_t &= \tfrac{1}{2}g_{t-1} + \tfrac{1}{2}j_{t-1}, \\ j_t &= \tfrac{1}{2}(\tfrac{1}{2} + \tfrac{1}{2}f_{t-1}) + \tfrac{1}{2}g_{t-1}. \end{aligned} \tag{3.8.12}$$

Letting $f = 1 - h$, $g = 1 - k$, and $j = 1 - m$ in 3.8.12, we obtain

$$\begin{aligned} h_t &= k_{t-1}, \\ k_t &= \tfrac{1}{2}k_{t-1} + \tfrac{1}{2}m_{t-1} \\ m_t &= \tfrac{1}{4}h_{t-1} + \tfrac{1}{2}k_{t-1}. \end{aligned} \tag{3.8.13}$$

Table 3.8.1. Decrease of heterozygosity with four mating systems, starting from a randomly mating population. The numbers are the ratio of the heterozygosity in generation t to the original heterozygosity, $H_t/H_0 = h_t$.

GENERATION t	SELF-FERTILIZATION $N=1$	SIB MATING $N=2$	DOUBLE FIRST-COUSIN MATING $N=4$	CIRCULAR HALF-SIB MATING $N=4$
0	1.000	1.000	1.000	1.000
1	.500	1.000	1.000	1.000
2	.250	.750	1.000	.875
3	.125	.625	.875	.813
4	.063	.500	.813	.750
5	.031	.406	.750	.695
6	.016	.328	.688	.644
10	.001	.141	.492	.477
15		.048	.324	.327
20		.017	.213	.224
30		.002	.092	.105
50			.017	.023
λ	.500	.809	.920	.927

These are now homogeneous and can be written in matrix form

$$\begin{pmatrix} h_t \\ k_t \\ m_t \end{pmatrix} = \begin{pmatrix} 0 & 1 & 0 \\ 0 & \frac{1}{2} & \frac{1}{2} \\ \frac{1}{4} & \frac{1}{2} & 0 \end{pmatrix} \begin{pmatrix} h_{t-1} \\ k_{t-1} \\ m_{t-1} \end{pmatrix}. \qquad 3.8.13a$$

The characteristic equation is

$$\begin{vmatrix} -\lambda & 1 & 0 \\ 0 & \frac{1}{2}-\lambda & \frac{1}{2} \\ \frac{1}{4} & \frac{1}{2} & -\lambda \end{vmatrix} = 0, \qquad 3.8.14$$

which, upon expansion, becomes

$$\lambda^3 - \tfrac{1}{2}\lambda^2 - \tfrac{1}{4}\lambda - \tfrac{1}{8} = 0, \qquad 3.8.14a$$

and the largest root is $\lambda = .9196$.

Some numerical values for heterozygosity with this mating system are given in Table 3.8.1.

Notice that for a population of size 4 this is the system of mating in which mated pairs are least related. A corresponding system in a population of size 8 would be quadruple second-cousin mating. Wright (1921) designated such systems as having maximum avoidance of inbreeding.

Such systems do, in fact, minimize the rate of approach to homozygosity during the initial generations, but somewhat surprisingly there are systems of mating that ultimately have a slower rate of decrease in heterozygosity. An example, for a population of 4, is half-sib mating, or circular mating, as illustrated in Figure 3.8.4.

Letting g_t be the coefficient of consanguinity of individuals one position apart and j_t be that for individuals two positions apart,

$$g_t = f_{AB} = f_{BC} = f_{CD} = f_{AD}$$

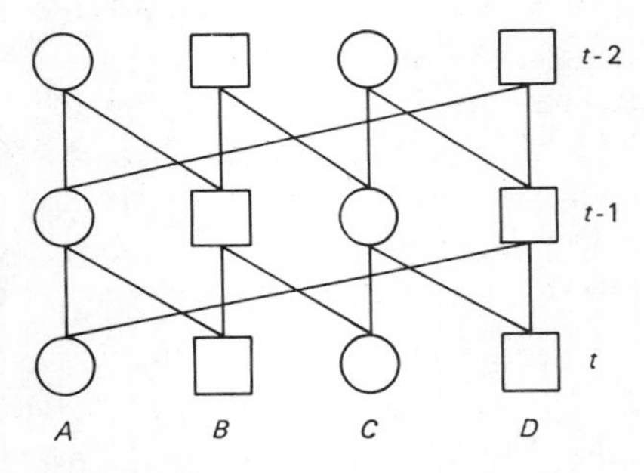

Figure 3.8.4. Half-sib mating in a population of 4, or circular mating. $\lambda = .927$.

and

$$j_t = f_{AC} = f_{BD},$$

we have

$$\begin{aligned} f_t &= g_{t-1}, \\ g_t &= \tfrac{1}{4}(\tfrac{1}{2} + \tfrac{1}{2}f_{t-1}) + \tfrac{1}{2}g_{t-1} + \tfrac{1}{4}j_{t-1}, \\ j_t &= \tfrac{1}{2}g_{t-1} + \tfrac{1}{2}j_{t-1}. \end{aligned} \qquad 3.8.15$$

Substituting $f = 1 - h$, $k = 1 - g$, and $m = 1 - j$ as before leads to

$$\begin{pmatrix} h_t \\ k_t \\ m_t \end{pmatrix} = \begin{pmatrix} 0 & 1 & 0 \\ \frac{1}{8} & \frac{1}{2} & \frac{1}{4} \\ 0 & \frac{1}{2} & \frac{1}{2} \end{pmatrix} \begin{pmatrix} h_{t-1} \\ k_{t-1} \\ m_{t-1} \end{pmatrix} \qquad 3.8.16$$

with the characteristic equation

$$\lambda^3 - \lambda^2 + \tfrac{1}{16} = 0 \qquad 3.8.17$$

and the largest root is

$$\lambda = .9273.$$

Notice that the eventual rate of decrease in heterozygosity is less in this system than with double first-cousin mating. Referring to Table 3.8.1, we see that the heterozygosity curves for the two systems cross at about the fifteenth generation. The general principle is that more intense inbreeding produces a lower ultimate rate of decrease in heterozygosity, provided that there is no permanent splitting of the population into isolated lines. Conversely, a system that avoids mating of relatives for as long as possible does so at the expense of a more rapid final approach to homozygosis. The breeder therefore may choose a different system of mating if he is more interested in maximum heterozygosity during the initial generations than in the long-time future population. An extension of the procedures of this section to larger populations than $N = 4$ has been given by Kimura and Crow (1963). Robertson (1964) and Wright (1965*a*) have shown that many of these results can be brought together very generally under a single point of view. For other types of mating systems see Wright (1921, 1951). Many of Wright's earlier results are summarized by Li (1955).

4. Partial Self-fertilization All the examples discussed thus far lead eventually to complete homozygosity. This is not always the case, and we shall now consider one such example. This is the simple, yet important, case where a certain fraction each generation are self-fertilized and the remainder are mated at random, a situation found in several plant species.

Let S be the fraction of the population that is produced by self-fertilization; then $1 - S$ is the fraction that is produced by random mating. From 3.8.1 we can write the expected recurrence relation for f as

$$f_t = S[(1 + f_{t-1})/2] + (1 - S)(0) = \frac{S}{2}(1 + f_{t-1}). \qquad \textbf{3.8.18}$$

This assumes that the plants to be self-fertilized each generation are a random sample of the population; for example, there is no tendency for the progeny of self-fertilized plants to be self-fertilized.

Substituting $f_t = (H_0 - H_t)/H_0$ from 3.2.4 into 3.8.18 we get

$$H_t = H_0(1 - S) + \frac{S}{2} H_{t-1}. \qquad \textbf{3.8.18a}$$

Subtracting $2(1 - S)H_0/(2 - S)$ from both sides and simplifying,

$$\begin{aligned} H_t - \frac{2(1-S)}{2-S} H_0 &= \frac{S}{2}\left[H_{t-1} - \frac{2(1-S)}{2-S} H_0\right] \\ &= \left(\frac{S}{2}\right)^2 \left[H_{t-2} - \frac{2(1-S)}{2-S} H_0\right] \qquad \textbf{3.8.19} \\ &= \left(\frac{S}{2}\right)^t \left[H_0 - \frac{2(1-S)}{2-S} H_0\right]. \end{aligned}$$

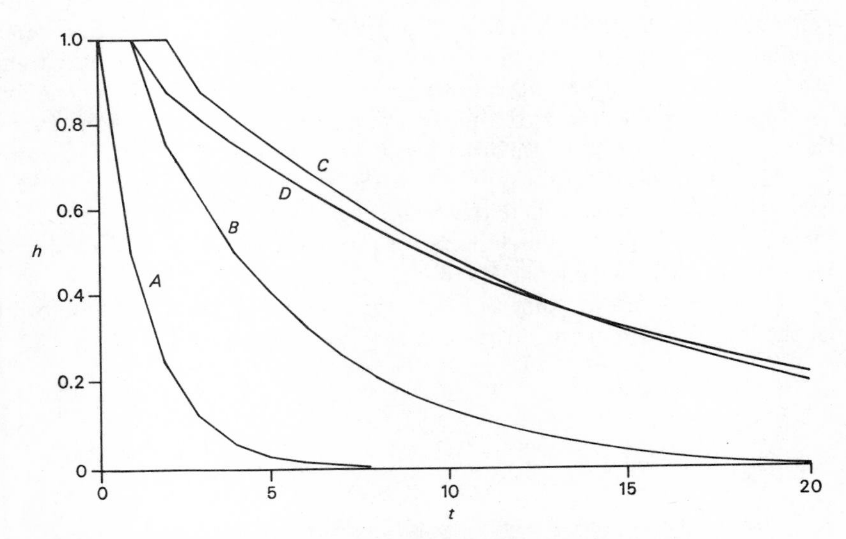

Figure 3.8.5. Change in heterozygosity with four mating systems. A. Self-fertilization; B. Sib mating; C. Double first-cousin mating; D. Circular half-sib mating. The ordinate is the heterozygosity relative to the starting population; the abscissa is the time in generations.

Since $(S/2)^t$ approaches 0 as t becomes large, the heterozygosity approaches a limit where the heterozygosity is a fraction $2(1 - S)/(2 - S)$ of its original value. The rate of approach is such that the departure from the equilibrium value is decreased by a fraction $1 - S/2$ each generation. Notice that when $S = 1$ we get the usual formula for self-fertilization.

This situation is striking in that unless S is large there is almost no cumulative effect; most of the effect occurs in the first generation. For example, with 10% self-fertilization, the initial heterozygosity is reduced by 5% in the first generation, but even when equilibrium is reached the reduction is only 5.3%!

5. Repeated Backcrossing to the Same Strain Frequently a plant breeder may wish to introduce one or more dominant genes from an extraneous source into a standard variety. For example he may have a highly desirable variety, except for its being susceptible to some disease. The resistant gene may exist in another strain which is less desirable in other respects. He can introduce this gene by crossing the two strains and then repeatedly crossing resistant plants to the susceptible strain. In this way the resistant gene is inserted into a genetic background that becomes more like the susceptible strain with each backcross. As another example, a mouse breeder may wish to introgress a new histocompatibility gene into a standard inbred strain.

It is clear that in recurrent backcrossing the number of loci that contain genes from both strains is reduced by half each generation. Thus, after t generations, a fraction equal to $1 - .5^t$ are from the recurrent parental strain. After seven generations less than 1% of the loci contain a gene from the other parental strain. If the recurrent parental strain is homozygous, the heterozygosity will reduce by half each generation, as with self-fertilization.

However, genes that are linked to the resistance or histocompatibility gene will tend to remain heterozygous. The question of how large a linked region will remain after a certain number of generations of backcrossing has been investigated by Haldane (1936) and Fisher (1949).

Consider a chromosome segment on one side of the selected factor and let the length of the segment be $100x$ map units in length (see Figure 3.8.6).

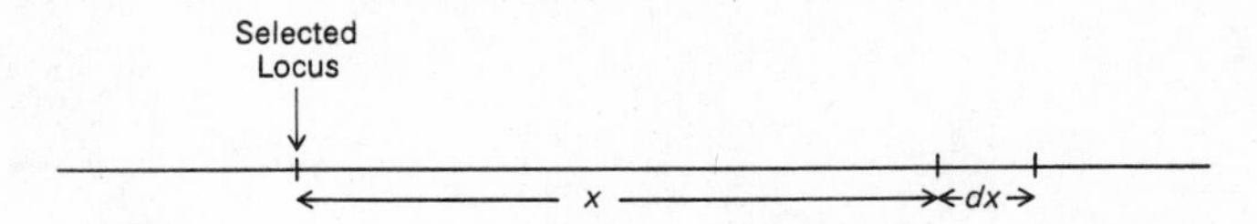

Figure 3.8.6. A chromosome segment. One locus is selected during recurrent backcrossing and the problem is to determine the length of chromosome to the right of this locus that will be intact after t generations.

If there is no interference, the probability of no crossover in this interval in one generation is e^{-x} (see Appendix A.5.6). The probability of no crossover in t generations is e^{-tx}. The chance of a crossover in the small interval x to $x + dx$ is dx, if we take this interval small enough that multiple crossovers can be ignored. The probability of a crossover in the interval dx sometime during t generations is tdx. Thus the probability after t generations of having had a crossover in the interval dx but not in the interval x is $e^{-tx}tdx$. Then the mean value of the intact interval x is

$$\bar{x} = \int_0^\infty e^{-tx} tx \, dx = \frac{1}{t}, \tag{3.8.20}$$

or $100/t$ map units.

For example, after 20 generations the average segment remaining intact would be 5 units on each side of the selected locus, or 10 units altogether. The derivation has assumed no interference, but for short regions the interference pattern makes very little difference.

There is a closely related question that can be answered very simply. We ask: What is the mean number of backcross generations that a gene will remain linked to the selected locus when the recombination probability is r? The number may be derived as follows. The probability that the genes will remain linked for t generations and then recombine in the next generation is $(1 - r)^t r$. The average number of generations until they become separated is

$$\begin{aligned} \bar{t} &= r + 2(1 - r)r + 3(1 - r)^2 r + 4(1 - r)^3 r + \cdots \\ &= r(1 + 2y + 3y^2 + 4y^3 + \cdots), \end{aligned}$$

where $y = 1 - r$.

Notice that the quantity in parentheses is the derivative of the series $1 + y + y^2 + y^3 + y^4 + \cdots = 1/(1 - y)$. Therefore

$$\bar{t} = r \frac{d}{dy}\left(\frac{1}{1 - y}\right) = r \frac{1}{(1 - y)^2} = \frac{1}{r}. \tag{3.8.21}$$

The value of x (the map distance in 3.8.20) and r (the recombination value in 3.8.21) will not in general be the same, but will become more and more similar as x becomes small enough that multiple crossovers can be neglected.

3.9 Inbreeding with Two Loci

The idea of the inbreeding coefficient can be extended to cover multiple loci. The principal item of interest is that inbreeding may cause the association of two or more recessive traits. This can happen in either of two ways: (1) with unlinked loci but nonuniform inbreeding, and (2) with uniform inbreeding and linkage—and, of course, because of both linkage and nonuniform inbreeding.

Our treatment follows very closely the method of Haldane (1949). We consider first unlinked loci in linkage equilibrium.

Let p_i be the frequency of allele A_i and r_k that of the independent allele B_k. The frequency of $A_i A_i$ homozygotes (from 3.2.2) is $p_i^2(1-f) + p_i f = p_i^2 + p_i(1-p_i)f$ with a similar formula for $B_k B_k$, $r_k^2 + r_k(1-r_k)f$.

Since the loci are in linkage equilibrium the frequency of $A_i A_i B_k B_k$ is the product, or

$$(p^2 + pqf)(r^2 + rsf) = p^2 r^2 + (pqr^2 + p^2 rs)f + pqrsf^2,$$

where $q = 1 - p$ and $s = 1 - r$, and the subscripts have been dropped for simplicity since we are discussing only one allele at each locus.

In a population with different values of f from individual to individual the frequency of the double homozygote, $P(AABB)$, is

$$p^2 r^2 + (pqr^2 + p^2 rs)\bar{f} + pqrs\overline{f^2}.$$

But, by the definition of the variance

$$\overline{f^2} = \bar{f}^2 + V_f.$$

Hence

$$P(AABB) = (p^2 + pq\bar{f})(r^2 + rs\bar{f}) + pqrsV_f. \qquad \text{3.9.1}$$

Suppose that in the human population two recessive genes each have a frequency of .01, and that 1% of the marriages are between cousins while the rest are random. Then $\bar{f} = 1/1600$, $V_f = 1/25600 - (1/1600)^2 = 3.87 \times 10^{-5}$, $(p^2 + pq\bar{f})(r^2 + rs\bar{f}) = (.00011)^2 = 11.28 \times 10^{-9}$, and $pqrsV_f = 3.79 \times 10^{-9}$. Hence the frequency of double homozygotes in the population is 15.07×10^{-9}, just about 4/3 what it would be if all families had the same inbreeding coefficient ($V_f = 0$).

To consider the second case we extend the inbreeding and consanguinity coefficient, f, to two loci. Let F be the probability that both of two loci carry identical alleles. F will be a function, not only of the pedigree, but also of the amount of recombination between the two loci.

A pair of uniting gametes can be: (1) identical for both loci with probability F; (2) identical for only the A locus, probability $f - F$; (3) identical for only the B locus, also $f - F$; or (4) identical for neither, with probability $1 - 2f + F$.

If the population has reached equilibrium with respect to the linkage phases, the frequency of AB gametes will be the product of the frequencies of the A and B genes.

Thus

$$P(AABB) = prF + pr^2(f - F) + p^2r(f - F) + p^2r^2(1 - 2f + F)$$
$$= pr(Fqs + fqr + fps + pr)$$
$$= pr[(qsf^2 + fqr + fps + pr) + Fqs - f^2qs] \quad \textbf{3.9.2}$$
$$= (p^2 + fpq)(r^2 + frs) + \phi pqrs,$$

where $\phi = F - f^2$. The frequency of any other genotype is obtainable the same way. For a summary table see Haldane (1949).

The quantity ϕ measures the degree of association in identity beyond that which would occur if the loci were independent. A comparison of 3.9.1 and 3.9.2 shows that ϕ and V_f enter the formula the same way. The effects are approximately additive, so that,

$$P(AABB) = (p^2 + \bar{f}pq)(r^2 + \bar{f}rs) + (\phi + V_f)pqrs \quad \textbf{3.9.3}$$

with analogous expressions for other genotypes.

There remains the problem of computing F (or ϕ) from a pedigree. There is no simple algorithm for F comparable to that for f. However, Denniston (1967) has discovered a method for computing F for any degree of relationship. The simple relationships are not hard to do by strong-arm methods.

Consider first two gametes produced from the same individual. They will be identical if both sets of alleles are derived from the same chromosome, there having been no recombination between A and B, or if the A's come from one chromosome and the B's from the other, which demands that each gamete be the result of a recombination. Letting c be the frequency of recombinant gametes, and d the frequency of nonrecombinant ($c + d = 1$), the probability of double autozygosity is

$$F = \tfrac{1}{2}(c^2 + d^2),$$
$$f = \tfrac{1}{2},$$
$$\phi = \tfrac{1}{4}(c - d)^2.$$

With parent-offspring mating, in order to have double autozygosity one of the identical gametes from the common ancestor must pass intact through another generation, which has probability $d/2$. Therefore

$$F = \tfrac{1}{4}(c^2 + d^2)d,$$
$$f = \tfrac{1}{4},$$
$$\phi = \tfrac{1}{16}(d - c)(3d^2 + c^2).$$

For half-sibs, each of two gametes must pass intact through a generation, which has probability $d^2/4$. Thus

$$F = \tfrac{1}{8}(c^2 + d^2)d^2,$$
$$f = \tfrac{1}{8},$$
$$\phi = \tfrac{1}{64}(8d^2 + 8c^2d^2 - 1).$$

With full-sib mating a complication arises since there must be consideration of the possibility that the A's come from one grandparent and the B's from the other. For the details we refer to Haldane's paper; his results for several kinds of relationship are given in Table 3.9.1.

Table 3.9.1. Values of the inbreeding coefficient for two loci, where c is the proportion of recombination and $d = 1 - c$. The parents are not inbred and are in linkage equilibrium. From Haldane (1949).

RELATIONSHIP OF PARENTS	F	f	$\phi = F - f^2$
Identical	$\frac{1}{2}(c^2 + d^2)$	$\frac{1}{2}$	$\frac{1}{4}(c - d)^2$
Parent-offspring	$\frac{1}{4}(c^2 + d^2)d$	$\frac{1}{4}$	$\frac{1}{16}(d - c)(3d^2 + c^2)$
Full sibs	$\frac{1}{8}(2d^4 + 2c^2d^2 + c^2)$	$\frac{1}{4}$	$\frac{1}{16}(4d^4 + 4c^2d^2 + 2c^2 - 1)$
Half-sibs	$\frac{1}{8}(c^2 + d^2)d^2$	$\frac{1}{8}$	$\frac{1}{64}(8c^2d^2 + 8d^4 - 1)$
Uncle-niece	$\frac{1}{16}(2d^4 + 2d^2c^2 + c^2)d$	$\frac{1}{8}$	$\frac{1}{64}(8d^5 + 8c^2d^3 + 4c^2d - 1)$
First cousins	$\frac{1}{32}(2d^4 + 2c^2d^2 + c^2)d^2$	$\frac{1}{16}$	$\frac{1}{256}(16d^6 + 16c^2d^4 + 8c^2d^2 - 1)$
Double half-cousins	$\frac{1}{128}(8d^6 + 8c^2d^4 + c^2)$	$\frac{1}{16}$	$\frac{1}{256}(16d^6 + 16c^2d^4 + 2c^2 - 1)$

It is interesting that several relationships with the same f may have different F's. Compare, for example, ordinary first cousins with double half-cousins.

As a numerical example, assume $p = r = .01$. The frequency of AA or BB in the progeny of cousins is .00072 compared with .00010 with random mating. In the absence of linkage the frequency of $AABB$ is $(7.2 \times 10^{-4})^2 = 5.2 \times 10^{-7}$; with 10% recombination the frequency is $5.2 \times 10^{-7} + \phi pqrs = 34.6 \times 10^{-7}$.

However, such a sharp rise is expected only with rare genes and close linkage. Some association of rare recessive traits should be expected as a consequence of inbreeding and linkage, but probably not of sufficient magnitude to be detected in ordinary circumstances.

The effect of inbreeding on quantitative traits involving two loci when epistasis and linkage disequilibrium are present may be obtained by using expressions like 3.9.3. for all the genotypes. Nei (1965) did this for the mean fitness, as an extension of 3.6.3, which was derived by assuming linkage equilibrium.

3.10 Effect of Inbreeding on the Variance

In Section 3.6 we considered the effect of inbreeding on quantitative traits, especially as the mean is affected. In addition to its effect on the population mean, inbreeding also has an effect on the variance.

Consider again the single-locus model of Section 3.6. It is convenient to let $Y = 0$, as this does not change the conclusion and saves some troublesome algebra. To assess the total effect we consider first a single locus.

GENOTYPE	A_1A_1	A_1A_2	A_2A_2
FREQUENCY	$p_1^2(1-f)+p_1 f$	$2p_1p_2(1-f)$	$p_2^2(1-f)+p_2 f$
CONTRIBUTION TO TOTAL PHENOTYPE	$-A$	D	A

We write the mean as a function of f,

$$\begin{aligned}\overline{Y}_f &= (1-f)(p_2^2 A + 2p_1p_2 D - p_1^2 A) + f(p_2 A - p_1 A)\\ &= (1-f)\overline{Y}_0 + f\overline{Y}_1,\end{aligned} \tag{3.10.1}$$

where $\overline{Y}_0$ is the mean value with random mating ($f = 0$) and $\overline{Y}_1$ is the mean value with complete homozygosity ($f = 1$).

Notice that this can be written as

$$\overline{Y} = \overline{Y}_0 + (\overline{Y}_1 - \overline{Y}_0)f, \tag{3.10.2}$$

illustrating once again, as was shown in Section 3.6, that the phenotype is a linear function of f. Equation 3.10.2 is the same as 3.6.1.

We now write the expression for the contribution of this locus to the total population variance.

$$\begin{aligned}V_f &= (1-f)(p_1^2A^2 + 2p_1p_2 D^2 + p_2^2 A^2) + f(p_1A^2 + p_2 A^2) - \overline{Y}^2\\ &= (1-f)(V_0 + \overline{Y}_0^2) + f(V_1 + \overline{Y}_1^2) - [(1-f)\overline{Y}_0 + f\overline{Y}_1]^2\\ &= (1-f)V_0 + fV_1 + f(1-f)(\overline{Y}_0 - \overline{Y}_1)^2,\end{aligned} \tag{3.10.3}$$

where V_0 and V_1 are the variance with random mating ($f = 0$) and complete inbreeding ($f = 1$). This formula comes from Wright (1951).

This shows that, unlike the mean, the variance is not a linear function of f but is quadratic.

Later, in considering the effects of selection, we shall see that the rate of gene frequency change depends on the additive component of the gene effect. Therefore it is of special interest to examine the effect of inbreeding on the mean and variance of a locus without dominance.

Without dominance there is no effect of inbreeding on the mean, as can be seen by letting $D = 0$ in 3.10.1 or 3.6.1. $\overline{Y}_0 = \overline{Y}_1 = A(p_2 - p_1)$,

$$V_0 = p_1^2 A^2 + p_2^2 A^2 - \overline{Y}_0^2 = 2p_1 p_2 A^2, \tag{3.10.4}$$

$$V_1 = p_1 A^2 + p_2 A^2 - \overline{Y}_1^2 = 4p_1 p_2 A^2 = 2V_0. \tag{3.10.5}$$

Thus, with no dominance the total variance is

$$V_f = (1 - f)V_0 + fV_1 = V_0(1 + f), \tag{3.10.6}$$

which in this case is linear rather than quadratic. A population within which there is some consanguineous mating has an increase in genetic variance proportional to f, provided there is no dominance.

If there is subdivision of the population into inbred strains, we can measure the variance within and between such strains. From 3.10.4. we see that the variance of a randomly mating population for a nondominant locus is proportional to the amount of heterozygosity, measured by $2p_1p_2$. In a strain of inbreeding coefficient f the heterozygosity is reduced by a fraction f; hence the variance is reduced by this amount.

The effect of any inbreeding which is of a type that divides the population into isolated groups will be to decrease the variance within groups, to increase the variance between groups, and increase the total variance. This happens, for example, in the development of a series of inbred lines of plants or livestock.

These conclusions apply only to the special case of no dominance. With varying degrees of dominance explicit formulae for variance within and between lines become more difficult (Robertson, 1952; Wright, 1952). This is discussed in Chapter 7.

A case of special interest arises when there are recessive alleles of low frequency. In this circumstance the variance within a subpopulation increases during the early stages of the inbreeding process, despite the fact that the population is becoming more homozygous. The reason is that inbreeding brings out previously hidden recessive factors which now can contribute to the phenotypic variance. This increase will continue until it is eventually offset by the increasing homogeneity in gene content from individual to individual within the subpopulation.

3.11 The Inbreeding Effect of a Finite Population

As was mentioned in the introduction to this chapter, there is a decrease of heterozygosity in a finite population even if there is random mating within the group. Each generation may be regarded as being made up of $2N$ gametes drawn from the previous generation and which combine to make the N individuals of this generation. The gene frequency will therefore change somewhat, the amount depending on the smallness of the population. The change will have a variance of $p(1-p)/2N$ where p is the frequency of the gene under consideration in the parents. We have assumed here that any particular successful gamete is equally likely to have come from any one of the parents, so that the situation is analogous to binomial sampling of $2N$ gametes (see A.5.5).

It might not seem obvious that random changes in gene frequency, which can be in either direction, will on the average cause a net decrease in heterozygosity and increase in homozygosity. One way to visualize this is to regard each of the $2N$ genes at a locus in the N parents as individually labeled. The effect of random processes in drawing (with replacement) from these is that some will be omitted entirely while others will be drawn more than once. Thus there will be a certain amount of identity next generation. If the process continues long enough, all the genes will be descended from a single individual gene and complete autozygosity will be attained.

The distribution of the probabilities of various gene frequencies during this process is a difficult problem. On the other hand, the average change in heterozygosity is uniform and easily derived. It can be expressed as a function of the inbreeding coefficient.

Consider first a population with completely random mating, including self-fertilization. As before, we consider a single locus. Imagine that the progeny are produced by drawing random pairs of gametes from an infinite pool to which each parent had contributed equally (or, if the pool is finite, the drawing is with replacement). Two gametes then have a chance $1/2N$ of carrying identical genes, since the N diploid parents have $2N$ genes at this locus. Two gametes have a chance of $1 - 1/2N$ of carrying different parental genes.

In the first case, the probability of the genes being identical is of course 1. In the second case the probability of their being identical is f_{t-1}, the inbreeding coefficient of an average individual in the previous generation. The reason for this is that the two alleles were drawn at random from the parent generation, and since the parents were the result of random mating, the probability of any two alleles being identical is the same as that for two in the same zygote; the latter is, by definition, f_{t-1}. Therefore,

$$f_t = \frac{1}{2N} + \left(1 - \frac{1}{2N}\right) f_{t-1}. \qquad 3.11.1$$

Recalling that H_t, the heterozygosity at time t, is $H_0(1 - f_t)$, we obtain by substitution in 3.11.1

$$H_t = \left(1 - \frac{1}{2N}\right)H_{t-1} = \left(1 - \frac{1}{2N}\right)^t H_0. \tag{3.11.2}$$

The result is the very simple one; despite the complexities in the changes in gene frequencies, the average heterozygosity decreases by a fraction $1/2N$ each generation.

As stated earlier, this formula assumes completely random mating (that is, random combination of gametes) including the possibility of self-fertilization. Note that, when $N = 1$, the results agree with the formula for self-fertilization, as expected (3.8.1 and 3.8.2).

No Self-fertilization We now assume that gametes are combined at random, but with the restriction that two uniting gametes cannot come from the same parent. We let f_t be the inbreeding coefficient at time t, as before, and let g_t be the coefficient of consanguinity of two different randomly chosen individuals in generation t. The inbreeding coefficient in generation t is clearly the same as the consanguinity coefficient in generation $t - 1$, since mating is at random. To get the consanguinity coefficient in generation t, we note that the two chosen genes (one from each individual) have come from the same individual in the previous generation with probability $1/N$ and from different individuals with probability $1 - 1/N$. In the first case the probability of identity is $(1 + f_{t-1})/2$ as explained earlier (see, for example, 3.4.1 and 3.8.1); in the second case it is g_{t-1}. Putting these together, we obtain

$$\begin{aligned} f_t &= g_{t-1}, \\ g_t &= \frac{1}{2N}(1 + f_{t-1}) + \frac{N-1}{N} g_{t-1}. \end{aligned} \tag{3.11.3}$$

Substituting the first into the second gives

$$f_{t+1} = \frac{1}{2N}(1 + f_{t-1}) + \frac{N-1}{N} f_t,$$

which, after going back one generation and rearranging, is

$$f_t = f_{t-1} + (1 - 2f_{t-1} + f_{t-2})/2N. \tag{3.11.4}$$

Separate Sexes If the individuals in generation $t - 1$ consist of N_m males and N_f females adding up to a total of N individuals, only a slight modification is required. The two sexes necessarily make the same contribution to later generations, since each fertilization event involves one maternal and one paternal gamete. Therefore the probability that two genes in different individuals in generation t are both derived from a male in generation $t-1$ is 1/4; and that they came from the same male is $1/4N_m$. Likewise, the probability of their coming from the same female is $1/4N_f$. Then the probability that the two genes came from the same individual in generation $t-1$ is

$$\frac{1}{4N_m} + \frac{1}{4N_f} = \frac{1}{N_e}, \qquad 3.11.5$$

where N_e is the *effective* number of individuals in generation $t-1$.

This means that equations 3.11.3 and 3.11.4 are correct with separate sexes; it is only necessary to use N_e instead of N in the formula. Notice that when $N_m = N_f = N/2$, then $N_e = N$. When the two sexes are equally frequent, the actual number and the effective number are the same.

The effective population number is used with a wider definition than this single example would suggest. If there are fluctuations in the population number from time to time, or if the distribution of number of progeny per parent is nonbinomial, or if there is any other kind of deviation from the idealized model that we have assumed, then it is conventional to define for that population an effective number. The effective number then, as in this example, is the size of an ideally behaving population that would have the same homozygosity increase as the observed population. We shall return to this subject in Section 3.13 and in Chapter 7.

Notice that, when the number of males and females is equal, there is no distinction (as far as rate of homozygosity change caused by random gene-frequency drift) between a population with separate sexes and a hermaphroditic population without self-fertilization.

Returning to equation 3.11.4 we can make the same substitution as before, $H_t = H_0(1 - f_t)$, and write an equation for the heterozygosity in successive generations. This yields

$$H_t = \frac{N-1}{N} H_{t-1} + \frac{1}{2N} H_{t-2}. \qquad 3.11.6$$

Notice that when $N = 2$ equations 3.11.4 and 3.11.6 reduce to the equations for sib-mating (3.8.4 and 3.8.5) as expected. This comparison may be misleading in one regard, however. Part of the increase in homozygosity in larger populations is due to the fact that different members of the population leave different numbers of descendants. The general formula takes this into account, and with larger N it turns out that about half of the increase in homozygosity is caused by the restricted number of parents and the other half by differential contribution of offspring. Equation 3.11.6 assumes under these circumstances that the distribution of progeny-number is binomial, as follows from the assumption that each progeny gene has an equal chance of having come from any parent. However, as is seen, the formula is still correct in the limiting case of sib mating, despite the fact that the progeny number per parent is necessarily constant in this case.

The ultimate ratio by which H_t decreases each generation is given by the larger root of the quadratic equation

$$\lambda^2 - \left(\frac{N-1}{N}\right)\lambda - \frac{1}{2N} = 0. \qquad \textbf{3.11.7}$$

This equation is easily obtained by setting $H_t/H_{t-1} = H_{t-1}/H_{t-2} = \lambda$ in 3.11.6.

The relevant solution to the equation is

$$\lambda = \frac{N - 1 + \sqrt{N^2 + 1}}{2N}, \qquad \textbf{3.11.8}$$

or approximately, unless N is very small,

$$1 - \lambda \sim \frac{1}{2N + 1}, \qquad \textbf{3.11.9}$$

or, when N is large,

$$1 - \lambda \sim \frac{1}{2N}. \qquad \textbf{3.11.10}$$

Hence, in a moderately large population the average heterozygosity is reduced by about $1/2N$ per generation whether there are separate sexes or not, provided that the two sexes are equal in frequency.

The accuracy of the approximation 3.11.9 for small N can be seen by comparison with the data in Table 3.1.2. The last column in this table gives the proportion by which the heterozygosity is reduced, which is to be compared with the approximation $1/(2N + 1)$. Within 6 generations the approximation is quite good: with $N = 2$, $1/(2N + 1) = 1/5$ or .20 whereas the exact answer is .19. So if the population is anywhere near a steady state, 3.11.9 is very good even for quite small numbers.

The results of this section were all obtained first by Wright (1931) by his method of path coefficients. A summary of this method and its applications to problems in population genetics is given in his Galton Lecture (Wright 1951).

3.12 Hierarchical Structure of Populations

At the beginning of this chapter we emphasized that inbreeding can occur under two quite different circumstances. In both cases there is a decrease in heterozygosity measured by f, but some of the consequences are quite different.

In the first case we can have a large population within which isolated consanguineous matings occur. The inbreeding coefficient measures the average decrease in heterozygosity and equations 3.2.1—3.2.3 give the expected genotypic frequencies for given gene frequencies. As soon as random mating occurs the inbreeding coefficient returns to zero. An example is the case of partial self-fertilization discussed in Section 3.8; as soon as self-fertilization is prevented, the original heterozygosity is restored. (We should perhaps mention that here, as throughout this chapter, we are ignoring the effects of selection and mutation.)

On the other hand, there may be inbreeding because of restriction of population number, even though mating is at random within the population. The average heterozygosity within the population is reduced and the individuals become more closely related, both measured by f (i.e., f_I and f_{IJ}). But, as emphasized in Section 3.11, the increased homozygosity within the population is not due to departure from Hardy–Weinberg ratios, but to changes in the gene frequencies. The change is such as to make the value of $2p_1p_2$ (or $\Sigma p_i p_j$ with multiple alleles) decrease in proportion to f. The formulae 3.2.1–3.2.3 hold only in the sense of giving the genotype frequencies averaged over a whole series of such populations.

In this situation the loss of heterozygosity within the population is permanent and could be restored only by crossing with other populations. Repeated self-fertilization and sib mating constitute extreme cases of small populations.

There are circumstances in which the two effects are combined. For example, we might inquire about the inbreeding coefficient of an animal whose parents were sibs in an isolated population of effective size N. His homozygosity will be greater than if his parents were sibs in a large population. We now derive a procedure to handle this situation. The conclusions were first reached by Wright (1943, 1951) by a different method.

Let S be a subpopulation derived by isolating a finite number of individuals from a large total population T. For example, S could be a breed or a strain isolated from a foundation stock T. Or, S could be one of a series of geographically isolated subpopulations of a large population T. Let I be an individual within subpopulation S.

We now define f_{IS} as the probability that two homologous genes in I are derived from the same gene in a common ancestor within the subpopulation. Let f_{ST} be the probability that two homologous genes, chosen at random from the subpopulation, are both descended from a gene in the subpopulation. We let f_{IT} be the overall probability of identity in individual I.

The probability of nonidentity is then the product of two terms: $(1 - f_{IS})$, the probability that the two genes do not both come from a gene in a known common ancestor in the population, and $(1 - f_{ST})$, the probability that if the two genes are randomly chosen from within the population they will not be identical because of some more remote relationship. Therefore,

$$1 - f_{IT} = (1 - f_{IS})(1 - f_{ST}), \qquad 3.12.1$$

or

$$f_{IT} = f_{ST} + (1 - f_{ST})f_{IS}. \qquad 3.12.1a$$

For example, what is the inbreeding coefficient of a child whose parents were cousins on an island whose population is descended from a shipwreck 10 generations ago? Assume for simplicity that there were 50 survivors, equally divided between the two sexes, and that the population has remained of this size and sex distribution since that time. From equation 3.11.2 (accurate enough, although it would be better to use $2N + 1$ instead of $2N$) we have

$$1 - f_{ST} = \left(1 - \frac{1}{100}\right)^{10} = 0.9045$$

and

$$1 - f_{IS} = 1 - \frac{1}{16} = 0.9375,$$

1/16 being the inbreeding coefficient of a child of a cousin marriage. From 3.12.1,

$$1 - f_{IT} = (0.9045)(0.9375) = 0.848$$

and

$$f_{IT} = 0.152,$$

compared with 0.0625 for cousin marriage in an infinite population.

Students of animal breeding will enjoy Wright's (1951 and earlier) application of these methods to the history of Shorthorn cattle. He showed that there was a substantial increase in the inbreeding coefficient f_{IT} of British Shorthorns, almost entirely due to f_{ST} and hardly at all because of consanguineous matings within the breed (f_{IS}).

Similar analyses are possible in human isolates if accurate pedigree records are available. If they are not, it is sometimes possible to get reasonably satisfactory information from marriage records. Since a person's name is inherited as if it were linked to the father's Y chromosome, marriages between persons of the same surname (isonymous) can be used as indications of consanguinity. Using this procedure on the Hutterite population, Crow and Mange (1965) were able to show that f_{ST} in this population is quite appreciable, about 4%, but that f_{IS} is not significantly different from 0.

In other words, the increased homozygosity is due almost entirely to the small effective size of the population and not at all to nonrandom marriage within the isolate.

Notice that 3.12.1 may be written as

$$1 - f_{IS} = \frac{H_0(1 - f_{IT})}{H_0(1 - f_{ST})}, \tag{3.12.2}$$

which shows that $1 - f_{IS}$ is a measure of the heterozygosity of an individual in the subpopulation relative to that of an individual derived from random mating in the same subpopulation.

As was mentioned earlier, the inbreeding coefficient can be described in terms of correlation as well as in terms of gene identity. In fact Wright's original definition and derivation was through correlation analysis. One advantage of the correlation interpretation, as opposed to the probabilistic, is that negative values have a meaning. This is especially useful in this section. F_{IS} is the correlation between homologous genes in an individual relative to genes chosen at random from his subpopulation. F_{IT} is the correlation between homologous genes in an individual relative to the whole population. F_{ST} is the correlation between randomly chosen genes in the subpopulation relative to the total population.

F_{ST} is necessarily positive, but the others need not be. For example, if there is specific avoidance of matings between related individuals within the subpopulation, F_{IS} may be negative.

Wright defined f_{ST} as the correlation between two random gametes from the same subpopulation and derived the relation

$$f_{ST} = \frac{V_p}{\bar{p}(1 - \bar{p})}, \tag{3.12.3}$$

where $\bar{p}$ and V_p are the mean and variance of the gene A among the subpopulations. We can derive this as follows, using 3.12.1.

Consider a pair of alleles, A and A', and let p be the frequency of A. Then the frequency of a heterozygote in the subgroup with frequency p is $2p(1 - p)(1 - f_{IS})$. Thus the frequency of heterozygotes in the total population is

$$H = E\{2p(1 - p)(1 - f_{IS})\}$$

where $E\{\ \}$ designates taking the expectation or average over all subpopulations. If the variation of gene frequencies among subgroups is independent of f_{IS}, then

$$H = 2(1 - f_{IS})E\{p(1 - p)\}.$$

But

$$E\{p(1-p)\} = E\{p - p^2\}$$
$$= \bar{p} - \bar{p}^2 - V_p$$

(see A.2.2 or 2.9.2). Thus

$$H = 2(1 - f_{IS})(\bar{p} - \bar{p}^2 - V_p). \qquad 3.12.4$$

On the other hand, from the definition of f_{IT},

$$H = 2(1 - f_{IT})\bar{p}(1 - \bar{p}). \qquad 3.12.5$$

Equating these two expressions gives

$$(1 - f_{IT}) = (1 - f_{IS})\left(1 - \frac{V_p}{\bar{p}(1-\bar{p})}\right).$$

Comparing this with 3.12.1 gives us the desired expression, 3.12.3.

Notice that if there is no inbreeding in the subpopulation ($f_{IS} = 0$), then

$$H = 2\bar{p}(1 - \bar{p}) - 2V_p. \qquad 3.12.6$$

This, although expressed in terms of heterozygotes rather than homozygotes, is the same as we described as Wahlund's principle in 2.9.2.

Nei (1965) extended this procedure to cover the case of multiple alleles and has shown that

$$f_{ST} = \frac{-\text{Cov}_{ij}}{\bar{p}_i \bar{p}_j}, \qquad 3.12.7$$

where $\bar{p}_i$ and $\bar{p}_j$ are the mean frequencies of the alleles A_i and A_j and Cov_{ij} is the covariance of their frequencies. (See problem 22, Chapter 1.)

Nei and Imaizumi (1966) applied these formulae to study the differentiation of *ABO* blood group gene frequencies among the prefectures of Japan. They obtained 6 values of f_{ST} that were in close agreement, suggesting that the local differentiation of gene frequencies was mainly random.

Equation 3.12.1. can easily be extended to include subdivisions of a subdivision in a hierarchy. For example,

$$(1 - f_{IT}) = (1 - f_{IR})(1 - f_{RS})(1 - F_{ST}), \qquad 3.12.8$$

where R is a subpopulation of S which is a subpopulation of T. Then f_{RS} is the probability that two randomly chosen homologues in R are identical because of common ancestry within R and f_{ST} is the probability that two randomly chosen genes in S are identical because of common ancestry within S.

3.13 Effective Population Number

In our discussion of random drift in gene frequencies and the resulting decrease in heterozygosity in a finite population we have assumed, in addition to random mating, that the expected number of progeny is the same for each individual. That is to say, we regarded a successful gamete as being equally likely to have come from any individual in the parent generation. We also assumed that the population size remains constant from generation to generation. Neither of these conditions is likely to be met in nature.

In equation 3.11.5 we introduced the concept of effective population number by noting that, if the numbers of males and females differ, we can replace N in the formulae such as 3.11.6 by N_e, the effective number, where

$$\frac{1}{N_e} = \frac{1}{4N_m} + \frac{1}{4N_f}$$

or

$$N_e = \frac{4N_m N_f}{N_m + N_f}. \qquad 3.13.1$$

Notice that the value of N_e is influenced much more by the smaller than by the larger of N_m and N_f. For example, if $N_m = 1$ and $N_f = 100$, N_e is about 4; so, in highly polygynous species the number of males is much more important than the number of females in determining the amount of random drift in the population. N_e is proportional to the harmonic mean of N_m and N_f, and the harmonic mean is more strongly influenced by the smaller values.

We can also use the concept of effective population number when the true size varies with time. From 3.11.2 we note that the heterozygosity after t generations, when N varies from generation to generation, is

$$\begin{aligned}\frac{H_t}{H_0} &= \left(1 - \frac{1}{2N_0}\right)\left(1 - \frac{1}{2N_1}\right)\left(1 - \frac{1}{2N_2}\right)\cdots\left(1 - \frac{1}{2N_{t-1}}\right) \\ &= \prod_{i=0}^{t-1}\left(1 - \frac{1}{2N_i}\right).\end{aligned} \qquad 3.13.2$$

We then ask: What population of constant size would have the same decrease in heterozygosity over the time period involved? We call this number the effective population number.

To find this, we write

$$\frac{H_t}{H_0} = \left(1 - \frac{1}{2N_e}\right)^t = \prod_{i=0}^{t-1}\left(1 - \frac{1}{2N_i}\right), \qquad 3.13.3$$

which can be solved for N_e.

Notice that if the N_i's are fairly large and t is small, this equation is roughly

$$1 - \frac{t}{2N_e} = 1 - \sum \frac{1}{2N_i}$$

or

$$\frac{1}{N_e} = \frac{1}{t} \sum \frac{1}{N_i}. \qquad 3.13.4$$

So, if the population size fluctuates, the effective population number is roughly the harmonic mean of the various values.

In Chapter 2 we showed that with varying population growth rates at different times the arithmetic mean is appropriate as a summarizing value if fitness is measured by the Malthusian parameter, m (1.2.5), and the geometric mean is used if fitness is measured by w (1.1.5). Here we find that still a third average, the harmonic mean, serves best as a single representative value.

Finally, if the individuals in the population do not have the same expected number of progeny, the effective number will be less than the census number. If the population size is constant

$$N_e = \frac{4N - 2}{\sigma_k^2 + 2} \qquad 3.13.5$$

where σ_k^2 is the variance in the number of progeny per parent. This is derived, along with more general formulae, later in Section 7.6.

This formula assumes complete random mating, including the possibility of self-fertilization. Notice that when the number of progeny per parent has a binomial distribution with mean 2 (the mean must be 2 in a sexual population that is neither increasing nor decreasing) this is equivalent to drawing a sample of $2N$ gametes randomly from the N parents. The binomial variance is

$$\sigma_k^2 = 2N\left(\frac{1}{N}\right)\left(1 - \frac{1}{N}\right) = \frac{2(N-1)}{N}.$$

Substituting this into 3.13.5 gives

$$N_e = N,$$

as it should. In this "ideal" population the actual and effective numbers are the same.

Notice that the ideal population model when $N_e = N$ does not imply a constant number of progeny per parent, but a randomly varying number. If the number has greater than binomial variance the effective number is smaller than the census number; if the variance is less than binomial, the effective number is larger. An extreme case is that where each parent is constrained to have the same number of progeny, as in many livestock or laboratory animal breeding systems. If $\sigma_k^2 = 0$ in 3.13.5 then $N_e = 2N - 1$, almost twice the census number. So, with random mating and binomial progeny distribution, about half the reduction in heterozygosity is due to consanguinity among mates; the other half is caused by variable numbers of progeny.

In most populations in nature the effective number is less than the census number. For a discussion of laboratory and census data on this question, see Crow and Morton (1955).

The concept of effective population number and the formulae in this section are all due to Sewall Wright (see especially his 1931 and 1938*b* papers). For extensions to more complicated situations see Crow (1954), Crow and Morton (1955), and Kimura and Crow (1963*a*). We shall return to this subject in Section 7.6.

3.14 Problems

1. In Table 3.1.1, if the initial frequencies are in Hardy–Weinberg ratios, show that in generation t, $D_t = p(1 - 2^{-t}q)$, where p and q are the frequencies of the dominant and recessive genes.
2. In deriving 3.8.19 we rather arbitrarily subtracted $2(1 - S)H_0/(2 - S)$ from both sides of the equation. Show, by equating H_t to H_{t-1}, that this is an equilibrium value (and that therefore the procedure was not arbitrary).
3. Is individual I inbred?

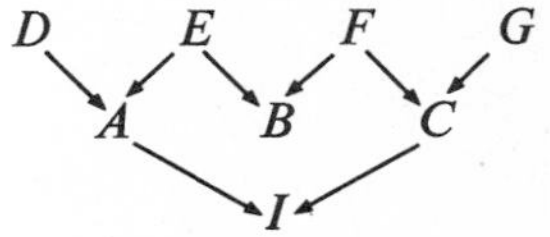

D E F G

A B C

I

4. The algorithm for computing the inbreeding (or consanguinity) coefficient (3.4.1) permits no individual to be counted twice in the same path. Invent an algorithm that can dispense with the $(1 + f_A)$ term by permitting the path to pass through some individuals twice. (Figure 3.4.3 provides a good example.)
5. What is the rate of decrease in heterozygosity from mother to daughter in mother-son mating in honeybees? Remember that the male bee is haploid, being derived parthenogenetically from the mother. Thus the rule is the same as for the X chromosome in ordinary species. See Figure 3.14.1.

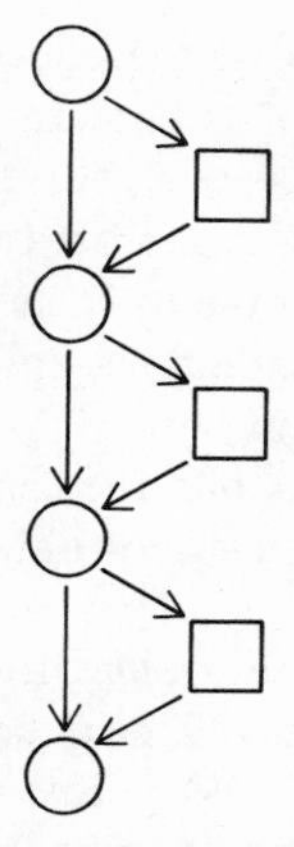

Figure 3.14.1. Repeated mother-son mating in honeybees.

6. Compare the results of parent-offspring mating for an autosomal locus with that for a sex-linked locus (or for bees) in problem 5. See Figure 3.14.2.

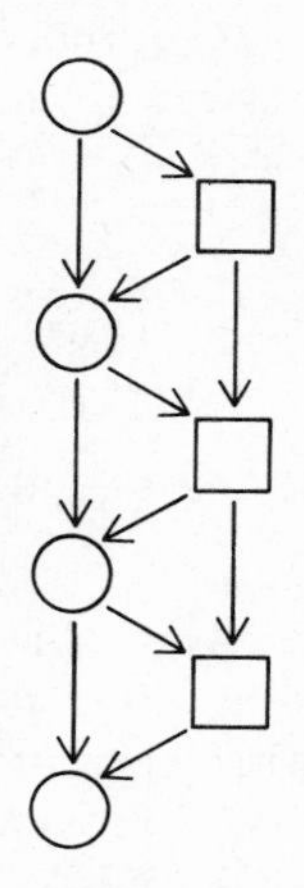

Figure 3.14.2. Repeated parent-offspring mating for autosomal locus.

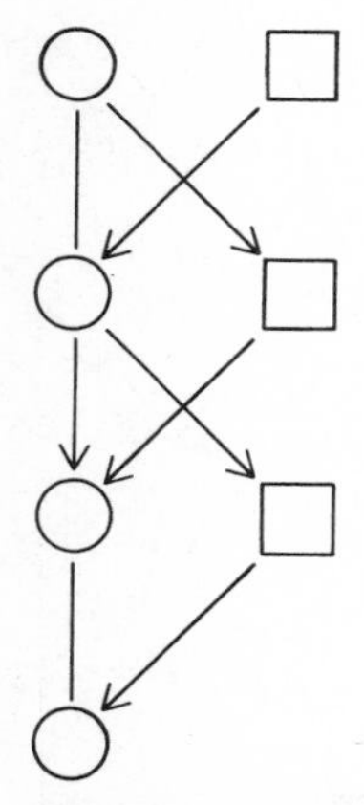

Figure 3.14.3. Repeated sib (or half-sib), mating in honeybees.

7. Show that continued sib mating leads to the same rate of decrease in heterozygosity in honeybees (or for an *X*-linked locus) as in ordinary diploid inheritance. See Figure 3.14.3.
8. Artificial insemination in bees requires pooled sperm from several males. If these males are brothers, will this alter the results of problem 7?
9. Show that, with partial self-fertilization, when self-fertilized plants produce fewer progeny than cross-fertilized the gene frequency does not change (although, of course, the rate of increase in homozygosity is less).
10. Is the rate of decrease in heterozygosity by random drift the same for a hermaphroditic species, such as earthworms, where individuals produce both eggs and sperm but do not self-fertilize, as for a bisexual species with equal numbers of males and females?
11. Show that with continued sib mating the heterozygosity, H_t, is given by $H_t = H_{t-1} - (1/8)H_{t-3}$.
12. How would you expect inbreeding to affect the rate of approach to linkage equilibrium?
13. Generalize the formula for maximum avoidance of inbreeding to a population of size 8. What is the value of λ?
14. Figure 3.14.4 gives an example of a mating system that reaches an equilibrium level of heterozygosity rather than approaching 0. Show that the eventual heterozygosity alternates between the original amount and 2/3 this amount. (You might enjoy generalizing this to two generations of sib mating between outcrosses. The inbreeding coefficient then cycles through the values 0, 5/14, and 3/7 at equilibrium.)

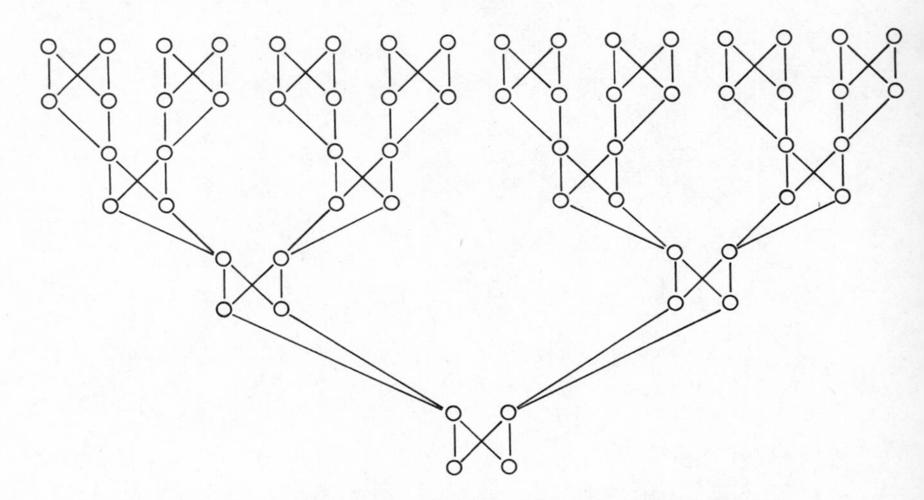

Figure 3.14.4. A mating system in which sib-mating is alternated with out-crossing.

15. Show that, starting with a randomly mating monoecious population of size N, the inbreeding coefficient after t generations is $1 - [(2N - 1)/2N]^t$.
16. What is the mean number of backcross generations required to separate a gene from the selected locus if the recombination frequency is 0.1?
17. If 10 generations of backcrossing are carried out, what is the probability that the genes are not yet separated?
18. How many generations of backcrossing must an experimenter carry out if he wants a probability of at least 0.95 that the genes have separated?
19. Show that the variance of the number of backcross generations that a gene will remain linked (see 3.8.21 for the mean number) is $(1 - r)/r^2$. *Hint*: Note that when $Z = 1 + y + y^2 + y^3 + \cdots$,

$$\frac{d^2Z}{dy^2} - \frac{dZ}{dy} = 1 + 4y + 9y^2 + 16y^3 + \cdots.$$

20. Show that with maximum avoidance of inbreeding in a population of size 4,

$$h_t = \tfrac{1}{2}h_{t-1} + \tfrac{1}{4}h_{t-2} + \tfrac{1}{8}h_{t-3}.$$

21. Comparing this to the formula for sib mating, which is maximum avoidance in a population of size 2 (3.8.5), and that for self-fertilization (size 1), you can probably guess what the generalization is to populations of size 8, 16, etc. Show this for a population of size 8.
22. Show that the variance of x in 3.8.20 is $1/t^2$.

4 CORRELATION BETWEEN RELATIVES AND ASSORTATIVE MATING

In this chapter we consider the analysis of population variance, the correlation between relatives, and the effect of inbreeding and assortative mating on both of these.

Since assortative mating is the mating of individuals with similar phenotypes, and since the phenotype is to some extent a reflection of the genotype, we should expect assortative mating to have qualitatively the same consequences as inbreeding, which is the mating of genotypically similar individuals. We shall see that this is true, but there are interesting quantitative differences.

A recurring problem in genetics is the separation of genetic from environmental effects. The first thing to do is to define the question more precisely, for clearly it gets us nowhere to try to answer such a question as "Is heredity or environment more important?" For one thing we must specify what trait is being discussed. We then ask questions of this type: Of the variability existing in a population, how much is caused by differences in the genetic composition and how much is caused by environmental differences? In principle we could hold the genotype constant and see how much variation remains in the population; this remaining variation would be environmental. We could also hold the environment constant; the variability then remaining would be due to genetic differences.

These two quantities should add up to the original variability, but only if we have chosen a proper measure of variability, and if the genetic and environmental factors are independent in occurrence and action. The fact that the variance of the sum of two independent quantities is the sum of the variances (see A.2.12) means that the variance is a suitable measure. We shall see in Chapter 5 that it is also naturally related to the way that selection acts on the population. We now consider ways in which the population variance may be subdivided into fractions assignable to various kinds of gene action and the environment.

4.1 Genetic Variance with Dominance and Epistasis, and with Random Mating

The variance of any trait in a population is in general determined partly by genetic factors, partly by environmental factors, and, when these are not independent, their covariance. We are now primarily interested in that part of the variance which is determined solely by genetic differences among the individuals in the population. This we call the *genotypic* variance, V_h; it is also sometimes called the total genetic variance. Later we shall bring in the environmental components as well.

Consider first a single locus with two alleles, A_1 and A_2. Suppose that the average phenotype or "yield" of A_1A_1 individuals is $Y_{11} = \bar{a} + a_{11}$, that of A_1A_2 is $Y_{12} = \bar{a} + a_{12}$, and of A_2A_2 is $Y_{22} = \bar{a} + a_{22}$, where $\bar{a}$ is the population mean. The phenotype may be any quantity of interest—height, weight, yield, fertility, or fitness, for example—measured in any convenient units. Ordinarily we are considering traits that are influenced by many gene loci so that the quantities a_{11}, a_{12}, and a_{22} are small deviations introduced by gene substitutions at this particular locus. The value of the Y's is determined mainly by other loci and by the environment. We call a quantity like a_{11} the average excess of the genotype (in this case, A_1A_1). In this example it is also the average effect of the genotype. Later, in Chapter 5, we shall show that when the genotypes are not in random-mating proportions, there is a difference between the effect and the excess; but for the moment they are indistinguishable because we are assuming Hardy–Weinberg ratios at the locus under consideration.

Since $\bar{a}$ is the mean value of the phenotype, letting p_1 and p_2 be the frequencies of alleles A_1 and A_2, we have

$$\bar{a} = p_1^2(\bar{a} + a_{11}) + 2p_1p_2(\bar{a} + a_{12}) + p_2^2(\bar{a} + a_{22}), \qquad 4.1.1$$

since p_1^2, $2p_1p_2$, and p_2^2 add up to 1.

We wish now to assign a value to each allele, α_1 for A_1 and α_2 for A_2. These will be called the genic values. These are measures of the contribution of each gene, averaged over all genotypes into which this gene enters, and measured as a deviation from the mean. More precisely and specifically, α_1 is the average effect of choosing an allele at the A locus at random and replacing it with A_1. If the allele chosen happens to be A_1, then of course the replacement has no effect. The genic value of A_1A_2, for example, will be $\bar{a} + \alpha_1 + \alpha_2$.

The genic and genotypic values (which, since we are ignoring the environment, are the same as the phenotypic) are given below.

GENOTYPE	A_1A_1	A_1A_2	A_2A_2
FREQUENCY	p_1^2	$2p_1p_2$	p_2^2
GENOTYPIC VALUE	$Y_{11} = \bar{a} + a_{11}$	$Y_{12} = \bar{a} + a_{12}$	$Y_{22} = \bar{a} + a_{22}$
GENIC VALUE	$\bar{a} + 2\alpha_1$	$\bar{a} + \alpha_1 + \alpha_2$	$\bar{a} + 2\alpha_2$

The genic value is also sometimes called the additive value, since it is a linear approximation to the genotypic value.

Soon we shall discuss the way in which the numerical values of the α's may be determined. For the moment they remain as abstract symbols. First, we want to define the genic and genotypic variances. To do this, note that 4.1.1 can be written

$$p_1^2 a_{11} + 2p_1p_2a_{12} + p_2^2 a_{22} = 0 \qquad 4.1.2$$

which illustrates the principle that we have used several times: The sum of the deviations from the mean is 0. For the same reason

$$p_1^2(2\alpha_1) + 2p_1p_2(\alpha_1 + \alpha_2) + p_2^2(2\alpha_2) = 0,$$

which is the same as

$$p_1\alpha_1 + p_2\alpha_2 = 0. \qquad 4.1.3$$

The *genotypic variance* is the total variance, since there is no environmental effect in this case. From the definition of the variance, that is

$$V_h = p_1^2 a_{11}^2 + 2p_1p_2 a_{12}^2 + p_2^2 a_{22}^2. \qquad 4.1.4$$

The *genic variance* is

$$\begin{aligned} V_g &= p_1^2(2\alpha_1)^2 + 2p_1p_2(\alpha_1 + \alpha_2)^2 + p_2^2(2\alpha_2)^2 \\ &= (2p_1^2\alpha_1^2 + 2p_1p_2\alpha_1^2) + (2p_1^2\alpha_1^2 + 4p_1p_2\alpha_1\alpha_2 + 2p_2^2\alpha_2^2) \\ &\quad + (2p_1p_2\alpha_2^2 + 2p_2^2\alpha_2^2) \\ &= 2p_1\alpha_1^2 + 2(p_1\alpha_1 + p_2\alpha_2)^2 + 2p_2\alpha_2^2. \end{aligned}$$

But, by making use of 4.1.3, we see that the middle term drops out, and we have

$$V_g = 2(p_1\alpha_1^2 + p_2\alpha_2^2). \tag{4.1.5}$$

There is considerable variability in the literature of population genetics and animal breeding in the words used for these expressions. What we have called the genic variance (so called because it reflects the direct contribution of the average values of the genes) is also called the additive genetic variance (because it measures the additive effects of the genes). What we have called the genotypic variance is also called the total genetic variance. Since it is the total hereditary contribution to the variance we have symbolized it V_h for mnemonic ease. We shall avoid using genetic variance when we intend a precise meaning, since it is used in the literature both for the genic and genotypic variance.

Equations 4.1.4 and 4.1.5 are written so as to suggest immediately the form of the equation when it is extended to multiple alleles. In both cases the summation is simply continued.

For two alleles it is for some purposes more convenient to write 4.1.5 as

$$V_g = 2p_1p_2(\alpha_1 - \alpha_2)^2. \tag{4.1.6}$$

We now are ready to determine the values of the α's. This is done, as first suggested by Fisher (1918), by the method of least squares. At the moment this seems arbitrary, but we shall show later in this chapter that this leads to a natural formulation for the correlation between relatives and in Chapter 5 that the genic variance measured in this way gives the rate at which the population changes under selection.

To get the least squares estimates, we choose the α's so as to minimize the quantity

$$Q = p_1^2(a_{11} - 2\alpha_1)^2 + 2p_1p_2(a_{12} - \alpha_1 - \alpha_2)^2 + p_2^2(a_{22} - 2\alpha_2)^2. \tag{4.1.7}$$

In other words, we minimize the weighted average of the squared deviations from the linear estimate, or the squared differences between the genic and genotypic values.

We get the minimum estimates in the standard way by differentiating Q and equating to 0. This gives

$$\begin{aligned} \frac{\partial Q}{\partial \alpha_1} &= -4[p_1^2(a_{11} - 2\alpha_1) + p_1p_2(a_{12} - \alpha_1 - \alpha_2)] = 0, \\ \frac{\partial Q}{\partial \alpha_2} &= -4[p_1p_2(a_{12} - \alpha_1 - \alpha_2) + p_2^2(a_{22} - 2\alpha_2)] = 0. \end{aligned} \tag{4.1.8}$$

If we add these two equations and recall 4.1.2 we verify the relation 4.1.3; the α's have been determined so as to be measured as deviations from the same mean $\bar{a}$ as the a's, as was intended in the model. Making use of this relation, the two equations are easily solved to yield

$$\alpha_1 = p_1 a_{11} + p_2 a_{12}, \qquad \text{4.1.9}$$
$$\alpha_2 = p_1 a_{12} + p_2 a_{22}.$$

Notice that these values for the α's make sense. If we write

$$\alpha_1 = \frac{p_1^2 a_{11} + p_1 p_2 a_{12}}{p_1},$$

this says that α_1 is the average deviation from the mean of all the genotypes containing the A_1 allele, weighted by the frequency of the genotype and by the fraction of its alleles that are A_1 (i.e., by one for A_1A_1, by one-half for A_1A_2).

You might wonder why, if we can obtain the α's in such a simple way, we bothered with the least squares method. The reason is that this simple value for α is true only if the population is in Hardy–Weinberg proportions. When the situation is more complex, as in some examples in Chapter 5, the least squares approach is needed. So we have introduced the procedure at this time, even though it is not needed for this example.

Returning to the problem, we substitute the values from 4.1.9. into 4.1.6 to get an expression for the genic variance,

$$V_g = 2p_1p_2[p_1(a_{11} - a_{12}) + p_2(a_{12} - a_{22})]^2. \qquad \text{4.1.10}$$

Likewise, subtracting this from the genotypic variance, 4.1.4, gives the residual variance, which is attributable to deviations from the linear model caused by dominance. This is

$$V_d = p_1^2 p_2^2 (a_{11} - 2a_{12} + a_{22})^2. \qquad \text{4.1.11}$$

Notice that, in the absence of dominance, a_{12} is the average of a_{11} and a_{22}. Under these circumstances $a_{11} - 2a_{12} + a_{22} = 0$, and the dominance variance disappears, as it should.

Our procedure has been to choose the α's to minimize the weighted sum of the squares of the deviations of the phenotypic values from the additive value. This sum is the same as the dominance variance. We got the dominance variance simply by subtracting the genic variance from the genotypic variance, a process which contains the hidden assumption that these two quantities are independent and therefore additive. That this is in fact correct can be verified by getting an expression for the dominance variance directly and comparing it with 4.1.11, as follows.

Notice, from equations 4.1.8, that

$$p_1^2(a_{11} - 2\alpha_1) = -p_1p_2(a_{12} - \alpha_1 - \alpha_2) = p_2^2(a_{22} - 2\alpha_2) = K, \text{ say}; \qquad \text{4.1.12}$$

then the dominance variance (which is Q in 4.1.7) is

$$V_d = K^2\left[\frac{1}{p_1^2} + \frac{2}{p_1 p_2} + \frac{1}{p_2^2}\right] = \frac{K^2}{p_1^2 p_2^2}. \tag{4.1.13}$$

From equations 4.1.9.

$$\begin{aligned} a_{12} - \alpha_1 - \alpha_2 &= a_{12} - p_1 a_{11} - p_2 a_{12} - p_1 a_{12} - p_2 a_{22} \\ &= -p_1 a_{11} - p_2 a_{22}. \end{aligned}$$

Since

$$p_1^2 a_{11} + 2p_1 p_2 a_{12} + p_2^2 a_{22} = 0 \quad \text{(from 4.1.2)},$$

we can add this quantity without changing the value. Doing so, and rearranging, gives

$$a_{12} - \alpha_1 - \alpha_2 = -p_1 p_2 (a_{11} - 2a_{12} + a_{22}). \tag{4.1.14}$$

Substituting this into 4.1.12 and 4.1.13 gives

$$V_d = p_1^2 p_2^2 (a_{11} - 2a_{12} + a_{22})^2,$$

in agreement with 4.1.11, as was to be shown.

When the effects of the different loci are independent (that is, no epistasis) the covariances between the contributions of the different loci are 0 and the variance for all loci is the sum of the variances for the individual loci. So for all loci together we can write

$$V_h = V_g + V_d, \tag{4.1.15}$$

where these symbols now refer to the genotypic, genic, and dominance variances for all loci combined.

Likewise, if the environmental and genetic effects are independent, their variance contributions are additive. Furthermore, just as we obtained dominance and genic variances which are independent, we can use the same procedures to compute an epistatic variance. We shall consider epistatic variance and its subdivision later in this section. Altogether, when genetic and environmental effects are independent

$$\begin{aligned} V_t &= V_g + V_d + V_i + V_e \\ &= V_h + V_e, \end{aligned} \tag{4.1.16}$$

where

V_t is the phenotypic (total) variance,
V_g is the genic (additive genetic) variance,
V_d is the dominance variance,

V_i is the interaction (epistatic) variance,
V_e is the environmental variance,
V_h is the genotypic (hereditary, or total genetic) variance.

The ratio, V_g/V_t, is often called the heritability. It is widely used by animal and plant breeders to predict the change produced by selection.

We should emphasize that the formulations in 4.1.16 make the assumption of no genotype-environment interactions. This is one of the greatest practical difficulties in using these formulae. It may be necessary to introduce covariance terms into the expression to deal with this; in order to determine the value of such terms one would need to know the effect of different genotypes in different environments. It is particularly difficult in human genetics, of course. Here the effects of genotype and environment are usually confounded and heritability estimates are difficult at best, and often impossible to interpret.

On the other hand, the procedure we have used guarantees that the genic and dominance variance components are independent. This is also true of the extension to epistasis.

Notice that the statistical procedure has been to determine the α's in such a way as to maximize the genic contribution and minimize the dominance variance. This gives a formula for predicting the progeny phenotype from knowledge of the parents. It also means that a low dominance variance does not mean that the genes are without dominance. It may simply mean that the particular combination of gene frequencies and dominance relations allocates most of the variance contribution of these genes to the additive component.

The basis for compartmentalization of epistatic variance is an extension of the same principle. As much variance as possible goes into the additive component, as much as possible of what is left goes into dominance, and what is left over goes into epistasis. Kempthorne has greatly extended the epistatic analysis and the details are in his 1957 book.

As an illustration of the importance of the gene frequency in determining the genic and dominance variance components, consider the case of complete dominance of the A_1 gene, which implies that $a_{11} = a_{12}$. We also assume that there is no environmental effect, so $V_e = 0$ and $V_t = V_h$. In this case V_g, V_d, and V_h are in the ratio $2p_1p_2^3 : p_1^2p_2^2 : p_1p_2^2(1 + p_2)$. The heritability, or V_g/V_t, is $2p_2/(1 + p_2)$. As p_2 approaches 0, that is when the recessive gene is rare, the heritability approaches 0. On the other hand, when the dominant gene is rare (p_2 approaching 1), the heritability approaches 1.

This accords with our intuitive knowledge from classical genetics. When the recessive gene is rare, selection is ineffective; when the dominant gene is rare, selection is just as effective as if the genes were additive.

For a particularly lucid discussion of this material, see Falconer (1960). Some examples are illustrated in Figures 4.1.1–4.1.3. See also Table 4.1.1.

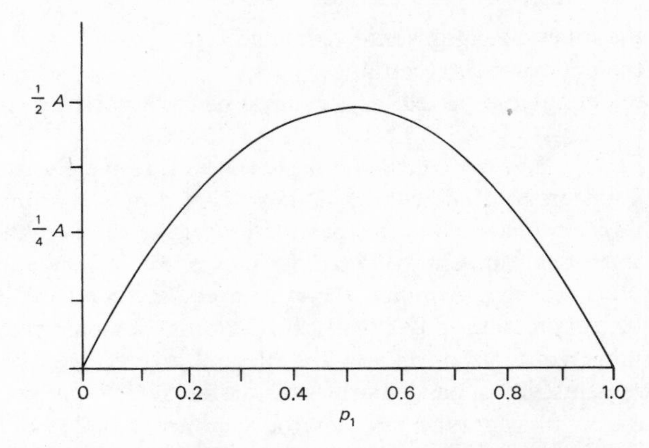

Figure 4.1.1. The relationship between genotypic variance and gene frequency, with no dominance, no epistasis, and with random mating. $A = [(a_{11} - a_{22})/2]^2$, $V = 2p_1p_2A$.

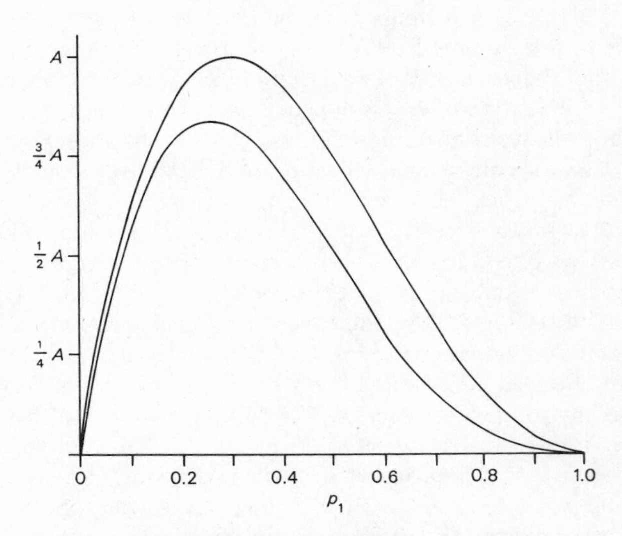

Figure 4.1.2. Genotypic variance (V_h) and genic variance (V_g) for a locus with complete dominance and random mating. The abscissa gives the frequency of the dominant gene. The dominance variance is the distance between the two curves. $A = [(a_{11} - a_{22})/2]^2$, $a_{12} = a_{11}$.

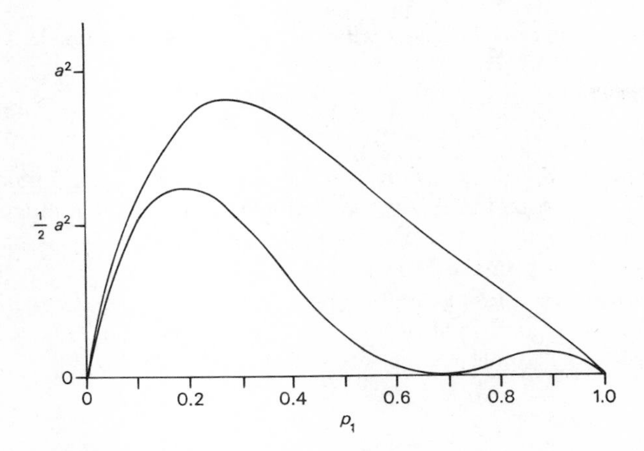

Figure 4.1.3. Genotypic variance (V_h) and genic variance (V_g) for an overdominant locus. The values assumed are $a_{12} - a_{22} = 2a$ and $a_{12} - a_{11} = a$. The abscissa is the frequency of A_1. The dominance deviation is the area between the two curves.

Figures 4.1.1 through 4.1.3 were adapted from Figure 8.1 of *Introduction to Quantitative Genetics,* by D. S. Falconer; Oliver and Boyd, Edinburgh, 1960; Ronald Press, New York, 1960.

Table 4.1.1. The heritability, V_g/V_t, with three levels of dominance and various gene frequencies. The environmental effect is assumed to be 0 and the effects of the different loci are additive.

FREQUENCY OF GENE A_1	NO DOMINANCE $a_{11} - 2a_{12} + a_{22} = 0$	COMPLETE DOMINANCE $a_{11} = a_{12}$	OVERDOMINANCE $a_{12} > a_{11} = a_{22}$
.01	1	.99	.98
.05	1	.97	.90
.10	1	.95	.78
.20	1	.89	.53
.50	1	.67	.00
.80	1	.33	.53
.90	1	.18	.78
.95	1	.10	.90
.99	1	.02	.98

Heritability is also defined in other ways. Sometimes it is defined as

$$H^2 = \frac{V_h}{V_t}.$$

This gives the fraction of the total variance that is attributable to differences among the genotypes. This is useful if the question is the relative influence of genotype and environment in determining phenotypic differences. This is sometimes called heritability in the broad sense.

On the other hand, the animal or plant breeder is less interested in this than he is in determining that part of the genotypic variability that is responsive to selection. This is V_g, so heritability is defined as

$$h^2 = \frac{V_g}{V_t},$$

and is often called heritability in the narrow sense. We have used the capital letter for the larger value and the small letter for the smaller. In some cases those components of epistatic variance which cannot be easily separated from the genic variance, and which contribute to the correlation between parent and offspring, are included.

In this book, unless the contrary is specified, we shall use the word heritability to mean V_g/V_t.

With epistasis the problems become more complicated. The extension of the procedures we have been using to multiple interacting loci was developed originally by Fisher (1918) and Wright (1935), but has been extended more recently by Cockerham (1954) and Kempthorne (1954, 1955).

The basic procedure is, as mentioned earlier, to take up as much variance as possible in the additive term, then as much of the remainder as possible with the dominance deviation, and what is left over is attributed to epistasis. The epistatic terms can be broken further into those that are due to pairs of loci, then after removing this the remaining variance is associated with loci taken three or more at a time. The 3-locus epistasis can in turn be removed, and so on. One would expect that, unless there are very intricate interactions (for example, a trait that is found only when three rare genes are simultaneously present), most of the variance is removed by 2-factor combinations.

Wright (1935) showed that for one form of epistasis this is exactly true. This is the kind of interaction that occurs when there is selection for an intermediate phenotype, as when animals of intermediate size are more likely to survive and reproduce than those that are too small or too large. This introduces epistasis of a rather extreme type, since a gene that increases size will be favored in genotypes where most of the other genes are for small size but selected against when most of the others are for large size. Such interaction must be quite frequent in occurrence. Wright assumed that the selective disadvantage is proportional to the square of the deviation from the optimum phenotype. With this model he showed that the total epistatic variance is simply the sum of all possible 2-locus components.

For more general situations 3-factor and higher interactions are undoubtedly involved, but they cannot ordinarily be measured and are probably usually small. In any case, we shall consider only the interactions of pairs of loci. Extensions to three or more are given by Kempthorne (1955, 1957).

The relationships are clearly seen in a two-way array (see Table 4.1.2). The items in the body of the table are the phenotypic measurements of each of the nine genotypes, each measured as a deviation from the population mean. The first two subscripts refer to the A locus and the second two to the B locus. Thus a_{1112} is the amount by which the phenotype of $A_1A_1B_1B_2$ exceeds the population mean. Below each of the a's is its frequency in the population.

Table 4.1.2. Basic calculations for subdividing the genotypic variance determined by two independent loci with two alleles each. The frequency of alleles A_1, A_2, B_1, and B_2 are p_1, p_2, q_1, and q_2. The a's are all measured as deviations from the average measurement in the population. The frequency of each class is given below the deviation.

	A_1A_1	A_1A_2	A_2A_2	MEAN	MEAN
B_1B_1	a_{1111} $p_1^2q_1^2$	a_{1211} $2p_1p_2q_1^2$	a_{2211} $p_2^2q_1^2$	$a_{..11}$ q_1^2	β_1 q_1
B_1B_2	a_{1112} $2p_1^2q_1q_2$	a_{1212} $4p_1p_2q_1q_2$	a_{2212} $2p_2^2q_1q_2$	$a_{..12}$ $2q_1q_2$	
B_2B_2	a_{1122} $p_1^2q_2^2$	a_{1222} $2p_1p_2q_2^2$	a_{2222} $p_2^2q_2^2$	$a_{..22}$ q_2^2	β_2 q_2
MEAN	$a_{11..}$ p_1^2	$a_{12..}$ $2p_1p_2$	$a_{22..}$ p_2^2	0 1	
MEAN	α_1 p_1		α_2 p_2		0 1

$$Z = V_g + V_d + V_i = p_1^2q_1^2a_{1111}^2 + 2p_1p_2q_1^2a_{1211}^2 + \cdots + p_2^2q_2^2a_{2222}^2$$

$$W = V_g + V_d \qquad = p_1^2a_{11..}^2 + 2p_1p_2a_{12..}^2 + p_2^2a_{22..}^2 + q_1^2a_{..11}^2 + 2q_1q_2a_{..12}^2 + q_2^2a_{..22}^2$$

$$U = V_g \qquad = 2(p_1\alpha_1^2 + p_2\alpha_2^2 + q_1\beta_1^2 + q_2\beta_2^2)$$

At the bottom of the table are the weighted averages of each column. For example

$$a_{11..} = \frac{p_1^2 q_1^2 a_{1111} + 2p_1^2 q_1 q_2 a_{1112} + p_1^2 q_2^2 a_{1122}}{p_1^2 q_1^2 + 2p_1^2 q_1 q_2 + p_1^2 q_2^2}$$

$$= q_1^2 a_{1111} + 2q_1 q_2 a_{1112} + q_2^2 a_{1122}.$$

Corresponding quantities for the B locus are on the next to the right column. The α's are obtained as before. Thus

$$\alpha_1 = p_1 a_{11..} + p_2 a_{12..},$$
$$\alpha_2 = p_1 a_{12..} + p_2 a_{22..}, \qquad 4.1.17$$

and the β's are corresponding quantities for the B locus.

Putting all this together, we have the variance components as given in the bottom part of Table 4.1.2.

Table 4.1.3 gives two numerical illustrations. In both cases epistasis and dominance are complete. In the left the dominant genes are complementary and in the right the dominant genes are duplicate, representing

Table 4.1.3. Numerical examples of two contrasting directions of epistatic deviations. On the left the dominant alleles are complementary; on the right, they are duplicate. These lead to the classical 9 : 7 and 15 : 1 Mendelian ratios. The gene frequencies are adjusted so that the two phenotypes are equally frequent in both examples; thus the deviations are equal and have been scaled to make the total variance 1.

	AA	Aa	aa
BB	1	1	−1
Bb	1	1	−1
bb	−1	−1	−1

$p_A = q_B = .459$

$V_g = .582$

$V_d = .247$

$V_i = .172$

$V_t = 1.000$

	AA	Aa	aa
BB	1	1	1
Bb	1	1	1
bb	1	1	−1

$p_A = q_B = .159$

$V_g = .757$

$V_d = .072$

$V_i = .172$

$V_t = 1.000$

extreme cases of diminishing and reinforcing epistasis. We have assumed that the two phenotypes are equally frequent. This necessitates that the dominant allele frequencies are 0.459 and 0.159 in the left and right tables, respectively.

Notice that in both cases, despite the complete epistasis, only 17% of the total variance appears in the epistatic term. The ratio of the additive or genic to the dominance variance is larger in the second case. This accords with the results in Table 4.1.1; the dominance component decreases as the dominant gene frequency decreases.

These examples are intended to illustrate the underlying principles. In practice one is dealing with quantitative traits such as size, weight, or fitness, and the effect of individual genes cannot be ascertained. What is observed is a set of cumulative effects of many genes reflected in correlations or covariances between relatives. From these the variance components can often be inferred.

The correlations between relatives will be considered later in the chapter, but we need to be able to subdivide the epistatic contributions further in order to study the correlations. This is necessary because the different epistatic components contribute differently to the covariances between individuals of different degrees of relationship.

Again we consider only two loci with two alleles at each locus. The epistasis may be broken down into interaction of the additive or genic components at the two loci, interaction between the additive component of one locus and the dominance component of the other, and interaction between the dominance components. We shall designate these as V_{AA}, V_{AD}, and V_{DD}. Thus

$$V_i = V_{AA} + V_{AD} + V_{DD}. \qquad \textbf{4.1.18}$$

The theory for such subdivision follows the principles of factorial experimental design (Fisher 1935) and is described by Kempthorne (1957). We shall not attempt a proof; but will show a simple procedure for obtaining these quantities. It is simpler to rearrange the phenotype values as in Table 4.1.4. Remember that the a's are deviations from the population mean. The quantities in parentheses are the weighted means of the two immediately adjacent values. For example,

$$a_{1.12} = p_1 a_{1112} + p_2 a_{1212}.$$

Likewise the values along the bottom are the means of the columns. For example,

$$a_{1.1.} = p_1 q_1 a_{1111} + p_2 q_1 a_{1211} + p_1 q_2 a_{1112} + p_2 q_2 a_{1212}.$$

Table 4.1.4. Basic calculations for subdividing the 2-allele, 2-locus epistasis into additive × additive, additive × dominance, and dominance × dominance components. The gametes and their frequencies are given at the upper and left margins and the phenotypes in the center. Values in parentheses are the weighted means of the two immediately adjacent values. All values are measured as deviations from the population mean.

	GAMETES					
	A_1B_1 p_1q_2		A_1B_2 p_1q_2	A_2B_1 p_2q_1		A_2B_2 p_2q_2
A_1B_1 p_1q_1	a_{1111}	$(a_{111.})$	a_{1112}	a_{1211}	$(a_{121.})$	a_{1212}
	$(a_{1.11})$		$(a_{1.12})$	$(a_{2.11})$		$(a_{2.12})$
A_2B_1 p_2q_1	a_{1211}	$(a_{121.})$	a_{1212}	a_{2211}	$(a_{221.})$	a_{2212}
A_1B_2 p_1q_2	a_{1112}	$(a_{112.})$	a_{1122}	a_{1212}	$(a_{122.})$	a_{1222}
	$(a_{1.12})$		$(a_{1.22})$	$(a_{2.12})$		$(a_{2.22})$
A_2B_2 p_2q_2	a_{1212}	$(a_{122.})$	a_{1222}	a_{2212}	$(a_{222.})$	a_{2222}
Mean	$a_{1.1.}$		$a_{1.2.}$	$a_{2.1.}$		$a_{2.2.}$

$$X = 4(p_1q_1a_{1.1.}^2 + p_1q_2a_{1.2.}^2 + p_2q_1a_{2.1.}^2 + p_2q_2a_{2.2.}^2)$$

$$Y = 2(p_1^2q_1a_{111.}^2 + p_1p_2q_1a_{121.}^2 + \cdots + p_2^2q_2a_{222.}^2 + p_2q_2^2a_{2.22}^2)$$

Now, we can put the various formulae together. Some are from Table 4.1.2 and the rest from 4.1.4.

$$\begin{aligned} U &= V_g, \\ W &= V_g + V_d, \\ X &= 2V_g + V_{AA}, \\ Y &= 3V_g + 2V_d + 2V_{AA} + V_{AD}, \\ Z &= V_g + V_d + V_{AA} + V_{AD} + V_{DD} = V_h. \end{aligned} \qquad 4.1.19$$

From these we readily obtain

$$\begin{aligned} V_g &= U, \\ V_d &= -U + W, \\ V_{AA} &= -2U + X, \\ V_{AD} &= 3U - 2W - 2X + Y, \\ V_{DD} &= -U + W + X - Y + Z. \end{aligned} \quad 4.1.20$$

The proof of these relations and extensions to multiple alleles and multiple loci are given by Kempthorne (1954, 1955).

An example is given in Table 4.1.5. Again, complementary dominant genes are assumed, but this time we have let the gene frequencies be 1/2 for each allele.

Table 4.1.5. A numerical example; two loci, two alleles at each, complete dominance, complete complementary epistasis. The phenotypic values are given at the left, the deviations from the population mean at the right. The gene frequencies are all equal to 1/2.

	AA	*Aa*	*aa*
BB	101	101	100
Bb	101	101	100
bb	100	100	100

	AA	*Aa*	*aa*
BB	7/16	7/16	−9/16
Bb	7/16	7/16	−9/16
bb	−9/16	−9/16	−9/16

	VARIANCE COMPONENT	FRACTION OF TOTAL VARIANCE	
V_g	.1406	.571	
V_d	.0703	.286	
V_{AA}	.0156	.064	
V_{AD}	.0156	.064	.143
V_{DD}	.0039	.016	

4.2 Variance Components with Dominance and Inbreeding

In Section 3.10 of the previous chapter we showed that the genotypic variance increases if there are consanguineous matings within the population. We are referring to the whole population, not to a subpopulation if the inbreeding is such as to break the population into groups. Equation 3.10.3 is,

$$V_f = (1-f)V_0 + fV_1 + f(1-f)(\overline{Y}_0 - \overline{Y}_1)^2, \qquad 4.2.1$$

where V_0 and V_1 are the variance of a randomly mating ($f = 0$) and completely inbred population ($f = 1$). $\overline{Y}_0$ and $\overline{Y}_1$ are the corresponding means. Unless the two means are the same, which would be true if there were no dominance, the variance is a quadratic function of f.

We now inquire as to how the genic and dominance components change with f. We generalize the model to include inbreeding.

GENOTYPE	A_1A_1	A_1A_2	A_2A_2
FREQUENCY	$p_1^2 + p_1p_2f$	$2p_1p_2(1-f)$	$p_2^2 + p_1p_2f$
GENOTYPIC VALUE	$Y_{11} = \bar{a} + a_{11}$	$Y_{12} = \bar{a} + a_{12}$	$Y_{22} = \bar{a} + a_{22}$
GENIC VALUE	$\bar{a} + 2\alpha_1$	$\bar{a} + \alpha_1 + \alpha_2$	$\bar{a} + 2\alpha_2$

Following exactly the same procedure as before, we see that again $p_1\alpha_1 + p_2\alpha_2 = 0$. The genic variance is

$$V_g = 2p_1p_2(\alpha_1 - \alpha_2)^2(1+f). \qquad 4.2.2$$

This, which corresponds to 4.1.6, would seem to imply that as the population is inbred the genic variance is simply multiplied by $1 + f$—as is true for a gene with no dominance (see 3.10.6). But this is misleading, for the α's also change as the genotype frequencies change with inbreeding. What we need is an expression for V_g in terms of quantities that do not change with f.

Continuing, if we go through the same least squares process as before, we have

$$Q = (p_1^2 + p_1p_2f)(a_{11} - 2\alpha_1)^2 + 2p_1p_2(1-f)(a_{12} - \alpha_1 - \alpha_2)^2 + (p_2^2 + p_1p_2f)(a_{22} - 2\alpha_2)^2. \qquad 4.2.3$$

$$\frac{\partial Q}{\partial \alpha_1} = -4[(p_1^2 + p_1p_2f)(a_{11} - 2\alpha_1) + p_1p_2(1-f)(a_{12} - \alpha_1 - \alpha_2)] = 0,$$

$$\frac{\partial Q}{\partial \alpha_2} = -4[p_1p_2(1-f)(a_{12} - \alpha_1 - \alpha_2) + (p_2^2 + p_1p_2f)(a_{22} - 2\alpha_2)] = 0.$$

From these equations we solve for α_1 and α_2, getting

$$\alpha_1 = \frac{(p_1 + p_2 f)a_{11} + p_2(1 - f)a_{12}}{1 + f}$$

and

$$\alpha_2 = \frac{p_1(1 - f)a_{12} + (p_2 + p_1 f)a_{22}}{1 + f}. \qquad 4.2.4$$

Substituting these into 4.2.2 gives

$$V_g = \frac{2p_1 p_2[(p_1 + fp_2)(a_{11} - a_{12}) + (p_2 + fp_1)(a_{12} - a_{22})]^2}{1 + f}. \qquad 4.2.5$$

This shows that the genic variance does not change in a linear way with f. Notice that a_{11}, a_{12}, and a_{22} are measured as deviations from a mean $\bar{a}$. But $\bar{a}$ is not constant, since it changes with f. Yet the quantities $(a_{11} - a_{12})$ and $(a_{12} - a_{22})$ do not change, since they are the same as $Y_{11} - Y_{22}$ and $Y_{12} - Y_{22}$, which are constant.

Note that when $a_{11} - a_{12} = a_{12} - a_{22}$ (no dominance), 4.2.5 becomes

$$V_g = 2p_1 p_2(a_{11} - a_{12})^2(1 + f), \qquad 4.2.6$$

linear in f as expected (compare 3.10.6).

We have not given an explicit formula for V_h comparable to 4.2.5 for V_g. This can be calculated from 4.2.1. Let

$$Y_{11} - Y_{12} = a_{11} - a_{12} = A,$$
$$Y_{12} - Y_{22} = a_{12} - a_{22} = B.$$

Then

$$V_0 = 2p_1 p_2(p_1 A + p_2 B)^2 + p_1^2 p_2^2(A - B)^2, \qquad 4.2.7$$

$$V_1 = p_1 p_2(A + B)^2, \qquad 4.2.8$$

$$\bar{Y}_0 - \bar{Y}_1 = -p_1 p_2(A - B). \qquad 4.2.9$$

These may be substituted directly into 4.2.1 which then gives V_h for any value of f. The heritability is given by the ratio of 4.2.5. to 4.2.1.

Equation 4.2.7 was obtained from 4.1.10 and 4.1.11; 4.2.8 and 4.2.9 were calculated from the table at the beginning of this section.

Notice that when $f = 1$, the genic variance becomes the same as the total variance and the dominance variance disappears. This is not surprising, for the dominance variance depends on the extent to which the heterozygote departs from the mean of the homozygotes; with no heterozygotes it has no meaning. Heritability increases with inbreeding because, although the total

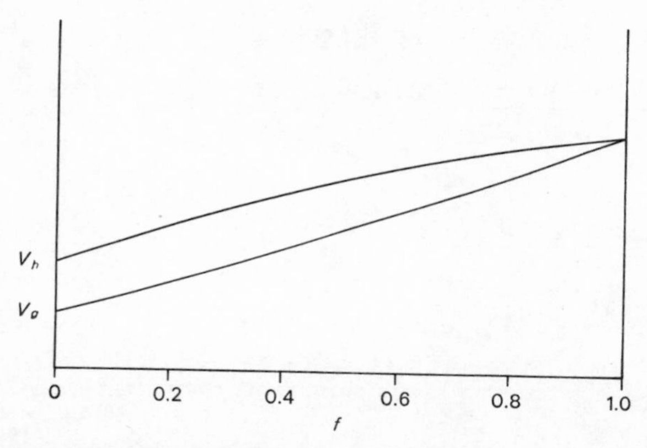

Figure 4.2.1. The changes in variance components with inbreeding. Notice that the dominance variance, which is the distance between the two lines, decreases to 0 as f approaches 1. Complete dominance is assumed, with the recessive gene frequency equal to 0.333.

variance increases, the genic variance increases more rapidly and becomes a larger fraction of the total. This is illustrated in Figure 4.2.1.

4.3 Identity Relations Between Relatives

We introduced in Section 3.2 the coefficients of inbreeding and of consanguinity. These measure the probability that two homologous genes drawn at random from an individual or from each of two individuals are identical by descent. We wish now to extend the ideas to measure different identity relations among diploid individuals who may be identical for neither, for one, or for both of their genes.

The method that we are using was first developed by Cotterman (1940). Cotterman worked only with the relationships between two individuals, neither of which was inbred. The extension of the method to include the relationships between two inbred individuals has been made by Denniston (1967). We shall consider only the simpler case where neither of the two individuals is inbred.

Consider two related individuals, I and J. We define the Cotterman k-coefficients as:

k_0 = the probability that no two genes at the locus are identical,
$2k_1$ = the probability that one gene in I is identical to one gene in J, but not both,
k_2 = the probability that both genes in I are identical to those in J.

For example,

GENOTYPE		k-PROBABILITY
I	J	
A_1A_2	A_1A_2	k_2
A_1A_2	A_1A_3	k_1
A_1A_2	A_3A_2	k_1
A_1A_2	A_3A_4	k_0

More precisely, if a and b are the two genes in I and c and d are the two in J, as shown in Figure 4.3.1, then (using = to mean "are identical by descent") the k-coefficients are:

$$\begin{aligned} k_2 &= \text{Prob}[(a = c) \text{ and } (b = d)] \text{ or } [(a = d) \text{ and } (b = c)], \\ 2k_1 &= \text{Prob}[(a = c) \text{ and } (b \neq d)] \text{ or } [(a = d) \text{ and } (b \neq c)] \\ &\quad \text{or } [(b = c) \text{ and } (a \neq d)] \text{ or } [(b = d) \text{ and } (a \neq c)], \\ k_0 &= \text{Prob}[a \neq c, a \neq d, b \neq c, \text{ and } b \neq d]. \end{aligned} \tag{4.3.1}$$

Notice that, if either or both of the two individuals I and J are inbred, there are other possible relations, such as $a = b = c \neq d$ or $a = b = c = d$. But without inbreeding of I or J the only possibilities are those measured by k_0, k_1, and k_2.

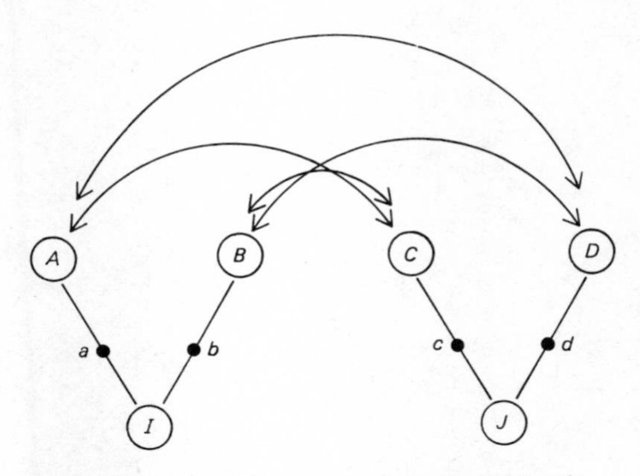

Figure 4.3.1. A diagram to show gene-identity relationships. The small letters refer to the genes in the gametes produced by the individuals designated by large letters. The curved arrows indicate possible consanguinity farther back in the pedigree.

To compute these k-coefficients we make use of the consanguinity coefficients of the parents. Referring again to Figure 4.3.1, we observe that

$$k_2 = f_{AC} f_{BD} + f_{AD} f_{BC}, \tag{4.3.2}$$

where f is the coefficient of consanguinity as defined in Chapter 3, Section 3. This follows immediately since the four f's are the probabilities that $(a = c)$, $(b = d)$, $(a = d)$, and $(b = c)$.

Furthermore,

$$\begin{aligned} 2k_1 &= f_{AC}(1 - f_{BD}) + f_{AD}(1 - f_{BC}) + f_{BC}(1 - f_{AD}) + f_{BD}(1 - f_{AC}) \\ &= f_{AC} + f_{AD} + f_{BC} + f_{BD} - 2(f_{AC} f_{BD} + f_{AD} f_{BC}) \\ &= 4f_{IJ} - 2k_2 . \end{aligned} \tag{4.3.3}$$

We can obtain k_0 by substraction since $k_0 + 2k_1 + k_2 = 1$.

As examples, consider the three sets of relationships shown in Figure 4.3.2. With ordinary single cousins

$$\begin{aligned} f_{BC} &= 1/4, \\ f_{AC} &= f_{AD} = f_{BD} = 0, \\ k_2 &= 0, \\ 2k_1 &= 4f_{IJ} = 1/4, \\ k_0 &= 3/4. \end{aligned}$$

With double first cousins

$$\begin{aligned} f_{AC} &= f_{BD} = 1/4, \\ f_{BC} &= f_{AD} = 0, \\ k_2 &= 1/16, \\ 2k_1 &= 6/16, \\ k_0 &= 9/16. \end{aligned}$$

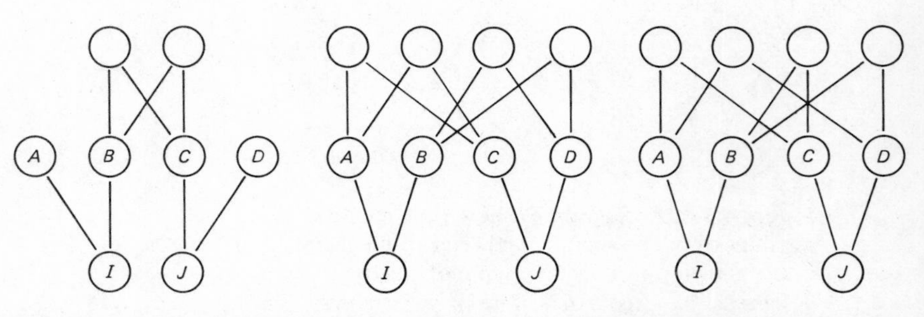

Figure 4.3.2. Single first cousins, double first cousins, and quadruple half-first cousins.

Finally, with quadruple half-first cousins,

$$f_{AC} = f_{AD} = f_{BC} = f_{BD} = 1/8,$$
$$k_2 = 1/32,$$
$$2k_1 = 14/32,$$
$$k_0 = 17/32.$$

The coefficient of consanguinity $f_{IJ} = (k_1 + k_2)/2 = 1/8$, the same as for double first cousins, as expected.

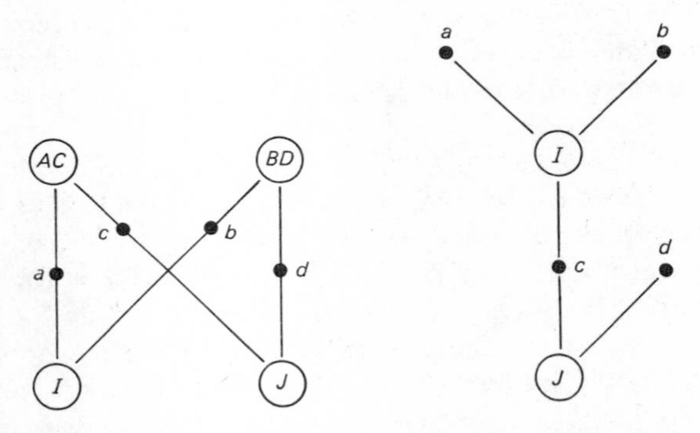

Figure 4.3.3. Full sibs, and parent and offspring.

For the k-coefficients of sibs we treat the pedigree as if two of the ancestors were collapsed into one, as shown in Figure 4.3.3, and write the f's in terms of the gametes. Thus

$$f_{ac} = f_{bd} = 1/2,$$
$$f_{ad} = f_{bc} = 0,$$

and

$$k_2 = 1/4,$$
$$2k_1 = 1/2,$$
$$k_0 = 1/4.$$

There is also a slight complication if one individual is an ancestor of the other, as in the parent-offspring relationship shown in Figure 4.3.3. In this case, we have drawn the relevant gametes; this makes the situation clear.

$$f_{ac} = f_{bc} = 1/2,$$
$$f_{ad} = f_{bd} = 0,$$
$$k_2 = 0,$$
$$2k_1 = 1,$$
$$k_0 = 0.$$

The results are as would be expected. A parent and child who are otherwise unrelated must share one and only one gene at a locus.

As stated earlier, we assume that neither I nor J is inbred. However, it is all right for other individuals in the pedigree to be inbred. The same rules apply as in the computation of the inbreeding coefficient under ordinary circumstances. Inbreeding is irrelevant for all members of a path except for the common ancestor; if the common ancestor is inbred the path is multiplied by $1 + f$, just as in computing the inbreeding and consanguinity coefficients.

Likewise the rules for X-linked traits are still good. Males in a path are not counted and any path with successive males is ignored. Of course the k-coefficients, since they apply to diploids, are meaningful only for females.

The k coefficients are of particular use in two contexts. One is for the computation of the solution to such problems as: Given that I and J are related and I is genotype aa, what is the probability that J is also? Suppose that p is the gene frequency. The probability, then, is $k_2 + 2k_1 p + k_0 p^2$. For example if I and J are double first cousins and I has a recessive disease of incidence p^2, then the probability of J being affected is $(1 + 6p + 9p^2)/16$.

The second use is in determining the correlation between relatives when there is dominance. That is the subject of the next section.

4.4 Correlation Between Relatives

The procedure for determining the correlations between relatives, with dominance, but restricted to individuals that are not inbred is now given. As in the previous section it is all right for other individuals in the pedigree to be inbred, but not for the two individuals under consideration. The necessary calculations are set forth in Table 4.4.1.

The covariance of X and Y is

$$\begin{aligned} C_{XY} = k_0(p_1^4 a_{11}^2 &+ 4p_1^3 p_2 a_{11} a_{12} + 4p_1^2 p_2^2 a_{12}^2 + 2p_1^2 p_2^2 a_{11} a_{22} \\ &+ 4p_1 p_2^3 a_{12} a_{22} + p_2^4 a_{22}^2) \\ &+ 2k_1(p_1^3 a_{11}^2 + 2p_1^2 p_2 a_{11} a_{12} \\ &+ p_1 p_2 a_{12}^2 + 2p_1 p_2^2 a_{12} a_{22} + p_2^3 a_{22}^2) \\ &+ k_2(p_1^2 a_{11}^2 + 2p_1 p_2 a_{12}^2 + p_2^2 a_{22}^2). \end{aligned}$$

Table 4.4.1. The calculation of the correlation between relatives in terms of gene effects, gene frequencies, and k-coefficients. The values a_{ij} are assumed to be measured as deviations from the mean to simplify the arithmetic.

GENOTYPES		PHENOTYPIC VALUES		FREQUENCY OF THIS COMBINATION
X	Y	X	Y	
A_1A_1	A_1A_1	a_{11}	a_{11}	$k_0 p_1^4 + 2k_1 p_1^3 + k_2 p_1^2$
A_1A_1	A_1A_2	a_{11}	a_{12}	$2[k_0 2p_1^3 p_2 + 2k_1 p_1^2 p_2]$
A_1A_2	A_1A_1	a_{12}	a_{11}	
A_1A_2	A_1A_2	a_{12}	a_{12}	$k_0 4p_1^2 p_2^2 + 2k_1 p_1 p_2 + k_2 2p_1 p_2$
A_1A_1	A_2A_2	a_{11}	a_{22}	$2[k_0 p_1^2 p_2^2]$
A_2A_2	A_1A_1	a_{22}	a_{11}	
A_1A_2	A_2A_2	a_{12}	a_{22}	$2[k_0 2p_1 p_2^3 + 2k_1 p_1 p_2^2]$
A_2A_2	A_1A_2	a_{22}	a_{12}	
A_2A_2	A_2A_2	a_{22}	a_{22}	$k_0 p_2^4 + 2k_1 p_2^3 + k_2 p_2^2$

But, the coefficient of k_0 is

$$(p_1^2 a_{11} + 2p_1 p_2 a_{12} + p_2^2 a_{22})^2,$$

which, by 4.1.2, is 0. The coefficient of k_1 is equivalent to

$$2[p_1(p_1 a_{11} + p_2 a_{12})^2 + p_2(p_1 a_{12} + p_2 a_{22})^2],$$

which, from 4.1.9, is

$$2(p_1 \alpha_1^2 + p_2 \alpha_2^2),$$

which, by 4.1.5, is V_g. Finally, the coefficient of k_2 is V_h, from 4.1.4.

Therefore, summing over all loci,

$$C_{XY} = k_1 V_g + k_2 V_h,$$

or, since $V_h = V_g + V_d$ in the absence of epistatic interaction,

$$C_{XY} = (k_1 + k_2)V_g + k_2 V_d. \qquad \textbf{4.4.1}$$

From the definition of the correlation coefficient (A.6.3) the correlation between two relatives, X and Y, neither of which is inbred, is

$$r_{XY} = (k_1 + k_2)\frac{V_g}{V_t} + k_2 \frac{V_d}{V_t}. \qquad \textbf{4.4.2}$$

Essentially the same results were derived by quite a different procedure by Fisher (1918).

As a numerical example, consider some correlations on human height. Fisher estimated $V_g/V_t = .74$ and $V_d/V_t = .26$. He found no appreciable environmental component in the population studied. From this, the correlation between parent and child, if marriages were random with respect to height, would be

$$r = \tfrac{1}{2}(.74) = .37$$

and, for sibs,

$$r = \tfrac{1}{2}(.74) + \tfrac{1}{4}(.26) = .44.$$

Actually, as we shall see later in this chapter, the correlations are considerably increased by the strong assortative marriage for height.

Notice that if there is no dominance, epistasis, or environmental effect, then the correlation becomes simply

$$r_{XY} = 2f_{XY}, \qquad 4.4.3$$

where f is the consanguinity coefficient.

The correlation is the covariance divided by the geometric mean of the two variances. For additively acting genes (no dominance) the effect of the two sets of genes in the zygote is the sum of the two haploid gametic sets. The covariance of $w + x$ with $y + z$ is not influenced by the correlation between w and x or that between y and z. On the other hand, the genotypic variance does increase with such a correlation, and in fact is multiplied by $1 + f$, where f is the inbreeding coefficient (or the correlation between alleles). This was shown in equation 3.10.6. Thus, if there is inbreeding, the denominator is increased and the correlation between two inbred individuals becomes

$$r_{XY} = \frac{2f_{XY}}{\sqrt{(1 + f_X)(1 + f_Y)}} \qquad 4.4.4$$

when there is no dominance, epistasis, or environmental effect.

Notice, by comparing with 3.3.1, that this is Wright's coefficient of relationship. In fact the original derivation of Wright's measure was through correlation analysis, and his intention was to have the relationship coefficient reflect the correlation between the gene values of the two individuals.

The extension of 4.4.1 to include epistasis is straightforward, but we shall simply give the results rather than the derivations. For two loci, when epistasis is considered, 4.4.1 becomes

$$C_{XY} = (k_1 + k_2)V_g + k_2 V_d + (k_1 + k_2)^2 V_{AA} + (k_1 + k_2)k_2 V_{AD} + k_2^2 V_{DD}. \qquad 4.4.5$$

The extension to more than two loci is direct. For an epistatic interaction between additive effects at r loci and dominance effects at s loci, the coefficient is

$$(k_1 + k_2)^r k_2^s . \tag{4.4.6}$$

Some examples are given in Table 4.4.2.

Table 4.4.2. Covariances between relatives of different degree in terms of variance components in a population mating at random. These quantities, when divided by the total (phenotypic) variance, V_t, give the correlations.

RELATIONSHIP	COVARIANCE
Sib	$\frac{1}{2}V_g + \frac{1}{4}V_d + \frac{1}{4}V_{AA} + \frac{1}{8}V_{AD} + \frac{1}{16}V_{DD}$
Parent-offspring	$\frac{1}{2}V_g + \frac{1}{4}V_{AA}$
Half-sibs, Uncle-niece, Parent-grandchild	$\frac{1}{4}V_g + \frac{1}{16}V_{AA}$
First cousins	$\frac{1}{8}V_g + \frac{1}{64}V_{AA}$
Double first cousins	$\frac{1}{4}V_g + \frac{1}{16}V_d + \frac{1}{16}V_{AA} + \frac{1}{64}V_{AD} + \frac{1}{256}V_{DD}$

It is often of interest to ask for the covariance or correlation between the offspring and the average of the parents, or the mid-parent, $\hat{P}$. Letting P_m stand for the measurement on the male parent, P_f for that of the female parent, and O for that of the offspring, we obtain from the definition of the covariance (A.2.9)

$$\begin{aligned} C_{\hat{p}o} &= \frac{\Sigma(1/2)(P_m + P_f - \bar{P}_m - \bar{P}_f)(O - \bar{O})}{N} \\ &= \frac{1}{2}\frac{\Sigma(P_m - \bar{P}_m)(O - \bar{O})}{N} + \frac{1}{2}\frac{\Sigma(P_f - \bar{P}_f)(O - \bar{O})}{N} \\ &= C_{po}, \end{aligned} \tag{4.4.7}$$

provided that the two sexes are equivalent. So, the covariance of offspring and mid-parent is the same as that between offspring and parent.

On the other hand, the variance of $\hat{P}$ is 1/2 the variance of P_m or P_f (if males and females are equally variable, and if they are independent as is the case with random mating). The regression of offspring on the average of the parents is

$$\begin{aligned} b_{o\hat{p}} &= \frac{C_{o\hat{p}}}{V_{\hat{p}}} \\ &= 2C_{op}/V_p \qquad \mathbf{4.4.8} \\ &= (V_g + \tfrac{1}{2}V_{AA})/V_t. \end{aligned}$$

Such a formula can be used to predict the rate of improvement by selection. The progeny are expected to deviate from the average by a fraction $b_{o\hat{p}}$ of the amount by which the mid-parent deviates, or more formally (see A.4.5),

$$O = \bar{P} + b_{o\hat{p}}(\hat{P} - \bar{P}), \qquad \mathbf{4.4.9}$$

where $\bar{P}$ is the population average.

Returning to the example of Table 4.1.5, the correlations are as follows:

Half-sibs	.147
Parent-offspring	.302
Full sibs	.382

If the epistatic factors are ignored, the half-sib correlation would be estimated as $\frac{1}{4}(.571) = .143$, not very different.

In practical problems, the breeder usually estimates the heritability and then uses this value as a guide to selection programs. His estimates usually come from various correlations between relatives. One of the most used measures of heritability is four times the half-sib correlation, particularly half-sibs with the same father and different mothers since this eliminates the confounding effects of a common uterine and early postnatal environment. Four times the half-sib correlation is .587. The correct prediction formula is $b_{o\hat{p}} = (V_g + V_{AA}/2)/V_t$, or .603. The error is about 2.5%! In this example the epistasis is quite large, since we have assumed completely complementary gene action. Yet it doesn't cause a very large error in heritability measurements or predictions based on these. It is these reasons, as well as the practical difficulty of measuring epistasis, that lead the breeder to ignore epistasis.

Variance Within and Between Groups of Relatives If the population is broken up into a series of groups we can relate the variance between and within the groups to the correlation coefficient. It is simpler to regard the groups as of equal size, but the general theory does not depend on this. We think of the quantitative trait or measurement as made up of the sum of a series of additive and independent components. A particular individual, the jth member of the ith group, has a measurement, y_{ij}, which is the sum of an overall mean (μ), a component common to all members of the group (b_i), and an additional component (w_{ij}) that is specific to the individual. Then if V_b and V_w are the variances of these quantities (the between-group and within-group variances), the correlation between members of a group is

$$r = \frac{V_b}{V_b + V_w} = \frac{V_b}{V_t}. \qquad \textbf{4.4.10}$$

For an explanation of these relationships, see A.4.10–A.4.15.

For example, the variance within families of full sibs is

$$V_w = (1 - r)V_t. \qquad \textbf{4.4.11}$$

From the information in Table 4.4.2, this is

$$V_w = \tfrac{1}{2}V_g + \tfrac{3}{4}V_d + \tfrac{3}{4}V_{AA} + \tfrac{7}{8}V_{AD} + \tfrac{15}{16}V_{DD}. \qquad \textbf{4.4.12}$$

There may also be environmental factors that are common to a sibship and others that differ for members of the sibship. Suppose that V_e and $V_{e'}$ are the environmental variance components within and between sibships; then the sib correlation is

$$r = \frac{V_b + V_{e'}}{V_t}. \qquad \textbf{4.4.13}$$

In animal and plant breeding experiments it is often possible to avoid such environmental correlations by randomization. In human genetics and any study of natural populations such difficulties are unavoidable.

4.5 Comparison of Consanguineous and Assortative Mating

Assortative mating means that mated pairs are more similar for some phenotypic trait than if they were chosen at random from the population. It may have either of two causes, or some combination of both. The tendency toward phenotypic similarity of mating pairs may be a direct consequence of genetic relationship. For example, in a subdivided population there will generally be a greater phenotypic similarity among the members of a subpopulation because they share a common ancestry. The genetic consequences in this case are the same as those of inbreeding.

On the other hand, there may be assortative mating based on similarity for some trait and any genetic relationship is solely a consequence of similar phenotypes. For example, there is a high correlation between husband and wife for height and intelligence, probably caused much more by nonrandom marriage associated with the traits themselves than by common ancestry.

There are also other situations. For example, there is a considerable correlation in arm length between husband and wife. This is probably a consequence of the fact that those factors, genetic and environmental, that increase height also increase the length of the arm. So, any assortative mating for height will be reflected in a similar assortative mating for arm length, diminished somewhat by the lack of perfect correlation between the two traits.

Assortative mating is between individuals of similar phenotypes; inbreeding is between individuals of similar genotypes. Since individuals with similar phenotypes will usually be somewhat similar in their genotypes, we should expect assortative mating to have generally the same consequences as inbreeding. An excess of consanguineous matings in a population has two effects: (1) an increase in the average homozygosity and (2) an increase in the total population variance. Assortative mating would be expected to produce the same general kinds of results.

In general, assortative mating causes less increase in homozygosity than inbreeding, especially if the trait is determined by several gene loci. On the other hand, assortative mating causes a large increase in the variance of a multifactorial trait, in contrast to that produced by a comparable amount of inbreeding. A further difference is that inbreeding affects all segregating loci, whereas assortative mating affects only those related to the trait involved.

Pure assortative mating, like inbreeding, does not change the gene frequencies. We shall refer in this book to any situation where different genotypes make different contributions to the next generation through differential survival, mating patterns, or fertility, as selection. Only when all genotypes make the same average contribution will we regard it as pure inbreeding or assortative mating. But we shall see later in the chapter that with many assortative-mating systems, and even more so with disassortative-mating systems, there are differential contributions of different genotypes; so the effects of assortment are often confounded with selection.

The variance-enhancing effect of assortative mating is apparent with a simple example. Suppose that an arbitrary quantitative trait is influenced by two loci without dominance. Let each gene with subscript 1 add one unit to the phenotype, whereas each gene with subscript 0 adds nothing. Then the genotype $A_1A_1\ B_1B_1$ represents one extreme phenotype and $A_0A_0B_0B_0$ the other, with $A_1A_1B_0B_0$, $A_1A_0B_1B_0$, and $A_0A_0B_1B_1$ being exactly intermediate. Inbreeding will increase the frequency of all four homozygous genotypes, $A_1A_1\ B_1B_1$, $A_0A_0\ B_1B_1$, $A_1A_1\ B_0B_0$, and $A_0A_0\ B_0B_0$. This will increase the variance; in fact, it will exactly double the variance if the population is changed from random-mating proportions to complete homozygosity.

On the other hand, with complete assortative mating, the population approaches a state where only the extreme homozygotes, $A_1A_1B_1B_1$ and

$A_0 A_0 B_0 B_0$, remain. This clearly causes a much greater enhancement of the variance, especially if the number of relevant loci is large. The variance increase with assortative mating has been shown experimentally in Nicotiana (Breese, 1956) and Drosophila (McBride and Robertson, 1963). The latter authors also found the expected decrease with disassortative mating and demonstrated that the rate of change under selection can be increased with assortative mating.

With inbreeding there is no systematic change in the frequencies of the gamete types A_1B_1, A_1B_0, A_0B_1, and A_0B_0. On the other hand, as the example shows, assortative mating causes a change in frequency of the gametic types, increasing two while decreasing the other two. So, another way of describing the effect of assortative mating and of understanding its variance-enhancing effect is to note that it causes gametic phase (or linkage) disequilibrium.

The simplest cases of assortative mating were worked out long ago. These involved mainly a single locus (Jennings, 1916; Wentworth and Remick, 1916). One example of this work is the simple case of two alleles where each genotype mates strictly assortatively, that is, only with another individual of the same genotype. The genetic consequences are obviously exactly the same as with self-fertilization. Heterozygosity is reduced by half each generation and the variance is eventually doubled.

It might be thought from this example that assortative mating leads eventually to complete homozygosity, as do many forms of inbreeding, but this is not the case. Partial assortative mating, like partial self-fertilization, leads to an equilibrium level of heterozygosity other than zero.

In the more general treatment of assortative mating two cases are of interest. At one extreme the individuals fall into two (or possibly more) discrete phenotypes with preference for mating within a phenotype. For example, deaf persons tend to marry others with the same trait. At the other extreme is a character like size, for which there is a correlation between mates, but the distribution of sizes is continuous and determined by multiple genetic and environmental factors.

Before dealing with more complex multifactorial models, we shall first consider a single-locus trait.

4.6 Assortative Mating for a Single Locus

With inbreeding the choice of a mathematical model is clear from knowledge of the relationships and from the Mendelian mechanism. With assortative mating the choice is not so obvious, as a different behavior pattern can produce a different consequence.

We shall measure the degree of assortative mating by the product-moment correlation between the parents, r. For a quantitative trait the correlation coefficient is directly measurable. For qualitative traits we measure the correlation coefficient as the decrease in the proportion of matings between dissimilar phenotypes, divided by that proportion which is expected with random pairs. We consider two situations.

Each Genotype with a Different Phenotype Assume that each genotype is distinct, the differences being determined by a series of alleles. No restriction is placed on the number of alleles.

Assume that in each genotype a fraction r select mates of their own genotype while the remainder mate at random. In this system, perfect assortative mating is equivalent to self-fertilization, so this model of imperfect assortative mating is formally equivalent to partial self-fertilization. This was considered in Section 3.8. From 3.8.19 the heterozygosity at equilibrium is given by

$$H_\infty = 2H_0\left(\frac{1-r}{2-r}\right). \qquad 4.6.1$$

This result was first obtained by Wright (1921). Notice that complete homozygosity is not approached unless the assortative mating is complete ($r = 1$). Otherwise, the population approaches a level of homozygosity which is equivalent to an inbreeding coefficient of $f = r/(2 - r)$.

We showed in Section 3.10 that when there is no dominance the variance is proportional to $1 + f$. Therefore, with partial assortative mating the population variance at equilibrium is

$$V_\infty = V_0(1 + f) = V_0\left(\frac{2}{2-r}\right), \qquad 4.6.2$$

where V_0 is the variance with random mating.

For example, if $r = 1/2$, the equilibrium heterozygosity is reduced by 1/3 and the variance is increased by the same fraction. Only if $r = 1$ does the population become homozygous, in which case the variance is eventually doubled.

A more important example, especially in human genetics, is the case where dominance is complete. We now consider this.

Complete Dominance We assume that there are only two alleles and, since dominance is complete, there are only two phenotypes. If there are more than two alleles it may be that they can be grouped into two sets as regards mating pattern. For example, it might be that the normal allele is dominant and that the wild types tend to mate among themselves leaving all the mutant types to mate with other mutants.

Let P_t be the frequency of AA in generation t, $2Q_t$ be the frequency of heterozygous Aa, and R_t that of the homozygous recessive aa. Let r be the correlation between mating individuals; that is to say, a fraction r mate strictly assortatively and the rest mate at random with respect to the trait considered.

To see the algebraic relationships we imagine the population as being divided into three groups: a randomly mating group comprising a fraction $(1 - r)$ of all matings and with the A gene frequency $P + Q = p$; a recessive assortatively mating group making up a fraction rR of matings and with the A gene frequency 0; a dominant assortative group comprising a fraction $r(1 - R)$ and with the A gene frequency $(P + Q)/(1 - R)$ or $p/(1 - R)$ and recessive gene (a) frequency $Q/(1 - R)$. From the randomly mated group the fraction of AA, Aa, and aa progeny will be p^2, $2pq$, and q^2, where $q = 1 - p$. The contribution of the dominant assortative group to the AA class next generation will be

$$r(1 - R)[p/(1 - R)]^2,$$

to the Aa class will be

$$r(1 - R)2[p/(1 - R)][Q/(1 - R)],$$

and to the aa class will be

$$r(1 - R)[Q/(1 - R)]^2.$$

The recessive assortative-mating group will make its entire contribution, rR, to the aa class.

Putting all this together the genotype frequencies next generation will be

$$P(AA) = P_{t+1} = (1 - r)p^2 + r(1 - R_t)\left(\frac{p}{1 - R_t}\right)^2 = (1 - r)p^2 + \frac{rp^2}{1 - R_t}, \tag{4.6.3}$$

$$P(Aa) = 2Q_{t+1} = (1 - r)2pq + r(1 - R_t)2\frac{p}{1 - R_t}\frac{Q_t}{1 - R_t} = 2(1 - r)pq + \frac{2rpQ_t}{p + Q_t}, \tag{4.6.4}$$

$$P(aa) = R_{t+1} = (1 - r)q^2 + rR_t + r(1 - R_t)\left(\frac{Q_t}{1 - R_t}\right)^2 = (1 - r)q^2 + r\left[\frac{q^2 + R_t(p - q)}{1 - R_t}\right]. \tag{4.6.5}$$

We have written p and q with no subscripts, since they do not change with time. This can be verified by summing 4.6.3 and half of 4.6.4. Recalling that $p = P + Q$ and $q = Q + R$, this simplifies to $p_{t+1} = p_t$, showing that the gene frequency does not change. As with inbreeding only the genotype frequencies change, not the gene frequencies.

When assortative mating is complete ($r = 1$), 4.6.4 becomes

$$2Q_{t+1} = \frac{2pQ_t}{p + Q_t}.$$

This approaches 0 as t increases, but extremely slowly. When $p = 1/2$, then $2Q_0 = 1/2$, and the frequency of heterozygotes in successive generations follows the simple harmonic series, 1/2, 1/3, 1/4, 1/5, ..., as first shown by Jennings (1916).

With any value of r except 1 the population never attains complete homozygosity but approaches an equilibrium. We can find the equilibrium heterozygosity by equating Q_{t+1} to Q_t, giving

$$\hat{Q}^2 + p^2(1 - r)\hat{Q} - p^2q(1 - r) = 0 \qquad 4.6.6$$

whose solution gives the equilibrium value, $\hat{Q}$, in terms of the correlation between mates and the gene frequency.

Alternatively, we can do as we did before and express the heterozygosity as a function of the inbreeding coefficient, f. Replacing $\hat{Q}$ by $pq(1 - f)$ in 4.6.6 gives

$$qf^2 + (r - 1 - q - rq)f + rq = 0. \qquad 4.6.7$$

When the correlation, r, is equal to 1, there is complete homozygosity ($f = 1$). Otherwise, there is equilibrium at an intermediate value of the inbreeding coefficient, given by the solution of the above equation lying between 0 and 1. Notice that, contrary to the results with inbreeding, the equilibrium value of the inbreeding coefficient with assortative mating is a function of the gene frequencies.

We note here that simply equating the frequency of heterozygotes in two successive generations does not prove that this equilibrium is actually approached, or that it is stable. That both of these are in fact true can be shown but the biological considerations make it quite clear that there must be a stable equilibrium, so we shall not demonstrate it more rigorously.

Of considerable interest is the extent to which assortative mating increases the frequency of homozygotes for recessive genes. The equilibrium proportion of recessive homozygotes is given by equating R_{t+1} and R_t. This gives the quadratic

$$\hat{R}^2 - \hat{R}(1 + q^2 - rp^2) + q^2 = 0 \qquad 4.6.8$$

with the solution

$$\hat{R} = \frac{1 + q^2 - rp^2 - \sqrt{(1 + q^2 - rp^2)^2 - 4q^2}}{2}. \qquad 4.6.9$$

Some numerical examples are given in Table 4.6.1.

Table 4.6.1. Proportion of recessive homozygotes with assortative mating, for various values of the recessive-allelle frequency (q) and the degree of assortative mating (r). The values given are the proportion of recessive homozygotes after one generation of assortative mating and at equilibrium.

	RECESSIVE-ALLELE FREQUENCY, q					
r	.01		.1		.5	
	R_1	R_∞	R_1	R_∞	R_1	R_∞
0	.00010	.00010	.010	.010	.250	.250
.125	.00011	.00011	.011	.011	.260	.261
.25	.00012	.00013	.012	.013	.271	.273
.50	.00014	.00020	.014	.017	.292	.305
.75	.00017	.00038	.016	.027	.312	.352
1.00	.00020	.01000	.018	.100	.333	.500

Several general conclusions emerge from examination of this table. First, with weak assortative mating there is little ultimate increase in homozygous-recessive genotypes, as seen in the values near the top of the table. However, the population goes a large fraction of the way to equilibrium in the first generation. On the other hand, as seen in the lower left part of the table, intensive assortative mating with a rare recessive gene can lead eventually to a considerable increase in recessive homozygotes, but this is approached very slowly.

Notice that when $r = 1$, the solution to 4.6.9 is $R = q$. As with inbreeding, the proportion of recessive homozygotes approaches the gene frequency. This is expected, of course, since there has been no change in gene frequency during the process. On the other hand, the rate at which the genotypes change, and the change in effective inbreeding coefficient, f, depend on the gene frequency.

Assortative mating is quite high for deafness and it might be thought that this is a major factor in increasing the incidence. It has been estimated

(Chung, Robison, and Morton, 1959) that there are at least 35 recessive genes, any one of which can cause deafness when homozygous, and with an average frequency of 0.002. Whatever the amount of assortative mating for deafness as a trait, it would be only about 1/35 of this amount for any one recessive gene—somewhat less because of other causes of deafness. Thus, even with strict assortative mating the incidence would not be increased by more than 2% or 3%. However, the tendency might be enhanced if there were a tendency for consanguineous marriages among the deaf.

4.7 Assortative Mating for a Simple Multifactorial Trait

There is strong assortative marriage for height and intelligence in the human population. These traits are determined by a large number of genes and are also influenced by the environment. We should expect that, if there are several genes acting somewhat cumulatively to produce the trait, assortative mating for the entire trait would have a very small effect on any one locus. On the other hand, we would expect that there would be an enhancement of the variability, more than with inbreeding.

The enhancement of variability can be seen by a simple example. Suppose that the trait depends on two pairs of factors, such that each substitution of an allele with a subscript 1 for an allele with subscript 0 adds one unit to the phenotype, as follows:

	GENOTYPE	PHENOTYPE ON ARBITRARY SCALE
1	$A_1A_1B_1B_1$	$Y+4$
2	$A_1A_1B_1B_0$, $A_1A_0B_1B_1$	$Y+3$
3	$A_1A_1B_0B_0$, $A_0A_0B_1B_1$, $A_1A_0B_1B_0$	$Y+2$
4	$A_1A_0B_0B_0$, $A_0A_0B_1B_0$	$Y+1$
5	$A_0A_0B_0B_0$	Y

Inbreeding will increase all four homozygotes. On the other hand, assortative mating will increase only the two extreme types, 1 and 5. This is easily seen to be true for complete assortative mating, for the extreme types can produce progeny only like themselves. Therefore the occurrence of an extreme type is an irreversible process; or, in a different vocabulary, types 1 and 5 represent absorbing barriers.

Assume that A_1 and B_1 have the same frequency, p and $q = 1 - p$. Then the frequencies will be as follows:

		EQUILIBRIUM FREQUENCY		
TYPE	CODED PHENOTYPE	RANDOM MATING	COMPLETE INBREEDING	COMPLETE ASSORTATIVE MATING
1	2	p^4	p^2	p
2	1	$4p^3q$	0	0
3	0	$6p^2q^2$	$2pq$	0
4	−1	$4pq^3$	0	0
5	−2	q^4	q^2	q

The mean phenotype, $\overline{Y}$, is $2(p - q)$. The three variances are

Random: $V = p^4(4) + 4p^3q(1) + 4pq^3(1) + q^4(4) - \overline{Y}^2 = 4pq;$
Inbred: $V = p^2(4) + q^2(4) - \overline{Y}^2 = 8pq;$
Assortative: $V = p(4) + q(4) - \overline{Y}^2 = 16pq.$

The inbred variance is a confirmation of the principle given in Chapter 3 (3.10.6) that without dominance or epistasis the variance when $f = 1$ is twice the variance when $f = 0$. This is true regardless of the number of factors involved in the trait. On the other hand, with assortative mating the variability increase depends on the number of factors.

The procedure can be extended to any number of loci. This was first done by Wright (1921*b*). We have modified his method somewhat, but follow the same general idea. The procedure comes from Felsenstein (see Crow and Felsenstein, 1968).

Consider a trait determined by n gene loci. At each locus is a gene with frequency p such that the substitution of this gene for its allele adds a constant amount α to the character under consideration. Later, this restriction to equal gene effects and equal frequencies at all loci will be removed.

Let

n = the (haploid) number of relevant gene loci,
f = correlation in value of homologous genes,
k = correlation of nonhomologous genes in the same gamete,
l = correlation of nonhomologues in different gametes,
m = correlation of homologues in different individuals, X and Y,
m' = correlation of nonhomologues in different invididuals.

These relations are shown in Figure 4.7.1.

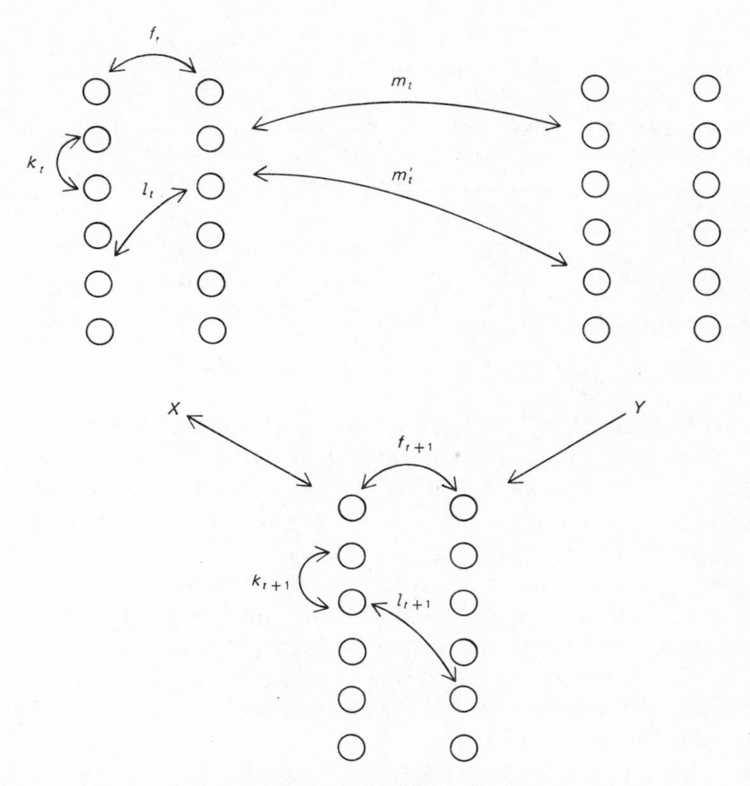

Figure 4.7.1. Correlations between the values of genes in two parents, X and Y, and their progeny. The circles represent individual genes. Homologous genes are opposite each other and genes from the same gamete are in a single vertical column.

An individual gene has a variance $pq\alpha^2$, where $q = 1 - p$. This can be shown as follows: For convenience, let the value of one allele be α and the other 0, with frequencies p and q. The mean value is $p\alpha + q0 = p\alpha$. The variance, v, is $p(\alpha - p\alpha)^2 + q(0 - p\alpha)^2 = pq\alpha^2$. Likewise the covariance, cov, of two genes, each with the same variance, is the variance times the correlation coefficient. For example, the covariance of two homologous genes is $pq\alpha^2 f$.

We can write the variance of the total value of individual X as the sum of the variances of the component genes. Thus

$$V(X) = \Sigma v_i + 2\Sigma\, cov_{ij}, \qquad \textbf{4.7.1}$$

where i and j designate individual genes.

The variance of an individual gene, v_i, is $pq\alpha^2$ and there are $2n$ of them, so $\Sigma v_i = 2npq\alpha^2$. The covariance of a pair of alleles is $pqf\alpha^2$, and there are n pairs. The covariance between nonalleles from the same gamete is $pqk\alpha^2$ and there are $n(n-1)$ combinations. Likewise, there are $n(n-1)$ pairs of nonalleles in different gametes with covariance $pql\alpha^2$. Putting all this together, we find that the variance of X at time t is

$$\begin{aligned} V(X)_t &= 2npq\alpha^2 + 2npqf_t\alpha^2 + 2n(n-1)pqk_t\alpha^2 + 2n(n-1)pql_t\alpha^2 \\ &= 2npq\alpha^2[1 + f_t + (n-1)(k_t + l_t)]. \end{aligned} \quad 4.7.2$$

Likewise the covariance of X and Y is

$$C(X, Y)_t = 4npqm_t\alpha^2 + 4n(n-1)pqm'_t\alpha^2. \quad 4.7.3$$

If the assortative mating is based solely on the phenotype, rather than being a by-product of common ancestry of the mates, and the gene frequencies are the same for all loci, there is no more reason for alleles in mates to be alike than nonalleles. Therefore $m_t = m'_t$ and we can drop the prime in equation 4.7.3, leading to

$$C(X, Y)_t = 4n^2pqm_t\alpha^2. \quad 4.7.4$$

From Figure 4.7.1 the following recurrence relations can be seen.

$$f_{t+1} = m_t, \quad 4.7.5$$

$$l_{t+1} = m'_t = m_t, \quad 4.7.6$$

$$k_{t+1} = (1-c)k_t + cl_t, \quad 4.7.7$$

where c is the proportion of recombination between the two loci concerned. In this case it is an average of the recombination between all pairs of loci concerned with the character, and for man is very nearly 1/2, since most pairs of loci are unlinked.

If r is the coefficient of correlation between the phenotypes of the two mates, X and Y, which have the same variance, the covariance is

$$C(X, Y) = rV(X). \quad 4.7.8$$

Substituting into this from 4.7.2 and 4.7.3 gives

$$4n^2pqm_t\alpha^2 = r[1 + f_t + (n-1)(k_t + l_t)]2npq\alpha^2. \quad 4.7.9$$

Now we substitute f_{t+1} for m_t (see 4.7.5) and f_t for l_t (4.7.5 and 4.7.6), which leads after some rearrangement to

$$f_{t+1} = \frac{r}{2n}[1 + nf_t + (n-1)k_t]. \quad 4.7.10$$

Using this and the relation

$$k_{t+1} = (1 - c)k_t + cf_t \tag{4.7.11}$$

(obtained from 4.7.6, 4.7.7, 4.7.8), we can compute f_t for any generation t, given the starting values, f_0 and k_0, which would both be 0 for a randomly mating population in gametic phase or "linkage" equilibrium.

At equilibrium there is no distinction between t and $t + 1$, so using carets to designate equilibrium values,

$$\hat{f} = \hat{l} = \hat{m} = \hat{k}. \tag{4.7.12}$$

Using these equilibrium relations 4.7.10 becomes

$$\hat{f} = \frac{r}{2n}[1 + n\hat{f} + (n - 1)\hat{f}], \tag{4.7.13}$$

leading to

$$\hat{f} = \frac{r}{2n(1-r) + r} \tag{4.7.14}$$

as first shown by Wright (1921).

If n is large, $\hat{f}$ is small unless r is very nearly 1. This shows that, unless the number of loci is small or the degree of assortative mating is very intense, there is only a very slight increase in homozygosity.

There is a much larger effect on the variance. From 4.7.2, substituting f_t for l_t from 4.7.5 and 4.7.6,

$$V(X)_t = V_0[1 + nf_t + (n - 1)k_t], \tag{4.7.15}$$

where $V_0 = 2npq\alpha^2$, the variance with random mating and linkage equilibrium. At equilibrium under assortative mating, substituting into 4.7.15 from 4.7.14 and 4.7.12,

$$\hat{V}(X) = \frac{V_0}{1 - r\left(1 - \dfrac{1}{2n}\right)} \tag{4.7.16}$$

(Wright, 1921), or, for large n,

$$\hat{V}(X) = \frac{V_0}{1-r}, \text{ approximately} \tag{4.7.17}$$

As a numerical example, let $r = 1/4$, which is roughly the correlation in height between husbands and wives. The homozygosity is increased only trivially if n, the number of factors, is large. After one generation $f = 1/8n$ and at equilibrium is $1/(6n + 1)$ or approximately $1/6n$. On the other hand, the variance is increased by 1/8 in the first generation and eventually by 1/3.

4.8 Multiple Alleles, Unequal Gene Effects, and Unequal Gene Frequencies

Still assuming no dominance and epistasis and no environmental effects, the assumption of only two alleles with equal effect and equal frequency will be dropped.

Let σ_i^2 be the variance of a gene at the ith locus. Thus $\sigma_i^2 = \Sigma p_k \alpha_k^2 - M_i^2$, where p_k and α_k are the frequency and effect on the trait of the kth allele, M_i is the mean effect of these alleles and the summation is over all alleles at the ith locus. The σ_i's remain constant under assortative mating since the gene frequencies do not change.

The covariance σ_{kl} between two genes is $\sigma_k \sigma_l r_{kl}$ where r_{kl} is the correlation between the two genes. The correlations f, k, l, m, and m' of Figure 4.7.1. are no longer constant for all pairs of genes. Equations 4.7.2 and 4.7.4 can be written more generally as

$$V(X) = 2\sum_{i=1}^{n} \sigma_i^2 + 2\sum_i \sigma_i^2 f_i + 2\sum_{i \neq j} k_{ij}\sigma_i\sigma_j + 2\sum_{i \neq j} l_{ij}\sigma_i\sigma_j, \qquad 4.8.1$$

$$C(X, Y) = 4\sum_{i,j} m_{ij}\,\sigma_i\,\sigma_j. \qquad 4.8.2$$

The recurrence relations 4.7.5, 4.7.6, and 4.7.7 still apply to individual gene pairs; that is,

$$f_{i,\,t+1} = m_{ii,\,t}, \qquad 4.8.3$$

$$l_{ij,\,t+1} = m_{ij,\,t}, \qquad 4.8.4$$

$$k_{ij,\,t+1} = (1 - c_{ij})k_{ij,\,t} + c_{ij}l_{ij,\,t}, \qquad 4.8.5$$

so that at equilibrium

$$f_i = m_{ii},\ k_{ij} = l_{ij} = m_{ij}. \qquad 4.8.6$$

Then at equilibrium 4.8.1 becomes

$$\hat{V}(X) = 4\sum_{ij} m_{ij}\sigma_i\sigma_j - 2\sum_i m_{ii}\sigma_i^2 + 2\sum_i \sigma_i^2. \qquad 4.8.7$$

We now let

$$n_e = \frac{\sum_{ij} m_{ij}\sigma_i\sigma_j}{\sum_i m_{ii}\sigma_i^2}. \qquad 4.8.8$$

Substituting this into 4.8.2 and 4.8.7 gives

$$\hat{V}(X) = \hat{C}(X, Y) - \frac{\hat{C}(X, Y)}{2n_e} + V_0 \qquad 4.8.9$$

where the carets indicate equilibrium values and

$$V_0 = V(X)_0 = 2\sum_i \sigma_i^2. \qquad \textbf{4.8.10}$$

V_0 is $V(X)$ before the assortative mating began ($f = l = k = m = 0$ in 4.8.1). Since $\hat{C}(X, Y) = r\hat{V}(X)$, we get

$$\hat{V}(X) = \frac{V_0}{1 - r(1 - 1/2n_e)}. \qquad \textbf{4.8.11}$$

At equilibrium the average inbreeding coefficient, weighted by the contribution of each locus to the variance, is

$$\hat{\bar{f}} = \frac{\sum_i m_{ii}\sigma_i^2}{\sum_i \sigma_i^2}. \qquad \textbf{4.8.12}$$

Substituting from 4.8.8, 4.8.2 and 4.8.10,

$$\begin{aligned} \hat{\bar{f}} &= \frac{\hat{C}(X, Y)}{2n_e} \cdot \frac{1}{V_0} \\ &= \frac{r}{2n_e} \cdot \frac{\hat{V}(X)}{V_0} \\ &= \frac{r}{2n_e(1 - r) + r}. \end{aligned} \qquad \textbf{4.8.13}$$

Comparing 4.8.11 and 4.8.13 with 4.7.16 and 4.7.14 shows the equivalence of n and n_e. When $m_{ii} = m_{ij} = m$, then from 4.8.8

$$n_e = \frac{\sum_{ij} \sigma_i \sigma_j}{\sum_i \sigma_i^2}, \qquad \textbf{4.8.14}$$

and if each locus has the same standard deviation ($\sigma_i = \sigma_j = \sigma$) then $n_e = n^2\sigma^2/n\sigma^2 = n$. We therefore call n_e the effective number of loci. It will be equal to the true number when there is free recombination and all loci contribute equally to the variance; otherwise it will be less.

Notice that when $n_e = 1$, 4.8.13 gives

$$\hat{f} = \frac{r}{2 - r}, \qquad \textbf{4.8.15}$$

the value mentioned earlier when we discussed a single locus. Likewise

$$\hat{V}(X) = V_0\left(\frac{2}{2 - r}\right). \qquad \textbf{4.8.16}$$

The variance after one generation of assortative mating is readily derived. From equations 4.8.1, 4.8.2, 4.8.3, 4.8.4, and 4.8.10 we can write

$$V(X)_1 = V_0 + 2\sum_i \sigma_i^2 m_{ii,0} + 2\sum_{i\neq j} k_{ij,1}\sigma_i\sigma_j + 2\sum_{i\neq j} \sigma_i\sigma_j m_{ij,0}. \qquad \textbf{4.8.17}$$

But $k_{ij,1} = 0$, from 4.8.5. Thus

$$V(X)_1 = V_0 + 2\sum_{i,j} m_{ij}\sigma_i\sigma_j$$

$$= V_0 + \tfrac{1}{2}C(X, Y)_0,$$

from 4.8.2, and therefore

$$V(X)_1 = V_0\left(1 + \frac{r}{2}\right), \qquad \textbf{4.8.18}$$

since $C(X, Y) = rV(X)$.

Table 4.8.1 gives numerical illustrations of the increase in homozygosity and variance after one generation of assortative mating and after equilibrium is reached.

Table 4.8.1. Effect of assortative mating on the average inbreeding coefficient, f, of relevant genes and the variance of the trait, V. Subscripts 0, 1, and ∞ refer to the randomly mating population, the population after one generation of assortative mating, and at equilibrium under assortative mating. Other symbols are: n_e = effective number of gene loci, r = correlation between mates, H = heritability.

	n_e	f_1	f_∞	$\frac{V_1}{V_0}$	$\frac{V_\infty}{V_0}$
$r = 1$	1	.500	1.000	1.500	2.00
$H = 1$	4	.125	1.000	1.500	8.00
$r = .5$	1	.250	.333	1.250	1.33
$H = 1$	4	.063	.111	1.250	1.77
	∞	0	0	1.250	2.00
$r = .25$	1	.125	.143	1.125	1.14
$H = 1$	4	.031	.040	1.125	1.28
	∞	0	0	1.125	1.33
$r = .5$, $H = .5$	∞	0	0	1.063*	1.21

* Exact only if $V_d = 0$.

We have not considered the effects of disassortative mating, but there is nothing about these formulae that demands that r be positive. Disassortative mating has opposite effects, a decrease of homozygosity and variance and a building up of linkage disequilibrium in the opposite direction (i.e., an association in the same gamete of genes of opposite effect).

4.9 Effect of Dominance and Environment

In a randomly mating population the variance can be divided into components

$$V_t = V_g + V_d + V_e \tag{4.9.1}$$

or

$$V_t = V_h + V_e, \tag{4.9.2}$$

where V_t is the total variance and V_g, V_d, V_h, and V_e are the genic (additive genetic), dominance, genotypic (or total genetic), and environmental components.

The equations above assume that the genetic and environmental factors are independent so that V_e is simply additive to the other components. This is a major limitation to precise quantitative prediction of the phenotypic effects of assortative mating, particularly in human populations. We are also ignoring the effects of epistasis. Finally, all the results from here on are only approximate.

According to Fisher (1918), assortative mating will increase V_g, but not V_d and V_e. This is not surprising, since with multiple factors only genic effects contribute to the correlation between parent and offspring (Reeve, 1961). However, it is not strictly true, for V_d does change. But, as noted earlier, with a large number of genes there is very little change in heterozygosity under assortative mating, and therefore V_d is not expected to change very much.

We let A be the correlation between the genic values of the mates. Thus,

$$A = r\frac{V_g}{V_t} = rH, \tag{4.9.3}$$

where H is the heritability. This is the same as h^2 of Section 4.1.

After one generation of assortative mating, we have approximately

$$V_t = V_g\left(1 + \frac{A}{2}\right) + V_d + V_e \tag{4.9.4}$$

from 4.8.18 after replacing r with A. Equation 4.8.18 is reasonably accurate if V_d is small. Otherwise the factor by which V_g is inflated may be appreciably in error. See Reeve (1961) for an exact expression for the 2-allele case. Equation 4.8.18 is strictly correct only in absence of dominance.

At equilibrium under assortative mating

$$\hat{A} = r\frac{\hat{V}_g}{\hat{V}_t} = r\hat{H}. \qquad 4.9.5$$

Substituting into 4.9.1 from 4.8.11 gives

$$\hat{V}_t = V_g\left(\frac{1}{1 - \hat{A}Q}\right) + V_d + V_e$$

$$= V_t + V_g\left(\frac{\hat{A}Q}{1 - \hat{A}Q}\right) = V_t\left[1 + H\left(\frac{\hat{A}Q}{1 - \hat{A}Q}\right)\right], \qquad 4.9.6$$

where $Q = 1 - \dfrac{1}{2n_e}\cdot$

We have used A instead of r because only the genic part of the correlation contributes significantly to the variance of future populations. Furthermore, we use the equilibrium value of A, since even with constant r there will be changes in A as the composition of the population changes.

The object is to express the population variance and the correlation between relatives after equilibrium under assortative mating, in terms of quantities that can be measured in the random-mating population before assortative mating began. To do this we must have a measure for $\hat{A}$. Note first the identity

$$V_t - V_g = V_g\left(\frac{1 - H}{H}\right) \qquad 4.9.7$$

which follows from the definition, $H = V_g/V_t$. But, since V_d and V_e do not change much with assortative mating,

$$V_t - V_g = \hat{V}_g\left(\frac{1 - \hat{H}}{\hat{H}}\right) = V_g\left(\frac{1}{1 - \hat{A}Q}\right)\left(\frac{1 - \hat{H}}{\hat{H}}\right). \qquad 4.9.8$$

Equating the right sides of 4.9.7 and 4.9.8 and recalling that $\hat{H} = \hat{A}/r$, we obtain after some algebraic rearrangement Fisher's equation for $\hat{A}$,

$$Q(1 - H)\hat{A}^2 - \hat{A} + Hr = 0. \qquad 4.9.9$$

H can be measured in the randomly mating population. Then, if Q is taken as 1 (i.e., the effective number of genes involved is assumed to be large, as it must be if other assumptions are to be correct), the equation can be solved for $\hat{A}$, and this value put into 4.9.6 to give the equilibrium variance.

As an example, let $H = .5$, $r = .5$, and $Q = 1$. Solving for $\hat{A}$ gives $(2 - \sqrt{2})/2 = .293$, Then, from 4.9.6,

$$\hat{V}_t = V_t\left[1 + .5\left(\frac{.293}{1 - .293}\right)\right] = 1.207V_t,$$

so the population variance is increased after equilibrium under assortative mating of this degree by about 21%.

This value is given in the bottom row of Table 4.8.1, along with the increase in variance after one generation of assortative mating. As noted above the latter value especially may be a poor approximation if V_d is large.

4.10 Effect of Assortative Mating on the Correlation Between Relatives

This was first done by Fisher (1918) and we follow his method.

Consider first parent-offspring correlation. The correlation is $V_g/2V_t$ in a randomly mating population. With equilibrium under assortative mating this will change for two reasons. One is that the variances increase, so we must replace V_g and V_t with their equilibrium values. The other reason is that the correlation between the two parents will, to the extent that this is reflected in genetic differences, add to the correlation of offspring with one parent through influences acting through the other.

If the chosen parent deviates by a unit amount from the population average, the other parent will deviate by r because of the correlation between the two mates. The mean deviation of the parents is thus $(1 + r)/2$, and the expected deviation of the children is the genic part of this, or $\hat{V}_g/\hat{V}_t$ times the parental mean deviation. Thus, the correlation between the chosen parent and the offspring at equilibrium under assortative mating is

$$\hat{r}_{po} = \frac{1}{2}\frac{\hat{V}_g}{\hat{V}_t}(1 + r), \qquad \text{4.10.1}$$

which in terms of the random-mating variances is

$$\hat{r}_{po} = \frac{1}{2}\frac{V_g + V_g\hat{K}}{V_t + V_g\hat{K}}(1 + r),\ \hat{K} = \frac{\hat{A}Q}{1 - \hat{A}Q}, \qquad \text{4.10.2}$$

as given by Fisher (1918).

Fisher also gives the grandparent-child correlation as

$$\hat{r}_{po} = \frac{\hat{V}_g}{\hat{V}_t}\frac{1 + r}{2}\frac{1 + \hat{A}}{2} \qquad \text{4.10.3}$$

and each additional descendant multiplies the correlation by $(1 + \hat{A})/2$, as expected, since only the genic component is transmitted and therefore $\hat{A}$ replaces r.

With full sibs the problem is more complicated because there are also correlations between the dominance components. Recall first the correlation between sibs under random mating, which is

$$r_{oo} = \frac{1}{2}\frac{V_g}{V_t} + \frac{1}{4}\frac{V_d}{V_t}. \qquad 4.10.4$$

The variance within a sibship with parents chosen at random is

$$\begin{aligned} V_s &= V_t(1 - r_{oo}) \\ &= V_t\left(1 - \frac{1}{2}\frac{V_g}{V_t} - \frac{1}{4}\frac{V_d}{V_t}\right) \qquad 4.10.5 \\ &= \tfrac{1}{2}V_g + \tfrac{3}{4}V_d + V_e. \end{aligned}$$

However, Fisher notes that this is also a good approximation to the variance within a sibship when the parents are mated assortatively, since the variance within a sibship depends only on genes for which the parents are heterozygous and, as we have learned, with a large number of genes the heterozygosity is only slightly decreased by assortative mating. Considering now the population at equilibrium under assortative mating, the correlation is a measure of the reduction of the variance within a sibship.

Thus

$$1 - \hat{r}_{oo} = \frac{\tfrac{1}{2}V_g + \tfrac{3}{4}V_d + V_e}{\hat{V}_t}.$$

But, from (4.9.6), $\hat{V}_t = V_t + \hat{K}V_g$. Making this substitution and rearranging we obtain

$$\hat{r}_{oo} = \frac{V_g(\hat{K} + \tfrac{1}{2}) + \tfrac{1}{4}V_d}{V_t + \hat{K}V_g}, \qquad 4.10.6$$

where

$$\hat{K} = \frac{\hat{A}Q}{1 - \hat{A}Q}. \qquad 4.10.7$$

$\hat{A}$ may be obtained from 4.9.9. Q is taken as 1.

Correlations for other relatives are given in Table 4.10.1.

Table 4.10.1. Correlations between relatives in a randomly mating population and in a population at equilibrium under assortative mating where r is the phenotypic correlation between mates, $H = V_g/V_t$, $D = V_d/V_t$, $A = Hr$. Equilibrium values under continued assortative mating are indicated by carets. The effective number of genes is assumed to be large, so that $(2n_e - 1)/2n_e$ may be regarded as 1.

Parent-offspring	$\frac{1}{2}H$	$\frac{1}{2}\hat{H}(1+r)$
Grandparent-offspring	$\frac{1}{4}H$	$\frac{1}{4}\hat{H}(1+r)(1+\hat{A})$
Great grandparent-offspring	$\frac{1}{8}H$	$\frac{1}{8}\hat{H}(1+r)(1+\hat{A})^2$
Sibs	$\frac{1}{2}H + \frac{1}{4}D$	$\frac{1}{2}\hat{H}(1+\hat{A}) + \frac{1}{4}\hat{D}$
Double first cousins	$\frac{1}{4}H + \frac{1}{16}D$	$\frac{1}{4}\hat{H}(1+3\hat{A}) + \frac{1}{16}\hat{D}$
Uncle-niece	$\frac{1}{4}H$	$\frac{1}{4}\hat{H}(1+\hat{A})^2 + \frac{1}{8}\hat{D}\hat{A}$
First cousins	$\frac{1}{8}H$	$\frac{1}{8}\hat{H}(1+\hat{A})^3 + \frac{1}{16}\hat{D}\hat{A}^2$

Fisher applied these methods to data on human stature. The data (obtained from earlier studies by Pearson and Lee) show

$$r = .2804,$$

$$r_{po} = .5066,$$

$$r_{oo} = .5433.$$

From 4.10.1 we calculate the equilibrium heritability

$$\frac{\hat{V}_g}{\hat{V}_t} = \hat{H} = .791,$$

from which

$$\hat{A} = \hat{H}r = .222,$$

assuming $Q = 1$.

Assuming the observed correlations represent equilibrium values we can ask what the heritability was before assortative mating began.

$$V_g = \hat{V}_g(1 - \hat{A}),$$

$$V_t = \hat{V}_t - V_g\left(\frac{\hat{A}}{1 - \hat{A}}\right),$$

$$H = \frac{V_g}{V_t} = .74.$$

So the assortative mating has increased the heritability from .74 to .79. From the sib correlation 4.10.6 we can estimate $\hat{V}_d/\hat{V}_t$, which turns out to be about the same as $1 - \hat{H}$. Hence, on the basis of these data, Fisher concluded that environment is of very little importance in determining variance in human stature.

The analysis of variance in a population at equilibrium under assortative mating would be

V_g	62%
V_d	21%
V_t	83%
Effect of assortative mating	17%
$\hat{V}_t$	100%

Fisher assumed that the environmental similarity between sibs was no greater than that between parent and offspring. This seems quite dubious; it is probable that genes for height are less dominant than he thought and the environmental influence greater.

To make it easier to go from this treatment to Fisher's 1918 paper, here is a list of equivalents:

V_g	τ^2	$\hat{V}_g/\hat{V}_h$	c_2
V_h	σ^2	$\hat{A}$	A
V_d	ε^2	Q	1
$\hat{V}_h/\hat{V}_t$	c_1	r	μ

4.11 Other Models of Assortative Mating

The correlation model that we have been discussing may not always be realistic. It is simple and natural, but of course there is an infinity of possible patterns of assortative-mating behavior. For quantitative traits the complexity may be such that it is not feasible to study more realistic models, except perhaps as special cases by computer simulation. Fortunately many traits of interest are normally distributed, or approximately normally, or may be transformed to be so, and the linear correlation and regression model works very well for most purposes.

On the other hand, there has been considerable discussion in the literature of specific models of assortative mating for single-locus traits (O'Donald, 1960*a*; Scudo, 1968; Parsons, 1962; Watterson, 1959; Scudo and Karlin, 1969; Karlin and Scudo, 1969).

As one considers the complexities of real populations there are many factors to take into account. The pattern may depend on whether the mating is monogamous or promiscuous, on the sex ratio, on which sex exerts the preference, on the nature of the inheritance of the trait, and on many other variables. Another complication is that the mating pattern may lead to a greater fertility of some genotypes than others. In other words, there may be selection in addition to pure assortative mating.

As stated in the beginning of this chapter, we shall ordinarily use the words assortative mating to describe pure assortative mating with no selection; that is, each genotype has the same expectation of surviving and fertility. When this is not so we shall speak of assortative mating with selection.

Even with this definition there will be difficulties in interpretation. For example, it may be that the same genotypes are more fertile in some mating combinations than others. In some instances it may be more convenient to designate the fertility of a mating combination than that of a genotype (see Bodmer, 1965).

We shall consider only a few of the many examples that could be used, first uncomplicated by selection and later with selection included.

Assortative-mating Models Without Selection We return first to the single locus with dominance, first discussed in Section 4.6. We assumed that the same level of preference existed among the recessive phenotypes as among the dominants, both measured by the correlation coefficient, r. But we would now like to be more general. For example, red-haired persons (or some red-haired persons) may prefer to marry others with red hair, but the rest of the population may have different preferences, or be indifferent to hair color. Consider the same model as before, but let the degree of assortment be r and r' among the recessives and dominants, instead of the same value for both. Again we assume that, after the designated fraction of assortative pairs is formed, the rest of the population mate at random.

The matings will then occur in the following ratios:

MATING	FREQUENCY		
	ASSORTATIVE	RANDOM	TOTAL
$A- \times A-$	$r'(1-R)$	$(1-r')^2(1-R)^2/D$	$1-R$
$A- \times aa$		$(1-r)(1-r')(1-R)R/D$	
$aa \times A-$		$(1-r)(1-r')(1-R)R/D$	R
$aa \times aa$	rR	$(1-r)^2R^2/D$	
	$r'+R(r-r') = 1-D$	$1-r'-R(r-r') = D$	1

The equations corresponding to 4.6.3–4.6.5 can be obtained from this table, giving

$$P_{t+1} = \frac{(1-r')^2p^2}{D} + \frac{r'p^2}{1-R_t}, \tag{4.11.1}$$

$$2Q_{t+1} = \frac{2(1-r')p[(1-r')Q_t + (1-r)R_t]}{D} + \frac{2r'pQ_t}{1-R_t}, \tag{4.11.2}$$

$$R_{t+1} = \frac{[(1-r')Q_t + (1-r)R_t]^2}{D} + rR_t + \frac{r'Q_t^2}{1-R_t}. \tag{4.11.3}$$

That the gene frequency does not change can be verified by adding 4.11.1 and half of 4.11.2 (or 4.11.3 and half of 4.11.2). This again shows that pure assortative mating does not change the gene frequency.

However, the final equilibrium and the rate of approach to this depend on r and r', as well as on the gene frequencies. The equilibrium value for Q is given by a cubic equation (see Scudo and Karlin, 1969), which of course reduces to 4.6.6 when $r = r'$.

As an example, suppose a certain fraction of deaf persons attend common schools and tend therefore to marry assortatively. The rest, say, are educated in public schools and join the population presumed to be marrying at random with respect to this trait. Then r' will be approximately 0.

In this case, the equilibrium equation corresponding to 4.6.8 and obtained by equating $R_{t+1} = R_t = \hat{R}$ is

$$r\hat{R}^2 - (1 - r + 2qr)\hat{R} + q^2 = 0, \tag{4.11.4}$$

with the solution

$$\hat{R} = \frac{1 - r + 2qr - \sqrt{(1 - r + 2qr)^2 - 4rq^2}}{2r}. \tag{4.11.5}$$

When r is less than 1 there is an equilibrium set of genotype frequencies. The proportion of recessive homozygotes is somewhat less than if $r = r'$, and the heterozygosity somewhat greater. When the mating preference is complete ($r = 1$), the population eventually becomes homozygous. Notice that in this case, 4.11.4 and 4.6.8 are equivalent, as they should be. If there is complete assortment within one phenotype, there must be within the other also.

Assortative Mating with Selection To continue with the same general model, suppose that those matings which are assortative differ in fertility from those which are random. This might happen, for example, if the assortment took place first; then the later random matings might have their fertility impaired by the delay.

As an example, suppose that in equations 4.11.1–4.11.3 the random matings have their fertility multiplied by a constant C, which may be greater or less than 1. This is equivalent to replacing D by D/C, which we shall call K. However, the equality signs must now become proportionality signs, since the three equations no longer add up to unity.

Consider first that $r = r'$. Then we can write, after some algebraic simplification (it will be helpful to recall that $P_t + Q_t = p_t$ and $Q_t + R_t = q_t = 1 - p_t$),

$$\frac{p_t}{q_t} = \frac{P_{t+1} + Q_{t+1}}{Q_{t+1} + R_{t+1}} = \frac{(1-r)^2 p + rpK}{(1-r)^2 q + rqK} = \frac{p}{q}, \tag{4.11.6}$$

regardless of the value of K.

In this case there is still no selection for, although different kinds of matings take place with different frequency, each genotype makes the same contribution of genes to the next generation. However, if $r \neq r'$, then the relationship 4.11.6 is no longer true. The gene frequencies change and assortative mating is complicated by an inherent selection in the process.

An interesting model that has been used for assortative mating with selection is the following. Suppose that matings occur at random but that disassortative matings are less fertile. This might happen if for some reason matings between unlike types were incompatible. We measure the extent of reduction in matings between different phenotypes by s. The model is specified in this way:

MATING	FREQUENCY RATIO	PROGENY RATIOS		
		AA	Aa	aa
$AA \times AA$	P^2	P^2		
$AA \times Aa$	$4PQ$	$2PQ$	$2PQ$	
$Aa \times Aa$	$4Q^2$	Q^2	$2Q^2$	Q^2
$AA \times aa$	$2PR(1\text{-}s)$		$2PR(1\text{-}s)$	
$Aa \times aa$	$4QR(1\text{-}s)$		$2QR(1\text{-}s)$	$2QR(1\text{-}s)$
$aa \times aa$	R^2			R^2
Total	$1 - 2R(1-R)s = D$			

Collecting the progeny of each genotype,

$$DP_{t+1} = (P_t + Q_t)^2 = p_t^2, \tag{4.11.7}$$

$$DQ_{t+1} = Q_t(P_t + Q_t) + R_t(P_t + Q_t)(1-s) = p_t q_t - p_t R_t s, \tag{4.11.8}$$

$$DR_{t+1} = (Q_t + R_t)^2 - 2Q_t R_t s = q_t^2 - 2(q_t - R_t)R_t s. \tag{4.11.9}$$

Adding 4.11.7 and 4.11.8 gives

$$p_{t+1} = P_{t+1} + Q_{t+1} = p_t\left[\frac{1 - R_t s}{1 - 2R_t(1 - R_t)s}\right]. \tag{4.11.10}$$

When $R_t < 1/2$, $p_{t+1} > p_t$, and the dominant gene increases in frequency. Whether the quantity in brackets is greater or less than 1 determines whether the gene frequency increases or decreases. So there is a tendency for the population to move away from the point where the two phenotypes are equal ($P + 2Q = R = 1/2$). The population tends to move toward fixation of whichever phenotype was more common in the first place. We end up eventually with a homozygous population, and which type it is depends mainly on the initial gene frequencies.

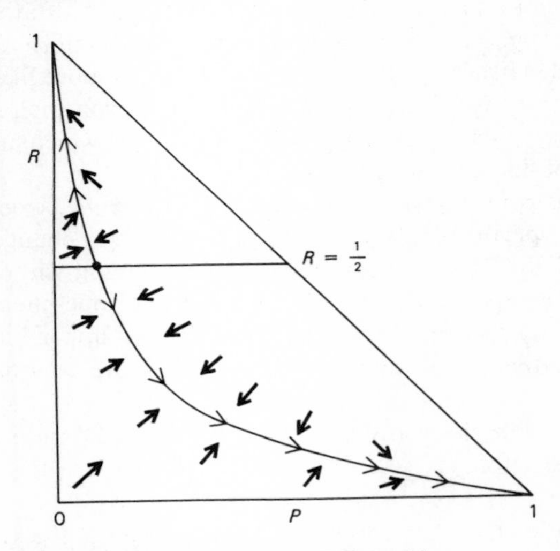

Figure 4.11.1. The paths followed by populations under assortative mating of the type described by equations 4.11.7–4.11.9. There is an unstable equilibrium at the point $q_t = q$, $R = 1/2$. Points corresponding to the Hardy–Weinberg proportions lie along the parabola, whose equation is $P^2 - 2PR + R^2 - 2P - 2R + 1 = 0$. When s is very small, the population tends to move quickly toward this curve and then proceed slowly along the curve to one or the other of the extreme points, $P = 1$, or $R = 1$, depending on which side of the horizontal line, $R = 1/2$, it started from.

Actually the situation is quite complex. There is a point of unstable equilibrium when $R = 1/2$ and the equilibrium value of the gene frequency q is obtained by solving 4.11.9 when $R = 1/2$ and $q_t = q$. Sometimes the gene frequencies will reach this equilibrium, but this point is unstable and the slightest displacement starts the process off toward fixation at one of the two extremes of fixation of the dominant or the recessive allele.

A discussion of this case has been given by Scudo and Karlin (1969). We have simply sketched the general picture in Figure 4.11.1. This shows the behavior when s is small.

4.12 Disassortative-mating and Self-sterility Systems

There has been much less attention paid to disassortative- than to assortative-mating systems, for the very good reason that with the great majority of traits the mating system, if departures from randomness occur, is more likely to be assortative than disassortative. There are, however, some conspicuous exceptions—one is the ordinary system of biparental reproduction, which may be regarded as an example of disassortative mating.

Even more than with assortative mating, disassortative tends to be confounded with selection. It is typically accompanied by gene frequency changes. In fact, it may be impossible to have strongly disassortative mating without selection. From the simple example of a population with 90% of one phenotype and 10% of the other, we can see that there is an obvious upper limit to the number of disassortative pairs that can occur. The more common type tends to get left out.

We shall consider two examples, both of which involve a mixture of disassortative mating and selection.

Self-sterility Alleles in Plants A number of plant species have this kind of system. The rule is that pollen is functional only on a plant, neither of whose two alleles at this locus is the same as that of the pollen. The system we are considering depends entirely on the haploid genotype of the pollen itself and not on the plant that produced it, although there are examples where the determination is by the genotype of the plant rather than the individual pollen.

The self-sterility allele system obviously prevents self-fertilization since neither of the two types of pollen produced by a plant can function on that plant. It is also clear that the system would work best in preventing self-fertilization, while permitting cross-fertilization, if there were a large number of alleles, so that a randomly chosen pollen is not likely to share its gene with the plant on which it lands. It is conventional to designate the alleles at this locus with the letter S, with individual alleles indicated by subscripts.

As an example, S_1 pollen would function on S_2S_3, S_2S_4, or S_2S_{50} plants, but not on S_1S_2 or S_1S_{47}. An immediate consequence of the system is that every plant is heterozygous. Thus, the total frequency of all genotypes carrying one S_i allele is $2p_i$.

We shall make two assumptions, both reasonable. One is that pollination is random. The second is that enough pollen is produced that every ovule has an equal chance of being fertilized. When there is no selection between different ovules, then each allele transmitted through the female has an equal chance.

The success of a pollen grain of genotype S_i depends on its falling on a plant not carrying this allele, and thus is proportional to $1 - 2p_i$. Thus the proportion of S_i pollen among all successful pollen will be $p_i(1 - 2p_i)/\Sigma p_i(1 - 2p_i)$. But the denominator is $\Sigma p_i - 2\Sigma p_i^2 = 1 - 2\Sigma p_i^2$, since the gene frequencies must add up to 1. Thus, dropping subscripts for simplicity of notation, the change in allele frequency due to pollen selection is in one generation

$$\frac{p(1-2p)}{1-2X} - p = -\frac{2p(p-X)}{1-2X},$$

where

$$X = \Sigma p^2.$$

But, there is no selection on ovules; all the selection is through the pollen. Thus, since the genes contributed to the next generation come equally from the male and female parents, the total change in gene frequency is only half as great. So, we write

$$\Delta p = \frac{-p(p-X)}{1-2X}, \qquad X = \Sigma p^2. \qquad \textbf{4.12.1}$$

This formulation is an excellent approximation, but it is not exact. It fails to take into account the exact nature of the other pollen grains with which any particular grain is competing on a particular stigma. For example, if most of the other grains carry one of the two alleles that the female has, this particular pollen has a better chance of being the successful one. However, if the number of alleles is large, most of the competing pollen will also fail to have an allele corresponding to that of the stigma on which they are competing, so that this factor can be ignored.

The more exact formulation has been given by Fisher (1958). It is also discussed by Moran (1962) and a rather similar approach was made by Wright (1939). The exact expression is in fact quite intractible, and the solutions obtained have been by approximations such as the one we have just given. Our formulation gives the correct equilibrium value, and the correct rate of approach unless the number of alleles is small or their frequencies grossly unequal.

It is clear from 4.12.1 that the frequency of an allele will increase if it is less than X, and will decrease if it is greater. The same is true of all alleles. Hence each allele frequency approaches the same value, X, and there is a stable equilibrium when all alleles are of the same frequency. If there are n alleles, then the equilibrium frequency of any one is

$$\hat{p}_i = 1/n. \quad \textbf{4.12.2}$$

Notice that equation 4.12.1 is equal to 0 when $p_i = 1/n$ for any number of alleles. That is, if one or more alleles have frequency 0 the rest again approach equality. Such a system will lead to the maximum number of alleles maintained. for every new mutant will tend to increase and, aside from loss due to sampling accidents, will be incorporated into the population, whereupon a new equilibrium is approached, with $p_i = 1/(n + 1)$.

It is no surprise then that actual investigations have revealed a very large number of such alleles persisting in plant populations. The only force tending to reduce the number of alleles (other than some limitation on the total range of mutational possibility) is the accidental loss of alleles from random changes. This problem has been discussed in great detail by Wright (1938, 1964, 1965), Fisher (1958), and others. We shall consider such random processes in the last three chapters.

Disassortative Mating; One Locus, Two Alleles We shall consider the problem briefly, and only for two alleles. The problem of disassortative mating for more than two alleles is quite complex (see Finney, 1952; Moran, 1962).

Imagine first a very simple case where the only matings are between $AA \times aa$ and $Aa \times aa$. It is obvious that, after the first generation, there will be no more AA homozygotes and the only remaining matings are $Aa \times aa$. This produces two kinds of progeny, like the parents, and in equal proportions. The equilibrium is immediately stable.

This situation is found in some plants where the dominant gene causes short style and the homozygous recessive is long. Fertilization normally occurs only between two different types. This is clearly expected to lead to a stable 1 : 1 polymorphism. A more familiar example is the ordinary sex-determining system in which all matings are XX by XY, again leading to a stable 1 : 1 sex-polymorphism.

Consider next a slightly more complicated example, and one with a rather interesting consequence. This time we still consider only two alleles, but each genotype is regarded as different. The rule is that each genotype can mate with any genotype but its own; otherwise mating is random. If, as usual, we let P, $2Q = H$, and R stand for the frequencies of the three genotypes AA, Aa, and aa, we can set forth the various possible matings as follows:

MATING	FREQUENCY RATIO	PROGENY RATIO		
		AA	Aa	aa
$AA \times Aa$	$2PH$	PH	PH	
$AA \times aa$	$2PR$		$2PR$	
$Aa \times aa$	$2HR$		HR	HR
Total	$D = 1 - P^2 - H^2 - R^2$			

The recurrence relations are easily written

$$P_{t+1} = P_t H_t / D_t, \qquad R_{t+1} = R_t H_t / D_t, \tag{4.12.3}$$

from which

$$\frac{P_{t+1}}{R_{t+1}} = \frac{P_t}{R_t}, \tag{4.12.4}$$

showing that the ratio of the two homozygotes does not change.

On the other hand,

$$\frac{R_{t+1}}{R_t} = \frac{P_{t+1}}{P_t} = \frac{H_t}{D_t} \tag{4.12.5}$$

and

$$D_t = 1 - P_t^2 - H_t^2 - R_t^2.$$

The homozygotes increase when $H > D$ and decrease when $H < D$. There is a stationary state when $H = D$, for then the genotypes have no tendency to change frequency. Setting $H = D$, and dropping subscripts since this is an equilibrium, gives the ellipse,

$$2P^2 + 2PR + 2R^2 - 3P - 3R + 1 = 0. \tag{4.12.6}$$

This is shown in Figure 4.12.1. As can be seen, the equilibrium is a rather peculiar one. Any point along the ellipse is stable with respect to perturbations changing the frequency of heterozygotes, for the population tends to return to the points along the ellipse. On the other hand, there is no tendency to return to the original point if there is a change (chance or otherwise) along the ellipse. Hence there are an infinity of points that are equilibria of this sort.

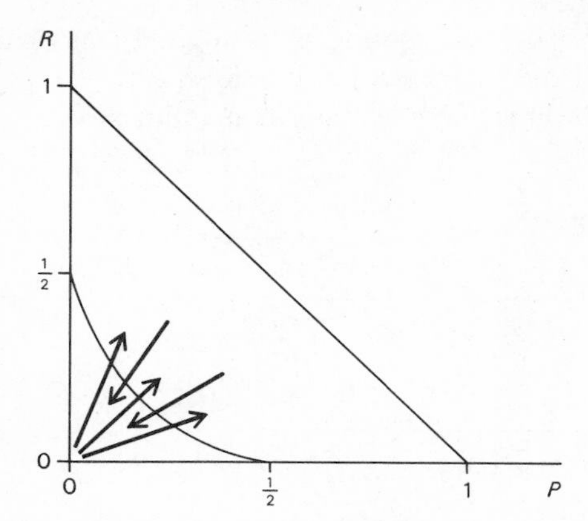

Figure 4.12.1. The case of complete disassortative mating with two alleles and three genotypes. The population follows the paths indicated by the arrows. Points along the ellipse represent equilibria. The arrows cross the ellipse, since the approach to equilibrium is oscillatory.

The only possible values lie within the triangle. The arrows indicate that there is no tendency for the P/R ratio to change. An actual population would drift randomly along the curve until one or the other of the two homozygotes is lost. Then the situation would reduce to the 2-phenotype polymorphism of the type discussed before, AA and Aa in equal proportions, or Aa and aa. These represent the two points at the end of the curve.

Another point of interest about this sytsem is that the approach to the ellipse is oscillatory. The population moves in the direction of the arrows in Figure 4.12.1, but overshoots each generation so that the value moves back and forth along the arrow, crossing the ellipse each time, and with decreasing amplitude until the equilibrium is reached.

In plants where there is an incompatibility system where the pollen function depends on the genotype of the plant that produces the pollen rather than the specific allele in the pollen grain itself, there are two possible mechanisms whose consequences differ slightly. It may be that incompatible pollen fails to fertilize the ovule (pollen elimination); alternatively, the fertilization may occur, but this particular ovule then fails to develop if the mating is incompatible (zygote elimination). The model we have just discussed is equivalent to zygote elimination. For a discussion, see Finney (1952) and Moran (1962).

4.13 Problems

1. Compute the mean and the genic, dominance, and genotypic variances at inbreeding coefficients 0, 1/2, and 1 for the following four examples.

	Y_{11}	Y_{12}	Y_{22}	P_1	P_2
(a)	99	100	101	p	q
(b)	99	100	100	.9	.1
(c)	100	100	99	.9	.1
(d)	99	100	99	.5	.5

2. For the system, $a_{11} \neq a_{12} = a_{22}$ (i.e., A_1 completely recessive), what gene frequency maximizes the genic variance? The dominance variance? The total variance? (Assume random mating.)
3. Give two reasons why the correlation between mother and daughter is likely to be lower for human weight than the correlation between sisters.
4. In terms of the model at the beginning of Section 3.10, show that the genic variance and dominance variance for $f = 0$ are $2p_1p_2[A + D(p_1 - p_2)]^2$ and $(2p_1p_2D)^2$.
5. What are the k-coefficients for a child and grandparent, for half-sibs, for uncle and niece, and for individuals D and H in pedigree 3.4.2?
6. Show that
$2k_0 = (1 - 2f_{AC})(1 - 2f_{BD}) + (1 - 2f_{AD})(1 - f_{BC}) - 2(f_{AC}f_{BD} + f_{BC}f_{AD})$.
7. Show that the dominance variance with random mating is the square of the difference in the population means at $f = 0$ and $f = 1$.
8. Relatives such that $k_2 = 0$ are sometimes called unilineal (Cotterman, 1941) and those with k_2 greater than 0 are bilineal. Give an example of a bilineal relationship other than identical twins, sibs, and double first cousins.
9. Show that for additive genes (no dominance, no epistasis, no environment effect) the correlation between parent and offspring is

$$\frac{1 + 2f_o + f_p}{2\sqrt{(1 + f_o)(1 + f_p)}}$$

where f_o and f_p are the inbreeding coefficients of the offspring and parent.
10. In deriving 4.4.4 we assumed that the covariance of the sums of two quantities is not changed if the quantities are correlated. Prove this.
11. Show from 4.11.9 that the equilibrium value of q for $R = 1/2$ is $(s + \sqrt{(2 - s)(1 - s)})/2$.
12. What is the limit of $\hat{q}$ in Problem 11 as s approaches 0? Show that the genotypes at this point are in Hardy–Weinberg ratios.

13. What is the heritability of a trait determined by a very rare recessive gene? What is the correlation between sibs?
14. Show that with "pure" overdominance ($a_{12} > a_{11} = a_{22}$) the heritability is $(p_1 - p_2)^2/(p_1^2 + p_2^2)$.
15. I and J are first cousins. I has phenylketonuria, the allele frequency being (say) 0.01. What is the probability that J has this recessive disease? Is the answer given by using k-coefficients exact?
16. Show that with three self-sterility alleles the proportion, P_t, of S_1S_2 heterozygotes in generation t is $(1 - P_{t-1})/2$. What is the equilibrium proportion? Is the approach direct or oscillatory?
17. Assuming the model of Table 4.4.2, what is the covariance of second cousins? Of individuals D and H in Pedigree 3.4.2?
18. Compare $\hat{R}$ from 4.11.5 for $r = r' = 0.5$ with $r = 0.5$, $r' = 0$ when $q = 0.01$. Do they differ appreciably?
19. Suppose that in a randomly mating population the heritability, $H\,(= h^2)$ of IQ is 0.6 and that the correlation between husband and wife is 0.5. By what fraction will the variance be increased when the population reaches equilibrium under this degree of assortative marriage? Compare this with the amount when the heritability is 1. If, prior to the beginning of assortative marriage, the IQ distribution had a mean of 100 and a standard deviation of 15, what fraction of the population would have IQ's above 130 before and after?

5 SELECTION

Selection occurs when one genotype leaves a different number of progeny than another. This may happen because of differences in survival, in mating, or in fertility. We are, as before, ignoring for the present differences that arise from random fluctuations. As mentioned in the preceding chapter, selection is distinguished from inbreeding and pure assortative mating in that under the latter systems the number of descendants is the same for all genotypes. Selection, along with migration and mutation, may alter the gene frequencies. However, it does not necessarily do so; it may be that the fitnesses of the different genotypes differ, but in such a way that opposing tendencies balance and the gene frequency is unchanged. Selection may be because of the greater fitness of some types, as in nature, or through artificial selection as practiced by the animal and plant breeders.

Sewall Wright (1931) has said:

> Selection, whether in mortality, mating or fecundity, applies to the organism as a whole and thus to the effects of the entire gene system rather than to single

> genes. A gene which is more favorable than its allelomorph in one combination may be less favorable in another. Even in the case of cumulative effects, there is generally an optimum grade of development of the character and a given plus gene will be favorably selected in combinations below the optimum but selected against in combinations above the optimum. Again the greater the number of unfixed genes in a population, the smaller must be the average effectiveness of selection for each one of them. The more intense the selection in one respect, the less effective it can be in others. The selection coefficient for a gene is thus in general a function of the entire system of gene frequencies. As a first approximation, relating to a given population at a given moment, one may, however, assume a constant net selection coefficient for each gene.

Selection involving both mortality and fertility is almost always complicated. One consequence of differential mortality is that a population counted at any stage except as zygotes will usually depart from Hardy–Weinberg ratios, even when mating is random. This would suggest that the proper time to census a population would be as soon as possible after fertilization. On the other hand, from the standpoint of assessing the effects of random gene frequency drift, it is more meaningful to count adults at the beginning of the reproductive period (Wright, 1931; Fisher, 1939*a*). When the probability of mating or the fertility is being considered, it may be more meaningful to measure the fertility of mating pairs than of individuals (Bodmer, 1965).

The systematic quantitative theory of natural selection came of age with a series of papers by Haldane (1924–1931). In the beginning of the first paper he said:

> A satisfactory theory of natural selection must be quantitative. In order to establish the view that natural selection is capable of accounting for the known facts of evolution we must show not only that it can cause a species to change, but that it can cause it to change at a rate which will account for present and past transmutations. In any case we must specify:
>
> (1) The mode of inheritance of the character considered,
> (2) The system of breeding in the group of organisms studied,
> (3) The intensity of selection,
> (4) Its incidence (e.g. on both sexes or only one), and
> (5) The rate at which the proportion of organisms showing the character increases or diminishes.
>
> It should then be possible to obtain an equation connecting (3) and (5).

Starting with the simplest cases—a single pair of alleles, random mating, discrete generations, constant selection coefficients equal in the two sexes—he proceeded to more and more complex cases. These included non-Mendelian inheritance, different intensities in the two sexes, within-family selection, *X*-linkage, inbreeding and assortative mating, multiple factors, linkage, polyploidy, sex-limited characters, reversal of dominance in the two sexes, gametic selection in one or both sexes, multiple-recessive and multiple-dominant traits, and overlapping generations. These early studies are summarized in his 1932 book *The Causes of Evolution*. The ways that selection can operate are uncountable, and many special cases are of genetic interest. However, we shall discuss only a few in order to illustrate general principles.

The effects of selection were also studied by R. A. Fisher and Sewall Wright, with a greater emphasis on quantitative traits. We shall also discuss the main generalities arising from these studies.

Natural selection, like classical mechanics, has both static and dynamic aspects. The statics of evolution will be dealt with in Chapter 6. This involves the relatively stable situation that results from the balance of various opposing forces—mutation, selection, migration, and random fluctuations. In this chapter we consider the dynamics—the way in which selection changes the composition of a population.

We shall consider two models, models 1 and 2 of Chapter 1. The first assumes that generations are discrete and nonoverlapping, as in annual plants. This is also applicable to many problems in animal breeding where pedigrees are known, and the generations can therefore be kept straight.

The second model applies strictly to organisms that reproduce and die continuously and with a constant probability of both, a situation approximated by some single-celled organisms. However, we should like to use it as an approximation to the situation in many organisms where generations overlap. The approximation is best when the population has reached stability of age distribution, as discussed in Chapter 1.

As expected, the two models become quite similar when the intensity of selection for the trait under consideration is small. For populations not in age-distribution equilibrium, it may be useful to weight each individual by its reproductive value (see Section 1.5). However, most of the time we shall use the equations in a simple form, regarding them as useful approximations from which we can reach interesting qualitative and semiquantitative generalizations.

5.1 Discrete Generations: Complete Selection

As a first example, we consider simple cases in which one class is lethal or sterile. In animal or plant breeding this corresponds to culling one class completely.

1. Selection Against a Dominant Allele If there is complete selection against a dominant factor, that is, if this phenotype is eliminated or fails to reproduce, the population next generation will be composed entirely of homozygous recessives (except for new mutations, incomplete penetrance, and such complications). So, one generation of selection is sufficient to eliminate the undesired type from the population.

2. Selection Against a Recessive Allele On the other hand, if all homozygous recessives are eliminated there are still recessive factors remaining in the population, hidden by heterozygosis with their dominant alleles. These can combine in later generations to produce zygotes that are homozygous for the recessive gene.

If the proportion of recessive genes in generation 0 is p_0 and mating is at random, there will be p_0^2 recessive homozygotes. When these are eliminated, the only possibility for a homozygous-recessive offspring is by the mating of two heterozygotes, in which case 1/4 of the offspring are expected to be homozygous recessives. Among the dominant phenotypes, a fraction $2p_0(1-p_0)/[(1-p_0)^2 + 2p_0(1-p_0)]$ will be heterozygous. The proportion of recessive homozygotes next generation will then be

$$\frac{1}{4}\left[\frac{2p_0(1-p_0)}{(1-p_0)^2 + 2p_0(1-p_0)}\right]^2 = \left[\frac{p_0}{1+p_0}\right]^2.$$

In the next generation, we can replace p_1 by $p_0/(1+p_0)$, and so on, leading to the following formulae for the change in gene frequency.

$$p_1 = \frac{p_0}{1+p_0},$$

$$p_2 = \frac{p_1}{1+p_1} = \frac{p_0}{1+2p_0}, \qquad 5.1.1$$

$$p_t = \frac{p_0}{1+tp_0},$$

and the zygotic frequencies are given by the squares of these quantities.

Another form of expression that might be mnemonically preferable is obtained by replacing p with $1/y$. This leads to

$$\frac{1}{y_t} = \frac{1}{1+y_{t-1}} = \frac{1}{t+y_0}. \qquad 5.1.2$$

The proportion of recessive homozygotes is the square of these quantities.

For example, if the present frequency of a rare recessive disease is 1/40,000 or $1/200^2$, and if none of these reproduce, the number next generation will decrease to $1/201^2$ or 1/40,301. This provides a numerical illustration of what is already well known—that selection against a recessive that is rare goes very slowly.

3. Selection Against a Sex-linked Recessive Trait If there is complete elimination of homozygous-recessive females and hemizygous-recessive males, then after the first generation there will be no more homozygous-recessive females. From that time on, all the recessive phenotypes will be males, who in turn came from heterozygous females. A heterozygous female may transmit the recessive gene to a son or to a daughter. In the former case it is eliminated; in the latter it is retained in the population. The proportion of heterozygotes among the daughters of affected females is 1/2. Since the affected males come from heterozygous females, they too are reduced by half each generation, the proportion of affected among males in any generation being exactly half the proportion of heterozygotes among females in the previous generation.

The results that we have just discussed are shown graphically in Figure 5.1.1. Note in particular the very slow rate of decrease of homozygotes for autosomal recessives, once the gene has become rare.

Figure 5.1.2 shows actual data on the decrease in frequency of a recessive lethal gene in Drosophila. In this experiment the generations were kept separated so the conditions of the model are fulfilled in this regard. In each generation the adults will be of two genotypes, *AA* and *Aa*, the *aa* type having died in the pre-adult stages. The two surviving types were classified by progeny

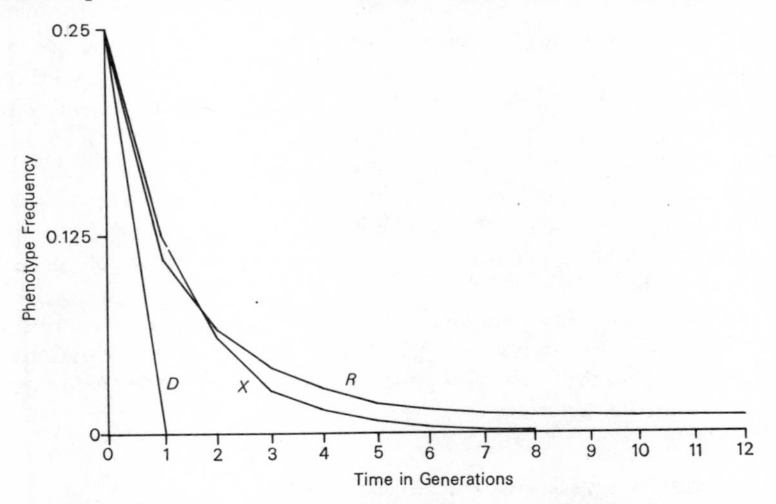

Figure 5.1.1. Selection against a recessive (*R*), dominant (*D*), and an *X*-chromosomal recessive (*X*). Selection is assumed to be complete, and in each case the starting frequency of the trait is 0.25. In the *X*-chromosome case it is assumed that there are no homozygous-recessive females, as would be the case after one generation of selection.

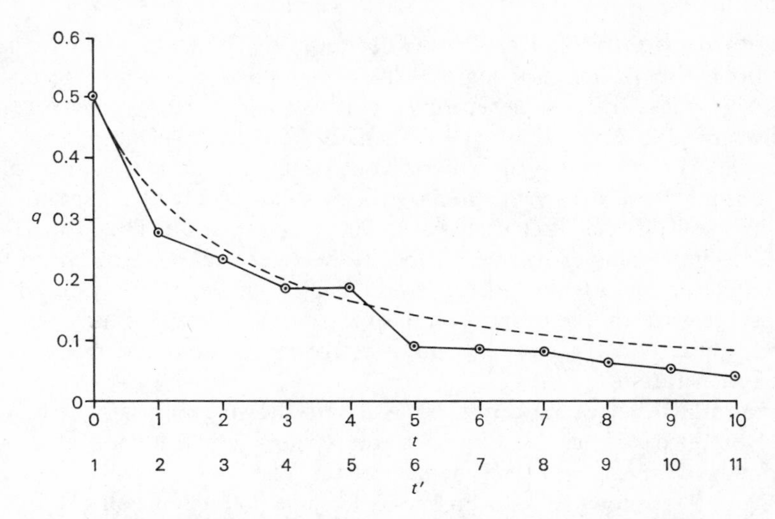

Figure 5.1.2. Selection against a recessive lethal gene. The allele frequency is given on the ordinate. In the abscissa t is the generation when adults are counted and t' is the generation for zygote gene frequency. Data from Wallace (*Amer. Natur.* 97: 65–66, 1963).

tests and the proportion of a genes among these is shown in the graph for each generation.

In generation 0 the adults were all Aa, since the experiment was started with flies all of the same heterozygous genotype. Thus $p_0 = 1/2$. This is also the expected proportion of a alleles among the zygotes in the next generation. At the adult stage the proportion of a alleles will have changed to 1/3, according to the relations in 5.1.1. In the figure the generations corresponding to the adult gene frequencies are indicated by t; those for zygote gene frequencies are indicated by t'.

The data agree approximately with the theoretical expectations, the standard errors of the points being large enough that most deviations are not significant. However, when all generations are considered there is some selection against the heterozygotes.

5.2 Discrete Generations: Partial Selection

We define the fitness, or selective value, as the expected number of progeny per parent. The parents and progeny must be counted at the same age, of course. The effects of differential mortality and fertility are kept within the same generation if each generation is enumerated at the zygote stage. In a biparental population, half of the progeny are credited to each parent.

Asexual Population Consider that there are many genetic types in the population, and that each reproduces its own type exactly. The model is also appropriate for a *Y*-chromosome factor (if only males are considered), for an *X*-chromosomal factor in either sex in an attached-*X* stock, or cytoplasmic inheritance transmitted through one sex.

Assume that the genotypes $A_1, A_2, A_3, \ldots$ have fitnesses $w_1, w_2, w_3, \ldots$ and are present in the population in frequencies $p_1, p_2, p_3, \ldots$. Then the proportion of A_i genotypes next generation will be

$$p_i' = \frac{p_i w_i}{p_1 w_1 + p_2 w_2 + \cdots} = \frac{p_i w_i}{\bar{w}},$$

where $\bar{w} = \sum p_i w_i$ is the average fitness.

The change in the proportion of A_i in one generation is

$$\Delta p_i = \frac{p_i w_i}{\bar{w}} - p_i = \frac{p_i(w_i - \bar{w})}{\bar{w}}. \qquad 5.2.1$$

The quantity $w_i - \bar{w}$ is the *average excess* in fitness of the genotype A_i. For only two alleles, this formula is conveniently written as

$$\Delta p_1 = \frac{p_1 p_2 (w_1 - w_2)}{\bar{w}} = \frac{s p_1 p_2}{\bar{w}}, \qquad 5.2.2$$

where s, the selection coefficient, is $w_1 - w_2$ and $p_1 + p_2 = 1$.

Notice that selection is most rapid when the two types are nearly equal in frequency and becomes slower when one is much more common than the other. For example, if w_1 is 1.1 and w_2 is .9, the frequency of A_1 will increase from .50 to .55 in one generation, but if the frequency is 0.10, it will only change to .1195 in one generation.

Diploid Sexual Population We now let w_{ij} stand for the average fitness of the genotype $A_i A_j$. As before we let P_{ii} be the frequency of the homozygous genotype $A_i A_i$ and $2P_{ij}$ the frequency of the heterozygote $A_i A_j$. Then the frequency of the gene A_i is (from equation 2.1.2)

$$p_i = \sum_j P_{ij}.$$

Next generation the proportion of A_i genes will be

$$p_i' = \frac{\sum_j P_{ij} w_{ij}}{\bar{w}} = \frac{p_i w_i}{\bar{w}}, \qquad 5.2.3$$

where

$$\bar{w} = \sum_i \sum_j P_{ij} w_{ij} = \sum_i p_i w_i \qquad 5.2.4$$

and

$$w_i = \frac{\sum_j P_{ij} w_{ij}}{p_i}. \qquad 5.2.5$$

Hence,

$$\Delta p_i = \frac{p_i(w_i - \bar{w})}{\bar{w}}. \qquad 5.2.6$$

The formula is the same as that for asexual selection, but w_i now has a more complex meaning. In this equation w_i is the average fitness of the A_i allele; more specifically, it is the average fitness of all genotypes containing A_i, weighted by the frequency of the genotype and by the number of A_i alleles (1 or 2). Then $\bar{w}$, the average fitness of the population, can be expressed in either of two ways: (1) the average of all the genotypes in the population, and (2) the average fitness of all the genes at this locus. These correspond to the two expressions given in 5.2.4.

Equation 5.2.6 shows that the rate of change of the gene frequency is proportional to:

(1) The gene frequency, p_i. Thus a very rare gene will change slowly, regardless of how strongly it is selected.
(2) The average excess in fitness of the A_i allele over the population average. If the excess, $w_i - \bar{w}$, is positive, the allele will increase; if negative, it will decrease. If this is large the frequency of A_i will change rapidly; if small, slowly.

Notice also that the gene frequency change will be slow when the allele becomes very common ($p_i \rightarrow 1$). In this case, w_i and $\bar{w}$ are not very different, since most of the population contains the A_i gene. This point can be brought out by rewriting 5.2.6 in another way:

$$\Delta p_i = \frac{p_i(1 - p_i)(w_i - w_x)}{\bar{w}}, \qquad 5.2.7$$

where w_x is the average fitness of all alleles other than A_i. This shows clearly that Δp_i approaches 0 as p_i gets near to either 0 or 1.

If the population is in random-mating proportions we can write 5.2.7 in still another way, often used by Wright (e.g., 1949). With random mating $P_{ij} = p_i p_j$ and

$$w_i = \frac{\sum_i \sum_j p_i p_j w_{ij}}{p_i} = \sum_j p_j w_{ij}, \qquad 5.2.8$$

or, representing all alleles that are not A_i as collectively A_x with frequency $p_x = 1 - p_i$,

$$w_{ix} = \frac{2\sum_j p_i p_j w_{ij}}{2p_i(1-p_i)} = \frac{\sum_j p_j w_{ij}}{(1-p_i)}, \quad j \neq i$$

$$w_{xx} = \frac{\sum_j \sum_k p_j p_k w_{jk}}{(1-p_i)^2}, \quad j, k \neq i$$

$$w_i = p_i w_{ii} + p_x w_{ix}, \qquad \textbf{5.2.9}$$

$$w_x = p_i w_{xi} + p_x w_{xx},$$

where $w_{ix} = w_{xi}$ is the average fitness of all heterozygotes where one allele is A_i and w_{xx} is the average of all genotypes where neither is A_i. In this notation, still assuming random-mating proportions,

$$\overline{w} = p_i^2 w_{ii} + 2p_i(1-p_i)w_{ix} + (1-p_i)^2 w_{xx}. \qquad \textbf{5.2.10}$$

In this formulation w_{ix} and w_{xx} are not constants, but depend on the relative frequencies of the non-A_i alleles (except when there are only two alleles). Notice that the partial derivative of $\overline{w}$ with respect to p_i is

$$\frac{\partial \overline{w}}{\partial p_i} = 2p_i w_{ii} - 2p_i w_{ix} + 2(1-p_i)w_{ix} - 2(1-p_i)w_{xx}$$

$$= 2(w_i - w_x). \qquad \textbf{5.2.11}$$

The two quantities w_{ix} and w_{xx} are treated as constants in this differentiation. Substituting this into 5.2.7 gives Wright's formula

$$\Delta p_i = \frac{p_i(1-p_i)}{2\overline{w}} \frac{\partial \overline{w}}{\partial p_i}. \qquad \textbf{5.2.12}$$

In analogy with physical theory we can regard $\overline{w}$ as a potential surface, in which case $\partial\overline{w}/\partial p_i$ becomes the slope of the surface with respect to p_i. Treating w_{ix} and w_{xx} as constants is equivalent to treating all allele frequencies except p_i as constant fractions of $1 - p_i$ (see 5.2.9). Thus $\partial\overline{w}/\partial p_i$ is the slope of $\overline{w}$ in the direction where the relative frequencies of the other alleles do not change. The gene frequency moves over the surface at a rate proportional to the slope, but governed by the term $p_i(1 - p_i)$. For extensions to populations not in random proportions, see Wright (1942, 1949).

In more complicated situations there may not be a fitness surface, $\overline{w}$, that is a function of the gene frequencies. For example, the fitnesses of the individual genotypes may not be constants, or there may be complications from linkage and epistasis. In such cases the fitness may not even increase at all. For a discussion, see Wright (1942, 1955, 1967) and Moran (1964).

When there are only two alleles, we can use equation 5.2.7 (for example) with $i = 1$ and $x = 2$. We shall now write explicit formulae for the 2-allele situation with random mating in several special cases. The fitnesses and frequencies of the three genotypes are given below:

GENOTYPE	A_1A_1	A_1A_2	A_2A_2
FITNESS	w_{11}	w_{12}	w_{22}
FREQUENCY	p_1^2	$2p_1p_2$	p_2^2

Then, from 5.2.9 and 5.2.7,

$$w_1 = w_{11}p_1 + w_{12}p_2,$$
$$w_2 = w_{12}p_1 + w_{22}p_2,$$
$$\Delta p_1 = \frac{p_1p_2[(w_{11} - w_{12})p_1 + (w_{21} - w_{22})p_2]}{\bar{w}}. \qquad 5.2.13$$

a. No dominance, $w_{12} = (w_{11} + w_{22})/2$.

$$\Delta p_1 = \frac{p_1p_2(w_{11} - w_{22})}{2\bar{w}} = \frac{sp_1p_2}{2\bar{w}} \qquad 5.2.14$$

where $s = w_{11} - w_{22}$

b. Dominant favored, $w_{11} = w_{12}$, A_1 dominant.

$$\Delta p_1 = \frac{sp_1p_2^2}{\bar{w}}. \qquad 5.2.15$$

c. Recessive favored, $w_{12} = w_{22}$, A_1 recessive.

$$\Delta p_1 = \frac{sp_1^2p_2}{\bar{w}}. \qquad 5.2.16$$

d. Asexual, haploid, or gametic selection (5.2.2).

$$\Delta p_1 = \frac{sp_1p_2}{\bar{w}}. \qquad 5.2.17$$

These four formulae illustrate several important facts about selection. One is that asexual selection is more effective than sexual when the whole range of gene frequencies is considered. A diploid sexual population with no dominance evolves half as fast as an asexual population in which the two types differ by the same amount as the two homozygotes in the sexual population. Haploid or gametic selection is equivalent to asexual for a single locus.

Comparison of b and c shows that selection is most effective when both the dominant and recessive genes are of intermediate frequency. In fact, the maximum rate of change is when the recessive allele is twice as frequent as the dominant. Selection becomes decreasingly effective as the recessive gene becomes rare and the p^2 term gets closer to 0. Notice, by comparison with Table 4.1.1, that the situation where selection is inefficient is that where the heritability is low, as expected.

A way of expressing partial dominance in a convenient manner is to use the terminology of Wright (1931 and later). We assign fitnesses and symbols to the genotypes as follows:

GENOTYPE	AA	Aa	aa
FREQUENCY	p^2	$2pq$	q^2
FITNESS (RELATIVE)	1	$1-hs$	$1-s$

In these terms,

$$\Delta q = \frac{-spq[q + h(p - q)]}{1 - sq(2hp + q)}. \qquad 5.2.18$$

The quantity, s, often is referred to as the selection coefficient. The quantity h is a measure of dominance. When the a allele is recessive, $h = 0$. When a is dominant, $h = 1$. When h is greater than 1 there is overdominance.

This formula brings out the importance of partial dominance when there is selection on a rare gene. If a is completely recessive and q is small the rate of change is proportional to q^2. When q is small and h larger than q, the corresponding term is hq (since p is very nearly unity). Thus, even a small amount of partial dominance (say, $h = .05$ or less) may make a great deal of difference in the nature of selection involving rare genes. This becomes especially important in the consideration of equilibrium gene frequencies when selection against a recessive genotype is balanced by new mutations; this will be discussed in the next chapter.

As mentioned earlier, the number of interesting special cases is almost endless. Many of them were worked out by Haldane and many more have been done since. We shall at this point mention only two.

It is probable that in nature much of selection is based on fertility. It is likely to be the rule, rather than the exception, that a gene has different fertility effects in the two sexes. Such differences are also frequent in mortality rates, as well. So, if we are to have a model that is at all realistic, it should take into account the possibility of different selection coefficients in the two sexes.

Fortunately, there is an easy solution; the considerations of equation 1.1.4 are applicable, and the combined fitness is the average of the two sexes. Furthermore, it is the simple average, since each individual is derived from one sperm and one egg. Therefore, except for X- and Y-linked genes, the contribution of the two sexes is the same, and the overall fitness of a genotype is the unweighted average of that in the two sexes. This is not exact, however, for the resulting differences in allele frequencies in the two sexes at the time of mating may lead to departures from Hardy–Weinberg proportions (see 2.5.1); but the formulae are usually satisfactory as an approximation.

X- and *Y*-linked genes cause no particular difficulty. Genes that are on the *Y* chromosome are found only in the heterogametic sex. The situation is entirely equivalent to asexual or gametic selection where only males (or females if they are *XY* in this species) are counted.

For an *X*-linked trait we note that the gene in a male is derived from his mother whereas those in a female are derived equally from the two parents. Therefore the frequency in males is given by the formula applied to the fitness and frequency in females of the previous generation. The gene frequency in females is given by the average for the two sexes in the previous generation. Specifically, if q_m and q_f stand for the frequencies of the gene of interest in males and females, and we use primes to designate the next generation,

$$q'_m = q_f + \frac{q_f(w_f - \overline{w}_f)}{\overline{w}_f} \qquad 5.2.19$$

and

$$q'_f = \frac{1}{2}\left[q_m + q_f + \frac{q_f(w_f - \overline{w}_f)}{\overline{w}_f} + \frac{q_m(w_m - \overline{w}_m)}{\overline{w}_m}\right]. \qquad 5.2.20$$

The symbols w_m and w_f stand for the average fitness of the allele in males and females, respectively. As mentioned above, these formulae are not exact because of departures from Hardy–Weinberg proportions; this may become important if selection is very intense.

Throughout this section we have spoken of gene frequency changes rather than changes of genotype frequencies. Formulae can be written for changes in the genotypes directly, but they are much more cumbersome. We effect a great simplification by working directly with the gene frequencies. Furthermore, the zygotic types are put together and taken apart every generation by the Mendelian processes of segregation and recombination, so that a zygote type (when many loci are considered) may never again be reconstituted. For these reasons, almost all of selection theory deals with changes in gene frequencies.

This has a price, however. There is usually some inaccuracy in going from gene frequencies to zygotic frequencies. We need to know something about the mating system. Even if mating is completely at random, there will be departures from Hardy–Weinberg ratios in all stages after mortality begins.

We therefore regard the procedures as useful approximations rather than exact formulae. We follow the changes in gene frequencies; then we get the genotype frequencies by the Hardy–Weinberg principle, or some modification thereof to include nonrandom-mating effects. If possible, we count the population at the zygote stage. Fortunately, much selection of evolutionary interest is relatively slow, and the Hardy–Weinberg ratios are very good approximations even for adult populations.

If we try to take into account the various complexities of populations in the real world the formulae naturally become more complicated. One obvious extension of what we have been doing is to consider survival and fertility as separate aspects of fitness. We illustrate with a special case.

Consider a locus with two alleles, A_1 and A_2, and with viabilities and fertilities as given below.

GENOTYPE	A_1A_1	A_1A_2	A_2A_2
VIABILITY	v_{11}	v_{12}	v_{22}
FERTILITY	f_{11}	f_{12}	f_{22}

The total fitness of a genotype will be the product of its viability and fertility.

The part of the life cycle in which survival is important (from the standpoint of natural selection) is that prior to reproduction, so we let v_{ij} be the survival to the time of reproduction. We are still assuming that generations are discrete and that matings take place at random among the adults.

If the proportions of the three genotypes are P_{11}, $2P_{12}$, and P_{22}, and the enumeration is made at the zygote stage, then the combined survival and fertility (or expected number of progeny, crediting half to each parent) of A_iA_j is $v_{ij}f_{ij}$. Letting $v_{ij}f_{ij} = w_{ij}$, the equations of the earlier parts of this section (e.g., 5.2.6) are applicable. Equation 5.2.6 gives the changes in gene frequency and the proportions of the three zygotic types are p_1^2, $2p_1p_2$, and p_2^2, where p_1 and p_2 are the new gene frequencies.

If, on the other hand, the population is enumerated at the adult stage the situation is more complicated. For one thing, the genotypes at this stage are no longer in Hardy–Weinberg ratios. Suppose the population is censused just before reproduction. Then v_{ij} is the viability up to this stage and f_{ij} is the fertility. (Deaths that occur during the reproductive period can be accommodated by regarding them as reducing f.) Let P_{11}, $2P_{12}$, and P_{22} stand for the proportions of the three genotypes at the stage of enumeration.

We can obtain the proportion of zygotes next generation as follows. The A_1 genes contributed will be proportional to $P_{11}f_{11} + P_{12}f_{12}$ and the A_2 genes proportional to $P_{12}f_{12} + P_{22}f_{22}$. With random mating the three zygotic types will be in the ratio

$$(P_{11}f_{11} + P_{12}f_{12})^2 : 2(P_{11}f_{11} + P_{12}f_{12})(P_{12}f_{12} + P_{22}f_{22}) : (P_{12}f_{12} + P_{22}f_{22})^2.$$

The adult frequencies next generation, indicated by primes, are

$$P'_{11} = \frac{(P_{11}f_{11} + P_{12}f_{12})^2 v_{11}}{K},$$

$$2P'_{12} = \frac{2(P_{11}f_{11} + P_{12}f_{12})(P_{12}f_{12} + P_{22}f_{22})v_{12}}{K}, \qquad 5.2.21$$

$$P'_{22} = \frac{(P_{12}f_{12} + P_{22}f_{22})^2 v_{22}}{K},$$

where K is the sum of the numerators and is introduced to make the frequencies total 1.

Successive application of these formulae gives the frequencies in later generations, but the equations are no longer simple functions of the allele frequencies. In most such cases and in more complicated ones the only way to get the results is to grind them out generation by generation. Of course high speed computers are a godsend for numerical results.

In some cases a simplifying transformation can be found. One such appears when one class is lethal or sterile (Teissier, 1944; Crow and Chung, 1967; Anderson, 1969). Suppose the A_2A_2 class dies before the age of enumeration. Then, since we are interested in relative rather than absolute frequencies, let us arbitrarily choose 1.0 as the viability and fertility of one class. Accordingly, let

$$v_{11} = 1,\ v_{12} = v,\ v_{22} = 0,$$
$$f_{11} = 1, f_{12} = f.$$

Then

$$P'_{11} = \frac{(P_{11} + P_{12}f)^2}{K},$$

$$P'_{12} = \frac{(P_{11} + P_{12}f)P_{12}fv}{K}, \qquad 5.2.22$$

$$P'_{22} = 0.$$

Note that the frequency, say q, of allele A_2 at the stage of enumeration is

$$q = P_{12},$$

since $2P_{12}$ is the proportion of heterozygotes, and $P_{11} + 2P_{12} = 1$.

We can simplify things by letting

$$y = \frac{P_{11}}{2P_{12}} = \frac{1-2q}{2q}. \qquad 5.2.23$$

Then the value of y next generation is

$$y' = \frac{P'_{11}}{2P'_{12}} = \frac{P_{11} + P_{12}f}{2P_{12}fv} = y\left(\frac{1}{fv}\right) + \frac{1}{2v}, \tag{5.2.24}$$

and we can write the simple recurrence relation

$$y_{t+1} = ay_t + b, \tag{5.2.25}$$

where

$$a = 1/vf \text{ and } b = 1/2v.$$

This gives the ratio of normal homozygotes to lethal heterozygotes at the adult stage. If we want the lethal-allele frequency, it is given each generation by

$$q_t = \frac{1}{2(1 + y_t)}. \tag{5.2.26}$$

An approximate expression for y_t for any t can be obtained by writing an expression for the rate of change of y and treating this as a differential equation.

$$\Delta y = y_{t+1} - y_t = \frac{1 - vf}{vf}\, y_t + \frac{1}{2v} \approx \frac{dy}{dt}.$$

This integrates into

$$y_t = (y_0 + C)e^{At} - C, \tag{5.2.27}$$

where

$$A = \frac{1 - vf}{vf}, \qquad C = \frac{f}{2(1 - vf)}.$$

Figure 5.2.1 shows some data from Drosophila population cages. The flies were of three genotypes, +/+ (normal), +/*Sb* (Stubble bristles), and *Sb*/*Sb* which is lethal in the larval stages and is not observed in the adult population. The data points are from weekly censuses and record the proportion of the *Sb* gene in the adult population. The data are an average of four populations, from each of which a sample of 200 adults was classified each week. The populations numbered several hundred.

The average generation length under these circumstances is estimated to be about 2.5 weeks. The dotted line is the expected proportion of *Sb* chromosomes if the gene is completely recessive and treating the situation as if the discrete model were appropriate (equation 5.1.1). The starting frequency, q_0, is taken as 0.3.

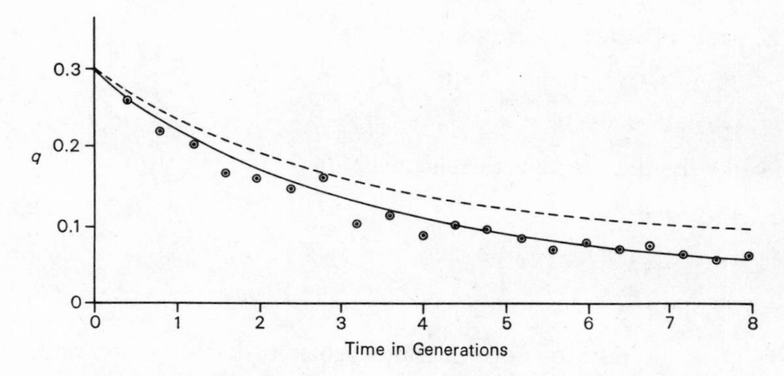

Figure 5.2.1. Selection against a recessive lethal gene that produces Stubble bristles when heterozygous. Abscissa: time in generations. Ordinate: *Sb* gene frequency at the adult stage. The dotted line shows expectation for fully recessive lethal; solid line, 12% disadvantage of heterozygote. (Data from W. Y. Chung.)

The solid line is obtained from 5.2.25 and 5.2.26, using $v = .970, f = .907$, $fv = .880$, and $y_0 = .667$ (corresponding to $q_0 = .30$). As can be seen, although the population is changing continuously and the generations overlap, the data fit the expectation very well.

The approximation 5.2.27, gotten by treating the process as if continuous, gives results that are almost indistinguishable from 5.2.22. Furthermore, it is not very important in this example to separate viability from fertility. For example, if we let $v = 0.88$ and $f = 1.00$, starting with a frequency of 0.30 we have after 10 generations a frequency of 0.035; with $v = 1.00$ and $f = 0.88$ we have 0.039, not very different. The product of v and f is more important than either of the components.

For a detailed discussion of the practical procedures for measuring v and f in actual populations, see Anderson (1969).

We can summarize this discussion of the time of enumeration by rewriting the formulae, and at the same time extending them to include multiple alleles. We are still assuming random mating.

Enumeration at the zygote stage:

$$P'_{ij} = \frac{(\sum_j P_{ij} v_{ij} f_{ij})(\sum_i P_{ij} v_{ij} f_{ij})}{C}. \qquad \textbf{5.2.28}$$

Enumeration at the adult stage:

$$P'_{ij} = \frac{(\sum_j P_{ij} f_{ij})(\sum_i P_{ij} f_{ij}) v_{ij}}{K} \qquad \textbf{5.2.29}$$

C and K are chosen to make the frequencies total to 1.

Note that, for zygote enumeration, $v_{ij}f_{ij} = w_{ij}$ and from 5.2.5

$$p_i w_i = \sum P_{ij} v_{ij} f_{ij}. \tag{5.2.30}$$

From the definition of gene frequency (2.1.2),

$$\begin{aligned} p_i' &= \sum_j P_{ij}' = \frac{\sum_j p_i w_i p_j w_j}{C} \\ &= \frac{p_i w_i \bar{w}}{C}, \end{aligned} \tag{5.2.31}$$

since $\sum p_j w_j = \bar{w}$.

If we sum both sides of 5.2.31, and note that $\sum p_i = 1$ and $\sum p_i w_i = \bar{w}$, we see that $C = \bar{w}^2$, so

$$p_i' = \frac{p_i w_i}{\bar{w}}, \tag{5.2.32}$$

corresponding to 5.2.3.

For adult enumeration, noting that

$$\begin{aligned} p_i f_i &= \sum_j P_{ij} f_{ij}, \\ p_i' &= \frac{\sum_j p_i f_i p_j f_j v_{ij}}{K} \\ &= \frac{p_i f_i v_i}{K}, \end{aligned} \tag{5.2.33}$$

where

$$v_i = \sum_j p_j f_j v_{ij}. \tag{5.2.34}$$

Summing, as before, we discover that $K = \tilde{w}$, where

$$\tilde{w} = \sum p_i f_i v_i. \tag{5.2.35}$$

Putting all this together, we have:

ZYGOTE ENUMERATION	ADULT ENUMERATION
$\Delta p_i = \frac{p_i(w_i - \bar{w})}{\bar{w}}$	$\Delta p_i = \frac{p_i(v_i f_i - \tilde{w})}{\tilde{w}}$
$w_i = \sum_j p_j v_{ij} f_{ij}$	$p_i f_i = \sum_j P_{ij} f_{ij}, \quad v_i = \sum_j p_j f_j v_{ij}$
$\bar{w} = \sum_i \sum_j p_i p_j v_{ij} f_{ij}$	$\tilde{w} = \sum_i \sum_j p_i p_j f_i f_j v_{ij}$

5.2.36

In these formulae, w_i is the weighted average of the product of v_{ij} and f_{ij}, while $v_i f_i$ is the product of the weighted averages of v_{ij} and f_{ij}. The gene frequencies refer, of course, to the frequencies at the time of enumeration.

As we saw in the numerical example, which involved rather strong selection (a lethal homozygote and a heterozygote with about 12% disadvantage), the results were approximated rather well by a model assuming that all the selection is through fertility differences and also by one assuming that all the selection is through viability differences of the heterozygote. Furthermore, a continuous approximation arrived at by treating the change by a differential equation also gave a good approximation to the results. This confirms our intuitive judgment that, unless selection is quite intense, there is not very much difference between the various models.

We shall sometimes use a discontinuous model and sometimes a continuous one, making the choice on the basis of which seems more natural or which is more manageable. We turn now to a discussion of the continuous model.

5.3 Continuous Model with Overlapping Generations

Many populations in nature have births and deaths occurring more or less continuously, with both reproduction and mortality at various ages. Under these circumstances the continuous models (models 2 and 4 of Chapter 1) are more appropriate. We shall develop formulae analogous to those of the preceding section.

Fitness is measured in terms of the Malthusian parameter, m. This is the rate of geometric increase such that the contribution of a class to the next generation is proportional to e^m. We expect equations of the general type discussed in Section 1.6.

Again, as in the last section, we concentrate on gene frequency changes. Because of mortality selection the Hardy–Weinberg (or other specified) ratios at birth will be changed as each cohort gets older. So there will ordinarily be no stage in which the entire population is in these ratios. One procedure that would at least partially mitigate this difficulty is to enumerate the population at birth, then get the proportion at different ages from life-table information; in other words, use genetic information only to predict the number of each genotype born at a particular instant.

However, we are mainly concerned with approximate results. So we assume that the conditions of model 2 are reasonably well met and that we are interested primarily in gene frequency changes.

As in Section 1.2 we let b and d stand for birth and death rates. For example, the genotype $A_i A_j$ would have a probability $b_{ij} \Delta t$ of giving birth and $d_{ij} \Delta t$ of dying during the infinitesimal time interval Δt. We let $m_{ij} = b_{ij} - d_{ij}$.

Let $2N$ stand for the total number of genes at the A locus in the diploid population and $2n_i$ for the number of A_i alleles. Then $p_i = n_i/N$. During the time interval Δt the increase in population number due to the contribution from $A_i A_i$ parents is $NP_{ii} m_{ii} \Delta t$, which is also a measure of increase in A_i genes due to $A_i A_i$ parents, since each parent contributes one A_i gene to each progeny. Likewise for any genotype $A_i A_j$, the increase in A_i genes due to contributions from this genotype is $NP_{ij} m_{ij} \Delta t$ (only half the total frequency of $A_i A_j$ is used since only half the contributed genes are A_i). Thus, when Δt becomes small,

$$\frac{dn_i}{dt} = \sum_j NP_{ij} m_{ij} = m_i n_i, \tag{5.3.1}$$

$$m_i = \frac{\sum_j NP_{ij} m_{ij}}{n_i} = \frac{\sum_j P_{ij} m_{ij}}{p_i}, \tag{5.3.2}$$

where m_i is the average fitness of gene A_i measured in Malthusian parameters. Likewise,

$$\frac{dN}{dt} = \sum_{ij} NP_{ij} m_{ij} = \overline{m}N, \tag{5.3.3}$$

$$\overline{m} = \sum_{ij} P_{ij} m_{ij} = \sum_i p_i m_i, \tag{5.3.4}$$

where $\overline{m}$ is the average fitness of the population, again measured in Malthusian parameters. A justification that the arithmetic mean of the individual m's is appropriate was given in equation 1.2.4.

From the ordinary rules of differentiation,

$$\begin{aligned} \frac{dp_i}{dt} &= \frac{d\left(\frac{n_i}{N}\right)}{dt} = \frac{N\frac{dn_i}{dt} - n_i\frac{dN}{dt}}{N^2} \\ &= \frac{Nn_i m_i - n_i N\overline{m}}{N^2} \\ &= p_i(m_i - \overline{m}), \end{aligned} \tag{5.3.5}$$

where $m_i - \overline{m}$ is the average excess in Malthusian parameters of the allele A_i.

The similarity of 5.3.5 and 5.2.6 is apparent. The continuous and discontinuous models become more nearly equivalent as the selective differences among the genotypes become smaller. If the w's are regarded as relative fitnesses and one genotype is assigned the value 1, then $\overline{w}$ is very nearly 1. Since $m_{ij} = \log_e w_{ij}$ (cf. Section 1.2),

$$w_i - \overline{w} \approx e^{m_i} - e^{\overline{m}} \to m_i - \overline{m},$$

and the two equations

$$\frac{dp_i}{dt} = p_i(m_i - \bar{m}) \quad \text{and} \quad \Delta p_i = \frac{p_i(w_i - \bar{w})}{\bar{w}}$$

become nearly equivalent.

We shall find that for some purposes one formula is more suitable than the other. Usually the qualitative results are very much the same so we shall often choose whichever model leads to the simplest results.

For two alleles and random mating, 5.3.5 can be written (in analogy with 5.2.7) approximately for slow selection as

$$\frac{dp_1}{dt} = p_1 p_2(m_1 - m_2). \tag{5.3.6}$$

This is only approximate because the adults may depart from Hardy–Weinberg ratios.

Now, consider the same special cases as before. With two alleles and random mating we can write the approximate formulae

$$m_1 = \frac{m_{11}p_1^2 + m_{12}\,p_1 p_2}{p_1} = m_{11}p_1 + m_{12}\,p_2\,,$$

$$m_2 = m_{12}\,p_1 + m_{22}\,p_2\,.$$

Substituting these into 5.3.6 gives

$$\frac{dp_1}{dt} = p_1 p_2[(m_{11} - m_{12})p_1 + (m_{21} - m_{22})p_2]\,. \tag{5.3.7}$$

a. No dominance, $m_{12} = (m_{11} + m_{22})/2$.

$$\frac{dp_1}{dt} = s p_1 p_2/2, \tag{5.3.8}$$

where $s = m_{11} - m_{22}$.

b. Dominant favored, $m_{11} = m_{12}$, A_1 dominant.

$$\frac{dp_1}{dt} = s p_1 p_2^2\,. \tag{5.3.9}$$

c. Recessive favored, $m_{12} = m_{22}$, A_1 recessive.

$$\frac{dp_1}{dt} = s p_1^2 p_2\,. \tag{5.3.10}$$

d. Asexual, haploid, or gametic selection.

$$\frac{dp_1}{dt} = s p_1 p_2\,. \tag{5.3.11}$$

The same general observations can be made as were made for the discontinuous model. The relative rates of gene frequency change are as before.

Notice that 5.3.11 is the equation of the logistic curve. This is apparent if we note the correspondence of p_1 with N/K and p_2 with $1 - N/K$ in equation 1.6.2. In integrated form, 5.3.11 becomes

$$p_t = \frac{1}{1 + \left(\frac{1 - p_0}{p_0}\right)e^{-st}}, \tag{5.3.12}$$

where p_t is the frequency of the favored gene at time t and p_0 is the initial frequency.

We can proceed as we did with equation 5.3.11 and write all the equations in integrated form. For our purposes it is more convenient to write them in the form $t = f(p)$, than with p as a function of time. Integrating and letting p_0 stand for the initial proportion (when $t = 0$) and $p_1 = p$, equations a, b, c, and d become:

a′. No dominance.

$$t = \frac{2}{s} \ln \frac{p_t(1 - p_0)}{p_0(1 - p_t)}. \tag{5.3.13}$$

b′. Dominant favored.

$$t = \frac{1}{s}\left[\ln \frac{p_t(1 - p_0)}{p_0(1 - p_t)} + \frac{1}{1 - p_t} - \frac{1}{1 - p_0}\right]. \tag{5.3.14}$$

c.′ Recessive favored.

$$t = \frac{1}{s}\left[\ln \frac{p_t(1 - p_0)}{p_0(1 - p_t)} - \frac{1}{p_t} + \frac{1}{p_0}\right]. \tag{5.3.15}$$

d.′ Asexual or haploid.

$$t = \frac{1}{s} \ln \frac{p_t(1 - p_0)}{p_0(1 - p_t)}. \tag{5.3.16}$$

In each case p designates the frequency of the favored gene, and t is the number of generations required to change the frequency from p_0 to p_t. The value of s is assumed to be small enough that the departure from Hardy–Weinberg proportions does not introduce serious errors.

The reason for writing the equations in this form is apparent. It emphasizes the fact that t is always inversely proportional to s. This is true, or approximately true, as long as s is not large enough to appreciably upset the

Hardy–Weinberg proportions. It is also true for slow selection with a discrete model. This can be seen from equations 5.2.14–5.2.17 by noting that when s is small Δp has about the same meaning as dp/dt, and $\bar{w}$ is approximately 1.

Table 5.3.1 shows the number of generations required to change the

Table 5.3.1. The rate at which gene and genotype frequencies change under selection.

NUMBER OF GENERATIONS REQUIRED WHEN $s = .001$ TO CHANGE GENE FREQUENCY	from .00001 to .01	from .01 to .5	from .5 to .99	from .99 to .99999
Asexual	6,921	4,592	4,592	6,921
No dominance	13,842	9,184	9,184	13,842
Dominant favored	12,563	5,595	102,595	99,896,918
Recessive favored	99,896,918	102,595	5,595	12,563
NUMBER OF GENERATIONS REQUIRED TO CHANGE THE PROPORTION OF DOMINANT (OR RECESSIVE) PHENOTYPES				
Dominant favored	6,920	4,819	11,664	309,780
Recessive favored	309,780	11,664	4,819	6,920
Asexual	6,921	4,592	4,592	6,921

gene or genotype frequency when $s = 0.001$. For any other value, say s', simply divide the numbers in the table by $s'/.001$. For example, with $s' = 0.01$, the times would be only 1/10 as large.

The values in this table are taken mostly from Haldane. For more extensive results and a variety of other cases, see his *The Causes of Evolution* (1932, 1966).

These equations are written as if s remains constant throughout the entire period of gene frequency change. That this should be strictly true is of course highly unlikely in any real case. But it gives us a general idea of the times involved in evolutionary change and the qualitative effects of dominance and recessivity. Furthermore, as emphasized in Section 1.6, equations of this type apply to changes in components of a population even though the whole population may be increasing, decreasing, or constant and under a variety of regulatory mechanisms. Again we see why it is usually convenient in population genetics to discuss proportions rather than numbers of individuals or of genes.

Obviously, some of the large values are unrealistic. 99,896,918 generations is probably longer than the life of the species, and certainly s is not going to be constant for that length of time. Furthermore, when the frequency of the gene is very near to 0 or to 1, random fluctuations in gene frequency can carry the gene to loss or fixation. A treatment of this problem taking chance factors into account has been given by Ewens (1967*d*). As might be expected, the values in the table are quite good for moderately large populations in the range of gene frequencies from 0.01 to 0.99, but the numbers at the tails of the distribution are often in serious error even in quite large populations. In particular, the largest values in the table are much too large.

Later, in Section 8.9, we shall consider the related problem of the length of time required for fixation of a mutant gene in a finite population.

5.4 The Effects of Linkage and Epistasis

When linkage and epistasis enter the problem, the situation immediately gets complicated. For one thing, whereas the Hardy–Weinberg ratios within each locus are attained within a single generation, gametic phase equilibrium is approached only asymptotically, as was discussed in Section 2.6. So we cannot be as free with the assumption of between-locus equilibrium as within a locus.

We can circumvent this to some extent by using the more generalized form of the Hardy–Weinberg principle that we discussed in Section 2.6. This states that the array of zygotic frequencies can be written as the square of the array of gametic frequencies. So we can deal with the problem by treating the chromosome, or the entire gamete, as the unit instead of the gene.

We will discuss the amount of linkage disequilibrium that is produced by selection with linkage and epistasis; in fact we shall find that there is "linkage" disequilibrium, even when there is no linkage. We shall also discuss the effect of epistasis on the rate of change by selection.

Many of the problems are still unsolved. It is not difficult to get reasonably good answers when linkage is loose and epistasis is weak. We can also treat the situation with very tight linkage as if the linked genes were a single gene. But the intermediate area, a small amount of crossing over and strong epistasis, is very difficult and no general theory exists. Individual examples have sometimes been worked out by computer. With more than two loci, the situation naturally gets still more complex.

Generation of Gametic Phase Disequilibrium with Epistasis

Assume that there are two loci, each with two alleles. There are therefore four gamete types. For the moment we shall treat the loci as if they were completely linked. We can then regard each chromosome as the formal equivalent of a gene and assign symbols in the same way. It is as if there were a single locus with four alleles.

CHROMOSOME	*ab*	*Ab*	*aB*	*AB*
FREQUENCY	p_1	p_2	p_3	p_4
AVERAGE FITNESS	m_1	m_2	m_3	m_4

Fitness is measured in Malthusian parameters; therefore the model is a continuous one. As before, we let b_{ij} stand for the birth rate and d_{ij} stand for the death rate of a particular genotype. The time, t, is measured in generations. The genotype *Ab*/*ab* would have a frequency $2p_1p_2$. It would have a probability $b_{12}\,\Delta t$ of giving birth and $d_{12}\,\Delta t$ of dying during the small time interval Δt. Then the fitness of *Ab*/*ab* is $m_{12} = b_{12} - d_{12}$.

In the absence of recombination we can use equation 5.3.5 of the previous section and write (assuming that there are Hardy–Weinberg proportions)

$$\frac{dp_i}{dt} = p_i(m_i - \overline{m}), \tag{5.4.1}$$

where

$$m_i = \sum_j p_j m_{ij}$$

and

$$\overline{m} = \sum_i p_i m_i = \sum_i \sum_j p_i p_j m_{ij}.$$

With crossing over the double heterozygote will contribute some gametes that are different from those it received from its parents. We measure crossing over by c, the recombination fraction. Unlinked genes will be treated as the special case when $c = 1/2$.

Consider the production of gametes of type *ab*. With homozygotes and single heterozygotes crossing over makes no difference, so their production of *ab* gametes is independent of c. There are $2p_2\,p_3$ *Ab*/*aB* double heterozygotes and a fraction $c/2$ of their gametes will be *ab*. There are $2p_1p_4$ *AB*/*ab* double heterozygotes and a fraction $(1 - c)/2$ of their gametes will be *ab*. Putting all this together, the change in the number of *ab* chromosomes in the time interval Δt will be

$$\begin{aligned} 2N[&p_1p_1b_{11} + p_1p_2\,b_{12} + p_1p_3\,b_{13} + p_1p_4\,b_{14}(1 - c) + p_2\,p_3\,b_{23}\,c \\ &- p_1p_1d_{11} - p_1p_2\,d_{12} - p_1p_3\,d_{13} - p_1p_4\,d_{14}]\Delta t \\ &= 2(Np_1m_1 - NbcD)\Delta t, \end{aligned} \tag{5.4.2}$$

where

$$D = p_1 p_4 - p_2 p_3 \qquad 5.4.3$$

and

$$b = b_{14} = b_{23},$$

on the entirely reasonable assumption that the two kinds of double heterozygotes have the same birth rate.

Meanwhile, the change in the number of all four kinds of gametes together in the same time interval is $2N\bar{m}\Delta t$. We then use the same method used in deriving 5.3.5 to get the rate of change in the proportions of the chromosome types, writing $dp_1/dt = d(n_1/N)/dt$, etc. This leads to

$$\frac{dp_1}{dt} = p_1(m_1 - \bar{m}) - cbD,$$

$$\frac{dp_2}{dt} = p_2(m_2 - \bar{m}) + cbD,$$

$$\frac{dp_3}{dt} = p_3(m_3 - \bar{m}) + cbD, \qquad 5.4.4$$

$$\frac{dp_4}{dt} = p_4(m_4 - \bar{m}) - cbD.$$

These equations were first derived by Kimura (1956).

There is another way of measuring departure from linkage equilibrium that is more useful than D for our purposes. Whereas D is the difference in frequency between the two types of heterozygotes, we define Z as the ratio of the two. Thus

$$Z = \frac{p_1 p_4}{p_2 p_3}, \qquad 5.4.5$$

and the relation between D and Z is given by

$$D = p_2 p_3(Z - 1). \qquad 5.4.6$$

Z has the property, first shown by Kimura (1965), of approaching a nearly constant value when mating is at random and recombination is large relative to the amount of epistasis. Such a slowly moving equilibrium is called quasi-equilibrium. The natural logarithm of Z is

$$\log_e Z = \ln Z = \ln p_1 - \ln p_2 - \ln p_3 + \ln p_4.$$

We are interested in determining what happens to the linkage disequilibrium as selection proceeds. To do this we inquire into the rate of change of log Z with time.

The time derivative of ln Z is

$$\frac{1}{Z}\frac{dZ}{dt} = \frac{1}{p_1}\frac{dp_1}{dt} - \frac{1}{p_2}\frac{dp_2}{dt} - \frac{1}{p_3}\frac{dp_3}{dt} + \frac{1}{p_4}\frac{dp_4}{dt}. \tag{5.4.7}$$

Substituting from 5.4.4 and simplifying leads to

$$\begin{aligned}\frac{1}{Z}\frac{dZ}{dt} &= m_1 - m_2 - m_3 + m_4 - cbD\left(\frac{1}{p_1} + \frac{1}{p_2} + \frac{1}{p_3} + \frac{1}{p_4}\right) \\ &= E - cbDP,\end{aligned} \tag{5.4.8}$$

where

$$E = m_1 - m_2 - m_3 + m_4 \tag{5.4.9}$$

and is a measure of epistasis, and

$$P = \sum_i \frac{1}{p_i}. \tag{5.4.10}$$

If c is larger than $|E|$ (more specifically, if cb is larger than $|E|$, but b is not far from 1 for a population of stable size), then the value of Z tends toward a value which is relatively stable. We start by writing 5.4.8 again and substituting for D from 5.4.6. This gives

$$\begin{aligned}\frac{dZ}{dt} &= EZ + cbZ(1 - Z)p_2 p_3 P \\ &= EZ + cb(1 - Z) - cb(1 - Z)^2(p_2 + p_3).\end{aligned} \tag{5.4.11}$$

Since $0 \le p_2 + p_3 \le 1$, we can write

$$\frac{dZ}{dt} \le EZ + cb(1 - Z), \tag{5.4.12}$$

$$\frac{dZ}{dt} \ge EZ + cb(1 - Z) - cb(1 - Z)^2 = EZ + cbZ(1 - Z). \tag{5.4.12a}$$

Consider first a haploid model, in which case the m_i's, and therefore E, are constants. We then integrate these two inequalities to obtain

$$Z \le \frac{cb}{cb - E}[1 + K_1 e^{-(cb-E)t}], \quad K_1 = Z_0\left(\frac{cb - E}{cb}\right) - 1, \tag{5.4.13}$$

$$Z \ge \frac{E + cb}{cb}\left[\frac{1}{1 + K_2 e^{-(E+cb)t}}\right], \quad K_2 = \frac{1}{Z_0}\left(\frac{E + cb}{cb}\right) - 1. \tag{5.4.13a}$$

Assume first that $cb > |E|$. Then as t gets large the value of Z is given by

$$\frac{cb + E}{cb} \le Z \le \frac{cb}{cb - E}. \tag{5.4.14}$$

For example, if $c = 10E$ and $b = 1$, Z comes to lie between 11/10 and 10/9. If E is negative, then Z becomes less than 1. For example, with $c = -10E$, Z comes to lie between 9/10 and 10/11.

This is equivalent to Felsenstein's (1965) statement that D becomes of the same sign as E. We can say that log Z becomes the same sign as E. Of course, if the selection is intense and c and E are small, the gene frequency change may be so rapid that the whole process is over (or nearly over) before the conditions of 5.4.14 are reached.

If one chromosome type (or pair of genes, if these are on independent chromosomes) is ultimately to be fixed, we can ask about the asymptotic behavior of Z. Assume, to be specific, that E is positive and that the ab or AB chromosome is favored by selection and will eventually prevail. As this stage is neared, $p_2 + p_3$ approaches 0, and 5.4.12 and 5.4.13 become equalities rather than inequalities. Thus, the limiting value of Z is given by

$$Z_\infty = \frac{cb}{cb - E}. \tag{5.4.15}$$

If Ab or aB is eventually fixed, then 5.4.12*a* and 5.4.13*a* are appropriate and

$$Z_\infty = \frac{cb + E}{cb}. \tag{5.4.15a}$$

This is illustrated in Figure 5.4.1. Three starting populations are assumed, one at linkage equilibrium ($Z = 1$), one with a large positive departure, and one with a large negative departure. E is taken to be 0.1 and $c = 0.5$. Therefore from 5.4.14 we would expect the value of Z to come eventually to lie between 1.20 and 1.25. Since the AB type is favored, it will ultimately be fixed in each of these examples, and the final value of Z will be 1.25, as indicated by the arrow in the figure.

Figure 5.4.2 shows an example that is the same in every regard except that the genes are linked with $c = 0.2$. In this case, the quasi-equilibrium will be between 1.5 and 2.0 with an eventual value of 2. The approach is slower because of the lesser recombination. Very roughly the rate of approach to quasi-equilibrium is the recombination rate, c.

Alternatively, if $cb < |E|$ there is not necessarily any limit on Z. Under these circumstances, when t gets large

$$\frac{cb + E}{cb} \leq Z, \qquad E/cb > 1, \tag{5.4.16}$$

$$0 \leq Z \leq \frac{cb}{cb - E}, \qquad E/cb < -1, \tag{5.4.16a}$$

with no upper limit in the first case and no positive lower limit in the second.

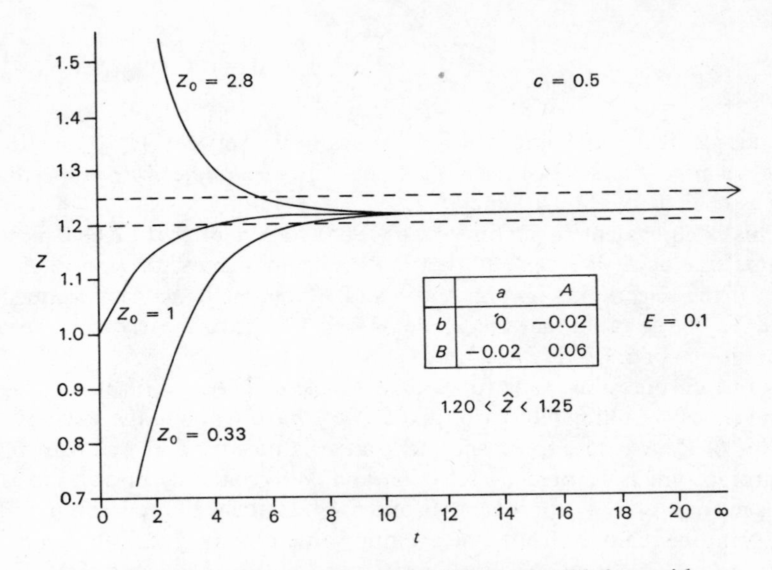

Figure 5.4.1. The attainment of quasi-linkage equilibrium with epistasis. This is a haploid model in which the fitnesses, measured in Malthusian parameters, are as given in the graph. Three starting values of Z are assumed. Note that value of Z comes to lie within the limits within about half a dozen generations and then very slowly changes to the upper limit. The recombination value is taken as 0.5.

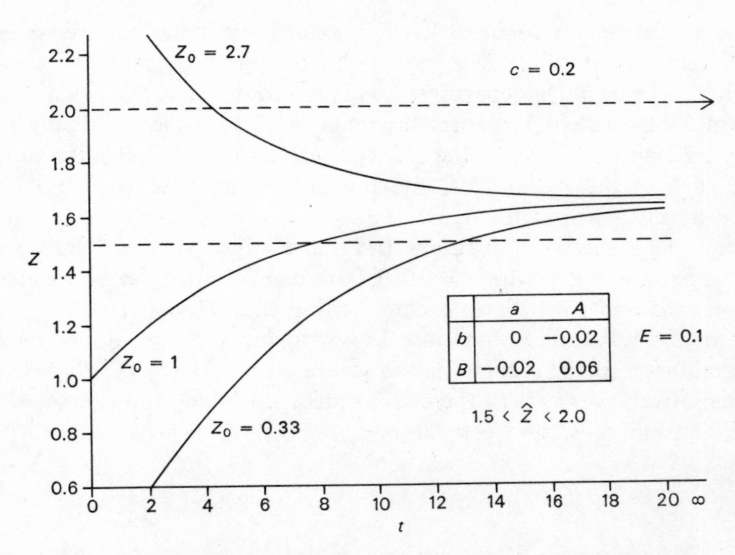

Figure 5.4.2. The same model except that the recombination value, c, is 0.2 instead of 0.5.

Further Analysis of the Diploid Case We can be specific only by considering particular epistatic models. As was pointed out in Section 3.6, four parameters are required to specify the possible kinds of 2-locus epistasis (two alleles at each locus). There we used I, J, K, and L; these parameters were chosen to bring out some of the effects of epistasis with inbreeding. For measuring the effects of directional selection another set of four parameters is more useful; at least it leads to a more transparent interpretation. The parameters were introduced by Fisher (1918). Our treatment follows Felsenstein (1965) whose paper may be consulted for a more detailed discussion. (Note: Felsenstein's Z is our $\log_e Z$.)
The four parameters are

$$\begin{aligned} E_1 &= m_{11} - m_{21} - m_{31} + m_{41}, \\ E_2 &= m_{12} - m_{22} - m_{32} + m_{42}, \\ E_3 &= m_{13} - m_{23} - m_{33} + m_{43}, \\ E_4 &= m_{14} - m_{24} - m_{34} + m_{44}. \end{aligned} \tag{5.4.17}$$

But, from the definition of m_i for diploids (5.4.1),

$$\begin{aligned} m_1 &= p_1 m_{11} + p_2 m_{12} + p_3 m_{13} + p_4 m_{14}, \\ m_2 &= p_1 m_{21} + p_2 m_{22} + p_3 m_{23} + p_4 m_{24}, \\ m_3 &= p_1 m_{31} + p_2 m_{32} + p_3 m_{33} + p_4 m_{34}, \\ m_4 &= p_1 m_{41} + p_2 m_{42} + p_3 m_{43} + p_4 m_{44}. \end{aligned} \tag{5.4.18}$$

Substituting these into 5.4.8 yields

$$\begin{aligned} \frac{1}{Z}\frac{dZ}{dt} &= p_1 E_1 + p_2 E_2 + p_3 E_3 + p_4 E_4 - cbDP \\ &= \bar{E} - cbDP, \end{aligned} \tag{5.4.19}$$

where

$$\bar{E} = p_1 E_1 + p_2 E_2 + p_3 E_3 + p_4 E_4. \tag{5.4.20}$$

We cannot proceed as we did with equations 5.4.13 and 5.4.13*a* because $\bar{E}$ is a function of the p_i's rather than a constant. However, we can obtain useful but less precise limits, when all the E_i's are less than cb in absolute value, by writing in analogy with 5.4.14

$$\frac{cb + E_{\min}}{cb} \leq Z \leq \frac{cb}{cb - E_{\max}}, \quad |E_i| < cb, \tag{5.4.21}$$

where $E_{\min}$ and $E_{\max}$ are the smallest and largest values of the E's.

Thus, as in the haploid model, if c is large relative to the absolute values of the E_i's, the value of Z eventually comes to lie within specified limits and the conditions for quasi-equilibrium are met. If the selective differences are small and c is large, the quasi-equilibrium is attained rapidly and the population is in this state when the gene frequencies are changing.

Table 5.4.1 shows the nature of the four epistatic parameters. The fitnesses of the genotypes are given in Malthusian parameters. Then, below, the table is decomposed into four parts, each comprising a quadrant of adjacent squares. If epistasis is "regular," the signs of the four epistatic parameters will tend to be the same. When E_i has the same sign in all four of the subtables, then the direction of departure from gametic phase balance is clear. When it is inconsistent and changes signs, then there are opposite tendencies and it is not obvious which will predominate. It is necessary to work out each case separately.

Table 5.4.1. The fitnesses in Malthusian parameters for the ten possible genotypes when there are two loci with two alleles each. Below are the parts of the figure that contribute to the four epistatic parameters. For example, the first small table is composed of the four upper left squares in the large table. E_i is positive when the average of the upper left and lower right square is greater than that of the other two.

	aa	*Aa*	*AA*
bb	m_{11}	m_{12}	m_{22}
Bb	m_{13}	$m_{14} = m_{23}$	m_{24}
BB	m_{33}	m_{34}	m_{44}

E_1		E_2		E_3		E_4	
m_{11}	m_{12}	m_{12}	m_{22}	m_{13}	m_{23}	m_{14}	m_{24}
m_{13}	m_{14}	m_{23}	m_{24}	m_{33}	m_{34}	m_{34}	m_{44}

Figure 5.4.3 shows the change in Z in a diploid model. The time period shown here is much longer than in the previous figures, and it can be seen that Z slowly changes during the period of selection. However, it remains between the limits set by 5.4.21, 1.00 and 1.29 during the entire time.

It is useful to go back to Table 3.7.1 at this point and see what could be predicted about selection in these systems. Notice that in models, 2, 6, and 8 the E_i's are all positive; therefore, selection in these cases will develop positive linkage disequilibrium ($Z > 1$). Model 5 is the opposite; all E_i's are negative. Therefore it would build up linkage disequilibrium in the opposite direction.

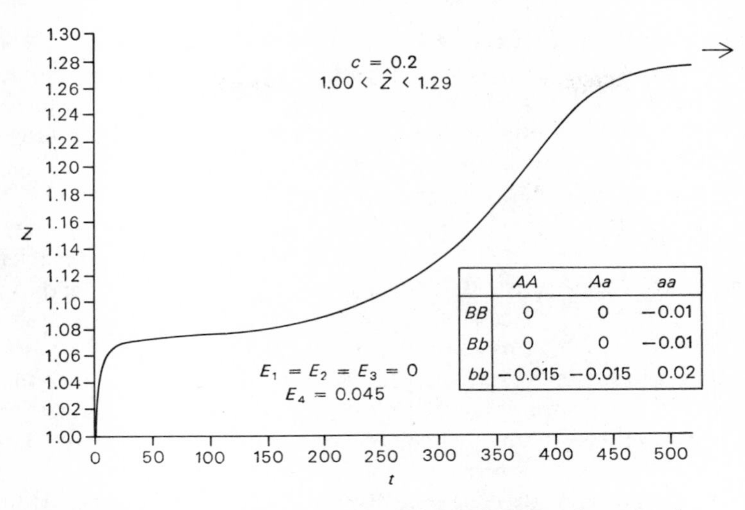

Figure 5.4.3. Another illustration of the fast approach to quasi-linkage equilibrium and the slow change in Z thereafter when epistasis is small and linkage is loose. This is a diploid model with recombination of 0.2 between the loci. The fitness and epistatic parameters are given in the figure.

Models 7 and 9 are mixed and the direction of departure depends on the gene frequencies.

Table 5.4.2 shows the change in Z when E is large and linkage is tight.

Table 5.4.2. Changes in chromosome frequencies and gametic phase unbalance (Z) in a diploid model with close linkage, $c = .01$. Fitnesses are: $w_{aabb} = 1.10$, $w_{A-bb} = w_{aaB-} = .95$, $w_{A-B-} = 1.00$.

GENERATION	CHROMOSOME FREQUENCIES			Z
	AB	$Ab = aB$	ab	
0	.250	.250	.250	1.00
10	.278	.218	.268	1.67
20	.294	.185	.336	2.87
40	.263	.114	.508	10.02
80	.025	.007	.961	483.

Notice that in this case there is no quasi-equilibrium and Z continues to change.

The Effect of Epistasis on the Rate of Natural Selection Unless linkage is close or epistasis strong (unless $\bar{E}/c$ is appreciable), the considerations that we have just discussed lead us to expect that epistasis would have rather little effect on selection. Although it is hard to make such a statement quantitative, we can give some indication of the direction of the effect. We shall return to the question of the magnitude of the effect in Section 5.7.

To be concrete, assume that the large letter genes generally increase fitness. Thus in Table 5.4.1, fitness tends to increase as we move from upper left to lower right. We have seen that if the E_i's are positive then Z tends to be greater than 1, that is, an excess of *ab* and *AB* chromosomes tends to develop. Those chromosomes that have the lowest fitness and those that have the highest, on the average, are the ones that are increased by epistasis. Thus the effect of positive epistasis is to make the population more variable. Since the rate of change by selection increases with the variability of the population, the effect of positive epistasis is to increase the speed of selection.

Conversely, if epistasis is negative (the E_i's are negative), the mediocre chromosomes accumulate in excess. The result is a less variable population and slower selection.

Linkage and the Establishment of Beneficial Mutant Combinations It is sometimes true that two or more genes that are individually deleterious interact to produce a beneficial effect. If these genes are newly arising mutants, and therefore rare, the population is in a troublesome situation. The essential situation is clear with a haploid model, so we shall consider this. Let the original population be *ab*. Suppose the fitnesses, w_{ab}, w_{Ab}, w_{aB}, and w_{AB} are in the ratio $1 : (1 - s_1) : (1 - s_2) : 1 + t$, where the s's and t are positive. If *ab* is the prevailing type, *Ab* and *aB* will be present in small numbers determined by their mutation rates and the magnitude of the s's. Very infrequently the *AB* type will arise, either by mutation or by recombination between *Ab* and *aB*. This individual will ordinarily mate with the most common type, *ab*. With free recombination less than half of the progeny will be *AB*, and unless this type possesses an enormous selective advantage it will not increase in frequency. Only if the fitness of the *AB* type is great enough to compensate for the loss of *AB* types through recombination will the genotype increase. However, if for any reason *AB* becomes common so that many of the matings are with individuals like themselves or with *Ab* and *aB*, then the genotype can increase. This type of situation has been extensively discussed by Haldane (1931), Wright (1959), and Bodmer and Parsons (1962).

One way in which the rare mutant combination might increase is when there is strong assortative mating. However, it is quite unlikely that the mutant combination that was favorable for some reason would also happen to predispose the individuals to mate assortatively.

The genes might also happen to be linked. If the rare *AB* individual mates with an *ab* type, which will usually be the case, the proportion of *AB* progeny will be proportional to $1 - c$, where c, as before, is the recombination fraction between the two loci. However, the *AB* type will increase from these matings if the extra fitness of the *AB* type is enough to compensate for this; that is, if $(1 + t)(1 - c) > 1$, or $t > c/(1 - c)$.

The conditions are actually a little less stringent, because some of the matings will be with *AB*, *aB*, or *Ab* types. Furthermore there is a small addition by recombination from $Ab \times aB$ matings. These do not change the direction of the inequality, however, so we can say that linkage of such intensity as to give a recombination of t or less is sufficient to insure the incorporation of the double mutant. Actually the same algebra works for diploids, where t is the advantage of the double heterozygote over the prevailing type (Bodmer and Parsons, 1962).

Whether such mutant combinations are important enough for this to be an important reason for having linkage is doubtful; but it does illustrate one situation where linkage introduces a qualitative change in the outcome, not just a change in rate.

5.5 Fisher's Fundamental Theorem of Natural Selection: Single Locus with Random Mating

Up to this point we have concentrated on the rates at which selection changes the genic and genotypic composition of a population. We are also interested in the way in which quantitative phenotypes are changed by selection. The selection may be natural or man-made. The character may be yield of a cereal, rate of gain in meat livestock, or whatever trait is of interest. In natural selection the trait of greatest significance is fitness—the capacity to survive and reproduce in the existing environment.

It is clear that natural selection tends to preserve those genes which, on the average, increase the fitness of their carriers, and therefore to increase the fitness itself. It is also clear that this is happening to all competing species at the same time, so that with increasing time it requires greater and greater intrinsic fitness for a species to survive the steadily increasing competition. We are looking for an expression describing the rate at which fitness is increasing, while realizing that this is not likely to be reflected in increasing rate of population growth because of limitations of the environment and competing species.

Intuition tells us that the rate at which selection changes the fitness will be related to the variability of the population. We shall show, as Fisher

(1930) first did, that the rate of increase in fitness is equal to the genic variance in fitness. If equations of the form of 5.3.5 are applicable, the theorem is precise. But we are primarily interested in it as an approximation applicable to a wide variety of population models.

We start first with a simple case: a single locus, two alleles, no environmental effect, and random mating.

GENOTYPE	A_1A_1	A_1A_2	A_2A_2
FREQUENCY	p_1^2	$2p_1p_2$	p_2^2
FITNESS	$m_{11} = \bar{m} + a_{11}$	$m_{12} = \bar{m} + a_{12}$	$m_{22} = \bar{m} + a_{22}$
GENIC VALUE	$\bar{m} + 2\alpha_1$	$\bar{m} + \alpha_1 + \alpha_2$	$\bar{m} + 2\alpha_2$

The average of the deviations from the mean must be 0; hence, as shown in 4.1.3,

$$2p_1^2\alpha_1 + 2p_1p_2(\alpha_1 + \alpha_2) + 2p_2^2\alpha_2 = 0$$

or, since $p_1 + p_2 = 1$,

$$p_1\alpha_1 + p_2\alpha_2 = 0. \tag{5.5.1}$$

The genic variance was shown earlier (4.1.5) to be

$$V_g = 2(p_1\alpha_1^2 + p_2\alpha_2^2) \tag{5.5.2}$$

with the obvious extension to multiple alleles,

$$V_g = 2\sum p_i\alpha_i^2. \tag{5.5.3}$$

The quantity α_i is called the average effect of the allele A_i. The term was introduced by Fisher (1930).

Alternatively, we can define α as the average effect of substituting A_1 for A_2, in which case $\alpha = \alpha_1 - \alpha_2$ and

$$V_g = 2p_1p_2\alpha^2. \tag{5.5.4}$$

However, we prefer 5.5.2 because it leads so naturally to an extension to multiple alleles.

In the model we are considering,

$$\bar{m} = p_1^2m_{11} + 2p_1p_2m_{12} + p_2^2m_{22}.$$

The rate of change in the average fitness will be

$$\begin{aligned}\frac{d\bar{m}}{dt} &= 2(p_1m_{11} + p_2m_{12})\frac{dp_1}{dt} + 2(p_1m_{12} + p_2m_{22})\frac{dp_2}{dt} \\ &= 2m_1\frac{dp_1}{dt} + 2m_2\frac{dp_2}{dt},\end{aligned} \tag{5.5.5}$$

where m_1 is the average fitness of the allele A_1 and m_2 is that of A_2 (5.3.7). But from 5.3.5,

$$\frac{dp_1}{dt} = p_1(m_1 - \overline{m}), \qquad \frac{dp_2}{dt} = p_2(m_2 - \overline{m}). \tag{5.5.6}$$

Hence

$$\begin{aligned}\frac{d\overline{m}}{dt} &= 2p_1m_1(m_1 - \overline{m}) + 2p_2m_2(m_2 - \overline{m}) \\ &= 2p_1(m_1 - \overline{m})^2 + 2p_2(m_2 - \overline{m})^2.\end{aligned} \tag{5.5.7}$$

The equivalence of the two expressions above may be shown by expanding the second to give (ignoring the factor 2)

$$\begin{aligned}&p_1(m_1 - \overline{m})(m_1 - \overline{m}) + p_2(m_2 - \overline{m})(m_2 - \overline{m}) \\ &\quad = p_1m_1(m_1 - \overline{m}) + p_2m_2(m_2 - \overline{m}) - \overline{m}(p_1m_1 + p_2m_2 - \overline{m}).\end{aligned}$$

But, from 5.3.4, $\overline{m} = p_1m_1 + p_2m_2$; so the quantity in the last parentheses is 0.

Now, note from 5.3.7 that $m_1 = p_1m_{11} + p_2m_{12}$, and therefore

$$\begin{aligned}m_1 - \overline{m} &= p_1m_{11} + p_2m_{12} - \overline{m} \\ &= p_1a_{11} + p_2a_{12} \\ &= \alpha_1\end{aligned} \tag{5.5.8}$$

from 4.1.9. Substituting into 5.5.7 gives

$$\frac{d\overline{m}}{dt} = 2p_1\alpha_1^2 + 2p_2\alpha_2^2 = V_g \tag{5.5.9}$$

from 5.5.2.

This example illustrates a special case of Fisher's Fundamental Theorem of Natural Selection. In his words: "The rate of increase in fitness of any organism at any time is equal to its genetic variance at that time." Fisher's genetic variance is what we are calling the genic or additive genetic variance.

When the genotypes are in random proportions, as in the model we have been discussing, the genic variance in fitness can be written in another form that is sometimes useful. Combining 5.3.5, 5.5.5, and 5.5.7, we can write

$$V_g = 2\left[\frac{1}{p_1}\left(\frac{dp_1}{dt}\right)^2 + \frac{1}{p_2}\left(\frac{dp_2}{dt}\right)^2\right], \tag{5.5.10}$$

or for multiple alleles,

$$V_g = 2\sum \frac{1}{p_i}\left(\frac{dp_i}{dt}\right)^2. \tag{5.5.11}$$

The discrete generation analog of 5.5.9, related to 5.2.6 in the same way that 5.3.5 is related to the continuous formula, is

$$\Delta\bar{w} = \frac{V_g}{\bar{w}}\left[1 + \frac{p_1 p_2}{2\bar{w}}(w_{11} - 2w_{12} + w_{22})\right]. \tag{5.5.12}$$

This can be derived easily as follows (Li, 1967*a*, *b*). We note first that with random mating

$$\bar{w} = w_{11}p_1^2 + 2w_{12}\,p_1 p_2 + w_{22}\,p_2^2 \qquad (p_2 = 1 - p_1),$$

$$\frac{d\bar{w}}{dp_1} = 2[p_1(w_{11} - w_{12}) + p_2(w_{12} - w_{22})]$$

$$= 2(\alpha_1 - \alpha_2) = 2\alpha,$$

by 5.5.8 when $a_{11} = w_{11} - \bar{w}$, etc.

$$\frac{d^2\bar{w}}{dp_1^2} = 2(w_{11} - 2w_{12} + w_{22}),$$

$$\Delta p_1 = \frac{p_1 p_2}{2\bar{w}}\frac{d\bar{w}}{dp_1}$$

from 5.2.13. We now expand $\Delta\bar{w} = \bar{w}' - \bar{w}$ into a Taylor series,

$$\Delta\bar{w} = \Delta p_1 \frac{d\bar{w}}{dp_1} + \frac{(\Delta p_1)^2}{2}\frac{d^2\bar{w}}{dp_1^2},$$

which is exact because $\bar{w}$ is a quadratic function of the p's and all derivatives beyond the second are 0. Substituting,

$$\Delta\bar{w} = \frac{2p_1 p_2 \alpha^2}{\bar{w}} + \frac{p_1^2 p_2^2 \alpha^2}{\bar{w}^2}(w_{11} - 2w_{12} + w_{22})$$

$$= \frac{V_g}{\bar{w}}\left[1 + \frac{p_1 p_2}{2\bar{w}}(w_{11} - 2w_{12} + w_{22})\right],$$

as was to be shown.

Unless selection is strong the quantity in brackets is very nearly 1. This is especially true if several loci contribute to the trait and dominance is in different directions, for then the quantity in parentheses is sometimes positive and sometimes negative with considerable cancellation resulting. So, to a good approximation in many instances,

$$\Delta\bar{w} = \frac{V_g}{\bar{w}}, \tag{5.5.13}$$

or perhaps more meaningfully,

$$\frac{\Delta \bar{w}}{\bar{w}} = \frac{V_g}{\bar{w}^2}.$$

That $\Delta \bar{w}$ is always positive may be shown by noting that

$$\begin{aligned} p_1 p_2(w_{11} - 2w_{12} + w_{22}) &= p_1 w_{11} + p_1 w_{22} - p_1^2 w_{11} \\ &\quad - 2p_1 p_2 w_{12} - p_2^2 w_{22} \\ &= \bar{w}_I - \bar{w}, \end{aligned}$$

where $\bar{w}_I$ and $\bar{w}$ are the means of a completely inbred population and a randomly mating population. Substituting this into 5.5.12 gives

$$\Delta \bar{w} = \frac{V_g}{2\bar{w}} \left(\frac{\bar{w} + \bar{w}_I}{\bar{w}} \right). \qquad 5.5.14$$

Since all the quantities in this expression are positive, or 0, the fitness must always increase or be at an equilibrium.

We shall not extend this to multiple alleles but the conclusion is still correct—the fitness with random mating and constant w_{ij}'s can never decrease. For references, see Mulholland and Smith (1959), Scheuer and Mandel (1959), and Kingman (1961, 1961*a*).

The interpretation of Fisher's theorem has been a matter of considerable discussion. Clearly, if the fitness of a genotype is defined as the intrinsic rate of increase, the average fitness cannot increase indefinitely, as the theorem would seem to say. The population growth rate may be 0, or its size may even decrease. One way in which this might happen is if the linkage relations change through recombination so that less favorable chromosomes increase in frequency despite natural selection. Another is if the fitnesses of individual genotypes change. For example, the fitness of a genotype may depend on its frequency or the frequency of other genotypes, or the environment may be deteriorating so that all genotypes become less fit.

One interpretation of the theorem is to say that it measures the rate of increase of fitness that would occur if the gene frequency changes took place, but nothing else changed. The theorem thus gives the effect of gene frequency changes alone, isolated from the other things that are happening. Fitness becomes a rather abstract quantity that continually increases while the population size is roughly stabilized by the various factors that cause the environment to change. This is what Fisher seems to be saying in at least one passage:

> For the majority of organisms, therefore, the physical environment may be regarded as constantly deteriorating, whether the climate, for example, is becoming warmer or cooler, moister or drier, and this will tend, in the majority

of species, constantly to lower the average value of m, the Malthusian parameter of the population increase. Probably more important than the changes in climate will be the evolutionary changes in progress in associated organisms. As each organism increases in fitness, so will its enemies and competitors increase in fitness; and this will have the same effect, perhaps in a much more important degree, in impairing the environment, from the point of view of each organism concerned. Against the action of Natural Selection in constantly increasing the fitness of every organism, at a rate equal to the genetic variance in fitness which that population maintains, is to be set off the very considerable item of the deterioration of its inorganic and organic environment.

Alternatively, we can interpret fitness more concretely as the actual rates of change and introduce corrective terms to include the effects of linkage and epistasis, nonrandom mating, changes in the fitnesses of individual genotypes, and the effects of overcrowding and deterioration of the environment. This we shall do in the next section.

5.6 The Fundamental Theorem: Nonrandom Mating and Variable Fitnesses

To derive the principle in its most general form, including the effects of epistasis and multiple alleles, is beyond the scope of this book. However, we shall give some indication of the nature of the extension to more complex situations at the end of this section.

Fisher's (1930, 1958) treatment of the subject is recondite. A clearer discussion of the circumstances under which the principle works was given in Fisher (1941). For a straightforward, general derivation, see Kimura (1958).

We shall remove the assumptions of random mating and constant fitness for each genotype. However, we shall continue to consider only a single locus with two alleles.

We assign values as follows:

GENOTYPE	A_1A_1	A_1A_2	A_2A_2
FREQUENCY	P_{11}	$2P_{12}$	P_{22}
FITNESS	$m_{11} = \bar{a} + a_{11}$	$m_{12} = \bar{a} + a_{12}$	$m_{22} = \bar{a} + a_{22}$
GENIC VALUE	$G_{11} = \bar{a} + 2\alpha_1$	$G_{12} = \bar{a} + \alpha_1 + \alpha_2$	$G_{22} = \bar{a} + 2\alpha_2$

We follow the same procedure as was used in Sections 4.1 and 4.2. There we measured departure from random-mating proportions by the inbreeding coefficient, f. Here we find it more convenient to measure it in another way, by a measure θ to be defined later. Although the inbreeding coefficient is a natural measure of the effects of inbreeding, random gene frequency drift, and some types of assortative mating, it is not the most natural measure to use when relating fitness change to the genic variance. For the rate of increase in fitness to be equal to the genic variance it is necessary, not that f be constant, but that θ be constant.

The phenotypic variance in fitness is

$$V = P_{11}a_{11}^2 + 2P_{12}a_{12}^2 + P_{22}a_{22}^2. \quad 5.6.1$$

However, we would expect selection to be more closely related to the genic variance, as defined in Section 4.1. In this case it is

$$V_g = P_{11}(2\alpha_1)^2 + 2P_{12}(\alpha_1 + \alpha_2)^2 + P_{22}(2\alpha_2)^2. \quad 5.6.2$$

The α's are to be determined by least squares, following the same procedures as before. We choose the α's so as to minimize the quantity

$$\begin{aligned} Q &= P_{11}(m_{11} - G_{11})^2 + 2P_{12}(m_{12} - G_{12})^2 + P_{22}(m_{22} - G_{22})^2 \\ &= P_{11}(a_{11} - 2\alpha_1)^2 + 2P_{12}(a_{12} - \alpha_1 - \alpha_2)^2 + P_{22}(a_{22} - 2\alpha_{22})^2. \end{aligned}$$

To minimize Q, we differentiate with respect to α_1 and α_2 and equate to 0, giving

$$\frac{\partial Q}{\partial \alpha_1} = -4P_{11}(a_{11} - 2\alpha_1) - 4P_{12}(a_{12} - \alpha_1 - \alpha_2) = 0,$$

$$\frac{\partial Q}{\partial \alpha_2} = -4P_{12}(a_{12} - \alpha_1 - \alpha_2) - 4P_{22}(a_{22} - 2\alpha_2) = 0.$$

Rewriting, we obtain

$$P_{11}a_{11} + P_{12}a_{12} = (P_{11} + P_{12})\alpha_1 + P_{11}\alpha_1 + P_{12}\alpha_2, \quad 5.6.3$$

$$P_{12}a_{12} + P_{22}a_{22} = (P_{12} + P_{22})\alpha_2 + P_{12}\alpha_1 + P_{22}\alpha_2. \quad 5.6.4$$

But, $P_{11}a_{11} + P_{12}a_{12} = p_1a_1$, where a_1 is the average excess of the allele A_1. Likewise $P_{12}a_{12} + P_{22}a_{22} = p_2a_2$. After making these substitutions in the left sides of the equations above, we then multiply the first by α_1 and the second by α_2. Adding the two equations we get

$$p_1a_1\alpha_1 + p_2a_2\alpha_2 = 2P_{11}\alpha_1^2 + P_{12}(\alpha_1 + \alpha_2)^2 + 2P_{22}\alpha_2^2.$$

The quantity on the right is half the genic variance (cf. 5.6.2). Therefore

$$V_g = 2p_1a_1\alpha_1 + 2p_2a_2\alpha_2. \quad 5.6.5$$

The extension to multiple alleles and to multiple loci is direct, so we can write, in general,

$$V_g = 2\sum\sum p_ia_i\alpha_i, \quad 5.6.6$$

where the inner sum is over the alleles at a locus and the outer is over the loci.

A comparison of 5.6.5 with 5.5.2, which was derived on the assumption of random mating, illustrates the fact that with Hardy–Weinberg ratios there is no distinction between a and α. With inbreeding, $a = \alpha(1 + f)$, where f is the inbreeding coefficient. This is clear from the comparison of 4.1.6 with 4.2.2. The difference between average excess and average effect of a gene is discussed in a different way by Fisher (1958, pp. 34–35).

We now proceed to write an expression for the rate of change in fitness,

$$\bar{m} = P_{11}m_{11} + 2P_{12}m_{12} + P_{22}m_{22}. \qquad 5.6.7$$

If the m's are no longer regarded as constant,

$$\frac{d\bar{m}}{dt} = P_{11}\frac{dm_{11}}{dt} + 2P_{12}\frac{dm_{12}}{dt} + P_{22}\frac{dm_{22}}{dt} + m_{11}\frac{dP_{11}}{dt} + 2m_{12}\frac{dP_{12}}{dt} + m_{22}\frac{dP_{22}}{dt}. \qquad 5.6.8$$

The first three terms to the right of the equality sign are the average of the rate of change of the fitnesses of the individual genotypes weighted by the frequency of each genotype. Thus

$$P_{11}\frac{dm_{11}}{dt} + 2P_{12}\frac{dm_{12}}{dt} + P_{22}\frac{dm_{22}}{dt} = \overline{\frac{dm}{dt}}. \qquad 5.6.9$$

Letting $m_{ij} = \bar{a} + a_{ij}$, the last three terms in 5.6.8 are

$$\bar{a}\left(\frac{dP_{11}}{dt} + 2\frac{dP_{12}}{dt} + \frac{dP_{22}}{dt}\right) + a_{11}\frac{dP_{11}}{dt} + 2a_{12}\frac{dP_{12}}{dt} + a_{22}\frac{dP_{22}}{dt}. \qquad 5.6.10$$

But, the quantity in parentheses must be 0, since it is equal to the total rate of change of the frequency of all genotypes. Because an increase in the proportion of one genotype must always be accompanied by a compensating decrease in the others, the net change is 0.

We measure the departure from random-mating proportions by the quantity

$$\theta_{ij} = \frac{P_{ij}}{p_i p_j},$$

which is 1 for Hardy–Weinberg ratios. Substituting $\theta_{ij} p_i p_j$ for P_{ij} in the least-squares equations 5.6.3 and 5.6.4 gives

$$p_1^2\theta_{11}a_{11} + p_1p_2\theta_{12}a_{12} = p_1\alpha_1 + \theta_{11}p_1^2\alpha_1 + \theta_{12}p_1p_2\alpha_2,$$

or

$$p_1\theta_{11}a_{11} + p_2\theta_{12}a_{12} = \alpha_1 + \theta_{11}p_1\alpha_1 + \theta_{12}p_2\alpha_2,$$

and in the same way

$$p_1\theta_{12}a_{12} + p_2\theta_{22}a_{22} = \alpha_2 + \theta_{12}p_1\alpha_1 + \theta_{22}p_2\alpha_2. \qquad 5.6.11$$

Likewise, since $P_{ij} = \theta_{ij} p_i p_j$,

$$\frac{dP_{11}}{dt} = 2\theta_{11} p_1 \frac{dp_1}{dt} + p_1^2 \frac{d\theta_{11}}{dt},$$

$$\frac{dP_{12}}{dt} = \theta_{12} p_1 \frac{dp_2}{dt} + \theta_{12} p_2 \frac{dp_1}{dt} + p_1 p_2 \frac{d\theta_{12}}{dt},$$

$$\frac{dP_{22}}{dt} = 2\theta_{22} p_2 \frac{dp_2}{dt} + p_2^2 \frac{d\theta_{22}}{dt}.$$

Substituting these into 5.6.8 and making use of equations 5.6.9, 5.6.10, and 5.6.11 leads to

$$\begin{aligned} \frac{d\bar{m}}{dt} = \overline{\frac{dm}{dt}} &+ 2(\alpha_1 + \theta_{11} p_1 \alpha_1 + \theta_{12} p_2 \alpha_2) \frac{dp_1}{dt} \\ &+ 2(\alpha_2 + \theta_{12} p_1 \alpha_1 + \theta_{22} p_2 \alpha_2) \frac{dp_2}{dt} \\ &+ a_{11} p_1^2 \frac{d\theta_{11}}{dt} + 2a_{12} p_1 p_2 \frac{d\theta_{12}}{dt} + a_{22} p_2^2 \frac{d\theta_{22}}{dt}. \end{aligned} \qquad 5.6.12$$

Now

$$\theta_{11} \frac{dp_1}{dt} + \theta_{12} \frac{dp_2}{dt} = -p_1 \frac{d\theta_{11}}{dt} - p_2 \frac{d\theta_{12}}{dt}, \qquad 5.6.13$$

and

$$\theta_{12} \frac{dp_1}{dt} + \theta_{22} \frac{dp_2}{dt} = -p_1 \frac{d\theta_{12}}{dt} - p_2 \frac{d\theta_{22}}{dt}. \qquad 5.6.14$$

This can be shown by noting that $p_1\theta_{11} + p_2\theta_{12} = 1$, which on differentiation leads to 5.6.13. Substituting 5.6.13 and 5.6.14 into 5.6.12 leads to

$$\begin{aligned} \frac{d\bar{m}}{dt} = \overline{\frac{dm}{dt}} &+ 2\alpha_1 \frac{dp_1}{dt} + 2\alpha_2 \frac{dp_2}{dt} + P_{11} D_{11} \frac{d\phi_{11}}{dt} + 2P_{12} D_{12} \frac{d\phi_{12}}{dt} \\ &+ P_{22} D_{22} \frac{d\phi_{22}}{dt}, \end{aligned}$$

where $D_{ij} = a_{ij} - \alpha_i - \alpha_j$, $\theta_{ij} p_i p_j = P_{ij}$, and $\phi_{ij} = \log_e \theta_{ij}$ (i.e., $d\theta/\theta = d\phi$). More concisely,

$$\frac{d\bar{m}}{dt} = 2\alpha_1 \frac{dp_1}{dt} + 2\alpha_2 \frac{dp_2}{dt} + \overline{\frac{dm}{dt}} + \overline{D \frac{d\phi}{dt}}.$$

Finally, from 5.3.5,

$$\frac{dp_i}{dt} = p_i(m_i - \overline{m}) = p_i a_i,$$

and since $2 \sum p_i a_i \alpha_i = V_g$,

$$\frac{d\overline{m}}{dt} = V_g + \overline{\frac{dm}{dt}} + \overline{D\frac{d\phi}{dt}}. \tag{5.6.15}$$

Although our derivation has considered only a single locus with two alleles, the formula is readily extended as was shown by Kimura (1958).

The three terms on the right can be interpreted as follows:

1. V_g is the genic variance.

2. $\overline{dm/dt}$ is the average rate of change in the fitness of the individual genotypes. If these are constant, as is frequently assumed, this term drops out. In a natural population the environment is continually deteriorating, primarily because of the evolutionary improvement of competing species. This term can be thought of as a measure of such deterioration.

3. The third term measures the effect of gene interaction and departure from Hardy–Weinberg ratios. The quantity D_{ij} is the difference between the fitness of a genotype and its best linear estimate. It is thus a measure of the effect of dominance. The quantity $\phi = \log_e \theta$ is a measure of departure from Hardy–Weinberg proportions, being the log of the ratio of the actual frequency of a genotype to its Hardy–Weinberg expectation.

Therefore, the term $\overline{Dd\phi/dt}$ will be 0 if there is no dominance or if the genotypes are in Hardy–Weinberg ratios. It will also be true under more general conditions. It is not necessary that ϕ be 0, only that its derivative be 0. In other words, there must be a constant amount of departure from random proportions, measured by ϕ.

Numerical examples where the mean fitness decreases because the third term in 5.6.15 (extended to include epistasis) is negative have been given by Kojima and Kelleher (1961) and Kimura (1965).

Notice that when the gene frequencies are changing, constant inbreeding coefficient (f) is not the same as constant ϕ. It is readily verified that when f is a constant other than 0, the third term in 5.6.15 is not 0, except in the absence of dominance (Crow and Kimura, 1956; Turner, 1967).

We have mentioned before that, even with random mating, there will generally be departures from random genotypic proportions if there is selection. The Hardy–Weinberg proportions will be found at the zygote stage, but if there is differential mortality this will ordinarily lead to deviations from random expectations. We should not expect this to alter greatly the fundamental theorem however. If gene frequency changes are slow, the proportions of each genotype in the population will be a constant multiple of the random-mating zygotic proportions; that is to say, the θ_{ij}'s will be approximately constants. Thus the condition $d \log \theta/dt = 0$ will be approximately correct with random mating, so the rate of fitness change is still given approximately by the genic variance.

For discrete generations the formula is similar. With random mating, the equation is

$$\Delta\bar{w} = \frac{V_g}{\bar{w}}\left[1 + \frac{p_1 p_2}{2\bar{w}}(w_{11} - 2w_{12} + w_{22})\right] + \overline{\Delta w}$$
$$= \frac{V_g}{\bar{w}} + \overline{\Delta w} \qquad 5.6.16$$

approximately. For a general discussion, see Wright (1955) and Li (1967).

Somewhat more generally, removing the assumption of random mating, we have

$$\Delta\bar{w} = \frac{V_g}{\bar{w}} + \sum \Delta P_{ij} D_{ij} + \overline{\Delta w}. \qquad 5.6.17$$

For a derivation of this, see Kempthorne (1957, p. 358). The close relationship of this to 5.6.15 can be seen by noting that

$$\overline{D\frac{d\phi}{dt}} = \sum P_{ij} D_{ij} \frac{d}{dt} \log \theta_{ij} = \sum D_{ij} \frac{d}{dt} P_{ij}, \qquad 5.6.18$$

which is equivalent to the middle right term in 5.6.17 for the continuous case.

The extension of 5.6.15 to multiple alleles and multiple loci with epistasis is not difficult in principle, but involves considerable algebra. It is given by Kimura (1958). The equation may be written

$$\frac{d\bar{m}}{dt} = V_g + \overline{\frac{dm}{dt}} + \sum \overline{\varepsilon \frac{d\phi}{dt}}, \qquad 5.6.19$$

where ε is a measure of the fitness of a genotype, expressed as a deviation from the linear least-squares expectation, and therefore is a measure of both dominance and epistasis, and $\phi = \log \theta$ is a measure of the departure from random-mating expectations of the frequency of each genotype.

An explicit expression for this term with two loci with linkage and epistasis, but with random mating, is given in the next section. It will be shown that with random mating the effects of linkage disequilibrium and epistasis effectively cancel, as long as epistasis is weak and linkage not too close, so that the last term in 5.6.19 becomes unimportant.

There has been considerable discussion in recent years as to whether Fisher's theorem is "exact." It is clear from the derivation as we have given it that the rate of change in fitness is equal to the genic variance if (1) the assumptions of population model 2 apply, that is, equations of the form $dp/dt = p(m - \bar{m})$ are appropriate, (2) the genotypic fitness coefficients are constant, and (3) the departures from random-mating proportions, as measured by θ, are constant.

However, no model can ever be an exact description of nature. We have not considered such complexities as arise when one pays careful attention to the pattern of deaths at different ages, the various ways in which mating combinations occur, different fitnesses of the genotypes in the two sexes, and such. We are mainly concerned with rather slow selection in which the details are of less influence. Fisher's principle, we believe, summarizes a great deal of biological complexity in a simple and elegant statement that relates population genetics to statistical theory in a very useful way. But of course, as a description of nature it (like all other descriptions) is only an approximation.

The effect of overcrowding can be included in the formulation. We let the actual rate of increase in a population of size N be given by M, so that

$$\frac{1}{N}\frac{dN}{dt} = M = \bar{m} - \psi(N)$$

where $\psi(N)$ is a function describing the reduction in population growth rate with overcrowding.

$$\begin{aligned} \frac{dM}{dt} &= \frac{d\bar{m}}{dt} - \frac{d\psi}{dt} \\ &= \frac{d\bar{m}}{dt} - \frac{d\psi}{dN}\frac{dN}{dt} \\ &= \frac{d\bar{m}}{dt} - MN\frac{d\psi}{dN} \\ &= V_g + \overline{\frac{dm}{dt}} + D\overline{\frac{d\phi}{dt}} - MN\frac{d\psi}{dN}, \end{aligned} \qquad 5.6.20$$

where $d\psi/dN$ describes the increase in resistance to population growth with overcrowding.

Fisher (1930, 1958, p. 46) writes

$$\frac{dM}{dt} = W - D - \frac{M}{C}. \qquad 5.6.21$$

Our formulation is similar. W is V_g, D is our $\overline{dm/dt}$, and $1/C$ is equivalent to $N(d\psi/dN)$, treated as a constant.

5.7 The Fundamental Theorem: Effects of Linkage and Epistasis

In most natural populations the assumption of random mating is a good approximation to reality. Furthermore, most pairs of genes are unlinked or loosely linked. In so far as a single locus is concerned, regardless of the number of alleles, the Hardy–Weinberg ratios are attained in a single generation; so this is a reasonable assumption. On the other hand, gametic phase equilibrium is approached only asymptotically, and with epistatic interactions there is permanent "linkage" disequilibrium even for unlinked genes, as was discussed in Section 5.4. For a discussion of this problem, see Bodmer and Parsons (1962), Kojima and Schaffer (1964), Lewontin (1964), Wright (1965), and Felsenstein (1965).

Therefore, it is necessary to consider the effects of linkage disequilibrium when there is epistasis, even when there is random mating. It is not immediately clear what this does to the term $\overline{Dd\phi/dt}$ (or rather, the corresponding term in the more general equation when epistasis is included, 5.6.19).

However, we shall now demonstrate that there is a remarkable property of randomly mating populations that is true unless linkage is very close or epistasis is very strong. This is that the amount of linkage disequilibrium maintained by selection is just enough to nullify the epistatic contribution to the variance, so that the genic variance remains the correct measure of the rate of change in population fitness.

We are assuming the same model as in Section 5.4, but we shall repeat the basic elements now for convenience.

Assume that there are two loci, each with two alleles. There are thus four gamete types. We shall assume that the loci are linked with recombination frequency c. If the loci are on independent chromosomes, then we regard this as the special case where $c = .5$. We assign symbols for the chromosomes, frequency, and average fitnesses as follows:

CHROMOSOME (GAMETE) TYPE	ab	Ab	aB	AB
FREQUENCY	p_1	p_2	p_3	p_4
AVERAGE FITNESS	m_1	m_2	m_3	m_4

Fitnesses are measured in Malthusian parameters. For example m_1 is the average fitness of all genotypes containing chromosome ab, weighted by their frequency and the number of ab chromosomes carried (1 or 2).

We make use of the gametic or marginal values in Table 5.7.1. All the necessary equations can be expressed in terms of these values because, under random mating, the zygotic frequencies are given by the square of the gametic frequencies. In this way we effect a considerable saving in algebraic manipulations.

We now show that under quasi-linkage equilibrium the Fisher theorem is correct. We would expect this to be the case, for quasi-equilibrium provides a nearly constant departure from gametic phase balance; in other words, the third term on the right side in equation 5.6.19 should disappear. However, we shall demonstrate this explicitly to illustrate how the epistatic-variance component may be isolated.

Table 5.7.1. Frequencies and fitnesses of the various zygotic combinations of two linked loci. The gametic values are given along the margins, and the zygotic values in the main body of the table.

	ab	*Ab*	*aB*	*AB*	
ab	p_1^2 m_{11}	p_1p_2 m_{12}	p_1p_3 m_{13}	p_1p_4 m_{14}	Frequency Fitness
Ab	p_2p_1 m_{21}	p_2^2 m_{22}	p_2p_3 m_{23}	p_2p_4 m_{24}	Frequency Fitness
aB	p_3p_1 m_{31}	p_3p_2 m_{32}	p_3^2 m_{33}	p_3p_4 m_{34}	Frequency Fitness
AB	p_4p_1 m_{41}	p_4p_2 m_{42}	p_4p_3 m_{43}	p_4^2 m_{44}	Frequency Fitness
Totals Averages	p_1 m_1 m	p_2 m_2 $m+\alpha$	p_3 m_3 $m+\beta$	p_4 m_4 $m+\alpha+\beta$	Frequency Fitness Genic value

$$\bar{m} = \sum_i \sum_j p_i p_j m_{ij} = \sum_i p_i m_i$$

$$m_i = \sum_j p_j m_{ij}$$

$$m_{ij} = m_{ji}$$

$$p_A = p_2 + p_4, \; p_B = p_3 + p_4 \text{ (gene frequencies)}$$

$$D = p_1p_4 - p_2p_3, \qquad Z = \frac{p_1p_4}{p_2p_3}$$

From Table 5.7.1 the mean fitness is

$$\bar{m} = \sum_i \sum_j p_i p_j m_{ij} = \sum_i p_i m_i \tag{5.7.1}$$

and the average fitness of the ith chromosome is

$$m_i = \frac{\sum_j p_i p_j m_{ij}}{p_i} = \sum_j p_j m_{ij}. \qquad 5.7.2$$

The rate of change in fitness is

$$\begin{aligned} \frac{d\bar{m}}{dt} &= \sum_i \sum_j m_{ij}\left(p_i \frac{dp_j}{dt} + p_j \frac{dp_i}{dt}\right) \\ &= \sum_j \frac{dp_j}{dt} \sum_i p_i m_{ij} + \sum_i \frac{dp_i}{dt} \sum_j p_j m_{ij} \\ &= \sum_j m_j \frac{dp_j}{dt} + \sum_i m_i \frac{dp_i}{dt} \\ &= 2 \sum m_i \frac{dp_i}{dt}, \end{aligned} \qquad 5.7.3$$

since the two expressions on the right are the same.

Substituting into 5.7.3 from 5.4.4 and rearranging gives

$$\frac{1}{2}\frac{d\bar{m}}{dt} = \sum p_i m_i^2 - \bar{m} \sum p_i m_i - cbD(m_1 - m_2 - m_3 + m_4). \qquad 5.7.4$$

But, from 5.7.1 and 5.7.2,

$$\sum p_i m_i = \bar{m}. \qquad 5.7.5$$

Recalling 5.4.9 and using 5.7.5, 5.7.4 becomes

$$\frac{1}{2}\frac{d\bar{m}}{dt} = \sum p_i m_i^2 - \bar{m}^2 - cbDE. \qquad 5.7.6$$

However, $\sum p_i m_i^2 - \bar{m}^2$ is the variance of the m_i's, which we shall call the marginal or gametic variance, V_{gam}. Refer to Table 5.7.1. This includes components from epistasis, but not from dominance. Thus

$$\frac{d\bar{m}}{dt} = 2(V_{\text{gam}} - cbDE). \qquad 5.7.7$$

We now use the familiar least-squares procedure to estimate the additive component of the gametic variance. The quantity to be minimized is, from Table 5.7.1,

$$\begin{aligned} Q = {} & p_1(m_1 - m)^2 + p_2(m_2 - m - \alpha)^2 + p_3(m_3 - m - \beta)^2 \\ & + p_4(m_4 - m - \alpha - \beta)^2. \end{aligned} \qquad 5.7.8$$

Taking derivatives and equating to 0 to minimize gives:

$$(1)\quad \frac{1}{2}\frac{\partial Q}{\partial m} = -p_1(m_1 - m) - p_2(m_2 - m - \alpha) - p_3(m_3 - m - \beta) - p_4(m_4 - m - \alpha - \beta) = 0$$

$$(2)\quad \frac{1}{2}\frac{\partial Q}{\partial \alpha} = -p_2(m_2 - m - \alpha) - p_4(m_4 - m - \alpha - \beta) = 0,$$

$$(3)\quad \frac{1}{2}\frac{\partial Q}{\partial \beta} = -p_3(m_3 - m - \beta) - p_4(m_4 - m - \alpha - \beta) = 0.$$

Let

$$p_1(m_1 - m) = K.$$

Then, after subtracting (3) from (1),

$$-p_2(m_2 - m - \alpha) = K.$$

Likewise, after subtracting (2) from (1),

$$-p_3(m_3 - m - \beta) = K.$$

Therefore, also

$$p_4(m_4 - m - \alpha - \beta) = K. \qquad 5.7.9$$

Dividing these four equations by p_1, p_2, p_3, and p_4, respectively, and adding, we obtain

$$m_1 - m_2 - m_3 + m_4 = K\sum\frac{1}{p_i}, \qquad 5.7.10$$

or, using 5.4.9 and 5.4.10,

$$E = KP. \qquad 5.7.11$$

Q, which is the nonadditive component of the marginal or gametic variance, is, from 5.7.8 and 5.7.9,

$$Q = K^2\sum\frac{1}{p_i} = K^2P. \qquad 5.7.12$$

Thus, $V_{\text{gam}} = V_a + K^2P$, where V_a is the additive component of the gametic variance. Substituting into 5.7.7 yields

$$\begin{aligned}\frac{d\bar{m}}{dt} &= 2(V_a + K^2P - cbDE)\\ &= 2\left(V_a + \frac{E^2}{P} - cbDE\right) \qquad 5.7.13\\ &= 2\left(V_a + \frac{E}{PZ}\frac{dZ}{dt}\right)\end{aligned}$$

from 5.4.8.

We showed earlier (Section 5.4) that with loose linkage ($cb > |E|$) Z attains quasi-equilibrium. So, when there is quasi-equilibrium the last term in 5.7.13 vanishes.

We can write 5.7.13 in two ways, recalling that the genic variance V_g is twice the additive component of the gametic variance, V_a, since we are assuming random mating, and that $2V_{\text{gam}} = V_g + \frac{1}{2}V_{AA}$ (see 4.1.19).

$$\frac{d\bar{m}}{dt} = V_g + \frac{2E}{PZ}\frac{dZ}{dt}, \tag{5.7.14}$$

$$\frac{d\bar{m}}{dt} = V_g + \frac{1}{2}V_{AA} - 2cbDE. \tag{5.7.15}$$

The first equation is appropriate when there is quasi-equilibrium and the second is more useful for strong epistasis and tight linkage, although both are correct and in fact fully equivalent. However, under the one circumstance the last term of the first equation tends to 0 and in the other the last term in the second equation does.

We therefore see that with free recombination natural selection operates in such a way that just enough gametic phase disequilibrium is generated ($-cbDE$) to cancel exactly the epistatic variance (K^2P). Therefore the rate of change in fitness is given by the genic component of the variance, as the Fisher principle says.

From this, it would appear that parent-offspring correlations do not involve any significant epistatic terms if the population is near quasi-linkage equilibrium. In fact, the assumption of quasi-equilibrium is probably closer to the truth than the conventional simple assumption of gametic phase equilibrium. It is therefore quite possible that the epistatic components of variance that are sometimes added to the expressions for parent-offspring correlations (e.g., Kempthorne, 1957) may be making the expressions less accurate than when only the genic variance is used.

A numerical example is shown in Table 5.7.2. The slow change in linkage disequilibrium, as measured by Z, is evident. Notice how, after the first few generations, the change in fitness is given very closely by the genic variance, despite great changes in the genic and the total gametic variance. However, in the very first generation, the change is given more accurately by the total gametic variance (doubled), because the proper level of linkage disequilibrium has not been attained.

A similar example, but with less recombination, is shown in Figure 5.7.1. Again, as soon as quasi-equilibrium is attained, the rate of change in fitness is given by the genic variance.

With very tight linkage the rate of change is given more closely by $2V_{\text{gam}}$ than by V_g. The reason is obvious; with very little crossing over the

Table 5.7.2. The attainment of quasi-equilibrium with random mating and free recombination. The genotypic fitnesses are: *aa bb*, 1.02; *aa B-*, .985; *A- bb*, .99; *A- B-*, 1.00. The recombination frequency is 1/2. The population starts in gametic phase equilibrium, and disequilibrium is very slowly generated. Discrete generation model. (From Kimura, 1965.)

GENERATION	CHROMOSOME FREQUENCIES AB	Ab	aB	ab	LINKAGE DISEQUILIBRIUM Z	CHANGE IN FITNESS $\times 10^5$	GENIC VARIANCE V_g $\times 10^5$	TWICE GAMETIC VARIANCE $2V_{gam}$ $\times 10^5$
0	.200	.200	.300	.300	1.000	2.93	0.73	2.91
10	.201	.204	.291	.304	1.028	.66	.66	2.91
50	.196	.224	.267	.313	1.029	.46	.45	2.85
100	.186	.243	.244	.328	1.030	.37	.37	2.98
200	.146	.259	.210	.385	1.035	.93	.92	4.24
300	.073	.229	.162	.536	1.049	5.72	5.64	10.09
400	.005	.085	.050	.860	1.081	15.29	15.42	16.78
500	.000	.006	.003	.991	1.095	1.68	1.72	1.73

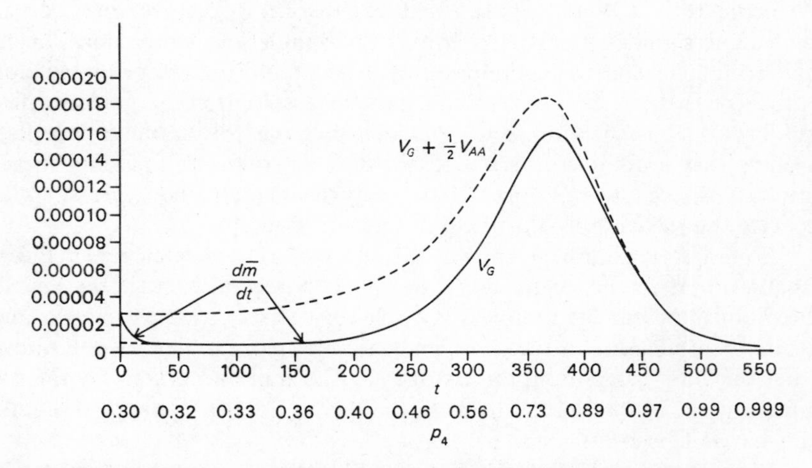

Figure 5.7.1. The genic variance (lower line) and the genic variance plus half the additive $\times$ additive epistatic variance (upper line). Note that the rate of change in fitness, after the attainment of quasi-equilibrium, follows almost exactly the lower line. The example is the same as Figure 5.4.3.

chromosomes behave as units. Notice, though, that in neither case does the dominance variance contribute to the rate of change in fitness.

Table 5.7.3 shows an example of the latter kind. The data are the same

Table 5.7.3. The same population as was shown in Table 5.4.2, showing that with very low recombination and high epistasis the gametic variance (×2) is a better predictor of the rate of change in fitness than the genic variance.

GENERATION	CHANGE IN FITNESS $\times 10^5$	GENIC VARIANCE $\times 10^5$	TWICE GAMETIC VARIANCE $\times 10^5$
0	31.84	.00	31.25
10	38.76	1.46	41.72
20	53.56	8.79	61.31
40	144.62	79.88	168.73
80	82.82	76.65	99.29
120	1.41	1.35	1.71

as in Table 5.4.2. Recombination is low ($c = .01$) and epistasis is large, so that E/c is much larger than 1. As is apparent in the table, the doubled gametic variance is a better predictor than the genic. (The doubling is simply because the diploid individual with random mating is derived from two independent gametes.) For several generations this is the best predictor, but later this turns out to be an overestimate. Notice, though, that there is no contribution from the dominance variance to the prediction—it is already an overestimate without dominance.

To return to the more usual situation—after all, most pairs of genes are not closely linked—we should not conclude that epistasis can have no effect on the rate of fitness change just because this is governed by the genic component. Epistasis, by altering the total variability in the population, can change all components of the variance, including the genic. This change could be in either direction.

Wright (1965 and many earlier papers) has emphasized that the most important kind of epistasis is probably the kind that arises from the fact that intermediate values for most metrical traits are optimum from the standpoint of fitness. For example, individuals far from the mean for size in either direction have lower survival and fertility rates. This would produce, generally, epistasis such as to make E negative. Hence the variance would be reduced. This would then make the population slower to respond to directional selection; so we can say that in general the effect of epistasis on such metrical traits is probably such as to slow somewhat the rate of progress by selection, progress being measured by the rate at which the mean for the trait is changed.

We have considered only that part of the total epistasis that is caused by interaction of pairs of genes. There are of course possible higher order interactions. It is generally believed that for quantitative traits these are not very important. Indeed, for one particular model, studied by all three pioneers—Haldane, Fisher, and Wright—they make no contribution. This is a model in which the fitness decreases in proportion to the square of the distance of the metrical trait from the mean. Wright (1935) showed that with this model all of the epistatic variance is contained in 2-locus interactions and that higher order interactions make no contribution. This is not true for most other models, but this case suggests that the magnitude of the contribution of higher order effects may be small.

One other possibility is being studied currently by several investigators. This is the effect of several linked genes on the chromosome. Even though the outer members may be only loosely linked with each other, there may be associations because they are both linked with intermediate genes. So far there is no general theory for this, although computer simulations suggest that there may be such effects, and that departures from gametic phase balance may thereby be generated (Lewontin, 1964).

Two final points connect this section with earlier discussions. One is that twice the gametic variance, as we have used it here, includes half the additive × additive component of the epistatic variance, V_{AA} (see Section 4.1).

The other is that, because of inevitable departures from Hardy–Weinberg proportions after selection, the genic variance is not exactly the sum of the additive portions of the two gametic variances. However, for the same reason as given before, if gene frequencies are changing slowly the departure from random proportions is roughly constant and the genic variance gives the rate of change in fitness.

A Note on Terminology We have used the word fitness as a synonym for the selective or reproductive value of a genotype. It may be either an absolute value, measured by the number of progeny per parent, or it may be relative to some reference genotype. For a discontinuous model we have used w and for a continuous model, m, as measures of fitness. The average fitness, $\bar{w}$ or $\bar{m}$, is a function of the individual genotypic fitnesses and the frequencies of the various genotypes, and again may be absolute or relative.

The word is also used more restrictively. In his recent discussions Wright (1955, 1969) used the word fitness for that population function that increases with time according to Fisher's Fundamental Theorem of Natural Selection. It is not in general the same as $\log_e \bar{w}$ or m. As we have seen, $d\bar{m}/dt = V_g$ only under certain conditions such as constant individual m's or w's, random mating or constant departure therefrom, loose linkage, and quasi-linkage equilibrium. In fact $d\bar{m}/dt$ may even be negative.

There is value in defining a quantity which always increases and which measures the theoretical evolutionary improvement in a population brought about by gene-frequency change, despite the fact that this is hardly ever experienced as a corresponding increase in population numbers because of such things as competing species, overcrowding, frequency-dependent selective values, and changes in the mating system and linkage relations. Such a quantity, called fitness by Fisher and the fitness function by Wright, always increases at a rate equal to the genic variance. For a discussion of this quantity and the way in which it can be defined, see Wright (1969, p. 121).

5.8 Thresholds and Truncation Selection for a Quantitative Trait

Some situations in nature approach a model in which nothing happens until a certain quantity is attained and then all values above this show the phenomenon. For example, it may be that doses of a drug may be harmless up to a point and that beyond this point they become harmful. There may be in some instances similar kinds of gene action; an effect of some kind appears only when a certain number of harmful genes are present. This has been suggested by Lerner as the basis for some congenital anomalies in chicks.

The existence of sharp thresholds in nature is open to discussion, but it is clear that this model is approached rather closely in some kinds of breeding experiments. All the individuals above a certain level are saved and reproduce; the rest are culled.

Since mass directional selection must be very similar to natural selection, we should be able to adapt the theory developed in these chapters to this subject.

We assume that the trait under consideration is determined by a large number of factors, genetic and environmental, the number being large enough that the effect of any one locus is small relative to the total variability. The quantitative trait (e.g., yield, Y) is assumed to be normally distributed (see A.5). If the data are not normally distributed they can often be transformed to be approximately normal. For example, one might work with the logs, or the square roots, or some other function of the original measurements that would have a symmetrical distribution approximating the normal.

The breeder saves for reproduction a certain fraction, S, of the population. All of these lie above the cutoff or truncation point, C, on the abscissa.

This is illustrated in Figure 5.8.1. This method of truncation selection is equivalent to a threshold at point C. We assume that all the animals or plants that are saved are equally fertile, or more generally that there is no correlation between fertility and the yield beyond the truncation point. Let z be the ordinate of the normal curve at the truncation point.

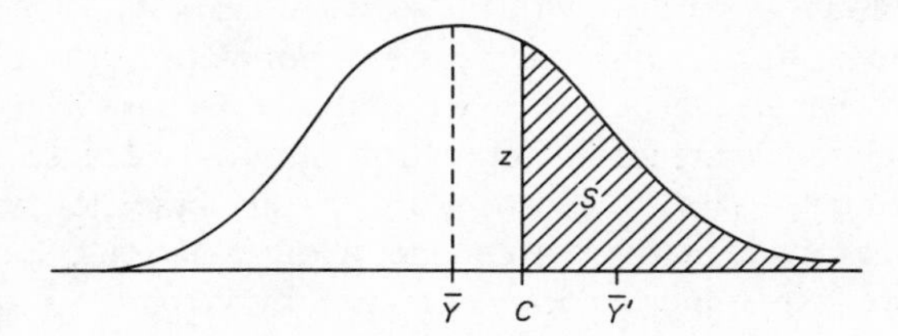

Figure 5.8.1. Truncation selection. The individuals in the shaded area are saved for breeding; the ones to the left are culled.

Assuming a normal distribution, the frequency $f(Y)$ of individuals with yield Y is given by

$$f(Y) = \frac{1}{\sigma\sqrt{2\pi}} \exp\left[-\frac{(Y-\bar{Y})^2}{2\sigma^2}\right] \tag{5.8.1}$$

where σ^2 $(=V_t)$ is the total variance in yield from all causes, genetic and environmental.

The proportion saved, S, is related to the cutoff point, C, by

$$S = \int_C^\infty f(Y)\,dY. \tag{5.8.2}$$

The mean value of the group chosen to be parents is given by

$$\bar{Y}' = \frac{1}{S}\int_C^\infty Yf(Y)\,dY. \tag{5.8.3}$$

Then, the difference between the mean of the selected parents and the population average (called the selection differential, I) is

$$\begin{aligned} I = \bar{Y}' - \bar{Y} &= \frac{1}{S}\left[\int_C^\infty Yf(Y)\,dY - \bar{Y}S\right] \\ &= \frac{1}{S}\left[\int_C^\infty Yf(Y)\,dY - \bar{Y}\int_C^\infty f(Y)\,dY\right] \\ &= \frac{1}{S}\int_C^\infty (Y-\bar{Y})f(Y)\,dY. \end{aligned} \tag{5.8.4}$$

We now let

$$\frac{Y - \bar{Y}}{\sigma} = y, \qquad \frac{C - \bar{Y}}{\sigma} = c, \qquad dy = \frac{dY}{\sigma}.$$

Then

$$I = \frac{\sigma}{S}\frac{1}{\sqrt{2\pi}}\int_c^{\infty} ye^{\frac{-y^2}{2}}\,dy = \frac{\sigma}{S}\frac{1}{\sqrt{2\pi}}e^{\frac{-c^2}{2}}$$

$$= \frac{\sigma}{S}z, \qquad 5.8.5$$

where z is the ordinate at the cutoff point.

Common sense tells us that the selection differential, I (or $\bar{Y}' - \bar{Y}$ in the original units), increases as the selection intensity is increased. But perhaps it exaggerates the amount by which the effectiveness of selection, as measured by I, increases with a decreasing proportion of individuals selected to be parents. Some numerical values are given below. These can be obtained from a table of areas and ordinates of the normal integral.

S	$\frac{I}{\sigma}$
.80	.34
.50	.80
.40	.97
.20	1.20
.10	1.76
.01	2.66

Thus, increasing the intensity by 10-fold (i.e., decreasing S from .1 to .01) changes I by only 2.66/1.76 or about 50%.

For example, suppose a plant breeder would like to select a strain for resistance to two independent virus diseases. Suppose also that the fertility is such that he can afford to discard 99% of the plants each generation, saving 1%. In two generations he could select intensively for one disease the first generation, selecting the most resistant 1%, and do the same thing for the second disease the second generation. He would thus have a selection differential, I, of 2.66σ for each disease. On the other hand, he could save the most resistant 10% for each disease each generation (thus saving 1% of the plants, as before). He would have a differential of 1.76σ for each disease each generation, or 3.52σ in two generations. Thus he could expect to make considerably more progress by selecting simultaneously for the two traits than by selecting one at a time.

We now want to relate this to the genetic composition of the selected individuals.

Let $\bar{Y}$ be the mean yield of the population and Y_1 be the mean yield of genotypes carrying A_1. The frequency distribution of the yield of these genotypes, assuming that the effect of this gene is small relative to the total variance, σ^2, is

$$f_1(Y) = \frac{1}{\sigma\sqrt{2\pi}} \exp\left[-\frac{(Y - Y_1)^2}{2\sigma^2}\right]. \tag{5.8.6}$$

Let

$$y_1 = \frac{Y_1 - \bar{Y}}{\sigma}, \qquad y = \frac{Y - \bar{Y}}{\sigma}, \tag{5.8.7}$$

and note that

$$\begin{aligned}(Y - Y_1)^2 &= [(Y - \bar{Y}) - (Y_1 - \bar{Y})]^2 \\ &= (Y - \bar{Y})^2 - 2(Y - \bar{Y})(Y_1 - \bar{Y}) + (Y_1 - \bar{Y})^2.\end{aligned}$$

Then

$$\begin{aligned}f_1(Y) &= \frac{1}{\sigma\sqrt{2\pi}} \exp\left[-\frac{y^2}{2} + y_1 y - \frac{y_1^2}{2}\right] \\ &= f(Y) \exp\left[y_1 y - \frac{y_1^2}{2}\right] \qquad (5.8.8) \\ &= f(Y)(1 + y_1 y + \cdots\cdots);\end{aligned}$$

since y_1 is assumed to be very small, terms involving higher powers of y_1 can be neglected.

The proportion saved of genotypes carrying A_1 is given by

$$\begin{aligned}S_1 &= \int_C^\infty f_1(Y)\,dY = \int_C^\infty f(Y)\,dY + y_1 \int_C^\infty \frac{Y - \bar{Y}}{\sigma} f(Y)\,dY \\ &= S + y_1 \frac{I}{\sigma} S\end{aligned} \tag{5.8.9}$$

(from 5.8.2 and 5.8.4).

Now the fitness of A_1 relative to the population as a whole is

$$\frac{W_1}{\bar{W}} = \frac{S_1}{S}$$

or

$$\frac{W_1 - \bar{W}}{\bar{W}} = \frac{S_1 - S}{S} = y_1 \frac{I}{\sigma}. \tag{5.8.10}$$

The average yield, $\overline{Y}$, in a randomly mating population is

$$\overline{Y} = p_1^2 Y_{11} + 2p_1 p_2 Y_{12} + p_1^2 Y_{22}.$$

For small increments, so that products of the Δp's can be neglected,

$$\begin{aligned} \Delta \overline{Y} &= 2(p_1 Y_{11} + p_2 Y_{12})\Delta p_1 + 2(p_1 Y_{12} + p_2 Y_{22})\Delta p_2 \\ &= 2(Y_1 \Delta p_1 + Y_2 \Delta p_2), \end{aligned} \qquad 5.8.11$$

where Y_i is the average yield of allele A_i.

From 5.2.1,

$$\Delta p_1 = \frac{p_1(W_1 - \overline{W})}{\overline{W}}, \qquad \Delta p_2 = \frac{p_2(W_2 - \overline{W})}{\overline{W}}. \qquad 5.8.12$$

Substituting into 5.8.12 from 5.8.10 gives

$$\Delta p_1 = \frac{p_1 y_1 I}{\sigma}, \qquad \Delta p_2 = \frac{p_2 y_2 I}{\sigma}. \qquad 5.8.13$$

But, recalling that $y_i = (Y_i - \overline{Y})/\sigma$, substituting 5.8.13 into 5.8.11 yields

$$\begin{aligned} \Delta \overline{Y} &= \frac{2I}{\sigma}\left[\frac{p_1 Y_1(Y_1 - \overline{Y})}{\sigma} + \frac{p_2 Y_2(Y_2 - \overline{Y})}{\sigma}\right] \\ &= \frac{2I}{\sigma^2}[p_1(Y_1 - \overline{Y})^2 + p_2(Y_2 - \overline{Y})^2]. \end{aligned} \qquad 5.8.14$$

But $Y_i - \overline{Y} = a_i$, the average excess of the allele A_i, so

$$\Delta \overline{Y} = \frac{2I}{\sigma^2}[p_1 a_1^2 + p_2 a_2^2], \qquad 5.8.15$$

which, from 5.5.2, recalling that with random proportions there is no difference between the average excess, a, and the average effect, α, yields

$$\Delta \overline{Y} = I\frac{V_g}{V_t}, \qquad 5.8.16$$

where $V_t = \sigma^2$, the total variance.

V_g/V_t is the heritability, as we have discussed before (Section 4.1). This is the familiar formula of the animal or plant breeder: The expected increase in yield from selection is the heritability multiplied by the selection differential.

As a numerical example, suppose the heritability of 6 months weight in swine is 0.25. Suppose the herd average is 100 pounds. If we select as parents a male who weighs 130 pounds at this age and a female who weighs 110, their average is 120. (In practice the weights would be adjusted for sex differences.) Then the progency would be expected to exceed the population mean by 0.25(120 − 100) or 5 pounds; so we would expect the litter to average 105 pounds.

Equation 5.8.16 is the same as derived earlier by an entirely different procedure (see 4.4.9). We have done it again in this way to show its close relation to the Fundamental Theorem of Natural Selection. Whether such terms as V_{AA} should be included with V_g depends on whether the population is in quasi-linkage equilibrium.

5.9 A Maximum Principle for Natural Selection

The effectiveness of selection in changing the composition of a population depends on the genic variance of the population, as we have shown in the previous sections. We can then ask: Are the gene frequency changes such as to produce the maximum increase in fitness that is possible within the restrictions imposed by the amount of genic variance? In simple cases the answer is yes. Furthermore, we can use this principle to deduce the equations for change in the genetic composition of the population.

In its simplest form the Maximum Principle says: Natural selection acts so as to maximize the rate of increase in the average fitness of the population. Stated more explicitly: With a given amount of genic variance in a short time interval the gene frequencies change in such a way that the increase in population average fitness is maximized.

This principle holds under the same conditions as the Fundamental Theorem of Natural Selection. These are: (1) Fitness is measured in Malthusian parameters; (2) the gene frequencies change continuously such that equations of the form of 5.3.5 are applicable; (3) the genotypes are in random proportions or the departure therefrom is constant (i.e., the rightmost term in 5.6.19 vanishes), and (4) the fitness coefficients are constant (i.e., the next to right term in 5.6.19 vanishes).

Consider now a single locus with genotypes in random-mating proportions. The average fitness (from 5.3.4) is

$$\overline{m} = \sum_{i,j} m_{ij} p_i p_j = \sum_i m_i p_i . \qquad 5.9.1$$

For any set of small changes in gene frequencies, δp_1, δp_2, etc., the change in mean fitness is

$$\begin{aligned} \delta\overline{m} &= \sum_{i,j} m_{ij} p_j \,\delta p_i + \sum_{i,j} m_{ij} p_i \,\delta p_j \\ &= 2 \sum_i m_i \,\delta p_i . \end{aligned} \qquad 5.9.2$$

To demonstrate the principle we shall show that maximization of $\delta\overline{m}$ leads to the correct equations of gene frequency change. This is maximized in the usual way, by equating the derivative to 0 subject to the appropriate conditions. There are two of these. The first is

$$\sum \delta p_i = 0, \tag{5.9.3}$$

which follows, since

$$\sum p_i = \sum (p_i + \delta\, p_i) = 1. \tag{5.9.4}$$

The second condition is that the genic variance is (momentarily) constant. With random proportions the genic variance is

$$V_g = 2 \sum p_i(m_i - \overline{m})^2 \tag{5.9.5}$$

from 5.5.7 and 5.5.9. Alternatively,

$$\tfrac{1}{2}V_g = \sum p_i m_i^2 - \overline{m}^2. \tag{5.9.6}$$

From 5.5.11

$$\frac{1}{2} V_g = \sum \frac{1}{p_i} \left(\frac{dp_i}{dt}\right)^2. \tag{5.9.7}$$

So, we write as our second condition

$$\sum \frac{1}{p_i} (\delta p_i)^2 = C(\delta t)^2, \tag{5.9.8}$$

where C is a constant, equal to $\sum p_i(m_i - \overline{m})^2$.

The most direct procedure for maximizing $\delta\overline{m}$ is by the use of Lagrange multipliers (A.10). We let

$$\psi = \delta\overline{m} + A\left[\sum \frac{1}{p_i} (\delta p_i)^2 - C(\delta t)^2\right] + B \sum \delta p_i, \tag{5.9.9}$$

where A and B are undetermined constants. Equating $\partial\psi/\partial(\delta p_i)$ to 0 gives

$$2m_i + \frac{2A}{p_i}\, \delta p_i + B = 0. \tag{5.9.10}$$

We now multiply each term by p_i and sum this over all alleles, $i = 1, 2, \ldots, k$. Noting 5.9.1 and 5.9.2 we find $B = -2\overline{m}$, and 5.9.10 becomes

$$\delta p_i = \frac{p_i(m_i - \overline{m})}{-A}. \tag{5.9.11}$$

Substituting this into 5.9.8 yields

$$\sum p_i(m_i - \bar{m})^2 = A^2C(\delta t)^2. \tag{5.9.12}$$

If we note the definition of C in 5.9.8 we see that $A^2(\delta t)^2 = 1$, and $1/A = \pm\delta t$. The appropriate value is $-1/A = \delta t$ (we want the favored gene to increase in frequency, not decrease), so, substituting this into 5.9.11, gives

$$\frac{\delta p_i}{\delta t} = p_i(m_i - \bar{m}), \tag{5.9.13}$$

which is equivalent to 5.3.5. Thus, the familiar equations for gene frequency change follow from the principle of maximum increase in fitness.

This is reminiscent of the principle of least action and Hamilton's principle in classical mechanics, whereby a minimization or maximization process leads to Newton's laws of motion.

The results for a single locus are readily extended to multiple loci if there is no epistasis and the loci are in gametic phase (linkage) equilibrium. This is discussed by Kimura (1958).

What if the fitness coefficients are not constant? We have seen earlier that under this circumstance the fitness does not change in a simple way and in fact may not increase at all. In this case we replace $\delta\bar{m}$ by the value that it would have if the fitnesses were constant; that is to say, we forget that the m_{ij}'s are variable and maximize $\delta\bar{m} = 2\sum m_i\,\delta p_i$.

With epistasis and linkage the principle works if the population is in quasi-linkage equilibrium. We shall illustrate this with the same model used before in Section 5.7. As we did then, we let p_1, p_2, p_3, and p_4 stand for the four gamete (chromosome) types and let $Z = p_1 p_4/p_2 p_3$ be the measure of linkage disequilibrium. We state the Maximum Principle as before: For a given genic variance, natural selection produces gene frequency changes δp_i so as to maximize $2\sum m_i \delta p_i$ (or, if the m_{ij}'s are constant, to maximize the change in fitness $\delta\bar{m}$).

In addition to the two side conditions used earlier,

$$\sum \frac{1}{p_i}(\delta p_i)^2 = \frac{1}{2}V_g(\delta t)^2 = K(\delta t)^2 \tag{5.9.14}$$

and

$$\sum \delta p_i = 0, \tag{5.9.15}$$

we have the additional condition imposed by the necessity for quasi-equilibrium. This asserts that Z (or $\log Z$) is constant, which is the same as saying that its differential is 0. Thus the third condition is

$$\delta \ln Z = \frac{1}{p_1}\,\delta p_1 - \frac{1}{p_2}\,\delta p_2 - \frac{1}{p_3}\,\delta p_3 + \frac{1}{p_4}\,\delta p_4 = 0,$$

or

$$\sum \frac{k_i}{p_i}\,\delta p_i = 0, \qquad 5.9.16$$

where

$$k_i = 1,$$

when $i = 1$ or 4,

$$k_i = -1,$$

when $i = 2$ or 3.

Proceeding as before, we let

$$\psi = 2\sum m_i\,\delta p_i + A\left[\sum \frac{1}{p_i}(\delta p_i)^2 - K(\delta t)^2\right] + B(\sum \delta p_i) + C\left(\sum \frac{k_i}{p_i}\,\delta p_i\right), \qquad 5.9.17$$

where A, B, and C are undetermined constant multipliers. Differentiating,

$$\frac{\partial \psi}{\partial(\delta p_i)} = 2m_i + \frac{2A}{p_i}\,\delta p_i + B + C\,\frac{k_i}{p_i} = 0. \qquad 5.9.18$$

First, multiply through by k_i and add. This gives

$$2\sum k_i m_i + 2A\sum \frac{k_i}{p_i}\,\delta p_i + C\sum \frac{1}{p_i} = 0. \qquad 5.9.19$$

But, the first term is equal to $2E$ (from 5.4.9), the second is 0 (from 5.9.16), and the third is CP (from 5.4.10). Thus

$$C = -\frac{2E}{P}. \qquad 5.9.20$$

Now, multiply 5.9.18 by p_i and sum, giving

$$2\overline{m} + 2A\sum \delta p_i + B - \frac{2E}{P}\sum k_i = 0$$

But, since $\sum \delta p_i = 0$ and $\sum k_i = 0$,

$$B = -2\overline{m}. \qquad 5.9.21$$

Substituting for B and C in 5.9.18 gives

$$\delta p_i = \frac{p_i(m_i - \overline{m}) - \dfrac{k_i E}{P}}{-A}. \tag{5.9.22}$$

The genic variance is

$$V_g = 2 \sum p_i(m_i - \overline{m})^2 - \frac{2E^2}{P} \tag{5.9.23}$$

(see 5.7.13 and earlier equations). Substituting these expressions for δp_i and V_g in the left and center terms of 5.9.14 gives

$$\frac{1}{A^2} \sum \frac{1}{p_i} \left[p_i(m_i - \overline{m}) - \frac{k_i E}{P} \right]^2 = \left[\sum p_i(m_i - \overline{m})^2 - \frac{E^2}{P} \right] (\delta t)^2.$$

Expanding the left side,

$$\frac{1}{A^2} \left[\sum p_i(m_i - \overline{m})^2 - 2 \frac{E}{P} \sum k_i m_i + \left(\frac{E}{P}\right)^2 \sum \frac{1}{p_i} \right]$$

$$= \frac{1}{A^2} \left[\sum p_i(m_i - \overline{m})^2 - \frac{E^2}{P} \right].$$

Therefore, $A^2(\delta t)^2 = 1$ and, choosing the right sign, $A\delta t = -1$. Equation 5.9.22 becomes

$$\delta p_i = \left[p_i(m_i - \overline{m}) - \frac{k_i E}{P} \right] \delta t. \tag{5.9.24}$$

But, at quasi-equilibrium, $E/P = cbD$, so we obtain

$$\delta p_i = [p_i(m_i - \overline{m}) - k_i cbD]\delta t. \tag{5.9.25}$$

Comparison of this with 5.4.4 shows that we have derived the equations for the rate of change of chromosome frequencies.

The principle can be extended to situations where there is not random mating and the population is not in quasi-equilibrium, but we shall not do so here. For a discussion, see Kimura (1958). For a single locus not in Hardy–Wienberg ratios, we state the Maximum Principle as follows:

Natural selection acts to change the gene frequencies in such a way that $\sum m_i \delta p_i$ is maximum for a given value of

$$\sum p_i(m_i - \overline{m})^2 = \sum \frac{1}{p_i} \left(\frac{\delta p_i}{\delta t} \right)^2 \tag{5.9.26}$$

during a small time interval δt.

The quantity on the left is not quite the same as half the genic variance. It is the variance of the average excess of the alleles, whereas the genic variance might be called the covariance of the average excess and effect of the alleles (cf. 5.6.6). Under some circumstances (for example, in a randomly mating population derived from an F_1 hybrid) fitness may actually decrease under natural selection. However, this principle gives the correct formula for gene frequency changes.

The principle (in the less general form first stated) is illustrated in Figure 5.9.1. We are assuming two loci, each with two alleles. The frequency

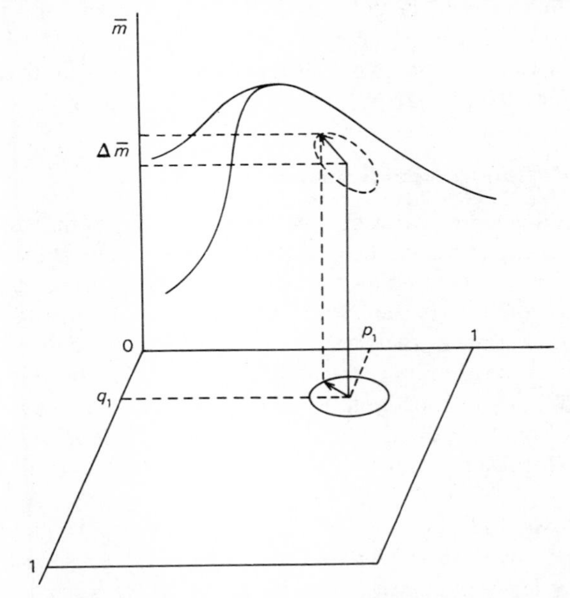

Figure 5.9.1. An illustration of the Maximum Principle. For discussion see the text.

of allele A_1 is given by p_1 and of B_1 by q_1. The changes in gene frequency, δp_1 and δq_1, lie on the ellipse given by 5.9.8. The fitness corresponding to each gene frequency combination is given by the ordinate so that there is a surface of fitnesses corresponding to the various gene frequencies at the two loci. The principle states that among the various possible gene frequency changes, determined by the ellipse on the plane of p_1 and q_1, the path taken by the gene frequency change is the one that leads to the largest increase in the ordinate. It is also clear that if the surface has more than one maximum the population will tend to move toward the nearest peak and not the one that is highest.

We have here been using a model that has been extensively developed by Sewall Wright. Corresponding to the various allele frequencies at the various loci there is an adaptive surface. If the fitnesses are constant and if there is equilibrium with respect to linkage and random-mating proportions (or a constant departure therefrom, as measured by the θ's and ε's of Section 5.6), the surface is uniquely determined. With loose linkage a population will tend to move rather rapidly to the surface as quasi-equilibrium is attained, and then tend to move upward on the surface toward the nearest peak. This model applies, obviously, only if there *is* such a surface. If, for example, the fitness coefficients are not constant the surface undulates with time and the population fitness does not necessarily increase maximally, or even at all in extreme cases. For a discussion of this, two of Wright's papers are (we think) especially relevant (Wright 1955, 1967).

5.10 The Change of Variance with Selection

If the number of genes influencing a trait or measurement is large the variation is essentially continuous, and superficially the pattern of inheritance is much the same with Mendelian heredity as with pre-Mendelian blending theories. For example, it is quite consistent with both hypotheses for progeny to be somewhere near the average of their parents, for selection to change the population in proportion to the amount of variability, and for the correlation between relatives to be roughly proportional to the amount of their common ancestry. In fact, much of natural selection and the progress made in animal and plant improvement by selection can be explained equally well on either assumption.

In one respect, though, the Mendelian particulate theory is greatly different. With particulate inheritance the variability is lost only very slowly; in fact, with an infinitely large population and no selection the variance is conserved indefinitely. In a finite population the variance is reduced by about $1/2N$ per generation. When N is large, this is a very slow rate of reduction and a small input of new mutations is sufficient to balance it.

With blending inheritance the basic nature of heredity is more analogous to a fluid than to a series of particles. Just as a mixture of red and white paint is forever pink and the red and white cannot be recovered, the variability of a population is reduced by each mating between differing types. Suppose that in a population of size N the ith individual deviates from the mean by an amount x_i for whatever quantitative trait is being measured. The variance then is $(\sum x_i^2)/N$. With blending inheritance the progeny of each mating are uniformly the average of the parents. Thus the progeny of a mating between two individuals deviating by amounts x_i and x_j will deviate by $(x_i + x_j)/2$. The average squared deviation of all progeny will be $\sum [(x_i + x_j)/2]^2/N$, or $\sum (x_i^2 + 2x_i x_j + x_j^2)/4N$. But, with random mating x_i and x_j are uncorrelated, so the middle term has an average value of 0 and drops out. Since there is no difference between the sum of the x_i^2's and the sum of the x_j^2's, the variance becomes $(\sum x_i^2)/2N$, just half the value of the previous generation. The reduction in variance under random mating with blending inheritance is the same as that with self-fertilization in a Mendelian system.

This means that there must be an enormous source of new variability if a blending system is to work. Darwin realized that natural selection requires variability in order to be effective. For this reason, and not knowing of Mendel's work, he gave considerable thought to what we now realize was wholly unnecessary—the devising of various schemes to generate new variability. Although inventive and showing Darwin's characteristic thoughtfulness and originality, they seem quite strange to contemporary biologists. Now we know that a small mutation rate is quite sufficient to offset the decay of variability from random drift in all but very small populations. The subject of blending vs. particulate inheritance is discussed in the first chapter of Fisher's *Genetical Theory of Natural Selection* (1930, 1958). We found it interesting reading and you probably will also.

Selection changes the variance. Whether the variance is increased or decreased depends on such things as gene frequencies, dominance, epistasis, linkage, and the mating system. Natural selection is based on total fitness and the effect on any measurable trait depends on the relation between that trait and the expectation of surviving and reproducing. For most measurements, such as size, the optimum from the standpoint of fitness is intermediate, so selection tends to remove extreme deviates. Ordinarily this tends to decrease the variance, but not necessarily at every locus since the favored type may be heterozygous at some loci.

If selection proceeds long enough with no input of new variability from mutation or migration the genic variability eventually becomes smaller and smaller. The gene frequencies change so that either (1) the favorable alleles all have a frequency of one, in which case there is no variance, or (2) there are stable equilibria, such as with a locus that is overdominant for fitness, at which point the genic variance is 0 (see, for example, the last column in Table 4.1.1). So eventually the genic variance becomes 0 for each locus, although it may increase temporarily. For additive loci with free recombination, the same is true for the combined genic variance from all loci.

Let us now consider the change of variance in more detail. In particular we want to ask what is the rate of change of the genic variance. We can do this by the same methods that have been used earlier in this chapter to study the change in the mean fitness.

As in Section 5.6 we shall consider a single locus with two alleles, but the formulae will be written so that the extension to multiple alleles is apparent.

With Hardy–Weinberg ratios the genic variance in fitness is

$$V_g = 2[p_1(m_1 - \overline{m})^2 + p_2(m_2 - \overline{m})^2]. \tag{5.10.1}$$

The rate of change of V_g is given by

$$\frac{dV_g}{dt} = 2\left[(m_1 - \overline{m})^2 \frac{dp_1}{dt} + (m_2 - \overline{m})^2 \frac{dp_2}{dt}\right] + 4\left[p_1(m_1 - \overline{m})\left(\frac{dm_1}{dt} - \frac{d\overline{m}}{dt}\right) + p_2(m_2 - \overline{m})\left(\frac{dm_2}{dt} - \frac{d\overline{m}}{dt}\right)\right]. \tag{5.10.2}$$

But,

$$\frac{dp_1}{dt} = p_1(m_1 - \overline{m}), \qquad \frac{dp_2}{dt} = p_2(m_2 - \overline{m}).$$

Making these substitutions,

$$\frac{dV_g}{dt} = 2[p_1(m_1 - \overline{m})^3 + p_2(m_2 - \overline{m})^3] + 4\left[\frac{dp_1}{dt}\frac{dm_1}{dt} + \frac{dp_2}{dt}\frac{dm_2}{dt} - \left(\frac{dp_1}{dt} + \frac{dp_2}{dt}\right)\frac{d\overline{m}}{dt}\right]. \tag{5.10.3}$$

From its original definition (e.g., 5.3.2),

$$m_1 = m_{11}p_1 + m_{12}p_2, \quad m_2 = m_{12}p_1 + m_{22}p_2$$

and

$$\frac{dm_1}{dt} = m_{11}\frac{dp_1}{dt} + m_{12}\frac{dp_2}{dt}, \qquad \frac{dm_2}{dt} = m_{12}\frac{dp_1}{dt} + m_{22}\frac{dp_2}{dt}.$$

Also

$$\frac{dp_1}{dt} + \frac{dp_2}{dt} = 0,$$

since the total change in all alleles must add up to 0. So, making these substitutions,

$$\frac{dV_g}{dt} = 2[p_1(m_1 - \overline{m})^3 + p_2(m_2 - \overline{m})^3] + 4\left[m_{11}\left(\frac{dp_1}{dt}\right)^2 + 2m_{12}\frac{dp_1}{dt}\frac{dp_2}{dt} + m_{22}\left(\frac{dp_2}{dt}\right)^2\right]. \tag{5.10.4}$$

Finally we let $m_{ij} = \bar{a} + \alpha_i + \alpha_j + D_{ij}$, as in Section 5.6. Substituting these and simplifying, we find that m_{11} is replaced by D_{11}, m_{12} by D_{12}, and m_{22} by D_{22}.

The generalization to multiple alleles is straightforward, so we can write

$$\frac{dV_g}{dt} = 2 \sum_i p_i(m_i - \bar{m})^3 + 4 \sum_{ij} \frac{dp_i}{dt} \frac{dp_j}{dt} D_{ij}. \qquad 5.10.5$$

This, summed over all loci, gives the rate of change of the genic variance for all loci. (We are ignoring the complications of close linkage and epistasis, and variable selection coefficients.)

The quantity D_{ij} is sometimes positive and sometimes negative, depending on the direction of dominance. With a large number of loci the positive and negative values cancel and, unless the direction of dominance is correlated with fitness, the average value will be close to 0. The same thing can be said for epistatic components that enter when the equation is extended to multiple loci. In many cases, probably most, the last term is small enough to be neglected.

The first term on the right side of the equation is the third moment about the mean of the distribution. The third moment is conventionally used as a measure of skewness. As the number of factors gets large there is a closer and closer approximation to the normal distribution, which is symmetrical and therefore has a zero value for the third moment. So, the larger the number of genes involved in the trait, the less important this term becomes.

We therefore reach the conclusion that with a large number of loci involved (as must always be true for fitness, and often is for artificial selection of quantitative traits in livestock), the variance is changed by selection only slowly. The population is not likely to exhaust its genic variance unless it is small, or the number of loci is small, or selection is very intense.

Recently, Turner (1969) has emphasized this point that with no input of new variability the genic (or with close linkage and epistasis, the gametic) variance eventually approaches 0. He suggests that this is the most fundamental statement that can be made about natural selection. In view of the very slow rate of change of variance that we have just discussed, we think that regarding this as *the* fundamental statement about selection is debatable, but in any case it is *one* statement that can be made about selection.

5.11 Selection Between and Within Groups

Our discussion of selection has dealt entirely with selection among individuals. Yet it is obvious that selection occurs at all levels—between individuals, between families, between communities, and between species. There may also be intercellular selection within an individual, although we shall not discuss this.

We should expect that the competition would be greatest between individuals who are close together geographically and ecologically, for they will then be competing most directly for the same resources. Likewise, the competition between individuals should be greater the more alike they are, for with greater similarity comes greater preference and need for the same resources. This gives an explanation for frequency-dependent selection whereby a genotype is favored when rare but becomes relatively less fit when common.

On the other hand, selection between similar genotypes has a lessened effectiveness in changing gene frequencies because there is less genetic variance. This is well illustrated by selection within families.

Within-family Selection This subject was discussed by Haldane (1924) and more recently by King (1965). They have given calculations and formulae for a number of special cases that we shall omit.

Competition between individuals within a family may be increased by any of the following: competition between members of a litter (for example, a female mouse producing 25% lethal zygotes often does not have a corresponding reduction in litter size); compensation in uniparous animals whereby a fetal or early childhood death leads to the next progeny's coming earlier than it would have otherwise, as we shall discuss in Section 6.10 in connection with selection on the sex ratio; the practice by animal breeders of culling litters down to a uniform size; family planning in humans whereby families tend to have the same number of surviving children.

If all selection is within families and none is between, we can get a general idea of what happens with selection as follows. The correlation between sibs for additive genetic effects in a randomly mating population is 1/2, as we showed in Chapter 4. If V_b and V_w are the genic variances between and within families, and the total genic variance is $V_g = V_b + V_w$, the correlation between members of a family is

$$r = \frac{V_b}{V_b + V_w}, \tag{5.11.1}$$

as shown in the appendix (A.4.12). So

$$\frac{V_w}{V_g} = 1 - r. \tag{5.11.2}$$

If $r = 1/2$, then the variance within sibships is just 1/2 that of a randomly mating population. So we would expect selection to change fitness or whatever trait is being selected only half as fast.

As for the change in gene frequencies, this too is only half as fast. It could hardly be otherwise, but here is a demonstration. From the Fundamental Theorem of Natural Selection,

$$V_g = \frac{d\overline{m}}{dt} = \sum_i \frac{\partial \overline{m}}{\partial p_i} \frac{dp_i}{dt}. \tag{5.11.3}$$

The genic variance within sibships is

$$V_w = \frac{1}{2} V_g = \frac{1}{2} \sum \frac{\partial \overline{m}}{\partial p_i} \frac{dp_i}{dt} = \sum \frac{\partial \overline{m}}{\partial p_i} \frac{1}{2} \frac{dp_i}{dt}. \tag{5.11.4}$$

Since $\partial \overline{m}/\partial p_i$ does not depend on the intensity of selection or rate of gene frequency change, the change in the p_i's is only half as fast as with ordinary selection.

This result will apply strictly only when the Fundamental Theorem holds, so it is not exact for strong selection with discrete generations. For example, rare recessive lethals are eliminated about 2/3 as fast with intrafamily selection as otherwise (King, 1965).

Completely Isolated Groups Somewhat the same thing must be going on in nature between populations as happens within populations. Some increase in frequency while others decrease, just as genes or chromosomes do within a population. Some become extinct; others enter into stable equilibria in which several populations coexist. Meanwhile changes are also taking place within each of the populations. If we consider a community of organisms of different competing species, there is a general overall increase in fitness (that is, intrinsic fitness, relative to a fixed standard). Such an increase is counteracted by crowding, deterioration of the environment, and such, just as within a species, so that the total biomass does not change much. But it is perhaps worthwhile to formulate a principle like that for within-population selection that states the rate of increase in overall fitness.

Consider a total population, divided into n subpopulations, not necessarily of equal size. We assume there is no gene interchange between the subpopulations. Let

$M = \sum_k P_k \overline{m}_k =$ the average fitness of the entire population,
$\overline{m}_k =$ the average fitness of the kth subpopulation, and
$P_k =$ proportion of all individuals that are in the kth subpopulation.

Then,

$$\frac{dM}{dt} = \sum P_k \frac{d\overline{m}_k}{dt} + \sum \overline{m}_k \frac{dP_k}{dt}. \tag{5.11.5}$$

But

$$\frac{d\overline{m}_k}{dt} = V_{gk},$$

the genic variance of the kth subpopulation, and

$$\frac{dP_k}{dt} = P_k(\overline{m}_k - M),$$

since $\overline{m}_k$ is the average rate of increase of the kth subpopulation, measured in Malthusian parameters (see, for example, 1.6.8a). Also

$$\sum \overline{m}_k P_k(\overline{m}_k - M) = \sum P_k(\overline{m}_k - M)^2 = V_{\overline{m}},$$

where $V_{\overline{m}}$ is the variance of the mean fitnesses of the subpopulations.

Substituting these into 5.11.5 gives

$$\frac{dM}{dt} = \overline{V_g} + V_{\overline{m}}. \qquad \textbf{5.11.6}$$

This says: The rate of increase in fitness of a structured population is equal to the average genic variance of the subpopulations plus the genotypic variance of the means of the subpopulations.

The effects of nonconstant departures from random-mating proportions, variable selection coefficients, and deterioration of the environment can be taken into account in the same way as in Section 5.6.

Since the selection between groups depends on the total genetic variance of the group means rather than only the additive portion thereof, this selection is somewhat more efficient than that within subpopulations in this respect. On the other hand, the group mean differences are not likely to be anything near as large as the differences among individuals within the groups.

The mechanism of intergroup selection is the increase or decrease, and eventual fixation or extinction, of whole populations or species. Such extinction and replacement by new species split off from existing ones is in no small part responsible for the continuing wide diversity of life. But how much of the overall increase in adaptability is developed this way is problematical. We are impressed by the much larger variance within a group than between group means, the larger numbers that minimize the "noise" from random events, and the slower effective "generation length" for intergroup selection.

The point has been made very forcefully by R. A. Fisher (1930, 1958, p. 49).

> The principle of Natural Selection . . . refers only to the variation among individuals (or co-operative communities), and to the progressive modification of structure or function only in so far as variations in these are of advantage

to the individual, in respect to his chance of death or reproduction. It thus affords a rational explanation of structures, reactions and instincts which can be recognized as profitable to their individual possessors. It affords no corresponding explanation for any properties of animals or plants which, without being individually advantageous, are supposed to be of service to the species to which they belong.

There would, however, be some warrant on historical grounds for saying that the term Natural Selection should include not only the selective survival of individuals of the same species, but of mutually competing species of the same genus or family. The relative unimportance of this as an evolutionary factor would seem to follow decisively from the small *number* of closely related species which in fact do come into competition, as compared to the number of individuals in the same species; and from the vastly greater *duration* of the species compared to the individual. Any characters ascribed to interspecific selection should of course characterize, not species, but whole genera or families, and it may be doubted if it would be possible to point to any such character, with the possible exception of sexuality itself, which could be interpreted as evolved for the specific rather than for the individual advantage.

Selection of Cooperative and Social Behavior How, then, do characters of benefit to the species evolve? For example, it has often been suggested that polymorphism is of benefit to the species, by making it more able to tolerate fluctuating environmental conditions. However, the mechanisms maintaining the polymorphism (heterosis, selection depending on frequencies, mimicry) are all intrapopulational. So we have no real assurance, solely from the existence of the polymorphism, that it is either good or bad for the species in addition to its value to the individuals in the population.

A great many seemingly altruistic behavior patterns may really benefit the individual himself. Much cooperative behavior must be mutually beneficial.

Altruistic, or truly self-sacrificing behavior, is we believe more easily explained by "kin selection." An extreme example of self-sacrifice is the worker bee who is likely to die in the process of stinging a person who threatens the hive. This is understandable, of course, in that the worker bee, being sterile, has hope of progeny only if the queen with whom the worker shares many genes is kept alive and fertile. This means that those genes which cause the worker bee to protect the queen will be transmitted disproportionately by the queen to future generations and the trait, among workers, will increase. In a less extreme way, factors tending to lead a mother to protect her children would have similar selective value; for in this way the mother enhances the contribution of her genes to the genetic makeup of future generations. Since her child shares 50% of her genes, and has more of its reproductive life span ahead, it is worth a large risk to the mother to keep the child alive; one would expect natural selection to favor such protective behavior. In general, and in a very rough way, one might expect natural selection to act in such a way as to perpetuate behavior patterns that would lead an individual to take a risk, r, to save the life of another individual, where r is the coefficient of relationship between the two individuals, who therefore share a fraction r of their genes.

The structure of the human population in the past, where there were many small, rather isolated groups, may have been important in developing social behavior. Individuals in a group are likely to be rather closely related genetically and this would be an opportunity for the kind of kin selection that we have been discussing. The important role that this may have played in the development of social systems in man has been emphasized by Muller (1967).

Partially Divided Populations and Wright's Theory of Evolution

There must be situations, and a great many of them, where the population is neither one large unit nor is it completely divided into subgroups. The population may be highly structured with subgroups that are isolated enough to prevent random exchange of genes, but between which there are some migrants. Such a structure lies at the base of the shifting balance theory of evolution that has been discussed extensively by Sewall Wright (1932, and many later papers elaborating the idea).

Wright argues that a partially subdivided population permits some local differentiation. This may take place for reasons of different selective forces in different localities. It also may happen because the local groups are small enough for random events to lead to substantial fluctuations. If, as a result of such processes, a particularly happy combination of genes or gene frequencies emerges, then this can spread to neighboring areas. Thus Wright thinks of natural selection as being a mixture of selection within and between populations; although he does not distinguish between traits that are of individual merit and those that have group merit.

The opportunity for local differentiation, with random processes playing an appreciable role, may help the population to get over a point of unstable equilibrium such as was discussed at the end of Section 5.4. In Wright's metaphor, the population may have an opportunity to move from one adaptive peak to another; something that would not happen in a strictly panmictic large population.

5.12 Haldane's Cost of Natural Selection

We have given in Section 5.3 formulae that tell what time is required to change the gene frequency under different intensities of selection. Haldane (1957) was the first to formulate the question of how much total selection is required for a gene substitution over the whole period of time involved. It is clear that there is a limit of some sort (although it might be quite elastic) on the number of genetic changes that can be carried out simultaneously. For example, this point was mentioned in the quotation from Sewall Wright at the beginning of Chapter 5. In his 1957 paper Haldane said: "In this paper I shall try to make quantitative the fairly obvious statement that natural selection cannot occur with great intensity for a number of characters at once unless they happen to be controlled by the same genes."

When we examine equations such as 5.3.13–5.3.16 we see that t and s are inversely proportional, and that for a particular situation the product st is a constant. If the selection is intense, the time is shorter. So, in a vague sense, we see that the total amount of selection during the time is constant for a given frequency change and type of gene action; or at least st is constant, whatever its meaning is.

Haldane formulated the question in a more meaningful way. Our treatment is slightly different from Haldane's, but the two lead to the same equations. We shall discuss the differences later.

Consider a single locus in a haploid species and with a discrete generation model. We shall extend this to diploids later, but the essential idea is present in the haploid situation and is less obscured by algebraic fog. Assume the following model:

GENOTYPE	A	A'	
FREQUENCY	p	$q = 1 - p$	$\bar{w} = w_A(1 - sq)$
FITNESS	w_A	$w_A(1 - s)$	
RELATIVE FITNESS	1	$1 - s$	

A is the favored gene. We assume that it is initially rare, either because it is a new mutation that hasn't occurred before, or, more likely, there has been a change in the environment such that a previously deleterious mutant is now favorable.

In the first generation (generation 0) the frequency of the A gene is p_0. This will be represented next generations by w_A descendants. The other genotype will be represented by $w_A(1 - s)$. Relative to the A genotype, a fraction s of the A' genotype will fail to survive or their reproduction will be lowered by an equivalent amount. In the whole population the ratio of individuals (or genes) not transmitted to those that are transmitted will be $w_A sq_0/w_A(1 - sq_0) = sq_0/(1 - sq_0)$. Next generation the proportion of A' genes will be $q_1 = q_0(1 - s)/(1 - sq_0)$, and so on in successive generations.

To be concrete, we may think of a population with a reproductive excess, but which because of mortality has an adult population that is roughly constant from generation to generation. In nature, as we discussed in Chapter 1, the population size stays stable despite gene frequency changes because its size is determined by other factors (e.g., food limitation). In this population, if $sq = .2$ and $1 - sq = .8$, the ratio of eliminated to noneliminated is .2 : .8 or .25. In other words, if selection is entirely through mortality (and all other factors are ignored), each parent would have to produce 1.25 progeny in order for the population size to remain stable.

If the selection is acting on fertility there must be enough variability in progeny number so that

$$\frac{\text{Average number from genotype } A - \text{Average of population}}{\text{Average of population}},$$

which is equal to

$$\frac{w_A - \bar{w}}{\bar{w}},$$

can, in this case, be .25.

We shall call the quantity $sq/(1 - sq)$, summed over all the generations involved in the gene substitution, the *cost of a gene substitution*. Designating this by C, we have

$$C = \sum \frac{sq_t}{1 - sq_t}. \qquad 5.12.1$$

The table below gives the value of C for several values of s, assuming that the initial frequency of A, p_0, is .01.

p_0	s	C
.01	1.00	99
.01	.99	52
.01	.50	6.2
.01	.10	4.8
.01	.01	4.63
.01	Limit	4.61

Three things stand out in this table. First, the values are strikingly high. For example, if $s = .1$, the total cost is 4.8 nonsurvivors per survivor, or in terms of numbers of individuals the total number of extinctions is 4.8 times the number of survivors in a single generation. This assumes, of course, that the number of adults is roughly constant from generation to generation. Second, the total cost is less if the selection takes place slowly, i.e., if s is small. Third, the value does not change much after s becomes as small as .10 and hardly at all when s is as small as .01.

It is not necessary for s to be constant during this process as long as it remains small and positive. It is quite likely, of course, that s changes as it is placed in combination with other genes, as the population density changes, and as environmental fluctuations occur. This does not make any difference unless the sign changes.

The third point is particularly remarkable. The fact that the cost is virtually independent of s when s is small means that we can calculate the total cost solely from the initial frequency p_0. This can be done as follows. We assume that the change in gene frequency is slow enough that the addition in 5.12.1 can be replaced by integration. Thus, to a satisfactory approximation,

$$C = \int_0^\infty \frac{sq}{1 - sq}\, dt. \tag{5.12.2}$$

From 5.2.2

$$\Delta p = \frac{spq}{1 - sq} \approx \frac{dp}{dt}$$

or

$$dt = \frac{1 - sq}{spq}\, dp. \tag{5.12.3}$$

Substituting this into 5.12.2 gives

$$C = \int_{p_0}^1 \frac{dp}{p} = -\ln p_0 . \tag{5.12.4}$$

Notice that $-\ln p_0$ is 4.61 when $p_0 = .01$, in agreement with the limiting value in the table above.

Actually, the upper limit should not be exactly 1 because the allele frequency will be kept from this value by reverse mutation; but this is a negligible correction, as can be seen by replacing the upper limit in the integration by a value close to 1.

Equation 5.12.4 shows that as long as s is small enough for the continuous model to be applicable the total load is independent of s and of the time required to make the gene substitution, but depends only on the initial frequency of the substituted allele, p_0.

Here are a few numerical examples.

p_0	$C = -\ln p_0$
10^{-6}	14
10^{-4}	9
10^{-2}	4.6
10^{-1}	2.3

These numbers are all quite large. This means that, if the typical allele has an initial frequency of 10^{-4}, a population of 1 million individuals will have to have 9 million genetic deaths each generation if it is to substitute an average of one allele per generation. Or more probably, if there is to be a gene substitution every 100 generations, the average fitness will be lowered by .09.

Notice the dependence of the total load on the initial frequency. Most of the cost is in the early generations when the favored gene is still rare. This means that genes that are initially common, either because of a high mutation rate, or because they were only mildly disadvantageous previously, are the easiest to substitute. However, it is probably such genes with only mild deleterious effect that are most important in evolution, not only because they are initially less rare, but because a gene that is only mildly deleterious in the old environment is more likely to be beneficial after an environmental change than one that was originally very harmful.

An alternative formulation that does not depend on s being small has been given by Felsenstein (see Crow, 1970).

The load becomes greater (and the algebra more difficult) when a diploid species is considered. The procedure follows the same line as before. We assume random mating and let h be a measure of dominance.

GENOTYPE	AA	AA_1	A_1A_1
FREQUENCY	$(1-p)^2$	$2p(1-p)$	p^2
RELATIVE FITNESS	$1-s$	$1-hs$	1

$$\begin{aligned}\bar{w} &= (1-s)(1-p)^2 + (1-hs)2p(1-p) + p^2 \\ &= 1 - s(1-p)(1-p+2hp).\end{aligned} \tag{5.12.5}$$

From 5.2.6

$$\frac{dp}{dt} \approx \frac{p(w_1 - \bar{w})}{\bar{w}}, \tag{5.12.6}$$

where w_1, the fitness of A_1, is $[p(1-p)(1-hs)+p^2]/p = 1 - hs(1-p)$. Substituting this and 5.12.5 into 5.12.6,

$$\frac{dp}{dt} = \frac{sp(1-p)[1-h-p(1-2h)]}{\bar{w}}. \tag{5.12.7}$$

The cost, by analogy with 5.12.2, is

$$C = \int_0^\infty \frac{(1-\bar{w})}{\bar{w}}\, dt. \tag{5.12.8}$$

Substituting from 5.12.5 for $\bar{w}$ and from 5.12.7 for dt in 5.12.8 leads to

$$C = \int_{p_0}^{1} \frac{s(1-p)(1-p+2hp)}{sp(1-p)[1-h-p(1-2h)]}\,dp$$
$$= \frac{1}{1-h}\left[\ln p_0 + h \ln \frac{h}{1-h-(1-2h)p_0}\right], \qquad 5.12.9$$

when $h \neq 1$.

$$C = -\left[1 + \ln p_0 - \frac{1}{p_0}\right], \qquad 5.12.9a$$

when $h = 1$.

Here are some approximate numerical values. These correspond very roughly to an initial frequency, p_0, determined by a mutation rate of 10^{-5} and homozygous disadvantage 0.1.

h	p_0	C	
0	10^{-4}	9	A_1 dominant
0.5	10^{-4}	18	
0.9	10^{-3}	50	
0.99	10^{-2}	70	
1.0	10^{-2}	100	A dominant

Thus, in a diploid system, the cost of substituting a moderately rare mutant (e.g., one previously maintained by a mutation rate of 10^{-5} to 10^{-4}) is from 10 to 100. That is to say, the total number of eliminated individuals is 10 to 100 times the number of adults in any single generation. Haldane suggests 30 as a representative number.

Haldane also suggests that a species might perhaps devote some 10% reproductive excess, or the equivalent in variability, to this process. With this cost value, a species could carry out one gene substitution on the average every 300 generations if each one involved a cost of 30. If a pair of species differ by 1000 loci and 10% reproductive excess were devoted to the process (with perfect efficiency), it would have required about 300,000 generations for this divergence to take place.

Our formulation differs from Haldane's in giving the cost as the ratio of zygotes eliminated by differential viability and fertility to those not eliminated. Haldane gave it as the ratio of those eliminated to all zygotes. When s is small these are practically equivalent, so he has the same equations. However, our formulation gives a more exact representation for larger values; and perhaps the meaning is more transparent.

Extensions to this problem to include epistasis and the effects of chance have been made (Kimura, 1967; Kimura and Maruyama, 1969). We have assumed here a sudden environment change so that a previously deleterious gene immediately becomes favorable, as might be the case if a species migrated to a new locale. On the other hand, the change may be gradual. It can be shown (Kimura, 1967) that this makes only a small change in the calculations; the general conclusions are essentially unchanged.

Most organisms have a reproductive rate far in excess of that required to maintain the population size in the absence of preadult mortality. Haldane was concerned with only that part of mortality and sterility differences that were selective. A million-fold reproductive excess, such as is found in some plants and invertebrates, is not necessarily more effective in selection than an excess of a few percent in which the differences are mainly genetic. Observations on the total reproductive rate set only an upper limit on the amount of gene substitution that can occur. We have no way, from this measurement alone, to determine whether or not Haldane's 10% is a reasonable value of the differential fitness that can be applied to evolutionary change.

According to King and Jukes (1969) the evolution of mammalian structural proteins has been at a rate averaging about 1.6×10^{-9} per year. Let's regard this as roughly 10^{-6} per cistron per generation. If the mutant phenotype had an initial frequency of 0.0001 the cost, as we have seen, is some 10–50 per gene substitution. The cost per generation is then about $10\text{–}50 \times 10^{-6}$ per generation for these proteins. If 10^4 loci were evolving at this rate the cost per generation would be 10% to 50%.

It is apparent from equations 5.12.9 and 5.12.9*a* that the cost is dependent very strongly on the initial frequency of the favored allele. One way in which the cost may be reduced is if the genes start at higher frequencies. Here are some representative values of the total cost required for a specified gene frequency change. These are for $h = 1/2$ in equation 5.12.9.

GENE FREQUENCY CHANGE FROM	TO	TOTAL COST
.0001	.9999	18.42
.001	.999	13.81
.01	.99	9.19
.1	.9	4.39
.2	.8	2.77
.3	.7	1.69
.4	.6	.81

This means that if the same phenotypic change can be accomplished by changing five genes from 0.4 to 0.6 as by changing one gene from .0001 to .9999 (as would seen reasonable), the cost is $5 \times 0.81 = 4.05$ as opposed to

18.42. Sewall Wright has repeatedly suggested that adaptive changes may depend more on shifting balances among genes with moderate frequencies than on substitution of rare mutants. In Wright's scheme the cost would appear to be less in this regard.

The remarkable property that the cost does not depend on s when the value of s is small implies that all gene substitutions involving the same dominance and the same initial frequency have the same cost. This suggests that if two mutants were closely linked they could both be substituted for half the cost of the two if substituted independently. A difficulty, however, is that the double mutant is likely to be rare. In simple cases these effects exactly compensate. If the single mutant has frequency p_0 and the double mutant p_0^2, then the cost for the linked double mutant is the same as for the two independent mutants, since $-\log p_0^2 = -2 \log p_0$. This relation is strictly true only when $h = 1/2$, but it gives the general idea.

The greatest difficulty in interpreting the Haldane cost principle is uncertainty about the importance of gene interaction. The Haldane argument that we have given assumes that the genes being substituted are independent in inheritance and in action. Sved (1968*a*) and Maynard Smith (1968) have both called attention to the change that is brought about when there is truncation selection. The consequences can be quite different. The number of loci at which selection can act simultaneously with the same total intensity of selection is many times as great.

Whether selection acts mainly on independent genes or in a threshold manner is not clear at present. Certainly some traits are quite independent of one another. For example, different genes are known to confer resistance to different insecticides or antibiotics. And presumably genes for feathers are independent of genes for laying more eggs. On the other hand, if the environment changes in such a way as to make increased size adaptive, a number of genes would be acting simultaneously on the same trait. Selection is not likely to act in a strict threshold manner in nature, but it might approach this nearly enough to reduce the cost substantially in comparison with the value if each allele were independent.

There is obviously a limitation somewhere on the number of genes—or at least on the number of independent genes—but our present ignorance as to the nature of the interactions of the genes concerned with fitness limits the practical utility of the principle. Nevertheless, we believe that this is an important idea and agree with Haldane's statement that "quantitative arguments of the kind here put forward should play a part in all future discussions of evolution." He also emphasized that this is only a beginning to a theory and added that "I am aware that my conclusions will probably need drastic revision."

For a review, see Crow (1970).

The Cost in Terms of Variance We have seen earlier that the rate of change in fitness and gene frequencies is related to the genetic variance of the population. So we might try a variance analogy to the Haldane cost principle. Again, we start with a simple haploid model, as before.

The genic variance (or since we are dealing with a single locus in a haploid, the genotypic variance as well) is pqs^2 for the model that we have just been discussing. Then the analogy to 5.12.2 would give the variance cost as

$$C_v = \int_0^\infty pqs^2 \, dt. \qquad \textbf{5.12.10}$$

This time it is more convenient, and perhaps more realistic, to consider a continuous model. Substituting from 5.3.6 gives

$$C_v = \int_0^1 s \, dp = s. \qquad \textbf{5.12.11}$$

So, the total variance required to produce a gene substitution, is the selective difference between the two alleles.

This is easily generalized to the diploid case, and the answer can be worked out quite generally by using the principles of Section 5.6. Suppose the difference in fitness between the original genotype and the one in which the substitution has been made is s; we can write for the total variance cost

$$C_v = \int_0^\infty V_g \, dt. \qquad \textbf{5.12.12}$$

But, from the Fundamental Theorem of Natural Selection,

$$\frac{d\bar{m}}{dt} = V_g. \qquad \textbf{5.12.13}$$

Substituting this into 5.12.12 and changing the limits of integration to correspond to the change in fitness by an amount s, we have

$$C_v = \int_A^{A+s} d\bar{m} = s. \qquad \textbf{5.12.14}$$

Thus to change the fitness of the population by an amount s every n generations demands a genic variance of s/n in the population. The actual variance of most populations is quite large, but it is hard to know what fraction is genic.

It would appear, however, that this principle is less restrictive than Haldane's; so Haldane's principle is likely to be the most useful in estimating the rate of gene substitution and comparing this with the actual values as they are beginning to be known from amino acid analysis. In any case, it appears to be more restrictive for the substitution of initially rare genes.

5.13 Problems

1. How many generations are required to change the homozygous recessive phenotype frequency from 0.5 to 0.01 when $s = 0.01$? When $s = 0.1$? Compare the value for $s = 1$ from Table 5.3.1 with the exact answer from 5.1.1.
2. Derive 5.2.7. Note that $\overline{w} = \sum_i p_i w_i = p_i w_i + \sum_{j \neq i} p_j w_j$ and $w_x = (\sum_{j \neq i} p_j w_j)/(1 - p_i)$.
3. Using 5.2.12 for the 2-allele case,

$$\Delta p_1 = \frac{p_1(1 - p_1)}{2\overline{w}} \frac{d\overline{w}}{dp_1},$$

derive 5.2.14 and 5.2.15.
4. Show that the rate of gene frequency change is maximum when the recessive allele is twice as frequent as its dominant allele.
5. What is the Malthusian parameter of a lethal genotype?
6. Show that with 2 alleles, no dominance, and a continuous model that $u_t = u_0 + st$, where t is time in generations, u is $\log_e [p/(1 - p)]$, and $2s$ is the selective difference between the two homozygous types.
7. With complete dominance and selection in favor of the dominant allele, define u_t in terms of p so that the simple linear relation of Problem 6 applies.
8. Show that with two alleles and constant inbreeding coefficient f the rate of gene frequency change is given by

$$\Delta p = \frac{p(1 - p)}{2\overline{w}} \frac{d\overline{w}}{dp} + \frac{d\overline{w}'}{dp} f$$

where $\overline{w}'$ is the average fitness of a completely homozygous population.
9. What transformation will linearize the data of Figure 5.2.1, assuming that Equation 5.2.27 is appropriate?
10. For a polygenic trait how fast (approximately) is selection within families of half-sibs compared with mass selection and random mating?
11. Show that the rate of elimination of a lethal mutant is about 2/3 as fast with intra-family selection as in general for a very rare recessive gene.
12. Show that for a single locus the average excess, a_i, of an allele is equal to $\alpha_i(1 + f)$, where α_i is the average effect and f is the inbreeding coefficient.
13. Why is it not generally possible to obtain the genic variance of a partially inbred population by measuring the genic variance before the consanguineous matings began and multiply this value by $1 + f$? When does this procedure give the right answer?

14. Show that when the ε's are constant in Section 8.7 ($\varepsilon_1 = p_1/p_a p_b$, etc.) then Z is a constant.
15. Show that constant θ's in Section 5.6 do not imply that the inbreeding coefficient f is constant, and therefore that when f is constant (except $f = 0$) the fitness increase is not equal to the genic variance.

6 POPULATIONS IN APPROXIMATE EQUILIBRIUM

In a natural population only a fraction, and surely only a very small fraction, of the selection is effective in causing the systematic changes in gene frequencies that we think of as evolution. Much differential survival and fertility is purely accidental—an animal may survive because it happens to be in the right place at the right time. This is especially true of organisms that produce a great excess of progeny of which only a few survive to maturity.

Of the survival and fertility that is not accidental, some is due to purely phenotypic differences rather than genetic, such as the effects of an earlier environment or an accident of development. Only that part of the selection that is based on genotypic, as opposed to environmental, differences has any influence on the genetic composition of future generations.

Much of the genotypic selection goes to eliminating recurrent mutations, to eliminating inferior genotypes that arise by segregation and recombination from better adapted parents, and to adjusting to changes in the environment.

Thus, most genetic selection goes to maintaining the status quo, rather than to making progressive evolutionary changes.

In Chapter 5 we considered systematic changes in gene frequency under selection, which might be called the *dynamics* of evolution. In this chapter we consider those factors that maintain the population in an approximate equilibrium—evolutionary *statics* (see Haldane, 1954*b*).

6.1 Factors Maintaining Gene Frequency Equilibria

There are many ways in which population variability can be maintained. They all depend on some kind of balance between opposing forces, or else the variability is transient. Here are some of the most important mechanisms. The items on this list are neither mutually exclusive nor collectively exhaustive.

1. Balance Between Mutation and Selection Each generation new mutant genes arise. Most of these are harmful in the environment of their occurrence and are eliminated sooner or later by natural selection, if not by chance. At any one time there will be a certain number that are not yet eliminated. The average number of such mutants will be determined by the balance of the mutation rate and the rate of selective elimination.

2. Selection Favoring Heterozygotes The genotypes of highest fitness may not be homozygous, but may be heterozygous at one or more loci. In a simple 2-allele situation where the heterozygote is favored it is clear that, since the favored genotype has two different alleles, there will be a tendency for both of these alleles to remain in the population. The classical example is the sickle-cell hemoglobin heterozygote which has neither the anemia of one homozygote nor the malaria susceptibility of the other. In complex multiple heterozygotes the same general principle applies; the population tends to retain more than one allele at each of the relevant loci.

3. Frequency Dependent Selection If the fitness of a genotype depends on its frequency and on what other genotypes are present—and to a greater or lesser extent this must very often be the case in nature—there are many ways in which stable equilibria can develop. The most obvious way, and perhaps the most important in maintaining variability, is selection in favor of rare genotypes. We have already discussed a situation that amounts to this—the self-sterility alleles found in some plants (Section 4.12).

There is one very general consideration that makes it reasonable that many genotypes are favored when they are rare. This comes from the fact that competition must be most keen between those individuals that are the most nearly alike. They will have the most similar food needs, for example, and will therefore compete most directly. To the extent that the genotypes differ in their needs a rare genotype will have less competition than a common one. Various experiments showing a greater yield or biomass in mixed than in homogeneous populations show that this is important, in at least some cases. The generality and importance of frequency-dependent selection as a factor maintaining population variability is largely unknown.

4. Heterogeneous Environment If there are different habitats and the various genotypes respond to these in different ways, individuals will be more favored in some environments than in others. Each type will be more favored when there is room in his best environment, but not when he has to live in another. This effect will be accentuated if each individual is able to select his environment. This leads to a situation in which genotypes are favored while they are rare but become disadvantageous when they outgrow their niches. This leads to a stable polymorphism for the same reason that the example just above does. The conditions under which multiple niches maintain a polymorphism have been discussed by Levene (1953) and Prout (1968).

The borders between the different environments may be sharp, or the transition may be gradual. There may be a geographical cline. For example, the amount of water might change from one end of the species range to the other. Organisms that do better in a dry environment are relatively favored at one end of the area and those that do best in a wet environment do better at the other end. This may maintain a gradual transition in gene frequencies over the range.

5. Conflicting Gametic and Zygotic Selection, or Meiotic Drive If a series of alleles are such that one gamete type is favored, either because of selection in the gametic stage (e.g., selection based on the genotype of the pollen tube) or because of meiotic events that lead to a segregation bias (meiotic drive as defined by Sandler and Novitski, 1957), this can lead to a stable polymorphism when the favored type is deleterious in the zygotic selection. For example, the *t*-alleles in mice and the Segregation Distorter factor in Drosophila are both maintained at a fairly high frequency in the population despite drastic effects on viability and fertility when the genes are homozygous.

6. Neutral Polymorphism If a series of alleles are nearly equivalent in their effect on fitness (near enough that the differences among them are of a lower order than the mutation rates or the reciprocal of the effective population number) there may be a neutral polymorphism in which the frequencies of the different types are determined largely by their mutation rates. In such a situation the effects of a finite population become especially important and the number of alleles actually maintained will be determined by the balance between mutation and random extinction through gene frequency drift. This will be discussed in Section 7.2.

7. Balance Between Selection and Migration Migrants that join a population from outside will usually be of a different frequency for one or more alleles from the local residents. Usually the outsiders will be less well adapted to local conditions, so there is a possibility of an equilibrium between the local selective forces which tend to favor some genotypes and the influx of other genotypes through migration.

8. Transient Polymorphism In addition to the various stabilizing mechanisms just mentioned, a population may be polymorphic because it is in the process of change. We saw in Chapter 5 that it may take many thousands of generations for a gene substitution to take place. This means that a transient polymorphism may persist for a long time and in nature this may be very difficult to distinguish from a stable polymorphism.

In considering equilibrium problems we usually use a discrete generation model. The reason is mainly convenience. We pick whichever model is most effective or the simplest, and for most of the questions that we ask in this chapter the discrete model is preferable. The conclusions are almost always directly transferable from one model to the other, more so than when one is considering evolutionary dynamics. As a population gets close to an equilibrium point, it makes less difference which model is chosen since the rate of change per generation (or other time unit) is ordinarily slowest as the population approaches a stationary value. We are referring here to equilibrium values at intermediate gene frequencies: typically these are approached asymptotically.

6.2 Equilibrium Between Selection and Mutation

Consider a discrete generation model. Each generation is counted at the same age; and we shall consider that enumeration is at the zygote stage.

We assume a single locus with two alleles, and with the three genotypic fitnesses and frequencies as follows:

GENOTYPE	A_1A_1	A_1A_2	A_2A_2
FITNESS	w_{11}	$w_{12} = w_{11}(1 - hs)$	$w_{22} = w_{11}(1 - s)$
FREQUENCY	P_{11}	$2P_{12}$	P_{22}

$$\bar{w} = P_{11}w_{11} + 2P_{12}w_{12} + P_{22}w_{22}.$$

If w_{12} and w_{22} are less than w_{11} (i.e., if h and s are between 0 and 1) the A_2 gene will decrease. This decrease will continue until the loss of A_2 genes is balanced by the rate at which new ones are created by mutation. We let u be the mutation rate from A_1 to A_2. If A_2 is rare, as it will be if selection is strong, reverse mutants from A_2 to A_1 will be very rare and can safely be ignored.

One generation later the frequency of allele A_1, in terms of the present frequency, will be

$$\frac{(w_{11}P_{11} + w_{12}P_{12})(1 - u)}{\bar{w}} \qquad 6.2.1$$

(see the derivation of 5.2.1). The factor $1 - u$ in the numerator represents the fraction of A_1 genes that have not mutated during the current generation and are therefore still to be counted as A_1.

If the population has reached equilibrium, the allele frequency does not change and 6.2.1 can be equated to p_1. Thus,

$$p_1\bar{w} = (w_{11}P_{11} + w_{12}P_{12})(1 - u) \qquad 6.2.2$$

at equilibrium.

We now consider several special cases.

1. A_2 Completely Recessive Letting $q = p_2$ be the frequency of the mutant gene and $p = p_1$ that of the normal allele, $w_{11} = w_{12}$, $w_{22} = w_{11}(1 - s)$, $\bar{w} = w_{11}(1 - sP_{22})$, $w_{11}P_{11} + w_{12}P_{12} = w_{11}p$, and

$$P_{22} = \frac{u}{s}. \qquad 6.2.3$$

We get the entirely reasonable result that the equilibrium frequency of recessive homozygotes is the ratio of the mutation rate to the selective disadvantage of the homozygous-mutant genotype.

If mating is at random, $P_{22} = q^2$ and

$$q = \sqrt{\frac{u}{s}}. \qquad 6.2.4$$

We need to mention a caution here. Although this formula is the correct average for an infinite population, it is not if the population is small. The distribution is highly skewed and the mean frequency of recessive genes, if these are strongly selected against, can be appreciably less than this formula gives even in quite large populations. For example, for a lethal mutant in a population of 1000 with a mutation rate of 10^{-5} the frequency of mutants is only about one-third its value in an infinite population. The distribution was first given by Wright (1937) and discussed more recently by Robertson (1962) and Nei (1968). This subject is taken up in Section 9.4 (see, for example, Figure 9.4.2).

2. Highly Inbred Population As $f \to 1$, $P_{11} = p^2(1-f) + pf \to p$, $P_{12} = pq(1-f) \to 0$, $P_{22} \to q$, $\bar{w} = w_{11}(1 - qs)$, and

$$P_{22} = q = \frac{u}{s}, \tag{6.2.5}$$

irrespective of h. We expect this, because with $f = 1$ there are no heterozygotes and the value of h is irrelevant.

3. Random Mating with Complete or Partial Dominance In this case $P_{11} = p^2$, $P_{12} = pq$, $P_{22} = q^2$, and $\bar{w} = w_{11}(1 - 2pqhs - q^2 s)$. The equilibrium equation 6.2.2 becomes $\bar{w} = w_{11}(1 - qhs)(1 - u)$.

This leads to the quadratic

$$s(1 - 2h)q^2 + hs(1 + u)q - u = 0, \tag{6.2.6}$$

whose solution gives the equilibrium value of q.

If $h = 1/2$, that is to say there is no dominance, then

$$q = \frac{2u}{s(1 + u)} \approx \frac{2u}{s}. \tag{6.2.7}$$

If $h = 1$, i.e., A_2 completely dominant,

$$q \approx \frac{u}{s}, \tag{6.2.8}$$

assuming that $s \gg u$. The frequency of the mutant phenotype is roughly $2u/s$. If s can be measured this gives a basis for estimating the mutation rate.

Many mutants, probably the majority, show some heterozygous expression but have a much more severe effect in homozygotes. For example, the average newly arisen "recessive" lethal in Drosophila has about a 5% decrease of viability in heterozygotes. So it is important to consider the population consequences when h is in the range 0.01 to 0.1.

When the frequency of q is determined mainly by selection against heterozygotes, mutant homozygotes are too rare to be of any consequence and, if u is small,

$$q \approx \frac{u}{hs}. \tag{6.2.9}$$

When the frequency is determined largely by homozygous deleterious effects,

$$\lim_{h \to 0} q = \sqrt{\frac{u}{s}}, \qquad 6.2.10$$

as we saw earlier for completely recessive mutants.

Figure 6.2.1 shows the relation graphically for transitional values of h. When heterozygous selection dominates, $qhs/u = 1$. Divergence downward from the horizontal line means that homozygous selection is playing a role. The graph shows that, for large s and small u, even a small heterozygous effect dominates the situation; homozygotes are too rare to have much influence statistically. For example, if the mutant is lethal ($s = 1$) and $u = .00001$, a mutant with $>1\%$ dominance has the same kinetics as if it were completely dominant, as the graph shows.

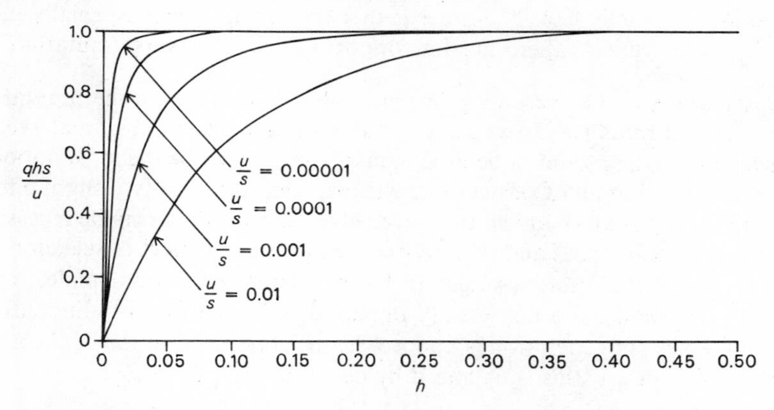

Figure 6.2.1. The equilibrium value of qhs/u as a function of h, where q is the mutant-gene frequency, s is its homozygous disadvantage, hs its heterozygous disadvantage, and u the mutation rate. When most of the elimination of mutants is by selection against heterozygotes, $qhs/u \to 1$ and the values lie along the horizontal line. If eliminations are partly through homozygotes the curve dips down and becomes 0 if all eliminations are this way.

Furthermore, there is strong evidence in Drosophila that h is larger for mutants with small s than when s is large (for a summary and references, see Crow, 1970). In general, the best evidence is that almost all the impact of natural selection on mutant genes is through their heterozygous effects.

If h is large enough that most selection is in the heterozygous state, then the difficulty in finite populations discussed above is no longer so important. With partial dominance the mean frequency of mutants is almost independent of the effective population number, although the variance is much greater in small populations of course.

4. Nonrandom Mating, Partial Dominance In this case, let $P_{11} = p^2(1-f) + pf$, $P_{12} = pq(1-f)$, and $P_{22} = q^2(1-f) + qf$. If h, f, and u are all small enough that products of three of them can be neglected, the solution of 6.2.2 becomes

$$q \approx \frac{u}{s(h+f+q)}. \qquad \textbf{6.2.11}$$

If $f = 0$, we have the situation just discussed. If $f > h$ then homozygous selection is most important. If $h > f$ (unless both are very small, or s is very small) most of the selection is in heterozygotes.

Reverse mutation is not likely to be important for two reasons. One is that the rate is usually less. The other is that the mutant gene, especially if hs is large, is very rare so there are few opportunities for reverse mutation.

X-linked Locus The case of primary interest is that where the mutant is deleterious and recessive. In contrast to the situation for autosomal recessives where a small amount of heterozygous selection may be the most important effect, a small amount of heterozygous expression has only a slight effect on the frequency of an X-linked recessive. The reason is that a single recessive gene is expressed in males and thereby exposed to the full effect of selection.

If the fitness of mutant males is $1 - s$, relative to normal males, and heterozygous females are not greatly impaired, then the rate of elimination of mutant genes per generation is about $s/3$, since one-third of the X chromosomes are in males. If this is balanced by new mutations, $u \approx qs/3$ or

$$q \approx \frac{3u}{s} \qquad \textbf{6.2.12}$$

at equilibrium. As expected, the frequency is higher than for autosomal dominants, but much less than for complete recessives.

If s is large the selective value of homozygous females is almost totally irrelevant as far as determination of gene frequencies is concerned. Very few such females exist; with a lethal there would be almost none at all. Likewise, the selection against heterozygous females is unimportant unless it gets to be comparable to s in magnitude.

6.3 Equilibrium Under Mutation Pressure

Ordinarily one regards selection as the strongest force influencing gene frequencies, with mutation providing a steady input of new variability. On the other hand, there is growing evidence that some, perhaps a large fraction, of DNA changes are nearly enough neutral that mutation rates become an appreciable factor. With such weak forces we should expect the consequences of random gene frequency drift to be important. That we shall consider later in Section 7.2. In this section we shall treat the subject deterministically, as if the population were infinite.

Consider a series of n alleles $A_1, A_2, A_3, \ldots, A_n$. The decrement in frequency of A_i due to mutation from A_i to other alleles is

$$\Big(\sum_j u_{ij}\Big) p_i,$$

where u_{ij} means the rate per generation of mutation from A_i to A_j and p_i, as usual, is the frequency of A_i. The increment in A_i alleles will be the sum of the mutation rates from other alleles to A_i, or

$$\sum_j u_{ji} p_j .$$

Thus the net increase in frequency of A_i per generation is

$$\Delta p_i = -p_i \sum_j u_{ij} + \sum_j u_{ji} p_j , \qquad 6.3.1$$

as given by Wright (1949, for example).

The equilibrium is obtained by setting $\Delta p_i = 0$ for each value of i and solving the series of simultaneous equations. For two alleles, $p_2 = 1 - p_1$ and we have

$$\Delta p_1 = -p_1 u_{12} + (1 - p_1) u_{21},$$

and when $\Delta p_1 = 0$,

$$p_1 = \frac{u_{21}}{u_{12} + u_{21}} = \frac{v}{u + v}, \qquad 6.3.2$$

where u and v are the mutation rates from and to allele A_1.

For three alleles we have the equilibrium equations

$$\begin{aligned} -(u_{12} + u_{13})p_1 + u_{21}p_2 + u_{31}p_3 &= 0, \\ u_{12}p_1 - (u_{21} + u_{23})p_2 + u_{32}p_3 &= 0, \\ u_{13}p_1 + u_{32}p_2 - (u_{31} + u_{32})p_3 &= 0. \end{aligned} \qquad 6.3.3$$

These three equations are not independent, since $p_1 + p_2 + p_3 = 1$. We can reduce this to two independent equations by replacing p_3 by $1 - p_1 - p_2$ in two of the equations and solving. A convenient way of getting the solution is to write the matrix of the coefficients

$$\begin{pmatrix} -(u_{12} + u_{13}) & u_{21} & u_{31} \\ u_{12} & -(u_{21} + u_{23}) & u_{32} \\ u_{13} & u_{23} & -(u_{31} + u_{32}) \end{pmatrix} = A. \qquad 6.3.4$$

Then

$$p_1 = \frac{\Delta_{11}}{D},$$

$$p_2 = \frac{\Delta_{22}}{D}, \tag{6.3.5}$$

$$p_3 = \frac{\Delta_{33}}{D},$$

where Δ_{11} is the 2×2 detriment obtained by deleting row 1 and column 1 from A. Likewise Δ_{22} is gotten by deleting the second row and column, and so on. D, the denominator, is the sum of the three numerators. D is also the value of the determinant obtained by replacing any row in A by 1's. (Some simple rules for evaluation of determinants are given in A.7.)

The extension to four or more alleles is straightforward. The determinant is larger, but the rules are the same.

With typically low mutation rates the approach to equilibrium is very slow. For example, with two alleles,

$$\Delta p_1 = -u_{12} p_1 + u_{21} p_2 . \tag{6.3.6}$$

Now let

$$p_1 = \hat{p}_1 + \xi,$$
$$p_2 = \hat{p}_2 - \xi,$$

where $\hat{p}_1$ and $\hat{p}_2$ are equilibrium values. Then substituting into 6.3.6 we get

$$\Delta\xi = \Delta p_1 = -(u_{12} + u_{21})\xi - u_{12}\hat{p}_1 + u_{21}\hat{p}_2 .$$

But the last two terms add up to 0 (see 6.3.2), so

$$-\frac{\Delta\xi}{\xi} = u_{21} + u_{12} = \sum u. \tag{6.3.7}$$

The approach to equilibrium is at a rate equal to the total mutation rate. If $\sum u = 10^{-5}$, about 70,000 generations are required to go halfway to equilibrium. So, to nobody's surprise, we learn that the approach to mutational equilibrium is a very slow process.

6.4 Mutation and Selection with Multiple Alleles

When discussing the equilibrium between mutation and selection we considered only two alleles and ignored reverse mutation. It is expected that the wild-type gene can mutate to many different states, as we have discussed in the previous section. If the mutant states are individually rare and are selected mainly as heterozygotes then there will be little chance for interaction, since hardly ever will more than one of a set of mutant alleles occur in the same individual. The total frequency will be governed mainly by the total mutation rate from the normal to all mutant alleles.

Experience with Drosophila mutants shows that the combinations of two recessive mutants with visible effects are often intermediate between the two homozygous mutants. In this case there will be little interaction. So we suspect that in most cases multiple allelism doesn't change the picture much.

We shall consider the question only briefly. For this it will be expedient to adopt a continuous model and assume that the gene frequency change can be written in the form

$$\frac{dp_i}{dt} = p_i(m_i - \overline{m}) - p_i \sum_j u_{ij} + \sum_j u_{ji} p_j. \qquad 6.4.1$$

We get this by combining 5.3.5 with 6.3.1.

For reasons mentioned in Section 6.2 reverse mutation can be ignored with little loss of accuracy if the deleterious effects of the mutants are large relative to the mutation rates. If the last term is 0, by setting $dp_i/dt = 0$ for equilibrium we obtain the pleasingly simple result

$$m_0 - \overline{m} = \sum_j u_{0j}, \qquad 6.4.2$$

where the subscript 0 designates the normal allele (which, of course, may be a population of indistinguishable isoalleles) and the j indicates any mutant allele.

In words: At equilibrium the average excess in fitness of the wild-type allele is equal to its total rate of mutation to all mutant alleles.

This depends on the fitness being measured in Malthusian parameters. It also assumes that reverse mutation to the wild-type allele can be neglected. However, no restriction is placed on the number of alleles, on the mating system, or on the rate of mutation from one to another mutant allele. We shall use this principle to find the equilibrium frequencies of mutant genes.

1. Fully Recessive Mutants We place no restriction on the interaction between mutant alleles, but all are recessive to the wild type, A_1. For algebraic simplicity we set the fitness of the wild type at 0, measured in Malthusian parameters. Let the fitness of the mutant type $A_i A_j$ be $-s_{ij}$. Then from 6.4.2,

$$m_0 - \overline{m} = \sum_{i,j} P_{ij} s_{ij} = \sum_i u_{0i}, \qquad 6.4.3$$

where the summation extends over the $n - 1$ mutant alleles. The average reduction in fitness caused by mutant phenotypes (homozygous mutants and combinations of two mutants) is

$$\bar{s} = \frac{\sum P_{ij} s_{ij}}{\sum P_{ij}}, \tag{6.4.4}$$

from which

$$\sum P_{ij} = \frac{\sum u_{0i}}{\bar{s}}. \tag{6.4.5}$$

This says that the total frequency of mutant phenotypes is the ratio of the total mutation rate (forward) divided by the (weighted) average selective disadvantage of the mutant phenotypes. The fitness is measured in Malthusian parameters.

For two alleles the discrete generation analogy is 6.2.3.

2. Partially or Completely Dominant Mutants Once again we let the wild-type ($A_0 A_0$) fitness be 0. The fitness of the mutant homozygote $A_i A_i$ is $-s_{ii}$, of the mutant heterozygote $A_i A_j$ is $-s_{ij}$, and of the normal-mutant heterozygote $A_0 A_i$ is $-h_i s_{ii}$. From 6.4.2,

$$-\sum_i \frac{h_i s_{ii} P_{0i}}{p_0} - \bar{m} = \sum_i u_{0i}. \tag{6.4.6}$$

But,

$$-\bar{m} = 2 \sum_i h_i s_{ii} P_{0i} + \sum_{ij} P_{ij} s_{ij}.$$

where, as before, the summation is over all mutant alleles. If there is appreciable dominance the last term is very small relative to the others. Omitting this, and noting that

$$\frac{\Sigma P_{0i} h_i s_{ii}}{\Sigma P_{0i}} = \overline{hs},$$

we obtain

$$\sum P_{0i} = \frac{\sum u_{0i}}{\overline{hs}} \times \frac{p_0}{2p_0 - 1} \approx \frac{\sum u_{0i}}{\overline{hs}} \times p_0. \tag{6.4.7}$$

With Hardy–Weinberg proportions $P_{0i} = p_0 p_i$, and

$$\sum p_i \approx \frac{\sum u_{0i}}{\overline{hs}}. \tag{6.4.8}$$

So, as a first approximation based on strong selection against heterozygotes, the total frequency of mutant alleles is the total forward mutation rate divided by the average selection against heterozygotes.

With weaker selection the situation is more complex, and we shall go no further with it.

6.5 Selection and Migration

The subject of migration and population structure has become very extensive in recent years. Various models of structure ranging from a continuous population with limited dispersal to completely isolated islands have been studied. We shall consider only one model, the island model of Sewall Wright (see, for example, his 1951 paper). Later, in Sections 9.2 and 9.9, this and other models will be treated stochastically.

The species is thought of as broken into a number of subpopulations, largely isolated from each other, but with some exchange of migrants. We let p_i be the frequency of an allele of interest in the ith subpopulation, and $\bar{p}$ be the frequency in the whole population. In Wright's model a fraction M of the population are replaced by migrants each generation. The migrants are assumed to have a gene frequency equal to that for the whole population $\bar{p}$.

The change in allele frequency in the ith subpopulation per generation is

$$\Delta p_i = -Mp_i + M\bar{p} = -M(p_i - \bar{p}). \qquad \text{6.5.1}$$

Since on this model the effect of migration is linear in the gene frequencies, there should be a correspondence with mutation which is also linear. If we rewrite 6.5.1,

$$\begin{aligned} \Delta p_i &= -Mp_i + M\bar{p} - Mp_i\bar{p} + Mp_i\bar{p} \\ &= -M(1-\bar{p})p_i + M\bar{p}(1-p_i), \end{aligned} \qquad \text{6.5.2}$$

we see that this corresponds in form to

$$\Delta p_1 = -u_{12}p_1 + u_{21}(1-p_1). \qquad \text{6.5.3}$$

So we can carry the same equations from mutation to migration by setting

$$\begin{aligned} M(1-\bar{p}) &= u_{12} = u, \\ M\bar{p} &= u_{21} = v. \end{aligned} \qquad \text{6.5.4}$$

We shall use this correspondence in Section 9.2.

In analogy with 6.4.1 we may regard selection and migration both as continuous processes and write

$$\frac{dp_i}{dt} = p_i(m_i - \bar{m}) - M(p_i - \bar{p}). \qquad \text{6.5.5}$$

We must keep straight that $\bar{m}$ is the average fitness of the allele of interest in the ith subpopulation whereas $\bar{p}$ is the average allele frequency in the entire population. We are using m for the Malthusian parameter of selection and M for migration.

It is to be expected that the immigrants coming into a subpopulation from the outside will be less well adapted to the local conditions since they have not had the benefit of previous selection in the local environment. This is quite analogous to an input of different genes by mutation (although the migrant genes have been pretested in an environment that is not wholly unrelated, and hence are likely to be less deleterious than new mutants).

To find equilibrium conditions we can equate 6.5.5 to 0. In general the selection term will be a cubic function of the gene frequencies. In the absence of dominance the expression is only quadratic, so we shall consider this. Using 5.3.8 and equating dp/dt to 0 we have at equilibrium

$$sp(1-p) - M(p-\bar{p}) = 0, \tag{6.5.6}$$

where, dropping the subscripts, p is the frequency of the gene (say A) that is locally favored, s is the selective advantage of AA, and $s/2$ that of Aa, both measured in Malthusian parameters. The relevant solution is

$$p = \frac{s - M + \sqrt{(M-s)^2 + 4Ms\bar{p}}}{2s}. \tag{6.5.7}$$

For example, if $M = s$, $p = \sqrt{\bar{p}}$.

It is clear that if migration is large and selection weak all the populations will tend to become alike. On the other hand, if $|s| \gg M$ and the selective values differ from one subpopulation to another, there will be considerable local differentiation. This model has been extensively discussed by Wright (1940, 1951).

6.6 Equilibrium Between Migration and Random Drift

If the population is broken up into subpopulations and these subpopulations are small there will be random drift of gene frequencies among the subpopulations so that they will drift apart. Migration from one to another will counteract this effect. We shall discuss this briefly now by elementary procedures and then return to it in Section 9.2 with a more sophisticated stochastic treatment.

As in the last section we let M be the amount of exchange each generation by migration. From equation 3.11.1 the increase in autozygosity in a subpopulation is given by

$$f_t = \frac{1}{2N_e} + \left(1 - \frac{1}{2N_e}\right) f_{t-1}, \tag{6.6.1}$$

where N_e is the effective number in the subpopulation. However, the genes will remain identical only if the individuals carrying them have not been replaced by migrants. The probability that neither of the two uniting genes has been exchanged for a migrant gene is $(1 - M)^2$. (This is not quite exact because our model assumes that self-fertilization is possible, but unless N_e is very small this is a trivial correction.)

The equation, including a correction for random exchange of genes with outsiders, is

$$f_t = \left[\frac{1}{2N_e} + \left(1 - \frac{1}{2N_e}\right)f_{t-1}\right](1 - M)^2. \qquad 6.6.2$$

Letting $f_t = f_{t-1} = f$ for equilibrium,

$$f = \frac{(1 - M)^2}{2N_e - (2N_e - 1)(1 - M)^2}. \qquad 6.6.3$$

When M is small so that M^2 can be neglected,

$$f \approx \frac{1 - 2M}{4N_e M + 1 - 2M} \approx \frac{1}{4N_e M + 1}. \qquad 6.6.4$$

We shall find exactly the same formula for mutation in Section 7.2. This is not surprising because of the mathematical equivalence of mutation and migration in Wright's migration model.

If $M \ll 1/4N_e$, then f becomes large and there is considerable local autozygosity. Contrariwise, if $M \gg 1/4N_e$, the migration swamps the local subpopulation and the whole thing becomes effectively one panmictic unit.

It is impressive that the amount of migration needn't be very great. A fraction $1/N_e$ of the population means one individual in a population of N_e, so if the number of migrants is much more than one per generation, there is little local differentiation. This is mitigated somewhat, however, because migrants tend to come from neighboring subpopulations, rather than being a random sample of the entire population, and neighboring genes are likely to be somewhat alike. Hence the swamping effect of migration may be less.

Equation 3.12.3 connects the inbreeding coefficient of a subpopulation with the variance in gene frequency caused by random variation among the subpopulations. From this

$$f = \frac{V_p}{\bar{p}(1 - \bar{p})}, \qquad 6.6.5$$

where V_p is the variance in gene frequency among subpopulations and $\bar{p}$

is the average frequency in the whole population. Substituting this into 6.6.4 gives

$$V_p = \frac{\bar{p}(1-\bar{p})}{4N_e M + 1}, \tag{6.6.6}$$

which is Wright's formula relating the amount of random differentiation among partially isolated subpopulations to the amount of migration.

We shall derive the same formula again later (9.2.10) by a quite different procedure.

6.7 Equilibrium Under Selection: Single Locus with Two Alleles

We start with a simple case—one locus, random mating, and two alleles. We shall use a discrete generation model; in general the conclusions are the same with a continuous model.

Suppose the generations are enumerated as zygotes and that the fitnesses of the three genotypes A_1A_1, A_1A_2, and A_2A_2 are w_{11}, w_{12}, and w_{22}. With random mating the zygote frequencies, counted before selection, will be p_1^2, $2p_1p_2$, and p_2^2, where p_1 and p_2 are the allele frequencies and $p_1 + p_2 = 1$.

From 5.2.13,

$$\Delta p_1 = \frac{p_1p_2[p_1(w_{11}-w_{12}) + p_2(w_{12}-w_{22})]}{\bar{w}}. \tag{6.7.1}$$

Equating this to 0 to find the equilibrium values gives

$$\hat{p}_1 = \frac{w_{22}-w_{12}}{w_{11}-2w_{12}+w_{22}}, \tag{6.7.2}$$

in addition to the two trivial values $p_1 = 0$ and $p_1 = 1$.

We can notice immediately that some restrictions are imposed by the fact that p_1 must lie in the range 0 to 1 and therefore cannot be negative. If we write the denominator as $(w_{11} - w_{12}) + (w_{22} - w_{12})$ we see that $0 < p_1 < 1$ only if both terms in the denominator have the same sign. So we see that one condition for an equilibrium is

$$w_{11} < w_{12} > w_{22}$$

or

$$w_{11} > w_{12} < w_{22}. \tag{6.7.3}$$

To have an equilibrium other than 0 or 1 the heterozygote must either be less fit or more fit than either homozygote. Otherwise the only equilibria are $p_1 = 0$ or $p_1 = 1$.

To investigate the stability of the equilibrium consider a small displacement ξ from the equilibrium $\hat{p}_1$, so $p_1 = \hat{p}_1 + \xi$. Then p_2 must be $\hat{p}_2 - \xi$ and, noting that $\Delta\hat{p}_1 = 0$,

$$\begin{aligned}\Delta\xi &= \Delta p_1 - \Delta\hat{p}_1 \\ &= \Delta p_1 \approx \hat{p}_1\hat{p}_2[(\hat{p}_1 + \xi)(w_{11} - w_{12}) + (\hat{p}_2 - \xi)(w_{12} - w_{22})]/\bar{w} \\ &= \hat{p}_1\hat{p}_2(w_{11} - 2w_{12} + w_{22})\xi/\bar{w},\end{aligned} \qquad \text{6.7.4}$$

since

$$\hat{p}_1(w_{11} - w_{12}) + \hat{p}_2(w_{12} - w_{22}) = 0.$$

When $w_{11} - 2w_{12} + w_{22} < 0$, $\Delta\xi/\xi$ is negative, and its absolute value is less than one. This means that $\hat{p}_1$ is a stable equilibrium in that a displacement from this point is followed by a tendency to return to the point.

When $w_{11} - 2w_{12} + w_{22} > 0$, then $\Delta\xi/\xi$ is positive. A displacement is followed by an even larger displacement in the same direction, so the equilibrium is unstable.

Selection favoring the heterozygote leads to a stable equilibrium, with the equilibrium value of p_1 given by 6.7.2. If the heterozygote is selected against, the population tends to move away from the unstable equilibrium toward $p_1 = 0$ or $p_1 = 1$. Whether the A_1 or the A_2 allele is fixed depends on which side of the equilibrium point the population starts from.

A locus with a superior heterozygote tends to persist in the polymorphic state so that it contributes quite disproportionately to the variability of the population. A locus with an inferior heterozygote ultimately is fixed at gene frequency 0 or 1 and does not contribute to the variability.

There is another way of testing the stability of an equilibrium. We saw in Section 5.5 that under natural selection with a single locus, random mating, and constant fitness coefficients the fitness always increases. With a continuous model this follows from the Fundamental Theorem of Natural Selection. With a discontinuous model it follows from 5.5.14.

The mean fitness is

$$\bar{w} = p_1^2 w_{11} + 2p_1 p_2 w_{12} + p_2^2 w_{22}, \qquad \text{6.7.5}$$

from which, recalling that $p_2 = 1 - p_1$,

$$\frac{d\bar{w}}{dp_1} = 2p_1(w_{11} - w_{12}) + 2p_2(w_{21} - w_{22}) = 0, \qquad \text{6.7.6}$$

leading to the same equation as before,

$$p_1 = \frac{w_{22} - w_{12}}{w_{11} - 2w_{12} + w_{22}}.$$

So, as expected, the equilibrium point is where the fitness is stationary—either maximum or minimum.

Differentiating again,

$$\frac{d^2\bar{w}}{dp_1^2} = 2(w_{11} - 2w_{12} + w_{22}). \tag{6.7.7}$$

When this is negative the equilibrium point is a relative maximum and therefore stable. When the second derivative is positive the equilibrium is unstable.

When there are multiple alleles the conditions for a maximum are more involved, as we shall see in the next section.

For a particularly lucid elementary exposition of the various kinds of equilibria under selection, see Li (1967*b*).

6.8 Selective Equilibrium with Multiple Alleles

Again we assume random mating and a discrete model. There are n alleles $A_1, A_2, \ldots, A_n$ with frequencies $p_1, p_2, \ldots, p_n$. Let w_{ij} be the fitness of genotype A_iA_j. Then, from 5.2.1, the change in frequency of allele A_i in one generation is

$$\Delta p_i = \frac{p_i(w_i - \bar{w})}{\bar{w}}, \tag{6.8.1}$$

where

$$\bar{w} = \sum_{ij} p_i p_j w_{ij}$$

and

$$w_i = \sum_j p_j w_{ij}.$$

At equilibrium when $\Delta p_i = 0$,

$$w_i = \bar{w}, \tag{6.8.2}$$

and we reach the quite reasonable conclusion that at equilibrium the average fitness of all alleles is the same and equal to the population average fitness. If this were not true, those alleles with the higher fitnesses would increase and the population would not be in equilibrium.

To investigate the stability of the equilibrium we should like to use the principle employed in the preceding section—that fitness is maximized at the point of stable equilibrium. For the continuous model we know this from 5.5.9, which extends readily to multiple alleles. We also know from 5.5.14 that this is true with a discrete generation model for random mating with two alleles.

We would expect this to be true for multiple alleles also, and this has been shown. For a discussion of this problem, see Scheuer and Mandel (1959), Mulholland and Smith (1959), and Kingman (1961, 1961*a*). We shall now demonstrate this. If you are willing to accept this without proof, you may wish to skip this section and proceed directly to equation 6.8.11.

Demonstration That Fitness Always Increases in the Neighborhood of the Equilibrium If we use a caret to designate equilibrium values,

$$\hat{w}_i = \sum_j w_{ij} \hat{p}_j = \hat{w} = \sum_{ij} w_{ij} \hat{p}_i \hat{p}_j. \tag{6.8.3}$$

To simplify the calculations, we express the fitnesses relative to the mean fitness at equilibrium, so we let

$$W_{ij} = \frac{w_{ij}}{\hat{w}}.$$

In this measure

$$\hat{W}_i = \sum_j W_{ij} \hat{p}_j = \hat{W} = 1. \tag{6.8.4}$$

The formula for Δp_i is the same in the new units;

$$\Delta p_i = \frac{p_i(W_i - \overline{W})}{\overline{W}}. \tag{6.8.5}$$

Our interest now is in the behavior of $\overline{W}$ near the equilibrium point. Let ξ_i be a small deviation from $\hat{p}_i$, as in the last section, so that

$$p_i = \hat{p}_i + \xi_i \qquad (\sum \xi_i = 0).$$

The change in $\hat{W}$ due to these displacements is

$$\delta \hat{W} = \sum_{ij} W_{ij}(\hat{p}_i + \xi_i)(\hat{p}_j + \xi_j) - 1$$

$$= 2 \sum_{ij} W_{ij} \hat{p}_j \xi_i + \sum_{ij} W_{ij} \xi_i \xi_j.$$

But since

$$\sum_{ij} W_{ij} \hat{p}_j \xi_i = \sum_i \xi_i = 0,$$

we have

$$\delta \hat{W} = \sum_{ij} W_{ij} \xi_i \xi_j. \tag{6.8.6}$$

Similarly, if we consider the change in W_i due to these displacements,

$$\begin{aligned}\delta \hat{W}_i &= \sum_j W_{ij}(\hat{p}_j + \xi_j) - \sum_j W_{ij}\hat{p}_j \\ &= \sum_j W_{ij}\xi_j \equiv \varepsilon_i,\end{aligned} \tag{6.8.7}$$

so

$$\delta \hat{W}_i = \varepsilon_i. \tag{6.8.8}$$

The change in gene frequencies in one generation near the equilibrium point, from 6.8.5, 6.8.6, and 6.8.8 is approximately

$$\Delta\xi_i = \hat{p}_i \varepsilon_i. \tag{6.8.9}$$

The change in mean fitness $\overline{W}$ in one generation due to natural selection is

$$\begin{aligned}\Delta \overline{W} &= \sum_{ij} W_{ij}(p_i + \Delta p_i)(p_j + \Delta p_j) - \sum W_{ij} p_i p_j \\ &= 2\sum_i W_i \,\Delta p_i + \sum_{ij} W_{ij}\,\Delta p_i\,\Delta p_j,\end{aligned}$$

which, in the neighborhood of the equilibrium point, reduces to

$$\begin{aligned}\Delta \overline{W} &= 2\sum_i (\hat{W}_i + \delta\hat{W}_i)\,\Delta\xi_i + \sum_{ij} W_{ij}\,\Delta\xi_i\,\Delta\xi_j \\ &= 2\sum_i \hat{p}_i \varepsilon_i^2 + \sum_{ij} W_{ij}\hat{p}_i\hat{p}_j\varepsilon_i\varepsilon_j,\end{aligned}$$

from 6.8.4, 6.8.8, and 6.8.9.

Now consider two quantities,

$$A = 2\sum_i \hat{p}_i \varepsilon_i^2$$

and

$$B = \sum_{ij} W_{ij}\hat{p}_i\hat{p}_j(\varepsilon_i + \varepsilon_j)^2,$$

both of which are nonnegative. Since

$$B = 2\sum_{ij} W_{ij} p_i p_j \varepsilon_i^2 + 2\sum_{ij} W_{ij}\hat{p}_i\hat{p}_j\varepsilon_i\varepsilon_j,$$

then

$$\begin{aligned}\sum_{ij} W_{ij}\hat{p}_i\hat{p}_j\varepsilon_i\varepsilon_j &= \frac{B}{2} - \sum_{ij} W_{ij}\hat{p}_i\hat{p}_j\varepsilon_i^2 \\ &= \frac{B}{2} - \sum_i \hat{W}_i\hat{p}_i\varepsilon_i^2 \\ &= \frac{1}{2}(B - A),\end{aligned}$$

since $\hat{W}_i = 1$.

Using these quantities, the change of $\overline{W}$ in one generation near the equilibrium is

$$\Delta\overline{W} = A + \tfrac{1}{2}(B - A) = \tfrac{1}{2}(A + B) \geq 0. \tag{6.8.10}$$

This shows for multiple alleles, as 5.5.14 did for two alleles, that (with random mating and constant w_{ij}'s) the fitness always increases unless the ε_i's are all 0, that is to say, unless the population is already at equilibrium.

We now use this principle to derive stability conditions for the multiple allele case.

Equilibrium Allele Frequencies The gene frequencies at equilibrium may be obtained directly by solving the simultaneous equations from 6.8.2 using the definitions of the w_i's (see 5.2.5).

$$\begin{aligned} w_{11}p_1 + w_{12}p_2 + \cdots + w_{1n}p_n &= \overline{w} \\ w_{21}p_1 + w_{22}p_2 + \cdots + w_{2n}p_n &= \overline{w} \\ w_{n1}p_1 + w_{n2}p_2 + \cdots + w_{nn}p_n &= \overline{w} \end{aligned} \tag{6.8.11}$$

The solutions, from Cramer's rule, are

$$\hat{p}_i = \frac{\hat{w}\,\Delta_i}{\Delta} = \frac{\Delta_i}{\sum \Delta_i}, \tag{6.8.12}$$

where

$$\Delta = \begin{vmatrix} w_{11} & w_{12} & \cdots & w_{1n} \\ w_{21} & w_{22} & \cdots & w_{2n} \\ \vdots & \vdots & & \vdots \\ w_{n1} & w_{n2} & \cdots & w_{nn} \end{vmatrix} \tag{6.8.13}$$

and Δ_i is the determinant gotton by substituting 1's for all the elements in the ith column of the determinant Δ. The latter part of 6.8.12 follows from the fact that $\sum p_i = 1$. The average fitness at equilibrium is $\hat{w} = \Delta/\sum \Delta_i$.

Stability of the Equilibrium We have seen that fitness always increases except at the equilibrium, so the stability condition is that $\overline{w}$ is a relative maximum. This means that $\delta\overline{w}$ must always be positive for small deviations from the equilibrium. Mathematically this is assured if the quadratic form 6.8.6 is *negative definite* (see below for definition).

Since $\sum \xi_i = 0$, then

$$\xi_n = -\sum_{i=1}^{n-1} \xi_i. \tag{6.8.14}$$

Substituting this into 6.8.6, and using the original fitness scale,

$$\delta\hat{w} = \sum_{i=1}^{n-1} \sum_{j=1}^{n-1} t_{ij} \xi_i \xi_j, \qquad \textbf{6.8.15}$$

where

$$t_{ij} = w_{ij} - w_{in} - w_{jn} + w_{nn} \qquad (w_{ij} = w_{ji}). \qquad \textbf{6.8.16}$$

The quadratic form on the right side of 6.8.15 is negative definite if and only if

$$t_{11} < 0, \qquad \begin{vmatrix} t_{11} & t_{12} \\ t_{21} & t_{22} \end{vmatrix} > 0, \qquad \begin{vmatrix} t_{11} & t_{12} & t_{13} \\ t_{21} & t_{22} & t_{23} \\ t_{31} & t_{32} & t_{33} \end{vmatrix} < 0, \qquad \textbf{6.8.17}$$

and so on up to order $n - 1$.

The second condition is that all gene frequencies be positive. This can be done as follows. Let $T = [t_{ij}]$ be the matrix of the quadratic form. It turns out that the determinant $|T|$ is equal to the denominator in the right term of 6.8.12. So

$$|T| = \sum_{i=1}^{n} \Delta_i .$$

However, 6.8.17 requires that

$$(-1)^{n-1}|T| > 0,$$

so

$$(-1)^{n-1}\,\Delta_i > 0 \qquad (i = 1, 2, \ldots, n) \qquad \textbf{6.8.18}$$

in order that all the gene frequencies are positive.

To summarize: The conditions for a stable equilibrium are 6.8.17 and 6.8.18.

It may be convenient mnemonically to remember that t_{ij} may be written symbolically as $(i - n)(j - n)$, which gives $ij - in - jn + nn$.

Exactly the same arguments apply to the continuous model. The necessary and sufficient conditions are the same, except that the w_{ij}'s are replaced by Malthusian parameters.

In fact the first report, by Kimura (1956*a*), was for a continuous model. Mandel's (1959) formulae for the discrete case are equivalent.

Those reading Mandel's papers can verify that, for example, his condition

$$\begin{vmatrix} a_{11} & a_{1n} & 1 \\ a_{n1} & a_{nn} & 1 \\ 1 & 1 & 0 \end{vmatrix} > 0$$

is the same as $t_{11} < 0$, since the expanded determinant is

$$-(a_{11} - a_{1n} - a_{n1} + a_{nn}) = -t_{11}.$$

In the case of two alleles the conditions become

$$(1) \quad t_{11} = w_{11} - 2w_{12} + w_{22} > 0,$$

$$(2) \quad -\Delta_1 = -\begin{vmatrix} 1 & w_{12} \\ 1 & w_{22} \end{vmatrix} = w_{12} - w_{22} > 0, \qquad \textbf{6.8.19}$$

$$-\Delta_2 = -\begin{vmatrix} w_{11} & 1 \\ w_{21} & 1 \end{vmatrix} = w_{21} - w_{11} > 0.$$

These are equivalent to the condition $w_{11} < w_{12} > w_{22}$ that we discussed earlier.

For three alleles

$$\Delta = \begin{vmatrix} w_{11} & w_{12} & w_{13} \\ w_{21} & w_{22} & w_{23} \\ w_{31} & w_{32} & w_{33} \end{vmatrix},$$

$$\Delta_1 = \begin{vmatrix} 1 & w_{12} & w_{13} \\ 1 & w_{22} & w_{23} \\ 1 & w_{32} & w_{33} \end{vmatrix}, \qquad \Delta_2 = \begin{vmatrix} w_{11} & 1 & w_{13} \\ w_{21} & 1 & w_{23} \\ w_{31} & 1 & w_{33} \end{vmatrix}, \qquad \Delta_3 = \begin{vmatrix} w_{11} & w_{12} & 1 \\ w_{21} & w_{22} & 1 \\ w_{31} & w_{32} & 1 \end{vmatrix},$$

$$t_{11} = (1-3)^2 = w_{11} - w_{13} - w_{31} + w_{33},$$

$$t_{12} = (1-3)(2-3) = w_{12} - w_{13} - w_{32} + w_{33} = t_{21},$$

$$t_{22} = (2-3)^2 = w_{22} - w_{23} - w_{32} + w_{33}.$$

$$(1) \quad t_{11} < 0, \qquad \begin{vmatrix} t_{11} & t_{12} \\ t_{21} & t_{22} \end{vmatrix} > 0,$$

$$(2) \quad \Delta_1 > 0, \qquad \Delta_2 > 0, \qquad \Delta_3 > 0.$$

But, since

$$\begin{vmatrix} t_{11} & t_{12} \\ t_{21} & t_{22} \end{vmatrix} = \Delta_1 + \Delta_2 + \Delta_3,$$

in the 3-allele case the necessary and sufficient conditions are:

$$\Delta_1 > 0, \qquad \Delta_2 > 0, \qquad \Delta_3 > 0, \qquad t_{11} < 0. \qquad \textbf{6.8.20}$$

With multiple alleles it is not necessary that each heterozygote be fitter than any homozygote; there can be a stable equilibrium without this. On the other hand, it is not sufficient for all heterozygotes to be superior; some such systems are unstable. However, the population fitness, $\bar{w}$, at equilibrium must exceed the fitness of any homozygote (Mandel, 1959). For a general review and extensions to inbred and polyploid populations, see Li (1967*b*).

6.9 Some Other Equilibria Maintained by Balanced Selective Forces

The number of possibilities for stable and unstable equilibria is enormous, even when only a single locus is considered, as we have just seen. When other complications are added, such as multiple loci, linkage, and variable selection coefficients, the possibilities become enormous. We shall mention only a few.

Sex-linked Polymorphism Consider a single locus with two alleles and with random mating. Let p_1^* and p_2^* be the frequencies of A_1 and A_2 in the male gametic output and, similarly, $p_1^{**}$and p_2^{**} be the corresponding frequencies in the female gametic output.

If we denote the fitnesses of A_1 and A_2 males by w_1 and w_2, and those of A_1A_1, A_1A_2, and A_2A_2 females by w_{11}, w_{12}, w_{22}, then the frequencies of these genotypes immediately after fertilization are as follows:

	MALE		FEMALE		
GENOTYPE	A_1	A_2	A_1A_1	A_1A_2	A_2A_2
FREQUENCY	p_1^{**}	p_2^{**}	$p_1^*p_1^{**}$	$p_1^*p_2^{**}+p_1^{**}p_2^*$	$p_2^*p_2^{**}$
FITNESS	w_1	w_2	w_{11}	w_{12}	w_{22}

Then the frequencies of A_1 in the male and female gametic outputs for the next generation are

$$p_1^{*\prime} = \frac{w_1 p_1^{**}}{\bar{w}^*},$$

$$p_1^{**\prime} = \frac{w_{11}p_1^*p_1^{**} + \frac{1}{2}w_{12}(p_1^*p_2^{**} + p_1^{**}p_2^*)}{\bar{w}^{**}}, \qquad 6.9.1$$

where $\bar{w}^*$ and $\bar{w}^{**}$ are respectively the mean fitness in male and females, and $p_2^* = 1 - p_1^*$, $p_2^{**} = 1 - p_1^{**}$.

In the following calculation, we will consider gene ratios rather than gene frequencies. Let

$$x = \frac{p_1^*}{p_2^*}$$

and

$$y = \frac{p_1^{**}}{p_2^{**}}$$

be respectively the gene ratios in male and female gametic outputs; then 6.9.1 is equivalent to

$$x' = \left(\frac{w_1}{w_2}\right) y$$

$$y' = \frac{w_{11}xy + \dfrac{w_{12}}{2}(x + y)}{\dfrac{w_{12}}{2}(x + y) + w_{22}}. \tag{6.9.2}$$

At equilibrium in which $x' = x$, and $y' = y$, letting

$$\frac{w_1}{w_2} = k,$$

we obtain from 6.9.2.

$$y = \frac{2w_{11}ky^2 + w_{12}(ky + y)}{w_{12}(ky + y) + 2w_{22}}.$$

By solving this for y and noting that $x = ky$, we obtain the gene ratios at equilibrium;

$$\hat{y} = \frac{(k + 1)w_{12} - 2w_{22}}{(k + 1)w_{12} - 2kw_{11}},$$

$$\hat{x} = k\hat{y}. \tag{6.9.3}$$

Since $\hat{y}$ is positive, it is necessary that either

$$(k + 1)w_{12} - 2w_{22} > 0, \qquad (k + 1)w_{12} - 2kw_{11} > 0,$$

or $\tag{6.9.4}$

$$(k + 1)w_{12} - 2w_{22} < 0, \qquad (k + 1)w_{12} - 2kw_{11} < 0.$$

Let us now investigate the conditions for stability of the equilibrium. From 6.9.2, considering the change in y over two generations, we get

$$y'' = \frac{2w_{11}kyy' + w_{12}(ky + y')}{w_{12}(ky + y') + 2w_{22}} \tag{6.9.5}$$

First, consider the situation in which the frequency of A_1 is very low in the population so that y is very near to 0. Neglecting higher order terms in 6.9.5, we get

$$y'' = \frac{w_{12}}{2w_{22}}(ky + y'),$$

or

$$2w_{22}y'' - w_{12}y' - w_{12}ky = 0. \tag{6.9.6}$$

For the equilibrium to be stable, it is necessary that y tends to increase. Using the procedures of Section 3.8, the characteristic equation of 6.9.6 is

$$2w_{22}\lambda^2 - w_{12}\lambda - w_{12}k = 0.$$

It has two roots,

$$\lambda = \frac{w_{12} \pm \sqrt{w_{12}^2 + 8w_{12}w_{22}k}}{4w_{22}},$$

one positive and the other negative and the former has the larger absolute value. For y to increase, it is necessary that this root be larger than unity. The condition for this is satisfied if

$$w_{12}(1 + k) - 2w_{22} > 0. \tag{6.9.7}$$

(Note that if $w_{12} > 4w_{22}$, the larger root is larger than unity but this condition is weaker than 6.9.7.)

Next, consider the situation in which the frequency of A_1 is very high, or the frequency of its allele A_2 is very low. In this situation it is necessary that A_2 tend to increase for the equilibrium to be stable. From the consideration of symmetry, it is easily seen that A_2 tends to increase if

$$w_{12}(1 + k^{-1}) - 2w_{11} > 0. \tag{6.9.8}$$

Conditions 6.9.7 and 6.9.8 satisfy one of the set of conditions in 6.9.4. Thus the conditions for the equilibrium to be stable are

$$\left.\begin{aligned} w_{12} &> \left(\frac{2}{1+k}\right)w_{22}, \\ w_{12} &> \left(\frac{2}{1+k^{-1}}\right)w_{11}, \end{aligned}\right. \tag{6.9.9}$$

or alternatively, noting that $k = w_1/w_2$,

$$\left.\begin{aligned} \frac{w_{12}}{w_{11}} &> \frac{2w_1}{w_1 + w_2}, \\ \frac{w_{12}}{w_{22}} &> \frac{2w_2}{w_1 + w_2} \end{aligned}\right. \tag{6.9.10}$$

(Kimura, 1960*a*). This agrees with the set of conditions first given by Bennett (1957). For a different derivation of the equivalent conditions and for the effect of inbreeding on the equilibrium, see Haldane and Jayakar (1964). Note here that the arguments leading to 6.9.7. and 6.9.8 are valid only when there is only one equilibrium point.

As an example of the application of formula 6.9.10, let us consider a mutant gene A_2 which is advantageous in males ($w_2 > w_1$) but deleterious in females ($w_{22} < w_{11}$). If A_2 is recessive in females ($w_{12} = w_{11}$), a stable polymorphism is possible provided that

$$\frac{2w_2}{w_1 + w_2} < \frac{w_{11}}{w_{22}}$$

is satisfied. On the other hand, if A_2 is dominant in females ($w_{12} = w_{22}$), the second inequality in 6.9.10 is violated and no stable polymorphism is possible.

Polymorphism Due to Selection of Varying Direction In this section, we will consider a situation in which selection fluctuates from generation to generation such that it sometimes favors one phenotype and sometimes another.

Let A_1 and A_2 be a pair of alleles at an autosomal locus and assume that A_2 is fully recessive with respect to fitness. Let w_t be the selective value of the recessive (A_2A_2) relative to the dominant (A_1A_1, A_1A_2), and, let p_t and $q_t(= 1 - p_t)$ be respectively the frequencies of A_1 and A_2, in the tth generation. Under random mating, the frequency of A_1 in the next generation is

$$p_{t+1} = \frac{p_t^2 + p_t q_t}{\bar{w}_t} = \frac{p_t}{\bar{w}_t}, \qquad 6.9.11$$

where

$$\bar{w}_t = p_t^2 + 2p_t q_t + w_t q_t^2.$$

First, consider the case in which A_1 is very rare. In this case, the ratio p_{t+1}/p_t is $1/w_t$ at the limit of $p_t \to 0$. This means that the proportional increase or decrease per generation of the frequency of the dominant allele (A_1) will be large and depends almost entirely on $1/w_t$. Thus

$$\frac{p_{t+1}}{p_0} = \frac{p_{t+1}}{p_t} \times \frac{p_t}{p_{t-1}} \cdots \frac{p_1}{p_0} \approx \frac{1}{w_t} \times \frac{1}{w_{t-1}} \cdots \frac{1}{w_0}. \qquad 6.9.12$$

In order that the polymorphism be maintained, it is necessary that the frequency of A_1 tend to increase when it is rare. Namely,

$$\prod_{i=0}^{t} \frac{1}{w_i} > 1,$$

or

$$\prod_{i=0}^{t} w_i < 1. \qquad 6.9.13$$

Next, consider the case in which the frequency of A_1 is very high, or equivalently, the frequency of A_2 is very low. In this case,

$$q_{t+1} = \frac{q_t(p_t + w_t q_t)}{\bar{w}_t} \quad 6.9.14$$

and the ratio q_{t+1}/q_t is unity at the limit $q_t \to 0$. This means that the proportional increase or decrease per generation of the frequency of the recessive allele (A_2) is extremely small when it is rare, and the consideration of the ratio is not very useful. Also the difference $q_{t+1} - q_t$ is of the order of q_t^2 and therefore tends to 0 very quickly as $q_t \to 0$. On the other hand, the reciprocal of q_t is very large and the difference of this quantity between two successive generations turns out to be finite and therefore more useful. That is to say we magnify the difference which actually is very small. Let $Q_t = 1/q_t$; then

$$Q_{t+1} - Q_t = \frac{(1 - w_t)(Q_t - 1)}{Q_t - 1 + w_t}, \quad 6.9.15$$

and, at the limit of $Q_t \to \infty$, this has the limit $1 - w_t$.

Therefore

$$Q_{t+1} - Q_0 \approx \sum_{i=0}^{t} (1 - w_i). \quad 6.9.16$$

In order that the polymorphism be maintained, A_2 must increase in this case, or the reciprocal of its frequency must decrease, so that

$$\sum_{i=0}^{t} (1 - w_i) < 0,$$

or

$$\frac{1}{t+1} \sum_{i=0}^{t} w_i > 1. \quad 6.9.17$$

This treatment shows that the necessary conditions for polymorphism are 6.9.13 and 6.9.17. Namely, the arithmetic mean of the fitness of the recessive is larger than unity while its geometric mean is less than unity. This elegant theorem was first proved by Haldane and Jayakar (1963).

These conditions would be fulfilled, for example, if the recessives were 5%–10% fitter than the dominants but an epidemic disease killed off all the recessives every twenty generations. The geometric mean is less than 1 while the arithmetic is larger. After each epidemic the frequency of recessive genes may decrease suddenly, followed by steady increase until the next epidemic.

Multiple-niche Polymorphism We shall consider briefly a simple model of Levene (1953). Suppose there are two alleles A_1 and A_2 and mating is at random. The zygote frequencies are p^2, $2pq$, and q^2. These genotypes are then distributed randomly into separate niches, each of which is different as regards the selective values of the three genotypes. In the ith niche the relative fitnesses are:

A_1A_1	A_1A_2	A_2A_2
$w_{11} = 1 - s_i$	$w_{12} = 1$	$w_{22} = 1 - t_i$

We make one more assumption—that the fraction that matures in the ith niche is k_i ($\sum k_i = 1$). After maturation the progeny from all niches are randomized before mating.

In this model the average fitness in the ith niche is

$$\begin{aligned}\bar{w}_i &= p^2(1 - s_i) + 2pq + q^2(1 - t_i) \\ &= 1 - p^2 s_i - q^2 t_i,\end{aligned} \tag{6.9.18}$$

and the average fitness of allele A_1 in the ith niche is

$$w_{1i} = p(1 - s_i) + q = 1 - ps_i. \tag{6.9.19}$$

From 5.2.6 the gene frequency change in the ith niche is

$$\Delta p_i = \frac{p(w_{1i} - \bar{w}_i)}{\bar{w}_i} = \frac{pq(qt_i - ps_i)}{\bar{w}_i}.$$

Averaging over all niches,

$$\Delta p = pq \sum_i k_i \frac{qt_i - ps_i}{\bar{w}_i}. \tag{6.9.20}$$

There will be at least one stable intermediate equilibrium if p increases when it is small and decreases when it is large. So we shall use this to determine sufficient conditions for a polymorphism. If $d(\Delta p)/dp > 0$ when $p \to 0$ and $d(\Delta p)/dp < 0$ when $p \to 1$ the conditions will be satisfied, since $\Delta p = 0$ at $p = 0$ or 1.

From 6.9.20,

$$\begin{aligned}\lim_{p \to 0} \frac{d(\Delta p)}{dp} &= \sum k_i \frac{t_i}{1 - t_i} \\ &= \sum k_i \frac{1 - w_{22,i}}{w_{22,i}} \\ &= \sum k_i \frac{1}{w_{22,i}} - 1,\end{aligned}$$

since $\sum k_i = 1$.

So one condition for a stable equilibrium is

$$\sum k_i \frac{1}{w_{22,i}} > 1,$$

or

$$1\Big/\left(\sum k_i \frac{1}{w_{22,i}}\right) < 1. \qquad \textbf{6.9.21}$$

Similarly, the second condition based on the limit as $p \to 1$ is

$$1\Big/\left(\sum k_i \frac{1}{w_{11,i}}\right) < 1. \qquad \textbf{6.9.22}$$

In words: If the weighted harmonic means of the fitnesses in the various niches show overdominance for fitness, there is a stable polymorphism.

Since the harmonic mean is generally smaller than the arithmetic, it is possible for there to be a stable polymorphism with the arithmetic means not showing heterozygote superiority in fitness. So, with multiple niches the conditions for a stable polymorphism are less stringent than with constant fitnesses.

Prout (1968) has shown that the necessary conditions are actually slightly weaker than we have indicated. We can actually replace $<$ by $\leq$ in equations 6.9.21 and 6.9.22. See Prout's paper for a discussion. This model has also been discussed by Dempster (1955*a*).

The Levene model that we have discussed was chosen deliberately to demonstrate the opportunity for polymorphism under a situation rather unfavorable to polymorphism. If the organisms are able to select their niches, then the opportunity for polymorphism is obviously greater. Another situation is when the opportunity for migration between niches is restricted so that there is more opportunity for maintenance of polymorphism by local differentiation. We shall not discuss these mathematically; in fact this subject has received very little attention. For a discussion of what has been done, see Moran (1959) and Levene (1967).

Two-locus Polymorphism When multiple alleles at multiple loci are considered, the problem quickly becomes very complicated. There are numerous possibilities for stable and unstable equilibria, particularly if the fitnesses of the different genotypes are not constant. We shall deal with only the simplest case, however. Assume that the population is mating at random and consider two loci with two segregating alleles at each.

With a single locus and random mating we saw that the fitness never decreases. If this is also true for two loci, then we can apply the kinds of stability criteria that were used in Section 6.8. If the linkage is loose so that the gametic types are in linkage equilibrium, a minor extension of these methods is directly applicable. It has been given by Kimura (1956*a*).

But it is not generally true that selection always leads to an increase in fitness. However, under one circumstance it does. We showed in Section 5.7 that the Fundamental Theorem of Natural Selection applies if the population is in quasi-linkage equilibrium. In Section 5.9 the same condition was shown to be sufficient for the Maximum Principle to work. So we should expect that under this condition there would be a stable equilibrium at the point where fitness is maximized. This is shown by doing almost exactly the same operation as we did in Section 5.9.

Let p_1, p_2, p_3, and p_4 be the frequencies of the four chromosomes *AB*, *Ab*, *aB*, and *ab* and c the proportion of recombination between the two loci. As before, we regard unlinked loci as a special case where $c = 1/2$. The equations of gene frequency change (5.4.4, 5.9.25) are

$$\frac{dp_i}{dt} = p_i(m_i - \bar{m}) - k_i cbD, \qquad \text{6.9.23}$$

where

$$D = p_1 p_4 - p_2 p_3$$

$$m_i = \sum_j m_{ij} p_j, \quad \bar{m} = \sum_{ij} m_{ij} p_i p_j$$

$$k_i = \begin{matrix} 1 \text{ when } i = 1, 4 \\ -1 \text{ when } i = 2, 3 \end{matrix},$$

and b is the birth rate of double heterozygotes.

At equilibrium, when $dp_i/dt = 0$.

$$p_1(m_1 - \bar{m}) = -p_2(m_2 - \bar{m}) = -p_3(m_3 - \bar{m}) = p_4(m_4 - \bar{m}) = cbD. \qquad \text{6.9.24}$$

The solution of these equations gives the equilibrium values.

Now we wish to show that the equilibrium is where fitness is maximum. We impose two side conditions,

$$\sum p_i = 1 \qquad \text{6.9.25}$$

and the quasi-linkage equilibrium relation

$$Z = \frac{p_1 p_4}{p_2 p_3} = \text{constant},$$

or, more conveniently,

$$\log Z = \log p_1 - \log p_2 - \log p_3 + \log p_4 = K, \qquad \text{6.9.26}$$

a constant.

Using A and B as Lagrange multipliers, as we did in Section 5.9, we maximize

$$\psi = \bar{m} - 2A(\sum p_i - 1) - 2B(\log Z - K) = 0. \quad 6.9.27$$

Differentiating,

$$\frac{\partial \psi}{\partial p_1} = \sum m_{1j} p_j + \sum m_{j1} p_j - 2A - \frac{2B}{p_1}$$

$$= 2m_1 - 2A - \frac{2B}{p_1} = 0,$$

$$\frac{\partial \psi}{\partial p_2} = 2m_2 - 2A + \frac{2B}{p_2} = 0,$$

$$\frac{\partial \psi}{\partial p_3} = 2m_3 - 2A + \frac{2B}{p_3} = 0,$$

$$\frac{\partial \psi}{\partial p_4} = 2m_4 - 2A - \frac{2B}{p_4} = 0. \quad 6.9.28$$

Multiplying the first equation by p_1, the second by p_2, the third by p_3, and the fourth by p_4 leads to

$$p_1(m_1 - A) = -p_2(m_2 - A) = -p_3(m_3 - A)$$
$$= p_4(m_4 - A) = B. \quad 6.9.29$$

To evaluate A and B, add the first and fourth equations of 6.9.28 and subtract the second and third. This gives

$$m_1 - m_2 - m_3 + m_4 - B \sum \frac{1}{p_i} = 0,$$

or in the symbols of 5.4.8,

$$B = \frac{E}{P}. \quad 6.9.30$$

Now, again multiply the four equations 6.9.28 by p_1, p_2, p_3, and p_4 and add. This gives

$$\sum p_i m_i - A \sum p_i = 0,$$

or

$$A = \bar{m}. \quad 6.9.31$$

Substituting into 6.9.29 and noting that at quasi-equilibrium

$$cbD = \frac{E}{P}$$

(see 5.7.13), we see that equations 6.9.29 are equivalent to 6.9.24.

So we have shown that the point where all the dp_i/dt's are 0 (6.9.24) and which is therefore an equilibrium is also the point at which fitness is maximum (or minimum) under quasi-linkage equilibrium. Once again we have an illustration that a population behaves simply (or at least *relatively* simply) when it is in quasi-equilibrium. Furthermore, as we showed in Sections 5.4 and 5.7, this is the state toward which the population tends to move—unless recombination is small relative to the epistatic parameters E. When the population is in quasi-linkage equilibrium any fitness change is attributable to changes in gene frequencies, not to changes in linkage relationships.

Selection proceeds to maximize fitness when there is quasi-linkage equilibrium, that is, with relatively free recombination and small amounts of epistasis. At the opposite extreme, with no recombination (or negligible amounts) the chromosomes behave as units and the situation is as if there were multiple alleles at a single locus. This, too, we have seen leads to a maximization of fitness by selection. So the stability criteria based on this assumption are valid.

The intermediate cases are more troublesome. When E is large enough relative to c that quasi-linkage equilibrium is not approached, then more complicated stability conditions are needed. One such procedure has been given by Kimura (1956*b*).

Ordinarily, epistasis is not sufficient to maintain a polymorphism: some overdominance is required. However, if one locus is overdominant, a second locus that is not overdominant can also be maintained in stable polymorphism provided there is the right kind of epistasis and tight enough linkage. A specific model, proposed by P. M. Sheppard, is this: Assume one locus with a pair of alleles, say A_1 and A_2, kept in balanced polymorphism by heterozygote superiority in fitness. Another pair of alleles, B_1 and B_2, are at another locus on the same chromosome, and interact with the genes at the first locus in such a way that A_1 is advantageous in combination with B_1 but is disadvantageous in combination with B_2, while the situation is reversed with A_2. Then the second locus will remain polymorphic if linkage between the two loci is sufficiently close. A special case was investigated by Kimura (1956*b*).

An interesting aspect of this situation is that it offers a possible explanation for modification of linkage intensity. In *The Genetical Theory of Natural Selection*, Fisher (1930, 1958) noted that "the presence of pairs of factors in the same chromosome, the selective advantage of each of which reverses that of the other, will always tend to diminish recombination, and therefore to increase the intensity of linkage in the chromosomes of that species." With a situation such as that of the paragraph above, where there is a stable polymorphism, recombination on the average increases the frequency of the less well-adapted chromosomes so any mechanism which reduces the amount of recombination will *cet. par.* have a selective advantage. Whether this is a major factor in increasing the intensity of linkage in actual evolution is not known. For further discussion, see Nei (1968).

6.10 Selection and the Sex Ratio

It is not immediately apparent how selection operates on the sex ratio, for a parent that gives rise to offspring with an unequal sex ratio still produces just as many offspring and transmits just as many of its autosomes to the next generation as if the sexes were produced in equal proportions. The *XX–XY* sex-determining mechanism assures approximate equality of the sexes in many species, but there are many examples of autosomal genes that produce minor deviations in the sex ratio. Thus, there is genetic variation on which selection could operate.

It is interesting that Darwin was unable to find a solution to the problem. In *The Descent of Man* he writes:

> In no case, as far as we can see, would an inherited tendency to produce sexes in equal numbers or to produce one sex in excess, be a direct advantage or disadvantage to certain individuals more than to others; for instance, an individual with a tendency to produce more males than females would not succeed better in the battle for life than an individual with an opposite tendency; and therefore a tendency of this kind could not be gained through natural selection.... I formerly thought that when a tendency to produce the sexes in equal numbers was advantageous to the species, it would follow from natural selection, but I now see that the whole problem is so intricate that it is safer to leave its solution to the future.

It is possible that if Darwin had considered three generations instead of only two he might have thought of a solution. For one can reason this way: Consider a population that for some reason has an excess of males. In such a population a female, on the average, has a larger number of offspring than a male. This follows from the simple fact that each individual progeny is the product of one egg and one sperm. Therefore the total contribution to the next generation is the same for males and females, and if there are fewer females the average female produces more progeny than the average male. Now, in such a population an individual that produces more than the average proportion of females will be producing more of that sex, which in turn contributes more to the next generation. Thus this individual will have more grandprogeny than the average of its contemporaries. To the extent that the tendency to produce progeny with an altered sex ratio is hereditary, the tendency will be transmitted and the average of the population will change in the direction of equality of the sexes. Whenever the sex ratio becomes equalized the selective advantage of the altered type will cease. The same argument holds in the opposite direction if the population has an initial excess of females. So, by the simple expedient of looking at one more generation than is customary, we arrive at an understanding of the problem.

This consideration helps us understand why the sex ratio is typically somewhere near 1 : 1. This would be expected to be true even of highly polygamous species, where the optimum ratio for perpetuation of the species might be thought to be far from equal. But this verbal analysis leaves several points vague. What happens if, as is almost always true, there is differential mortality in the two sexes? At what age is the sex ratio equalized by selection—at the zygote stage, at the reproductive age, or some other? For answers to these points we shall have to inquire in a more rigorous way.

This problem was first solved by R. A. Fisher (1930). His treatment involved the concept of "parental expenditure," the meaning of which is not entirely transparent (to us, at least). The question has been discussed further from similar points of view by Bodmer and Edwards (1960), Kolman (1960), MacArthur (1965), and Emlen (1968*a*, *b*). However, we shall treat the problem in a more mechanistic way. This is rather similar to the methods used by Shaw and Mohler (1953) and Shaw (1958).

Suppose that there is a population of adults (generation 0) producing a total of N progeny, of which $N\bar{p}$ are male and $N\bar{q}$ are female. Among the population is a special group, say A, who produce a fraction p of males and q of females. Suppose that this group contributes a total of n offspring. Let s_m and s_f be the probability of survival to adulthood for males and females and let f_m and f_f be the fertility of the two sexes (the f's will not be constants, for their ratio will be related to the proportion of the two sexes in the population). Then in the progeny (generation 1), we will have:

GENERATION	STAGE	PROGENY OF A'S MALES	PROGENY OF A'S FEMALES	ENTIRE POPULATION MALES	ENTIRE POPULATION FEMALES
1	zygote	np	nq	$N\bar{p}$	$N\bar{q}$
	adult	nps_m	nqs_f	$N\bar{p}s_m$	$N\bar{q}s_f$
	gametes	$2nps_mf_m$	$2nqs_ff_f$	$2N\bar{p}s_mf_m$	$2N\bar{q}s_ff_f$
		TOTAL		TOTAL	
2	zygote	$n(ps_mf_m + qs_ff_f)$		$N(\bar{p}s_mf_m + \bar{q}s_ff_f)$	
$W = \dfrac{\text{gen. 2 zygotes}}{\text{gen. 1 zygotes}}$		$ps_mf_m + qs_ff_f$		$\bar{p}s_mf_m + \bar{q}s_ff_f$	

The average excess of W for type A over the population average, expressed as a fraction of the population average, is

$$E_A = \frac{(W_A - \overline{W})}{\overline{W}} = \frac{ps_m f_m + qs_f f_f - \bar{p}s_m f_m - \bar{q}s_f f_f}{\bar{p}s_m f_m + \bar{q}s_f f_f}. \qquad 6.10.1$$

But, recall that the total contribution of males must equal that of females. So

$$\bar{p}s_m f_m = \bar{q}s_f f_f, \qquad 6.10.2$$

and with this substitution we obtain

$$\begin{aligned} E_A &= \frac{p}{2\bar{p}} + \frac{q}{2\bar{q}} - 1 \\ &= \frac{(p - \bar{p})(\bar{q} - \bar{p})}{2\bar{p}\bar{q}}. \end{aligned} \qquad 6.10.3$$

If the sex-ratio property of type A is completely heritable, then the rate of change per generation of this type is

$$\Delta P_A = P_A \frac{(W_A - \overline{W})}{\overline{W}} = P_A E_A, \qquad P_A = \frac{n}{N}, \qquad 6.10.4$$

where P_A is the proportion of A in the population. With less than complete heritability the rate is correspondingly less. However, the change is always delayed by one generation, since it is the excess of grandprogeny over progeny that we are measuring.

There are two ways for E_A to be 0. One is the trivial case where $p = \bar{p}$. This simply means that A produces the same sex ratio as the average of the population and therefore has no tendency to increase or decrease. More interesting is the case where $\bar{p} = \bar{q}$. This says that when male and female zygotes are produced in equal numbers, there is no differential contribution of a genotype with a different sex ratio.

That the 1 : 1 sex ratio is stable can be seen by noting that when $\bar{q} > \bar{p}$, E_A is positive if $p > \bar{p}$; that is to say, if there is an excess of females in the population, a genotype that produces more than the average proportion of males leaves more descendants than the average. Therefore, if the trait is heritable the proportion of males in the population increases. This will continue until $\bar{p} = \bar{q}$, in which case the A genotype no longer enjoys a selective advantage.

Likewise, if the population initially has an excess of males, a genotype producing a more than average proportion of females will leave more descendants. So, in either case the population tends to move in the direction of equal numbers of the two sexes.

The equality of the two sexes is at the zygote stage. Notice that the factors for survival cancel out and do not appear in the final equation, so differential mortality does not affect the 1 : 1 zygotic sex ratio. The number at later stages is determined by the relative mortality of the two sexes. At adulthood the sex ratio would be $\bar{p}s_m : \bar{q}s_f$.

It may seem as if differential mortality should have some effect on the zygotic sex ratio, but clearly this is not so for the model that we have been discussing. However, there is at least one way in which mortality can have such an effect. This is applicable to uniparous mammals and especially to man. If an embryo or child dies it may be replaced, or there may be some compensation in that the next zygote will be conceived earlier than if the first had survived, thus making possible a larger number of conceptions. In man, males have a higher death rate than females, both prenatally and postnatally. Thus if there is compensation, a woman producing an excess of males will produce a larger than average number of zygotes, although a smaller number of surviving children (unless the compensation is complete).

As a simple model, assume that compensation is complete so that, despite differential mortality prior to a certain age, the number of children that reach this age (say T) is independent of the sex ratio in the sibship. Let S_m and S_f be the fraction of survivors in each sex from conception until age T. Let s_m and s_f now stand for the probability of surviving from age T to adulthood. Then, starting with generation 0 as before, the table is modified as follows:

		PROGENY OF A'S		ENTIRE POPULATION	
GENERATION	STAGE	MALES	FEMALES	MALES	FEMALES
1	zygote	np	nq	$N\bar{p}$	$N\bar{q}$
	age T	$\dfrac{npS_m}{pS_m + qS_f}$ $= nx$	$\dfrac{nqS_f}{pS_m + qS_f}$ $= ny$	$\dfrac{N\bar{p}S_m}{\bar{p}S_m + \bar{q}S_f}$ $= N\bar{x}$	$\dfrac{N\bar{q}S_f}{\bar{p}S_m + \bar{q}S_f}$ $= N\bar{y}$
	adult	nxs_m	nys_f	$N\bar{x}s_m$	$N\bar{y}s_f$
	gametes	$2nxs_mf_m$	$2nys_ff_f$	$2N\bar{x}s_mf_m$	$2N\bar{y}s_ff_f$
		TOTAL		TOTAL	
2	zygote	$n(xs_mf_m + ys_ff_f)$		$N(\bar{x}s_ms_f + \bar{y}s_ff_f)$	

Note that, since $y = 1 - x$, the total number of progeny at age T is the same as the number of zygotes. This leads to the same equations as before (6.10.1–6.10.4), except that we replace p, q, $\bar{p}$, and $\bar{q}$ with x, y, $\bar{x}$, and $\bar{y}$. Since $\bar{x}$ and $\bar{y}$ are the proportions of males and females at age T, it follows that with complete compensation the sex ratio is adjusted to equality at that age when compensation ceases.

Compensation is almost certain to be less than complete (if for no other reason than that the replacements occur when the parents are older, thereby increasing the interval between generations). In this case the sex ratio would be adjusted to be equal at some age prior to the age when replacement no longer occurs. In the human population with excess mortality of males and partial compensation we would expect an excess of males at the zygote stage with a steadily decreasing proportion, fewer males than females at the end of the age of compensation, and with an intermediate age where the sex ratio is unity. There is no implication that compensation is conscious.

As we mentioned earlier, Fisher (1930) gives a different argument, but he reaches the same conclusion:

> The condition toward which Natural Selection will tend will be one in which boys are more numerous at birth, but become less numerous, owing to their higher death rate, before the end of the period of parental expenditure. The actual sex-ratio in man seems to fulfill these conditions somewhat closely, especially if we make allowance for the large recent diminution in the deaths of infants and children; and since this adjustment is brought about by a somewhat large inequality in the sex-ratio at conception, for which no a priori reason can be given, it is difficult to avoid the conclusion that the sex-ratio has really been adjusted by these means.

In this model the population sex ratio tends to equality, either at the zygote stage, or at a later stage if there is compensation for lost progeny. At the same time there is nothing in the model that says that each individual in the population will produce progeny with a 1 : 1 sex ratio. It only says that the frequencies of the different genotypes in the population will be adjusted so that the average sex ratio of their progeny is unity. A population in which each individual produces an equal number of male and female progeny would have one obvious advantage over a population composed of equal numbers of male-producing and female-producing types; it would be less dependent on maintaining the proper ratio of the two types and would therefore be better buffered against chance fluctuations in their numbers. On the other hand, this selection, depending on selection between groups rather than of individuals within groups, is slower and less effective. Therefore, we should not be surprised if there is considerable genetic variability among individuals in the sex ratio of their progeny in a population that maintains an average ratio of 1 : 1.

There are many reasons why one would not expect perfect agreement with this principle. It may well be that genes affecting the sex ratio of the progeny have direct effects on the individual, in which case the selection on the individual effect may be more important and the sex ratio would be modified as a by-product. A second reason is that sex-ratio modifiers may be located on the *X* or *Y* chromosome. It is especially clear in the *Y* case that if there are two kinds of *Y* chromosomes, one of which leads to a higher proportion of sons than the other, the one with the higher proportion of sons will have a selective advantage and, in the absence of other influences, eventually replace the competing *Y*. In this case, the sex ratio of the population would simply become the ratio associated with the behavior of this particular *Y*.

6.11 Stabilizing Selection

Several times in this book we have noted that for almost any metrical trait (except fitness itself) the most fit individuals are likely to have an intermediate value for the trait. If size is the trait, the species typically has a characteristic size and individuals that are too large or too small are less viable or fertile, and similarly for other quantitative traits.

This means that selection is directed toward elimination of deviants, and the effect should tend to reduce the variance and stabilize the population at an intermediate value. Of course the population, especially if it is Mendelian, will not become absolutely uniform. There are all sorts of variability-generating mechanisms, many of which we have discussed.

An immediate question is whether selection favoring intermediate values tends to perpetuate homozygous or heterozygous combinations. In either case the average will be much the same, but if the former is true the variance will be less.

We can get a rough answer by considering a very simple model, such as was used in Section 4.7. Suppose that alleles A_0 and B_0 decrease size while A_1B_1 increase it. Assume further that the genes are without dominance and that there is additivity between loci.

Selection for an intermediate measurement will favor genotypes $A_0A_0B_1B_1$, $A_1A_1B_0B_0$, or $A_1A_0B_1B_0$. Which type will prevail?

Note first that gametes A_0B_1 and A_1B_0 will more often enter into intermediate zygotes than A_1B_1 or A_0B_0 gametes, and therefore are favored. Then, if one of these (say A_1B_0) is more frequent, either A_1B_0 or A_0B_1 gametes can produce intermediate phenotypes with A_1B_0, but the latter will produce by recombination the two unfavored types A_1 B_1 and A_0 B_0. So we would expect that A_1B_0 type of gamete will increase and eventually the population will be homozygous $A_1A_1B_0B_0$.

This intuitive argument that selection favoring an intermediate type leads to an intermediate homozygous genotype rather than an intermediate heterozygous genotype was originally given, in a more convincing way than we have, by Wright (1935). Wright assumed that the genes were additive in producing the metrical trait and that fitness decreased as the square of the deviation from the optimum measure.

This quadratic model has been studied by several workers (e.g., Robertson, 1956; Kojima, 1959*a*; Jain and Allard, 1965; Lewontin, 1965*a*; Levene, 1967). The general conclusion is that with complete or partial dominance on the quantitative trait, selection for intermediate values leads to homozygosity. If the optimum phenotype is necessarily heterozygous at one or more loci, for example if the optimum falls between two homozygotes, heterozygosity for one (or perhaps more) loci may persist. But, if the number of genes affecting this trait is large, it is unlikely that more than a tiny fraction will remain heterozygous. The larger the number of loci, the smaller the fraction that remain segregating.

In summary, selection for an intermediate phenotype ordinarily leads to homozygosity. If we are to explain much variability we must look elsewhere.

The two most obvious sources of variation are mutation and heterozygote superiority in fitness. With the quadratic model, small deviations from the mean do not cause much change in the fitness. So, a moderate mutation rate can maintain a substantial amount of variability of the metrical trait because of the slow rate of elimination of mutants (see Wright, 1935, 1952).

The second possibility is that some loci are favored as heterozygotes for reasons other than their effect on the quantitative trait. The ubiquity of pleiotropy argues that genes affecting size (say) also have other effects. Some such effects may lead to heterozygote superiority. If so, this provides a ready mechanism for maintenance of variability (see Robertson, 1956).

An example—rather a specialized one—of selection for an intermediate is provided by selection for the amount of redundant DNA, such as ribosomal DNA. There is good evidence that either too small or too large an amount is deleterious. New variability can be generated by unequal crossing over, for which there is evidence in Drosophila where the *bobbed* mutant is the phenotypic manifestation of a reduced amount of ribosomal DNA. See Kimura (1965) for a similar model.

This example will show one more instance of the generation of a normal distribution. We shall assume that fitness is maximum for an intermediate length of chromosome and decreases with the square of the deviation from this optimum amount.

Let K and L be the number of repeats in the maternal and paternal chromosomes, respectively. Let the optimum fitness, M, be that of an individual with a total of N repeats, counting both homologs. Assume that fitness is measured in Malthusian parameters. We then have:

NUMBER OF REPEATS	N	$K+L$
FITNESS	M	$M-c(K+L-N)^2$

where c is a proportionality constant. Let p_K be the frequency of a chromosome with K repeats.

The average fitness of the population is

$$\overline{m} = M - c\sum_{K,L}(K + L - N)^2 p_K p_L. \tag{6.11.1}$$

Likewise, the average fitness of a chromosome with K repeats is

$$m_K = M - c\sum_{L}(K + L - N)^2 p_L. \tag{6.11.2}$$

From our familiar equation for frequency change (5.3.5),

$$\frac{dp_K}{dt} = p_K(m_K - \overline{m}). \tag{6.11.3}$$

It is convenient to rescale and measure the number of repeats as deviations from the optimum number (which is $N/2$ per chromosome), so we let $k = K - N/2$ and $l = L - N/2$. Then 6.11.3 becomes

$$\frac{dp_k}{dt} = -p_k c\left[\sum_l (k + l)^2 p_l - \sum_{k,l}(k + l)^2 p_k p_l\right]. \tag{6.11.4}$$

With random mating, k and l are independent and, since $\sum p_k = \sum p_l = 1$,

$$\begin{aligned}\sum_l (k + l)^2 p_l &= \left(\sum_l k^2 + \sum_l l^2\right)p_l \\ &= k^2 + \overline{l^2}.\end{aligned} \tag{6.11.5}$$

Thus 6.11.4 becomes

$$\begin{aligned}\frac{dp_k}{dt} &= -cp_k\left[k^2 + \overline{l^2} - \sum_k (k^2 + \overline{l^2})p_k\right] \\ &= -cp_k[k^2 - \overline{k^2}].\end{aligned} \tag{6.11.6}$$

That is to say, selection is as if it were acting on a haploid.

Now consider recombination. Let u be the rate at which a chromosome with k repeats is converted into something else. Then the change in frequency from this is

$$\frac{dp_k}{dt} = -up_k + u\sum_x p_{k-x}f(x), \tag{6.11.7}$$

where $f(x)$ is the proportion of changes that increase the length by an amount x. Let us assume that x is small and expand 6.11.7 into a Taylor series, which gives

$$\begin{aligned}\frac{dp_k}{dt} &= -up_k + u\sum_x\left[p_k - x\frac{dp_k}{dk} + \frac{x^2}{2}\frac{d^2p_k}{dk^2} - \cdots\right]f(x)\\ &= -u\bar{x}\frac{dp_k}{dk} + \frac{u\overline{x^2}}{2}\frac{d^2p_k}{dk^2},\end{aligned} \qquad 6.11.8$$

where $\bar{x} = \sum xf(x)$, $\overline{x^2} = \sum x^2f(x)$, and $\sum f(x) = 1$.

Combining 6.11.6 and 6.11.8 and letting $dp_k/dt = 0$ for equilibrium, we get

$$\frac{u\overline{x^2}}{2}\frac{d^2p_k}{dk^2} - u\bar{x}\frac{dp_k}{dk} - c(k^2 - \overline{k^2})p_k = 0. \qquad 6.11.9$$

The appropriate solution to this equation (you may verify the result by differentiating, noting that $\overline{k^2} = \bar{k}^2 + V$) is

$$p_k = \frac{1}{\sqrt{2\pi V}}e^{\frac{-(k-\bar{k})^2}{2V}}, \qquad 6.11.10$$

where

$$V = V(k) = \sqrt{\frac{u\overline{x^2}}{2c}},$$

$$\bar{k} = \frac{u\bar{x}}{\sqrt{2u\overline{x^2}c}}.$$

So k is normally distributed with mean $\bar{k}$ and variance V.

In the original units

$$\bar{K} = \bar{k} + \frac{N}{2}, \qquad 6.11.11$$

$$V(K) = V(k) = V. \qquad 6.11.12$$

With random mating the zygotic mean is $2\bar{K}$ and the variance is $2V$.

The mean and optimum will coincide if $\bar{x} = 0$, that is, if recombination is equally likely to increase or decrease the length—an assumption that is reasonable if the unequal exchange events are reciprocal.

6.12 Average Fitness and Genetic Loads

The change of average fitness associated with maintaining the variability in a population has come to be called the genetic load. The word "load" was first used by Muller (1950) in assessing the impact of mutation on the human population. In Muller's thinking the load was clearly a burden, measured in terms of reduced fitness, but felt in terms of death, sterility, illness, pain, and frustration. In most evolutionary and population genetics considerations, however, it is used as a measure of the amount of natural selection associated with a certain amount of genetic variability. It is not necessarily a burden; on the contrary, it may be the *sine qua non* for further evolution without which the population could become extinct. For example, consider a population, previously of genotype AA, which undergoes a mutation to A' such that $A'A$ is superior to AA, but $A'A'$ is inferior. This creates variability and therefore a genetic load, but the polymorphic population has a higher average fitness than if it were monomorphic for the original AA genotype.

The genetic load also applies to traits other than fitness. For example the load may be measured as an increased death rate, decreased fertility, or increase in some specific disease. The *expressed* load is the load that is expressed under the mating system of the population, usually approximated by random-mating theory. The total load includes also the *hidden* load that is concealed by heterozygosis and is brought out by inbreeding.

6.12.1 Kinds of Genetic Loads and Definitions

There are many kinds of genetic loads. Any factor that influences gene frequencies can lead to a change in the average fitness. We give here a list of some of the factors that may produce a genetic load, together with a short discussion of each. The first two were introduced by Haldane in his original paper (1937), from which the genetic load concept came.

1. Mutation Every population is exposed continuously to newly occurring mutations, the average effect of which is to lower the fitness in the environment where the population lives. This decrease in fitness, Haldane suggests, is the price paid by a species for its capacity to evolve.

2. Segregation The segregation load, or balanced load (Dobzhansky, 1955, 1957), occurs when the fittest genotype is heterozygous. In a sexual population homozygous progeny are regularly produced by Mendelian segregation and the segregation load is the decrease in fitness compared to the most fit heterozygote—the type that would prevail if it were not for segregation.

3. Recombination As pointed out by Fisher (1930), if two chromosomes, say *AB* and *ab*, are on the average favored by natural selection in the various zygote combinations into which they enter, while the two types *Ab* and *aB* are in disfavor, the average effect of recombination will be to increase the inferior types. Thus recombination produces a decrease in fitness.

4. Heterogeneous Environment If there are different environments and the various genotypes respond to these in different ways, it is advantageous if those genotypes that fare best in an unusual environment are rare while those which prefer an abundant environment are common. This means that there will be an optimum allotment of genotype frequencies corresponding to the numbers of available spaces in the various environments. Any actual population will depart from this ideal, and to the extent that this is true there is a reduced fitness, and therefore a load. Haldane (personal communication) suggested that such a load might be called the "dysmetric" load.

5. Meiotic Drive or Gamete Selection A gene that is favored by meiotic drive or gamete selection, or a gene that is closely linked to such a locus, may persist in the population despite being harmful in various zygotic combinations. The population fitness will therefore be lowered and this again creates a genetic load.

6. Maternal-fetal Incompatibility A genotype may have a lowered fitness in association with certain maternal genotypes. A well-known example is an Rh+ child with an Rh− mother. The genetic load from this cause has been called the incompatibility load.

7. Finite Population In any finite population the gene frequencies drift away from the values that they would have if they had reached equilibrium in an infinite population. The equilibrium proportions are usually the values that maximize the population fitness. The drift away from the equilibrium frequencies will usually reduce the fitness, thereby creating a genetic load.

8. Migration Immigrants to a new area ordinarily carry genes that were selected in their original habitat and are therefore not always optimum in the new area. The extent to which the average fitness of the population is reduced by the continued introduction of less favorable immigrant genes is the migration load.

Definitions In general we define the genetic load as the fraction by which the population mean is changed as a consequence of the factor under consideration in comparison with a population, supposed to be identical otherwise, in which the factor is missing. Usually the measurement of interest is fitness, but the concept can be used for other measurements. For the mutation load the reference population is a population with no mutation. For the segregation load the reference is a population at equilibrium without segregation, that is, an asexual population. In such a population the fittest genotype attains a frequency 1, so this is the reference.

The operational definition of the load will then be

$$L = \frac{w_{max} - \overline{w}}{w_{max}}, \qquad 6.12.1.1$$

where w_{max} is the fitness of the reference population, usually taken as the one with maximum fitness.

We shall treat four examples—mutation, segregation, incompatibility, and meiotic drive.

6.12.2 The Mutation Load

We shall assume that the generations are discrete and use a model based on this assumption. Let p be the frequency of the normal allele (or population of isoalleles) and q the frequency of the mutant. If the combination of two mutants is approximately the average of the two mutant homozygotes, as is often the case, q can stand for the sum of the frequencies of all such mutants. We use the inbreeding coefficient, f, as a measure of departure from random-mating proportions.

GENOTYPE	A_1A_1 or AA	A_1A_2 or Aa	A_2A_2 or aa
FREQUENCY	$P_{11} = p^2(1-f) + pf$	$P_{12} = 2pq(1-f)$	$P_{22} = q^2(1-f) + qf$
FITNESS	$w_{11} = w_{max}$	$w_{12} = w_{max}(1 - hs)$	$w_{22} = w_{max}(1 - s)$

The expressed genetic load in a population with inbreeding coefficient f is

$$L = \frac{w_{max} - \overline{w}}{w_{max}} = hs[2pq(1-f)] + s[q^2(1-f) + qf]. \qquad 6.12.2.1$$

When $f = 0$ the *random-mating* mutation load is

$$L_0 = 2hspq + sq^2, \qquad 6.12.2.2$$

and when $f = 1$ the *inbred* mutation load is

$$L_1 = sq. \qquad 6.12.2.3$$

From 6.2.11 the equilibrium frequency of q is approximately

$$q \approx \frac{u}{s(h + f + q)} = \frac{u}{sz}, \tag{6.12.2.4}$$

where $z = h + f + q$.

Substituting this into equation 6.12.2.1 gives the expressed load. We consider three special cases of interest:

1. Complete Recessive ($h = 0$) **and Random-mating Proportion** ($f = 0$)
Substituting from 6.12.2.4 into 6.12.2.1 and letting $h = f = 0$ gives

$$L = sq^2 = s\frac{u}{s} = u. \tag{6.12.2.5}$$

2. No Dominance ($h = 0.5$) **and Random-mating Proportions**
When $h = 1/2$ and $f = 0$, 6.12.2.1 becomes $L = qs$. Substituting from 6.2.7

$$L = \frac{2u}{1 + u}. \tag{6.12.2.6}$$

The striking fact, first noted by Haldane (1937), is that the effect of mutation on fitness is related to the mutation rate and not to the amount of deleterious effect of the mutant. From these two examples we see that the mutation load is equal to the mutation rate for a recessive mutant and to about twice this with partial dominance. What about other cases? In general, values of h larger than 0.5 give a larger value than $2u/(1 + u)$ and values less than 0.5 give a lesser value. But this statement does not do justice to the true picture. The value of L is very insensitive to changes of h over a wide range of values, so that unless h is quite close to 0 the load is very nearly $2u$. Table 6.12.2.1 gives some numerical examples that illustrate this.

3. Arbitrary Dominance, Random-mating Proportions If h and f are both small, but s is large enough to keep q small, then we can write 6.12.2.1 as approximately

$$L \approx sq(2h + q + f), \tag{6.12.2.7}$$

where all terms that involve a square or product of any of the quantities h, q, and f are ignored. Comparing this with 6.12.2.4 shows that when $h \to 0$, $L = u$, but when q and f are nearly 0 so that h is the dominant term, $L = 2u$.

The general principle is that if the mutant is completely recessive or if there is enough inbreeding that most mutant genes are eliminated as homozygotes, the load is equal to the mutation rate. On the other hand, if h is

Table 6.12.2.1. Values of C where Cu is the mutation load per locus. Diploidy is assumed, u is the mutation rate, s the homozygous selective disadvantage of the mutant, and hs is the disadvantage in heterozygotes. The mutation rate is taken to be 10^{-5}. (From Crow, 1970.)

		h							
f	*s*	0	.01	.02	.05	.1	.5	1.0	1.5
0	.001	1.00	1.10	1.18	1.41	1.66	2.00	2.01	2.01
	.01	1.00	1.27	1.47	1.78	1.93	2.00	2.00	2.00
	.1	1.00	1.62	1.83	1.97	1.99	2.00	2.00	2.00
	.5	1.00	1.86	1.96	1.99	2.00	2.00	2.00	2.00
	1.0	1.00	1.92	1.98	2.00	2.00	2.00	2.00	2.00
.05	.001	1.00	1.07	1.14	1.31	1.51	1.90	1.96	1.98
	.01	1.00	1.13	1.24	1.45	1.63	1.90	1.95	1.97
	.1	1.00	1.16	1.27	1.48	1.65	1.90	1.95	1.97
	.5	1.00	1.16	1.27	1.49	1.65	1.90	1.95	1.97
	1.0	1.00	1.16	1.27	1.49	1.65	1.90	1.95	1.97

appreciable, then most eliminations are through heterozygotes as we have discussed earlier in Section 6.2; in this case the load is twice the mutation rate. This is shown in Table 6.12.2.1.

For an X-linked recessive with equal mutation rates in the two sexes, Haldane showed that the mutation load is $1.5u$. If the mutation rates in the two sexes are not equal the load is approximately

$$L = \frac{u_m + 2u_f}{2}, \qquad 6.12.2.8$$

where u_m and u_f are the mutation rates in the two sexes. For a general discussion of the mutation load for an X-linked locus, see Kimura (1961).

That partially dominant mutants produce approximately twice the load that recessives do is not surprising. It can be seen intuitively by using the genetic extinction or "genetic death" argument of Muller (1950). A genetic death is a preadult death or failure to reproduce, or the statistical equivalent in reduced fertility. In a diploid population of constant size N (therefore having $2N$ genes), a genetic extinction caused by death or failure to reproduce eliminates one mutant among the $2N$ genes at this locus. If the mutant is recessive, two mutants are eliminated at once so there are two eliminations among the $2N$ genes per extinction. The 2 : 1 ratio will be correct as long as the dominant mutant is rare enough so that it is never eliminated as a homozygote. If the mutant is an X-linked recessive, practically all the elimininations are through genetic extinction of males. There are $3N/2$ genes in the population so the extinction eliminates one among the $3N/2$. Thus the number of genetic extinctions per gene eliminated for dominant, recessive, and X-linked recessives is in the ratio $2:1:1.5$. Likewise, if the inbreeding coefficient is high, the eliminations will be through homozygotes and therefore the situation is like that of a recessive mutant.

If the effects of different loci are independent, the load for all loci combined is given by

$$1 - L = \prod(1 - L_i) \approx e^{-\sum L_i}, \qquad \textbf{6.12.2.9}$$

where L_i is the load at the ith locus. Since the individual L_i's are very small, this is very close to $L = \sum L_i$. The mutation load is then, to a very good approximation,

$$L = C\sum u_i = CU, \qquad \textbf{6.12.2.10}$$

where C is a constant between 1 and 2 (usually much nearer 2), u_i is the total mutation to all deleterious mutants at the ith locus, and the summation is over all relevant loci. In other words, $U = \sum u_i$ is the total mutation rate to harmful genes per gamete per generation.

For a review and discussion of the mutation load with epistasis, see Crow (1970).

If a population, previously mating at random, is inbred without selection, the mutation load for $f = 1$ from 6.12.2.1 is

$$L_1 = qs, \qquad \textbf{6.12.2.11}$$

where q is the frequency before inbreeding began. Substituting from 6.12.2.4 and assuming that before inbreeding f was 0,

$$L_1 = \frac{u}{z}. \qquad \textbf{6.12.2.12}$$

For all loci

$$L_1 = \sum \frac{u}{z} = \frac{1}{\tilde{z}} \sum u, \qquad \textbf{6.12.2.13}$$

where $\tilde{z}$ is the harmonic mean of z, weighted by the mutation rate, that is,

$$\frac{1}{\tilde{z}} = \frac{\sum \frac{1}{z} u}{\sum u}.$$

If $h = 0$, the ratio of the inbred load ($f = 1$) to the random-mating load ($f = 0$) is

$$\frac{L_1}{L_0} = \frac{sq}{sq^2} = \sqrt{\frac{s}{u}}. \qquad 6.12.2.14$$

The more realistic case probably includes partial dominance so that h is appreciable. If h is the dominant term in z,

$$\frac{L_1}{L_0} = \frac{1}{2h}, \qquad 6.12.2.15$$

for a single locus, or if many loci are considered,

$$\frac{L_1}{L_0} = \frac{1}{2\tilde{h}}, \qquad 6.12.2.16$$

where $\tilde{h}$ is the harmonic mean of the individual h values. The general conclusion is that if h is small the mutation load is greatly increased by inbreeding. We are referring to inbreeding of a population that has come to equilibrium under random mating and is then inbred without selection.

6.12.3 The Segregation Load

The decreased fitness, in comparison with the heterozygote in situations where the heterozygote is favored, arises from the fact that in a Mendelian population there will always be inferior homozygotes arising by segregation.

To consider the question quantitatively we consider first a simple, 2-allele model.

GENOTYPE	A_1A_1	A_1A_2	A_2A_2
FITNESS	$w_{11} = w(1 - s)$	$w_{12} = w$	$w_{22} = w(1 - t)$
FREQUENCY	$P_{11} = p^2$	$2P_{12} = 2pq$	$P_{22} = q^2$

Assume that s and t are large enough, relative to the mutation rates, that mutation can be ignored. Following the procedure used above for the mutation load, we write the frequency of allele A one generation later in terms of the present genotypic frequencies and the selection coefficients. Thus, writing p' for the frequency next generation,

$$p' = \frac{p^2w(1 - s) + pqw}{\bar{w}},$$

where $\bar{w} = p^2w(1 - s) + 2pqw + q^2w(1 - t)$. Equating p to p' to specify an equilibrium and solving leads to

$$p = \frac{t}{s + t}, \qquad q = \frac{s}{s + t}, \tag{6.12.3.1}$$

in agreement with 6.7.2. From this, $\bar{w}$ at equilibrium is

$$\bar{w} = w\left(1 - \frac{st}{s + t}\right), \tag{6.12.3.2}$$

and from the definition 6.12.1.1, the segregation load is

$$L = \frac{st}{s + t}. \tag{6.12.3.3}$$

Notice that the formula can also be written as

$$L = sp^2 + tq^2 \tag{6.12.3.4}$$

$$= sp = tq, \tag{6.12.3.5}$$

where p and q are the equilibrium values of the allele frequencies.

This example illustrates two principles of the segregation load. The first, illustrated by 6.12.3.4, is that the allele with the least selective disadvantage makes the largest contribution to the segregation load. For example, if $s = 0.01$ and $t = 1$, which means that A_2A_2 is either lethal or sterile, $sp^2 = 100/101^2 = 0.01$ and $tq^2 = 1/101^2 = 0.0001$. So the more drastic allele contributes only 1% of the total segregation load.

The second principle, first noticed by N. E. Morton, is illustrated by equation 6.12.3.5. This is that one can compute the segregation load from information on only one allele. This is also true for multiple alleles.

If there are multiple alleles there are numerous possibilities for stable equilibria, as discussed earlier in Section 6.8. In the following discussion, we assume that the conditions for a stable equilibrium have been met. We still assume random-mating proportions.

Let the genotypes and fitnesses be:

GENOTYPE	A_iA_i	A_iA_j	
FITNESS	$w_{ii} = w(1 - s_{ii})$	$w_{ij} = w(1 - s_{ij})$	6.12.3.6
FREQUENCY	p_i^2	$2p_ip_j$	

From the general equation for gene frequency change,

$$\Delta p_i = \frac{p_i(w_i - \bar{w})}{\bar{w}}. \qquad 6.12.3.7$$

At equilibrium $\Delta p_i = 0$, and therefore $w_i = \bar{w}$, as was shown in Section 6.8. So we are led to the very reasonable conclusion that, at equilibrium, each allele has the same average fitness as any other, or as the average of all. If this were not true, then the allele with a higher average fitness would tend to increase and there would be no equilibrium.

The segregation load, assuming that for some heterozygote $s_{ij} = 0$, is

$$L = \frac{w - \bar{w}}{w} = \frac{w - w_i}{w} = s_i \qquad 6.12.3.8$$

where

$$w_i = w(1 - s_i).$$

This can also be written as

$$L = \sum_j p_j s_{ij}. \qquad 6.12.3.9$$

A more useful form is

$$L > p_i s_{ii}. \qquad 6.12.3.10$$

This follows from 6.12.3.9 *a fortiori*, since the right side of 6.12.3.9 contains several nonnegative terms, of which $p_i s_{ii}$ is only one.

If all heterozygotes have equal fitnesses and all homozygotes are inferior, but not necessarily equal to each other, then the inequality becomes an equality. This is because with equal heterozygotes $s_{ij} = 0$ for all combinations where $i \neq j$ and therefore $p_i s_{ii}$ is the only nonzero term in 6.12.3.9. Writing 6.12.3.10 as

$$p_i = \frac{L}{s_{ii}},$$

and noting that $\sum p_i = 1$, we obtain

$$L \sum \frac{1}{s_{ii}} = 1,$$

and

$$L = \frac{\tilde{s}}{n}, \qquad 6.12.3.11$$

where $\tilde{s}$ is the harmonic mean of the homozygous disadvantages, s_{ii}, and n is the number of alleles. This shows that for comparable fitness coefficients, the segregation load is inversely proportional to the number of alleles maintained in the population.

Equation 6.12.3.10 extends the principle previously mentioned to multiple alleles. Regardless of the individual fitnesses, if the population is at equilibrium under selective balance, the segregation load (or at least a minimum estimate thereof) can be gotten from information on a single allele and its homozygous effect on fitness. The minimum estimate of the load is the product of the frequency of that allele and its homozygous selective disadvantage relative to the best genotype.

For example, the frequency (q) of the recessive gene for phenylketonuria is about 0.01. The fitness of persons homozygous for this gene is very nearly 0 (or at least it has been until very recently), so $s = 1$. Thus the minimum segregation load if this allele is maintained by selective balance is sq or 0.01. The population fitness is at least 1% less than that of the heterozygote, or the best heterozygote if there are multiple alleles. In contrast, if this is a fully recessive mutant maintained by recurrent mutation, the necessary mutation rate is 0.0001. The mutation load is only 0.0001. So the genetic load in this case is 100 times as large if the abnormal gene is maintained by heterozygote advantage as if it were determined by recurrent mutation.

A group of independent loci will have a collective segregation load that is roughly the sum of the individual loads until the number gets large. Suppose there were 100 loci, independent in inheritance and in their effect on fitness, and each with a load as large as that just mentioned. The total load would then be 100×0.01, or 1. This means that, with independence, the average fitness of the population relative to the best heterozygote is $(1 - 0.01)^{100}$, or roughly $e^{-1} = 0.37$. With 400 loci, $e^{-4} = 0.02$.

In general, the average fitness and load are

$$\bar{w} = we^{-\Sigma l}, \qquad L = 1 - e^{-\Sigma l}, \tag{6.12.3.12}$$

where l is the load for an individual locus.

The load can quickly get to be very nearly one if the number of polymorphic loci is large. That this creates problems in accounting for large numbers of segregating loci in a population has been discussed repeatedly.

There are several ways in which a large amount of polymorphism can be maintained without a large segregation load.

One possibility is that the selective differences are very small. If these are much less than the reciprocal of the effective population size, the allele frequencies will be largely determined by random drift and mutation as we shall discuss in Chapter 8. Of course, with neutral genes there is no load.

Whether any large number of neutral or nearly neutral mutants exist is an open question, but the evidence for them is increasing (e.g., King and Jukes, 1969).

A second possibility is that some polymorphic loci are maintained by frequency-dependent selection. If each allele is favored when rare, but not when common, there is a stable equilibrium when each is present in intermediate frequencies. The selective differences are minimized at or close to the equilibrium point, so that a population at equilibrium can be polymorphic with very little load. However, in any real population there will be drift away from the equilibrium point because of random processes and the population is returned to the equilibrium only by selection. So the load is not 0.

Linkage can reduce the segregation load by holding together a group of heterotic genes, at least under some circumstances. An extreme example is the case where there are several loci, each occupied by two alleles, and with both homozygotes lethal at every locus. Under this model the segregation load is 1/2 at each locus, and for n loci will be $1 - .5^n$. However, if all the loci were linked into two chromosomes, complementary at each locus, the load would be reduced to 1/2.

Sved, Reed, and Bodmer (1967), King (1967), and Milkman (1967) have all suggested that a threshold or truncation-selection model can greatly decrease the segregation load. The model assumes that beyond a certain level of heterozygosity additional heterozygous loci make no increased contribution to the average fitness. In an extreme form, all individuals beyond a number x of heterozygous loci have fitness 1 and those below this number have fitness 0. An alternative model with much the same consequences is that a certain fraction p of the individuals are selected and the rest are culled. Those individuals that are selected are the ones with the largest number of heterozygous loci.

It is doubtful, we think, that natural selection acts by counting the number of heterozygous loci and then sharply dividing the population into two groups based on the number of such loci. But the animal breeder practices truncation selection with respect to phenotypes, and it may be argued that natural selection approximates this pattern sufficiently well to alter substantially the number of polymorphisms maintained by a certain amount of selection. The question is one that requires empirical answers; clearly not enough is known about gene interactions to judge the realism of such models at present.

The introduction of linkage into threshold models leads to mathematical difficulties, but computer simulations have shown that it is possible to devise systems in which a great amount of polymorphism is maintained. Wills, Crenshaw, and Vitale (1969) have studied one such model. They assume truncation selection of a certain intensity (e.g., 10% selective elimination, where eliminated individuals are those with the smallest number of heterozygous loci). With close linkage the population tends to retain particular chromosomes—generally those that are mutually complementary—and a moderate amount of selection can retain a large number of polymorphisms. Again, how realistic the assumptions are is unknown.

Which of these mechanisms are the more important in determining natural polymorphisms and whether additional mechanisms play a role are among the most intriguing questions of population genetics. The combined efforts of experimental studies, natural population censuses, and computer simulation studies will probably be required for any real understanding.

The inbred load for a locus with k alleles is

$$\begin{aligned} L_1 &= \sum p_i s_{ii} \\ &< kL, \end{aligned} \qquad 6.12.3.13$$

since $\sum p_i s_{ii}$ is the sum of k terms, each of which (by 6.12.3.10) is less than the random-mating load. If all heterozygotes are the same the inbred load is simply k times the random load.

Thus, for a locus maintained by selective balance, or for a group of such loci if they are independent, the inbred load is not greater than the random load multiplied by the number of alleles maintained in the equilibrium population. In principle, it should be possible to determine whether inbreeding effects in man are due mainly to mutationally maintained loci or loci maintained by selective balance—provided that the average dominance, $\tilde{h}$, and the number of alleles, k, are both small. But the uncertainty about these and other assumptions and the absence of reproducible data have kept this from being an informative approach to the problem.

6.12.4 The Incompatibility Load

The only well-understood cause of an incompatibility load is maternal-fetal incompatibility for antigenic factors. For example, an A child with an O mother has a certain risk of dying as an embryo or neonatally due to anti-A agglutinins of the mother.

Because of the rule that an individual can produce antibodies only against antigens he does not possess, and because antigens are (with perhaps a few exceptions) the result of dominant genes, it follows that any increased death rate will always be in heterozygotes.

We can write the possibly incompatible types by writing maternal genotypes and the allele contributed to the embryo by the father. With random mating, they occur with the following frequencies, letting p, q, and r stand for the frequencies of A, B, and O alleles.

MOTHERS GENOTYPE	FREQUENCY (1)	SPERM GENOTYPE	FREQUENCY (2)	(1) × (2)	PROB-ABILITY OF DEATH
OO	r^2	A	p	pr^2	d_A
OO	r^2	B	q	qr^2	d_B
AA	p^2	B	q	p^2q	d_B
AO	$2pr$	B	q	$2prq$	d_B
BB	q^2	A	p	pq^2	d_A
BO	$2qr$	A	p	$2pqr$	d_A

The total incompatibility load due to this locus is equal to

$$\begin{aligned} L &= d_A(pr^2 + pq^2 + 2pqr) + d_B(qr^2 + p^2q + 2pqr) \\ &= d_A\, p(1-p)^2 + d_B q(1-q)^2. \end{aligned} \qquad 6.12.4.1$$

More generally, if p_i represents the frequency of A_i and d_i is the probability of death due to the antigen resulting from allele A_i, we can write the load as

$$L = \sum d_i\, p_i(1-p_i)^2, \qquad 6.12.4.2$$

where d_i may, of course, be 0 for some alleles as is probably the case for O in the ABO system.

This assumes that d_i is the same irrespective of the mother's genotype (assuming she has no A_i allele) and of the other (non-A_i) allele in the child, which is a reasonable a priori assumption, but not always true. If it is not, a separate d for each maternal-fetal genotype combination has to be introduced.

If either the mother or the child is inbred the incompatibility load is changed.

1. Mother Inbred The load due to the antigen produced by allele A is now

$$\begin{aligned} & d_i\, p_i\Big[(1-f_m)\Big(\sum_{j\neq i} p_j\Big)^2 + f_m \sum_{j\neq i} p_j\Big] \\ &\quad = d_i\, p_i[(1-p_i)^2(1-f_m) + f_m(1-p_i)] \\ &\quad = d_i\, p_i(1-p_i)(1-p_i+p_i f_m), \end{aligned}$$

where f_m is the inbreeding coefficient of the mother.

Summed over all alleles, this is

$$D = \sum d_i\, p_i(1-p_i)(1-p_i+p_i f_m). \qquad 6.12.4.3$$

Thus the incompatibility load increases in proportion to the inbreeding coefficient of the mother.

2. Child Inbred Consider the diagram in Figure 6.12.4.1 where M is the mother and C is the child. The letters a and b are the two gametes that united to form the mother while c is the gamete contributed to the child by the father, F.

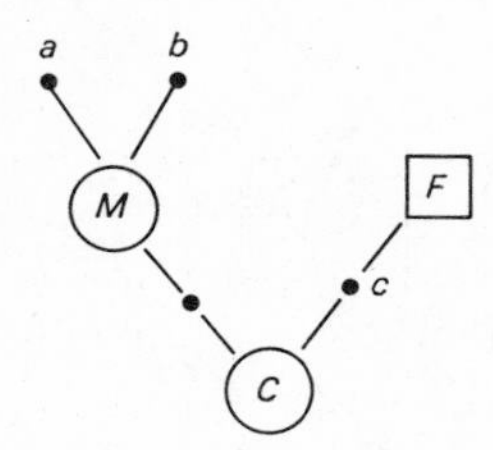

Figure 6.12.4.1. A diagram of gametic connections between mother (M), father (F), and child (C). The solid circles are gametes, designated by a, b, and c.

Let E be the probability that c is identical to either a or b, in which case there can be no incompatibility. Then, letting $=$ stand for identity by descent,

$$E = \text{prob}\ (a = c) + \text{prob}\ (b = c) - \text{prob}\ (a = b = c).$$

But the probability that $a = c$ is twice the contribution to the inbreeding coefficient of the child made by the path going through the grandparent that contributed a, and likewise for the probability that $b = c$. Therefore

$$E = 2f_c - \text{prob}\ (a = b = c), \qquad \textbf{6.12.4.4}$$

where f_c is the inbreeding coefficient of the child.

The prob $(a = b = c)$ is 0, unless there is an ancestor to which all three gametes, a, b, and c, can be traced back. There seems to be no simple general rule for computing this, but it is easily done for individual cases. In most human pedigrees this term is 0, of course.

The incompatibility load is then

$$L = D(1 - E). \qquad \textbf{6.12.4.5}$$

Hence, except for the possibility of a common ancestor of *a*, *b*, and *c*, we can say that the incompatibility load increases in proportion to the inbreeding coefficient of the mother and decreases in proportion to the inbreeding coefficient of the child. For further discussion, see Crow and Morton (1960).

6.12.5 The Load Due to Meiotic Drive

As the final example of a genetic load, consider the effect of meiotic drive (Sandler and Novitski, 1957). Typical examples are the *t*-alleles in mice and the SD factor in Drosophila. We shall consider only a simple example.

Assume that the homozygote is lethal or sterile (which is often true). Let h be the selection against heterozygotes. Assume further that the ratio of a to A genes contributed by males is $K : 1 - K$, but that the contribution from heterozygous females is the regular 1 : 1 of Mendelian heredity. Meiotic drive is typically found in one sex only, so this is a good assumption.

The genotypes and fitnesses can be designated as

GENOTYPE	AA	Aa	aa
FITNESS	1	$1-h$	0
FREQUENCY	$p_f p_m$	$p_f q_m + p_m q_f$	$q_f q_m$

6.12.5.1

where p_f and p_m are the frequency of allele A in the gametes of females and males; $q_m = 1 - p_m$ and $q_f = 1 - p_f$. With meiotic drive we cannot make the simplifying assumption of Hardy–Weinberg zygote ratios, since the gamete frequencies are different in the two sexes (recall Section 2.5).

The allele frequencies next generation are given by

$$q'_m = \frac{K(1-h)(p_m q_f + p_f q_m)}{1 - h(p_m q_f + p_f q_m) - q_m q_f}, \qquad 6.12.5.2$$

$$q'_f = \frac{q'_m}{2K}. \qquad 6.12.5.3$$

To specify the equilibrium conditions, let $q'_m = q_m = \hat{q}_m$ and $q'_f = q_f = \hat{q}_f = \hat{q}_m/2K$. This leads immediately to the quadratic

$$\hat{q}_m^2(1-2h) + \hat{q}_m[h(1+4K) - 2K] + K[2K - 1 - h(1+2K)] = 0 \qquad 6.12.5.4$$

or

$$A\hat{q}_m^2 + B\hat{q}_m + C = 0, \qquad 6.12.5.5$$

where

$$A = 1 - 2h$$

$$B = h(1+4K) - 2K,$$

$$C = K[2K - 1 - h(1+2K)].$$

The relevant solution is

$$\hat{q}_m = \frac{-B - \sqrt{B^2 - 4AC}}{2A}. \tag{6.12.5.6}$$

The load is

$$L = h(\hat{p}_m \hat{q}_f + \hat{p}_f \hat{q}_m) + \hat{q}_m \hat{q}_f = \frac{h\hat{q}_m}{2K}(1 + 2K - 2\hat{q}) + \frac{\hat{q}_m^2}{2K}. \tag{6.12.5.7}$$

The value of the meiotic drive load for several representative values of K and h are given in Table 6.12.5.1.

Table 6.12.5.1. The load due to meiotic drive. The gene favored by segregation distortion is assumed to be lethal when homozygous and to have a disadvantage of h relative to the normal homozygote. K is the proportion of gametes with the driven gene in one sex; the other sex is assumed to have normal Mendelian segregation. (From Crow, 1970.)

K	h							
	0.00	0.01	0.02	0.05	0.10	0.20	0.30	0.50
.5	0	0	0	0	0	0	0	0
.6	.010	.010	.010	.007	0	0	0	0
.7	.042	.042	.041	.039	.029	0	0	0
.8	.100	.100	.100	.097	.087	.033	0	0
.9	.200	.200	.200	.197	.187	.129	0	0
.95	.282	.282	.282	.279	.267	.205	.031	0
.98	.360	.360	.359	.356	.342	.270	.083	0
.99	.401	.400	.400	.396	.379	.298	.103	0
1.00	.500	.495	.490	.472	.438	.333	.125	0

When $h = 0$, as shown by Bruck (1957),

$$\hat{q}_m = K - \sqrt{K(1 - K)}, \tag{6.12.5.8}$$

$$L = \frac{\hat{q}_m^2}{2K} = \frac{1}{2} - \sqrt{K(1 - K)}. \tag{6.12.5.9}$$

This system has the interesting and unusual property that the load is decreased when the lethal gene is partially dominant. This property is brought out in the table, which also gives an idea of the range of values for K and h that maintain the polymorphism.

6.13 Evolutionary Advantages of Mendelian Inheritance

The ubiquity of Mendelian inheritance attests to its evolutionary value. For such an elaborate mechanism to be contrived implies that it must confer a great advantage on the population possessing it.

It is not at all apparent that sexual reproduction is of any selective advantage to the individual. Its value clearly lies in gene-shuffling, the value of which is more likely to be for the long time benefit of the population as a whole than for the individual. In fact, as we noted before, Fisher (1930, 1958) went so far as to suggest that sexual reproduction is perhaps the only trait that has evolved by intergroup competition.

We shall not try to discuss ways in which Mendelian inheritance may have evolved, but we shall discuss briefly some of the ideas about its value to the population.

It is true that the Mendelian system is capable of producing an enormous number of genotypes by recombination of a relatively small number of genes. The number of potential combinations is indeed great, but the number produced in any single generation is limited by the population size, and gene combinations are broken up by recombination just as effectively as they are produced by it. Furthermore, for a given amount of variability, the efficiency of selection is greater in an asexual population, for here the rate of progress is determined by the genotypic variance rather than by the genic variance, as it is in a sexual system. However, if the environment changes so that drastic change in phenotype is needed, an asexual population (in the absence of new mutants) is limited to the best genotype in the current population. Selection of Mendelian recombinants can produce strains that far transgress the former variability of the population. Numerous selection experiments have demonstrated that in a few generations the mean can come to lie outside the range of what were the most extreme deviants in the population before selection began.

The evolutionary advantage of recombination has often been discussed. The two principal ideas are: (1) that recombination makes it possible for favorable mutants that arose in different individuals to get into a single individual, and (2) that recombination permits the species to respond more effectively to an ever-changing environment. The first idea was developed originally by Fisher (1930) and by Muller (1932). The second was most clearly stated among early writers by Wright (1931 and later) and by Sturtevant and Mather (1938).

To consider the first idea, we shall rely on a model first proposed by Muller. In an asexual population two beneficial mutants can be incorporated into the population only if the second occurs in a descendant of the individual in which the first mutation took place. The limiting factor will be the time required for the descendants of the first mutant to increase to such numbers that a second mutant becomes reasonably probable. In a Mendelian population all the favorable mutants that occur during this interval can be incorporated. (We are ignoring the loss of a new mutant by random processes, to be discussed later in Chapter 8, for this is not essentially different in the two systems.) An asexual system will be as efficient as a sexual system only if the mutation rate is so low, the selective advantage of the mutant so great, or the population so small that the first mutant is established before another favorable mutant occurs.

We have discussed this situation quantitatively (Crow and Kimura, 1965) but we shall give only a qualitative summary here. The situation is illustrated in Figure 6.13.1, which is adapted from Muller's 1932 paper. The three

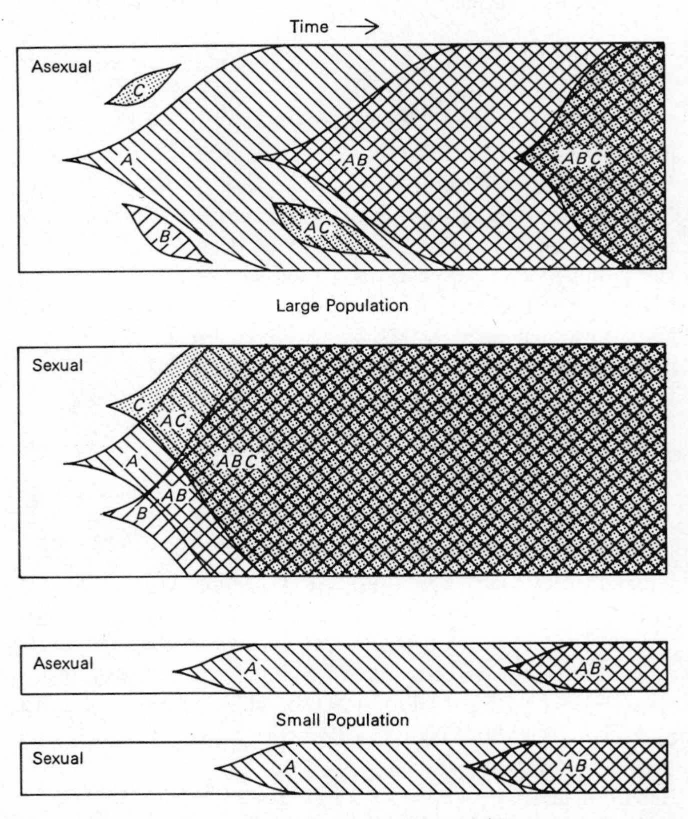

Figure 6.13.1. Evolution in sexual and asexual populations. For explanation, see the text.

mutants, A, B, and C, are all beneficial. In the asexual population when all three arise at roughly the same time only one can prevail. In this case A is more fit than B or C; or A may simply be luckier in happening to occur in an individual that was for other reasons unusually well adapted. B can be incorporated only when it occurs in an individual that already has mutant A, and this will not happen on the average until the descendants of the original A mutant have grown to numbers roughly the reciprocal of the mutation rate. C is finally incorporated, but only after AB individuals are in appreciable numbers. So, in an asexual population the mutants are incorporated in series.

The sexual population, on the other hand, permits the incorporation of mutants in parallel. The ratio of the evolutionary rate in a Mendelian to that in an asexual population is the number of mutants that occur in the interval of time between the occurrence of a mutant and the occurrence of a second mutant in a descendant of the first mutant. Numerical calculations have been presented in the paper referred to above.

The qualitative conclusions are clear from the diagrams however. The relative advantage of sexuality will be greatest when the population is large, when the mutation rate is high, and when the selective advantage of the mutant is small—for each of these favors the occurrence of more mutants than can be incorporated in series. In other words, the advantage of sexuality is greatest when the system is evolving by very minute steps in a large population.

On the other hand, the incorporation of new mutant is not the only evolutionary factor of importance. Evolution may also occur by the shifting of frequencies of genes that are already present in fairly high frequencies in the population, as Wright has emphasized. In this case the argument of Sturtevant and Mather is relevant. We can do no better than quote directly from their article:

> The simplest system that we have been able to devise, having the required property of favoring recombination, is as follows: Two gene pairs, A, a and B, b, exist in a population subjected frequently to three different sets of environmental conditions, D, E, and F. Condition D favors A but acts unfavorably on B, whereas E favors B but lowers the frequency of A. These, if properly adjusted as to intensity of selection and number of generations over which they operate, will insure the perpetuation of both allelomorphs at each locus. Recombination will not, however, be of importance. For this it seems necessary to introduce the third condition, F, favoring the combination AB (with or without a similar effect on ab), but acting adversely on the single types A and B. Under such conditions it would appear that there will be a selective action favoring recombination, of the order of magnitude of the selection of the AB type, as opposed to Ab or aB, under condition F. Other combinations of two or more loci may be exerting similar action in similar

or disimilar environmental conditions, and so the net effect will probably be to favor recombination for the majority of loci under the majority of condition changes.

Similar arguments apply when the population is divided into partially isolated subpopulations with different gene combinations favored in different subpopulations and occasional migration between subpopulations (Wright, 1931 and later).

Virtually the same argument has been stated recently (and independently) by Maynard Smith (1968). We do not know which of these arguments, the Fisher–Muller or the Wright–Sturtevant, is the more important. Very likely, both are valid, and which is more important depends on whether the incorporation of new mutants or the adaptation of existing variants to variable environments is more often the limiting factor in evolutionary advance. It is also possible, of course, that neither of these has identified the really crucial advantage, and that some third reason is still more important.

The Advantages of Diploidy The evolutionary advantages of recombination can be obtained in species that spend most of their lives as haploids as well as in those that are predominantly diploid. Yet there has often been evolution toward diploidy. What is the reason?

At first glance it would appear that there is an obvious advantage of diploidy in that dominant alleles from one haploid set can prevent the expression of deleterious alleles in the other. However, as soon as a new equilibrium is reached the mutation load will be the same; in fact, it will be twice as large unless the mutants are completely recessive. From this standpoint diploidy is a disadvantage, not an advantage.

However, when the population that is previously haploid suddenly changes to diploid there is an immediate advantage. To be sure, when the equilibrium is reached the advantage is gone; but by this time the diploid condition may be established and there is no way of going back to haploidy without all the deleterious effects of exposing recessive genes. So it may be that diploidy is not conferring a lasting benefit, but is the result of a temporary advantage that cannot easily be gotten rid of.

There are two other possible advantages of diploidy. One very obvious one is overdominance. If such loci are common, there is a genuine advantage to diploidy provided such effects are important enough to compensate for the greater mutation load from partially dominant loci.

A second possibility is the protection it affords from the effects of somatic mutation. The zygote in a diploid species or the gametophyte in a haploid plant may have approximately the same equilibrium fitness, but the effects of somatic mutation would be quite different. Diploidy would protect against recessive, or partially recessive, somatic mutants. If the soma were large and complicated, as in higher plants and especially animals, a diploid soma may provide a significant protection against the effects of recessive mutants in critical cells.

6.14 Problems

1. A mutant has a selective disadvantage hs in the heterozygote. Assume that hs is large enough that selection in mutant homozygotes can be ignored. Show that the mean number of generations that a mutant persists is $1/hs$.
2. Assume that X-linked hemophilia has an incidence of one in 20,000 males and a selective disadvantage, s, of about 0.8. Estimate the mutation rate of the gene.
3. Give two or more reasons why the mutation rate estimates for autosomal dominant and X-linked recessive mutants are more accurate if s is large.
4. What is the genic variance of an overdominant locus at equilibrium under selection?
5. Show that for equilibrium under selection the equilibrium gene frequencies remain the same when w_{ij} is replaced by $a + bw_{ij}$, where a and b are constants.
6. Show that the segregation load when $f = 1$ for one of a series of multiple alleles is the same as the load for all alleles when $f = 0$. (Assume that all heterozygotes are equal in fitness.)
7. Prove that in equation 6.7.4, when $w_{11} - 2w_{12} + w_{22} < 0$, $|\Delta\xi/\xi| < 1$.
8. Show, for the model of equation of 6.5.6, that the migration load is $M(p - \bar{p})/p$, where p is the equilibrium frequency in the subpopulation and $\bar{p}$ the gene frequency in the immigrants. Is the load the same when the favored gene is completely dominant?
9. Derive the characteristic equation of 6.9.6 by letting $y'/y = y''/y' = \lambda$.
10. Show how 6.12.2.8 may be obtained.
11. What would be the effect of polygamy on the zygotic sex ratio? Of infanticide where one sex is preferentially killed? Would you expect a different zygotic sex ratio for mice than for monkeys?
12. Assume that weight in mice is determined by a large number of loci, that there is no dominance or epistasis, and no environmental variance. Assume further that the fitness of a mouse is proportional to the square of its deviation from the average weight in the population. Show that the homozygous load is twice the randomly mating load.

13. Given the array of fitnesses

	A_1	A_2	A_3
A_1	0.90	1.00	1.15
A_2	1.00	1.00	1.10
A_3	1.15	1.10	0.80

will this lead to a stable equilibrium? Is there a stable equilibrium if allele A_3 is lost?

14. What frequency of the Rh^- gene (d) will maximize the incidence of hemolytic disease in Rh^+ embryos with Rh^- mothers?
15. What frequency of the A, B, and O alleles will maximize the ABO incompatibility load? Are these frequencies independent of d_A and d_B?

7 PROPERTIES OF A FINITE POPULATION

In preceding chapters we have considered mutation, migration, and selection as factors causing deterministic changes in gene frequencies and in such population properties as average fitness or performance. There was the implicit assumption that the population is large enough that random sampling of gametes does not introduce an appreciable amount of noise into the system and that the migration and selection coefficients are either constant or change in a predictable way.

Actual populations are finite so there is some random drift in gene frequencies as we mentioned briefly in Chapter 3. Also there may be fortuitous changes in the other factors, particularly in selection coefficients.

In this chapter we will continue the discussion begun in Chapter 3 on the effects of a finite population number, ignoring the influence of migration and selection. Later, in Chapters 8 and 9, we consider how these factors interact with random processes.

7.1 Increase of Homozygosity Due to Random Gene Frequency Drift

As was pointed out in Chapter 3, Section 11, there is a fluctuation in gene frequency from generation to generation in a finite population. We can regard each generation of N diploid offspring as being derived from a sample of $2N$ gametes from the parental generation. As time goes on the gene frequency will tend to depart more and more from its original value. Finally one of the alleles is fixed while all others are lost and the population becomes homozygous for this locus. Since the fixation of genes is an irreversible process (if we exclude mutation or migration) the number of fixed loci will tend to increase with time.

We also emphasized that, although the consequences of random gene frequency drift and consanguineous mating are both such as to lead to an increase in homozygosity and we can measure either by Wright's inbreeding coefficient, in one regard the two are quite different. In a large population with some consanguineous matings there is a departure from Hardy–Weinberg ratios, as given by equations 3.2.2 and 3.2.3, and the proportion by which the heterozygosity is reduced is equal to f. In a finite population within which mating is at random the population remains in approximate Hardy–Weinberg ratios. The reduction in heterozygosity comes from the random changes in gene frequency, the net effect of which is to reduce $\sum p_i p_j$ (the proportion of heterozygotes, $i \neq j$) by a fraction f.

In a population of N diploid individuals mating completely at random, including the possibility of self-fertilization, the heterozygosity decreases at a rate $1/2N$ where N is the effective population number (see 3.11.2). In terms of the inbreeding coefficient

$$f_t = \frac{1}{2N} + \left(1 - \frac{1}{2N}\right) f_{t-1} \qquad \textbf{7.1.1}$$

as was shown earlier (3.11.1).

Also, when there is no self-fertilization the change is given by

$$f_t = g_{t-1}, \qquad \textbf{7.1.2}$$

$$g_t = \frac{1}{2N}(1 + f_{t-1}) + \left(\frac{N-1}{N}\right) g_{t-1}, \qquad \textbf{7.1.3}$$

where f_t is the inbreeding coefficient in generation t and g_t is the coefficient of consanguinity of two randomly chosen individuals in generation t (see 3.11.3).

From this

$$f_t = \frac{1}{2N} + \left(1 - \frac{1}{N}\right) f_{t-1} + \frac{1}{2N} f_{t-2}. \tag{7.1.4}$$

If there are two sexes of unequal numbers, N_m males and N_f females,

$$N = \frac{4N_m N_f}{N_m + N_f} \tag{7.1.5}$$

in 7.1.4 (see 3.13.1).

We also introduced Wright's concept of effective population number in Section 3.13, an idea that will be developed further later in this chapter.

Random gene frequency drift, either because of a finite population or because the selection and migration coefficients vary, plays a central role in the "shifting balance" theory of evolution proposed by Sewall Wright. As we discussed in Chapter 5, deterministic selection tends to increase the frequency of those genes which enhance the fitness of the individual. So the population fitness tends to increase (subject to some qualification if there are complicating factors such as frequency-dependent selection and linkage disequilibrium). This was discussed in Sections 5.6, 5.7, and 5.9.

If the mean relative fitness is plotted as ordinate and the various gene frequencies as the various abscissae in multidimensional space, we can think of the fitness as a hypersurface. In this metaphor we can speak of the surface, as Wright has, as an adaptive surface. See Wright (1967 and earlier) for a discussion of this concept, and for the relation of the "existence" of such a surface to quasi-linkage equilibrium. If the surface has multiple peaks and valleys a population, which can be thought of as a point on the surface, will tend to climb the nearest peak, which is not necessarily the highest. There may be no deterministic way in which a population can change from one peak to a higher one.

Wright suggests that such evolutionary bottlenecks may be frequent and that it is important that there be ways by which a population can move away from the stable equilibrium represented by one peak in order to come under the sphere of influence of another, higher peak. This might happen if, for any of the reasons mentioned before, there is some random shifting of gene frequencies. Such a change might be sufficient to permit the population to wander randomly over the surface (still speaking metaphorically) and come under the influence of a higher peak. Of course, there is a loss in average fitness as the population drifts from the highest point. But Wright argues that this is a necessary price for letting evolution find new gene combinations. Multiple peaks and valleys will be prevalent if there are complex dominance and epistatic relations among the various loci, especially those of a type where two or more alleles that are individually deleterious are collectively beneficial.

The views of Fisher and Wright contrast strongly on the evolutionary significance of random changes in the population. Whereas, to Fisher, random change is essentially noise in the system that renders the deterministic processes somewhat less efficient than they would otherwise be, Wright thinks of such random fluctuations as one aspect whereby evolutionary novelty can come about by permitting novel gene combinations.

Whether random gene frequency drift is a way of creating new, favorable epistatic combinations or is more like background noise, there is increasing evidence that it is prevalent. Molecular biology has shown dramatically the wide range of mutational possibility at a single gene locus with its several hundred nucleotides. The possibility that many nucleotide substitutions may cause inconsequential changes has become increasingly apparent. Some, or perhaps many, such changes may alter fitness by an amount that is of the same order as the mutation rate or the reciprocal of the effective population number, or less. The fate of such mutants is determined largely by random processes. The possibility that such random changes may account for a substantial part of the amino acid changes observed in the evolution of hemoglobin and other proteins has been discussed by several authors recently (see Kimura, 1968, 1969b; King and Jukes, 1969; Crow, 1969). In the next chapter we shall discuss the rate of evolution that would be expected from neutral or near-neutral mutations.

The other aspect of near-neutral genes and a great multiplicity of potential mutations at each locus is the possibility that this may account for polymorphisms, particularly those having no overt effect and detected only by electrophoresis or other chemical trickery. This is the subject of the next section.

7.2 Amount of Heterozygosity and Effective Number of Neutral Alleles in a Finite Population

It has often been suggested in the past that the wild-type allele is not a single entity, but rather a population of different isoalleles that are indistinguishable by any ordinary procedure. Since each gene consists of several hundred, or perhaps thousands, of nucleotide pairs, the range of mutational possibility is enormous, especially when one considers combinations. That some of these are essentially equivalent seems reasonable and is reinforced by chemical studies showing that amino acid substitutions often do not affect in any detectable way the function of certain enzymes. In any case the probability that such alleles may exist in substantial numbers seems great enough to warrant an inquiry into the population consequences. The procedure given here is an extension of one presented earlier (Kimura and Crow, 1964).

In this chapter we shall consider only neutral alleles. Those with a slight selective advantage or disadvantage (especially those that are overdominant) are of interest, but require more advanced methods. These will be discussed in Chapter 9.

To isolate the essential question, we consider an extreme situation in which the number of possible isoallelic states at a locus is large enough that each new mutant is of a type not preexisting in the population. This provides an estimate of the upper limit for the number of different alleles actually maintained by mutation.

Let u be the average rate of mutation of the alleles existing in the population of effective size N_e. If the population consists of N individuals, there will be $2Nu$ new mutant genes introduced per generation. The probability of two uniting gametes carrying identical alleles is given by 7.1.1. The two genes which are identical in state will remain so in the next generation only if neither of them has mutated since the previous generation, the probability of which is $(1-u)^2$. Thus we can write

$$f_t = \left[\frac{1}{2N_e} + \left(1 - \frac{1}{2N_e}\right) f_{t-1}\right](1-u)^2. \tag{7.2.1}$$

When equilibrium is reached, $f_t = f_{t-1}$. The solution is

$$f = \frac{(1-u)^2}{2N_e - (2N_e - 1)(1-u)^2}. \tag{7.2.2}$$

Ignoring terms containing u^2, we obtain

$$f = \frac{1-2u}{4N_e u - 2u + 1} \approx \frac{1}{4N_e u + 1}. \tag{7.2.3}$$

The proportion of heterozygous loci, H, is $1 - f$ or

$$H \approx \frac{4N_e u}{4N_e u + 1}. \tag{7.2.4}$$

This gives the probability that an individual chosen at random will be heterozygous for this locus. If the effective population number is much smaller than the reciprocal of $4u$, the average individual will be homozygous for most such loci; on the other hand, if N_e is much larger than $1/4u$ the individuals in population will be largely heterozygous for neutral alleles.

It is convenient to define the *effective* number of alleles (n_e) maintained in the population by the reciprocal of the sum of the squares of allelic frequencies. In the present model, $n_e = 1/f$ (since the proportion of homozygotes $= f = \sum p_i^2$) and therefore

$$n_e = 4N_e u + 1. \tag{7.2.5}$$

Note that this will be less than the actual number of alleles, unless all alleles are of the same frequency.

If p_i is the frequency of the ith allele, then the effective number of alleles is $1/\sum p_i^2$. On the other hand, the actual number of alleles is $1/\bar{p}$, where $\bar{p}$ is the mean allele frequency. For example, with three alleles with frequencies 2/3, 1/6, and 1/6, $\bar{p} = 1/3$ (of course, since there are three alleles), but $\sum p_i^2 = 1/2$. Thus the number of alleles is three, but the amount of heterozygosity is the same as would be found in a population with two alleles of equal frequency.

Most of the time in population genetics we are more interested in the effective number of alleles than in the actual number. Many of the alleles will be represented only once or twice in the population and contribute very little to the average heterozygosity or genetic variance. Since it is ordinarily the latter quantities that we are interested in, the effective number is the more useful. Fortunately, the effective number is much more easily computed, as we have just shown. The actual number requires a knowledge of the distribution of allele frequencies; this will be deferred until Chapter 9.

The derivation of 7.2.3 assumed that each mutant is new; that is, that it is an allele not currently represented in the population. We can remove this restriction. Assume that there are k possible alleles and the rate of mutation from one to one of the $k - 1$ others is $u/(k - 1)$; that is, the total mutation rate is u. Then we modify 7.2.1 to become

$$f_t = \left[\frac{1}{2N_e} + \left(1 - \frac{1}{2N_e}\right) f_{t-1}\right](1 - u)^2 + \left(1 - \frac{1}{2N_e}\right)(1 - f_{t-1})2\frac{u}{k - 1}(1 - u), \tag{7.2.6}$$

ignoring the possibility that both alleles mutate simultaneously to the same new mutant state.

Ignoring terms in u^2 and letting $f_t = f_{t-1} = f$, we get for the equilibrium value,

$$f = \frac{4N_e u\left(\frac{1}{k - 1}\right) + 1}{4N_e u\left(\frac{k}{k - 1}\right) + 1}, \tag{7.2.7}$$

or

$$H = \frac{4N_e u}{4N_e u\left(\frac{k}{k-1}\right) + 1}. \qquad 7.2.8$$

As expected, when $k \to \infty$, 7.2.8 becomes the same as 7.2.4.

We have been assuming where the population includes self-fertilization. We can readily remove this restriction. Including the possibility of mutation and returning to our model where each mutant is new to the population, 7.1.2 and 7.1.3 are modified to become

$$f_t = g_{t-1}(1-u)^2, \qquad 7.2.9$$

$$g_t = \left[\frac{1}{2N_e}(1 + f_{t-1}) + \left(1 - \frac{1}{N_e}\right)g_{t-1}\right](1-u)^2. \qquad 7.2.10$$

Putting these together and eliminating the g's gives

$$f_t = \frac{1}{2N_e}(1 + f_{t-2})(1-u)^4 + \left(1 - \frac{1}{N_e}\right)f_{t-1}(1-u)^2. \qquad 7.2.11$$

Table 7.2.1. The average proportion of homozygosity, f (upper figure), and the effective number of neutral alleles (lower figure) in a randomly mating population of effective number N_e and mutation rate u. (From Kimura and Crow, 1964.)

	EFFECTIVE POPULATION NUMBER, N_e					
MUTATION RATE, u	10^2	10^3	10^4	10^5	10^6	10^7
10^{-4}	.96 1.04	.71 1.4	.20 5.0	.024 41	.0025 401	.00025 4001
10^{-5}	.996 1.004	.96 1.04	.71 1.4	.20 5.0	.024 41	.0025 401
10^{-6}	.9996 1.0004	.996 1.004	.96 1.04	.71 1.4	.20 5.0	.024 41
10^{-7}	.99996 1.00004	.9996 1.0004	.996 1.004	.96 1.04	.71 1.4	.20 5.0

At equilibrium, $f_t = f_{t-1} = f_{t-2} = f$, leading to

$$f \approx \frac{1}{4N_e u + 1} \tag{7.2.12}$$

and

$$H \approx \frac{4N_e u}{4N_e u + 1}, \tag{7.2.13}$$

just as before.

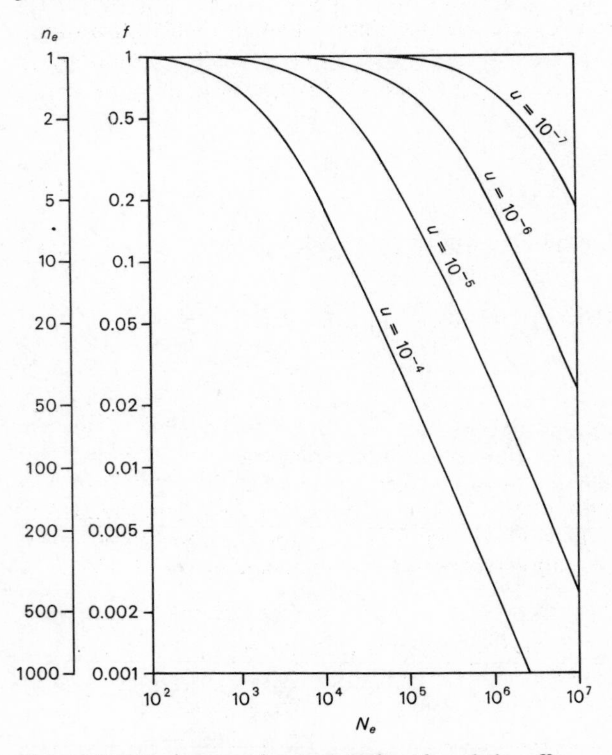

Figure 7.2.1. Average homozygosity, f, and the effective number of neutral alleles, n_e, as a function of the effective population number, N_e, for various mutation rates, u. The range of mutational possibility is assumed to be large enough that each mutant is an allele not currently in the population. Populations where f is between 0.5 and 1 are homozygous for the majority of loci; those below 0.5 are heterozygous for the majority.

Table 7.2.1 gives the average proportion of homozygosity, f, and the effective number of alleles, $n_e = 1/f$, for populations of various effective number and mutation rate. If $4N_e u = 1$ the effective number of alleles is 2 and increases as $4N_e u$ gets larger. As $4N_e u$ gets smaller the population becomes more nearly monomorphic. For populations with effective numbers of order 10^5 or larger, it should not be surprising to find considerable neutral polymorphism, provided that neutral mutations are occurring at rates of the order 10^{-5} or higher. This is shown graphically in Figure 7.2.1.

Later in this chapter we shall discuss effective population number in more detail. At that time we will show that there are two ways of defining the effective population number. For many cases these lead to the same consequences, but not always. To anticipate, the effective number that we have been discussing in this section is the *inbreeding* effective number.

7.3 Change of Mean and Variance in Gene Frequency Due to Random Drift

Consider a population of N breeding individuals in which the frequencies of a pair of alleles A_1 and A_2 are p and $1-p$ respectively. We assume that mating is such that the next generation is produced from union of N male and N female gametes, each extracted as a random sample from an infinitely large hypothetical population of gametes produced by the parents. Thus the probabilities that the number of gene A_1 will be $0, 1, 2, \ldots, i, \ldots, 2N$ in the pooled sample of $2N$ gametes is given by

$$(1-p)^{2N},\ 2Np(1-p)^{2N-1},\ \frac{2N(2N-1)}{2}p^2(1-p)^{2N-2},\ \cdots,$$

$$\frac{2N(2N-1)\cdots(2N-i+1)}{i!}p^i(1-p)^{2N-i},\ \cdots,\ p^{2N}.$$

In other words, the probabilities follow the binomial distribution which is obtained by expanding

$$[p+(1-p)]^{2N}.$$

Since the mean and variance of this distribution are respectively $2Np$ and $2Np(1-p)$, the mean and variance of the gene frequency x will be given by

$$\bar{x} = E(x) = p, \tag{7.3.1}$$

$$V_x = E[(x-p)^2] = E(x^2) - p^2 = \frac{p(1-p)}{2N}, \tag{7.3.2}$$

where E stands for the operation of taking expectations with respect to the number (i) of gene A_1 and $x = i/2N$. For a derivation of the binomial variance formula, $p(1-p)/2N$, see A.5.3–A.5.5.

In the previous chapters we used the letter p, or sometimes q, to designate allele frequencies. Now, as we shift the emphasis to random effects and stochastic processes, we shall designate the allele frequency by x to emphasize that this is a random variable. Usually, we use p for the initial value, which is what the frequency would be in the absence of random effects.

We will call the newly formed generation the first generation. Formulae 7.3.1 and 7.3.2 show that in going to the first generation, the mean of the gene frequency is unchanged, while the variance increases by an amount inversely proportional to the population number N. Suppose that the second generation is produced again from N male and N female gametes taken as random samples from the first generation. Let x' be the frequency of A_1 in the second generation; then

$$\overline{x'} = E(x') = E_1 E_2(x') = E_1[E_2(x')] = E_1(x) = p,$$

where E_1 and E_2 designate expected value operators in the first and the second generation. It will be seen that the mean is again unchanged. Similarly, the variance is calculated as follows: Since

$$\begin{aligned} V_{x'} &= E[(x'-p)^2] = E_1E_2[(x'-x)+(x-p)]^2 \\ &= E_1E_2[(x'-x)^2] + 2E_1[(x-p)E_2(x'-x)] + E_1[(x-p)^2], \end{aligned}$$

in which

$$E_2[(x'-x)^2] = \frac{x(1-x)}{2N} = \frac{1}{2N}[x - p^2 - (x^2 - p^2)],$$

and

$$E_2(x'-x) = 0,$$

we obtain

$$\begin{aligned} V_{x'} &= \frac{1}{2N}\left[p - p^2 - \frac{p(1-p)}{2N}\right] + \frac{p(1-p)}{2N} \\ &= \frac{p(1-p)}{2N}\left(1 - \frac{1}{2N}\right) + \frac{p(1-p)}{2N}. \end{aligned}$$

If the same pattern of reproduction is continued until the tth generation, the mean and the variance become respectively

$$\bar{x}^{(t)} = p, \tag{7.3.3}$$

$$\begin{aligned} V_x^{(t)} &= \frac{p(1-p)}{2N}\left[\left(1-\frac{1}{2N}\right)^{t-1} + \cdots + \left(1 - \frac{1}{2N}\right) + 1\right] \\ &= p(1-p)\left[1 - \left(1 - \frac{1}{2N}\right)^t\right]. \end{aligned} \tag{7.3.4}$$

After a large number of generations, the mean is still unchanged, i.e.,

$$\bar{x}^{(\infty)} = p, \tag{7.3.5}$$

but the variance increases with time and finally becomes

$$V_x^{(\infty)} = p(1-p), \tag{7.3.6}$$

since

$$\lim_{t\to\infty}\left(1-\frac{1}{2N}\right)^t = 0.$$

This suggests that at this limit, the gene frequency will become either 1 or 0 with probabilities p and $1-p$. The process leading to this state starting from an arbitrary gene frequency will be studied in detail in Chapter 8. The state $x=1$ corresponds to fixation of gene A_1 in the population and $x=0$ corresponds to its loss. The increase of variance with time as shown in 7.3.4 shows that the probability of belonging to one of these states increases with time.

We are referring here to the variance in gene frequency between a group of populations, each starting with the same gene frequency. This variance increases with time as the populations diverge from their initial value. At the same time the average genic variance within each population decreases as the individual gene frequencies move toward 0 or 1. In parallel with this the average heterozygosity of an individual decreases with time, as we shall now show.

Let H_t be the probability that a randomly chosen individual in the population at the tth generation be heterozygous. Since this probability is $2x(1-x)$ for a given gene frequency x, we have

$$\begin{aligned} H_t &= E[2x(1-x)] = 2E[x-(x^2-p^2)-p^2] \\ &= 2[\bar{x}^{(t)} - V_x^{(t)} - p^2], \end{aligned}$$

which reduces to

$$H_t = 2p(1-p)\left(1-\frac{1}{2N}\right)^t, \qquad (t=1,2,\ldots), \tag{7.3.7}$$

if we apply 7.3.3 and 7.3.4.

We see that the average heterozygosity decreases at a constant rate of $1/2N$ per generation. If the initial population (generation 0) is produced by random mating, the expected frequency of heterozygotes at that time is $2p(1-p)$, so we have

$$H_t = H_0\left(1-\frac{1}{2N}\right)^t \qquad (t=0,1,2,\ldots), \tag{7.3.8}$$

as we obtained earlier by another method (see 3.11.2).

When N is sufficiently large, this is approximated by

$$H_t = H_0 e^{-\frac{t}{2N}} = 2p(1-p)e^{-\frac{t}{2N}} \tag{7.3.9}$$

(see also formula 8.4.11 in the next chapter).

At this point, a few remarks are in order. First it should be noted that the homozygosity or heterozygosity of an individual within a population is a concept distinct from genetic homogeneity or heterogeneity of the population itself. For the latter Wright used the terms homallelic and heterallelic; a population is homallelic if it contains only one allele, but is heterallelic if it contains two or more alleles. Secondly, as shown above, the probability of heterozygosity decreases at the rate of exactly $1/2N$ per generation under random mating, while the probability of coexistence of both alleles within a population, though it diminishes each generation, does not decrease at a constant rate. Its rate approaches $1/2N$ only asymptotically. This will be demonstrated at the end of the next section and in more detail in Chapter 8.

Wright (1931) gave an approximate formula,

$$L_t = L_0 e^{-\frac{t}{2N}}, \tag{7.3.10}$$

for the number of unfixed loci at the tth generation. The approximation is satisfactory only for a large t and an intermediate initial gene frequency. The number of alleles introduces another complication into the process of decrease in genetic heterogeneity, but not in the decrease of heterozygosity: Formulae 7.3.8 and 7.3.9 are correct not only for two alleles but for any number.

Let $x_1, x_2, \ldots, x_n$ be respectively the frequencies of alleles $A_1, A_2, \ldots, A_n$ in the random-mating population at the tth generation $(x_1 + x_2 + \cdots + x_n = 1)$. The distribution of the frequencies of these alleles in the next generation produced by N male and N female gametes taken as random samples from generation t follows the multinomial distribution with means, variances, and covariances given by

$$E(x_i') = x_i, \tag{7.3.11}$$

$$E(x_i')^2 - x_i^2 = \frac{x_i(1-x_i)}{2N}, \tag{7.3.12}$$

$$E(x_i' x_j') - x_i x_j = -\frac{x_i x_j}{2N} \qquad (i \neq j). \tag{7.3.13}$$

The expected total frequency of heterozygotes in generation $t+1$ is

$$H_{t+1} = E\Big(\sum_{i \neq j} x_i' x_j'\Big) = \sum_{i \neq j} E(x_i' x_j'),$$

and if we apply 7.3.13 in the right side of the above expression, we obtain

$$H_{t+1} = \left(1 - \frac{1}{2N}\right) \sum_{i \neq j} x_i x_j$$

or

$$H_{t+1} = \left(1 - \frac{1}{2N}\right) H_t. \qquad 7.3.14$$

The expected total frequency of heterozygotes in a randomly mating population decreases at the rate of $1/2N$, regardless of the number of alleles or the allele frequencies. If we note that the expected frequency of heterozygotes is proportional to $1 - f_t$ (see Chapter 3), we obtain

$$1 - f_{t+1} = \left(1 - \frac{1}{2N}\right)(1 - f_t)$$

or

$$f_{t+1} = \frac{1}{2N} + \left(1 - \frac{1}{2N}\right) f_t, \qquad 7.3.15$$

which is equivalent to equation 7.1.1 and also 3.11.1. The above relation may also be written as

$$\frac{\Delta f_t}{1 - f_t} = \frac{f_{t+1} - f_t}{1 - f_t} = \frac{1}{2N}. \qquad 7.3.16$$

Starting from $f_0 = 0$, the inbreeding coefficient in generation t is given by

$$f_t = 1 - \left(1 - \frac{1}{2N}\right)^t = \frac{V_x^{(t)}}{p(1 - p)}. \qquad 7.3.17$$

When N is large we have, with good approximation,

$$f_t = 1 - e^{-\frac{t}{2N}} \qquad 7.3.18$$

7.4 Change of Gene Frequency Moments with Random Drift

This section is an extension of the previous one. Here we will study more generally the change of gene frequency moments. First, a discrete model will be employed to derive a few moments and the results will be compared with the corresponding values obtained by a continuous model. The latter model will then be used with advantage to derive a general formula for the moments when N is large. Some of the results obtained will be found useful in later sections. In the next section, we will find that the first four moments are required to describe the change of variance in quantitative characters in finite populations. Furthermore, by knowing all the moments for large N, we get extensive knowledge of the process of random genetic drift which will be treated in detail in Chapter 8, Section 4. Actually, it was through this method in conjunction with the method of partial differential equations that the entire process of random genetic drift was first constructed (Kimura, 1955*a*).

Let us consider an isolated population of N breeding diploid individuals. Let A_1 and A_2 be a pair of alleles with frequencies x and $1 - x$ respectively. We assume that mating is random and that the mode of reproduction is such that N male gametes and N female gametes are drawn as a random sample from the population to form the next generation. Here x may take any one of the sequence of values:

$$0, \frac{1}{2N}, \frac{2}{2N}, \ldots, 1 - \frac{1}{2N}, 1.$$

We designate by $f(x; t)$ the propability that the gene frequency is x at the tth generation ($t = 0, 1, 2, \ldots$). Let x_t be the frequency of A_1 in the tth generation and let δx_t be the change of x_t due to random sampling of gametes such that

$$x_{t+1} = x_t + \delta x_t. \tag{7.4.1}$$

Let

$$\mu_n^{(t+1)} = E(x_{t+1}^n) \tag{7.4.2}$$

be the nth moment of the gene frequency distribution about 0 in the generation $t + 1$. Conventionally the nth moment around 0 is designated by μ_n', but since we often use the prime (′) to designate values in the next generation, we will save the prime for this purpose.

Now, we note that the distribution in the $(t + 1)$th generation is the result of convolution of the distribution in the tth generation and the sampling "error" in reproduction. Thus, in calculating the expectation of x_{t+1}^n in terms of $(x_t + \delta x_t)^n$, we take the expectation in two steps: first taking the expectation for a random change δx_t which we shall denote by E_δ, and then taking the expectation for the existing distribution of gene frequency which we shall denote by E_ϕ, such that

$$E_\phi(x_t^n) = \sum_{x=0}^{1} x^n f(x; t) = \mu_n^{(t)} \tag{7.4.3}$$

and

$$E_\delta[(\delta x_t)^n] = \sum_{i=0}^{2N} \left(\frac{i}{2N} - x_t\right)^n \binom{2N}{i} x_t^i (1 - x_t)^{2N-i}, \tag{7.4.4}$$

since δx_t can be assumed to follow the binomial distribution.

For $n = 1$, 2, 3, and 4, 7.4.4 becomes

$$E_\delta(\delta x_t) = 0, \tag{7.4.5}$$

$$E_\delta[(\delta x_t)^2] = \frac{x_t(1 - x_t)}{2N} \tag{7.4.6}$$

$$E_\delta[(\delta x_t)^3] = \frac{x_t(1 - x_t)(1 - 2x_t)}{(2N)^2} \tag{7.4.7}$$

and

$$E_\delta[(\delta x_t)^4] = \frac{3x_t^2(1 - x_t)^2}{(2N)^2} + \frac{x_t(1 - x_t)(1 - 6x_t + 6x_t^2)}{(2N)^3}. \tag{7.4.8}$$

The mean, variance, and higher moments of the distribution can be obtained as follows.

From 7.4.1,

$$E(x_{t+1}) = E(x_t + \delta x_t).$$

The left side is $\mu_1^{(t+1)}$ by 7.4.2, while the right side is

$$E(x_t + \delta x_t) = E_\phi[x_t + E_\delta(\delta x_t)] = E_\phi(x_t) = \mu_1^{(t)},$$

since $E_\delta(\delta x_t) = 0$ from 7.4.5. Therefore

$$\mu_1^{(t+1)} = \mu_1^{(t)} = \mu_1^{(0)},$$

i.e., the mean stays constant:

$$\mu_1^{(t)} = \mu_1^{(0)}. \tag{7.4.9}$$

To obtain the second moment we square both sides of 7.4.1 and take expectations.

$$E(x_{t+1}^2) = E[x_t^2 + 2x_t\,\delta x_t + (\delta x_t)^2].$$

The left side is $\mu_2^{(t+1)}$ by definition, while the right side is

$$E_\phi[x_t^2 + 2x_t E_\delta(\delta x_t) + E_\delta(\delta x_t)^2] = E_\phi\left[x_t^2 + \frac{x_t(1 - x_t)}{2N}\right],$$

which becomes

$$\left(1 - \frac{1}{2N}\right)E_\phi(x_t^2) + \frac{1}{2N}E_\phi(x_t) = \left(1 - \frac{1}{2N}\right)\mu_2^{(t)} + \frac{1}{2N}\mu_1^{(t)},$$

by noting 7.4.5, 7.4.6, as well as 7.4.3. Since $\mu_1^{(t)} = \mu_1^{(0)}$, the above equation gives

$$\mu_2^{(t+1)} = \left(1 - \frac{1}{2N}\right)\mu_2^{(t)} + \frac{1}{2N}\mu_1^{(0)}. \tag{7.4.10}$$

The solution of this finite difference equation is

$$\mu_2^{(t)} = (\mu_2^{(0)} - \mu_1^{(0)})\left(1 - \frac{1}{2N}\right)^t + \mu_1^{(0)}. \tag{7.4.11}$$

Let $V_x^{(t)}$ be the variance of gene frequency in generation t, i.e.,

$$V_x^{(t)} = \mu_2^{(t)} - (\mu_1^{(t)})^2;$$

then

$$V_x^{(t)} = \mu_1^{(0)}(1 - \mu_1^{(0)}) + (\mu_2^{(0)} - \mu_1^{(0)})\left(1 - \frac{1}{2N}\right)^t. \tag{7.4.12}$$

To find the third moment, we use a similar procedure; cubing both sides of 7.4.1 and taking expectations,

$$\begin{aligned} E(x_{t+1}^3) &= E_\phi[x_t^3 + 3x_t^2 E_\delta(\delta x_t) + 3x_t E_\delta(\delta x_t)^2 + E_\delta(\delta x_t)^3] \\ &= E_\phi\left[x_t^3 + \frac{3x_t x_t(1 - x_t)}{2N} + \frac{x_t(1 - x_t)(1 - 2x_t)}{(2N)^2}\right], \end{aligned}$$

where the last term is obtained from 7.4.7. Thus, we obtain the recurrence relation,

$$\mu_3^{(t+1)} = \left(\frac{1}{2N}\right)^2 \mu_1^{(t)} + \frac{3}{2N}\left(1 - \frac{1}{2N}\right)\mu_2^{(t)} + \left(1 - \frac{1}{2N}\right)\left(1 - \frac{2}{2N}\right)\mu_3^{(t)}, \tag{7.4.13}$$

the solution of which is

$$\begin{aligned} \mu_3^{(t)} &= \mu_1^{(0)} + \frac{3}{2}(\mu_2^{(0)} - \mu_1^{(0)})\left(1 - \frac{1}{2N}\right)^t \\ &+ \left(\mu_3^{(0)} - \frac{3}{2}\mu_2^{(0)} + \frac{1}{2}\mu_1^{(0)}\right)\left[\left(1 - \frac{1}{2N}\right)\left(1 - \frac{2}{2N}\right)\right]^t. \end{aligned} \tag{7.4.14}$$

Carrying out the same procedure, we obtain the following recurrence formula for the fourth moment:

$$\begin{aligned} \mu_4^{(t+1)} &= \left(\frac{1}{2N}\right)^3 \mu_1^{(t)} + 7\left(\frac{1}{2N}\right)^2\left(1 - \frac{1}{2N}\right)\mu_2^{(t)} \\ &+ \frac{6}{2N}\left(1 - \frac{1}{2N}\right)\left(1 - \frac{2}{2N}\right)\mu_3^{(t)} \\ &+ \left(1 - \frac{1}{2N}\right)\left(1 - \frac{2}{2N}\right)\left(1 - \frac{3}{2N}\right)\mu_4^{(t)}, \end{aligned} \tag{7.4.15}$$

the solution of which is

$$\begin{aligned}\mu_4^{(t)} = \mu_1^{(0)} &+ \frac{18N-11}{10N-6}(\mu_2^{(0)} - \mu_1^{(0)})\left(1-\frac{1}{2N}\right)^t \\ &+ 2\left(\mu_3^{(0)} - \frac{3}{2}\mu_2^{(0)} + \frac{1}{2}\mu_1^{(0)}\right)\left[\left(1-\frac{1}{2N}\right)\left(1-\frac{2}{2N}\right)\right]^t \\ &+ \left[\mu_4^{(0)} - 2\mu_3^{(0)} + \frac{12N-7}{10N-6}\mu_2^{(0)} - \frac{2N-1}{10N-6}\mu_1^{(0)}\right] \\ &\times \left[\left(1-\frac{1}{2N}\right)\left(1-\frac{2}{2N}\right)\left(1-\frac{3}{2N}\right)\right]^t.\end{aligned} \qquad \text{7.4.16}$$

If we start from a population with gene frequency p, we have $\mu_1^{(0)} = p$, $\mu_2^{(0)} = p^2$, etc., and the above results are expressed as follows:

$$\mu_1^{(t)} = p, \qquad \text{7.4.17}$$

$$\mu_2^{(t)} = p - p(1-p)\left(1-\frac{1}{2N}\right)^t, \qquad \text{7.4.18}$$

$$V_x^{(t)} = p(1-p)\left[1 - \left(1-\frac{1}{2N}\right)^t\right], \qquad \text{7.4.19}$$

$$\begin{aligned}\mu_3^{(t)} = p &- \frac{3}{2}p(1-p)\left(1-\frac{1}{2N}\right)^t \\ &- \frac{1}{2}p(1-p)(2p-1)\left[\left(1-\frac{1}{2N}\right)\left(1-\frac{2}{2N}\right)\right]^t,\end{aligned} \qquad \text{7.4.20}$$

$$\begin{aligned}\mu_4^{(t)} = p &- \frac{18N-11}{10N-6}p(1-p)\left(1-\frac{2}{2N}\right)^t \\ &- p(1-p)(2p-1)\left[\left(1-\frac{1}{2N}\right)\left(1-\frac{2}{2N}\right)\right]^t \\ &+ p(1-p)\left[p(1-p) - \frac{2N-1}{10N-6}\right] \\ &\times \left[\left(1-\frac{1}{2N}\right)\left(1-\frac{2}{2N}\right)\left(1-\frac{3}{2N}\right)\right]^t.\end{aligned} \qquad \text{7.4.21}$$

The result for variance agrees with 7.3.4; see also Wright (1942) and Crow (1954) for alternative derivations. The moment formulae are important in finding the process of change in the genotypic variance within lines, between lines, and the additive component of the variance within lines in the case of inbreeding due to restricted population size. This was first worked out for a completely recessive gene by Robertson (1952). Though his expressions are less explicit, the present results seem to be in complete agreement with his, which were obtained by the use of matrix algebra.

In order to find the general formula for the moments of the distribution, we shall now make the assumption that the population size N is large enough that terms of the order $1/N^2$. $1/N^3$, etc., can be omitted without serious error. The recurrence formula for the nth moment is obtained as follows:

$$E(x_{t+1}^n) = E[(x_t + \delta x_t)^n]$$

$$= E\left[x_t^n + \binom{n}{1} x_t^{n-1}\delta x_t + \binom{n}{2} x_t^{n-2}(\delta x_t)^2 + \cdots\right],$$

$$E_\phi(x_{t+1}^n) = E_\phi\left[x_t^n + \frac{n(n-1)}{2}\frac{x_t^{n-1}(1-x_t)}{2N} + O\left(\frac{1}{N^2}\right)\right],$$

where $O(1/N^2)$ denotes a term of the order of $1/N^2$.

Neglecting higher order terms, we have

$$\mu_n^{(t+1)} = \left[1 - \frac{n(n-1)}{4N}\right]\mu_n^{(t)} + \frac{n(n-1)}{4N}\,\mu_{n-1}^{(t)} \qquad (n = 1, 2, \ldots). \tag{7.4.22}$$

For a large N, the moments change very slowly with generations, and we can replace the above system of finite difference equations by the following system of differential equations.

$$\frac{d\mu_n^{(t)}}{dt} = \frac{n(n-1)}{4N}\,[\mu_n^{(t)} - \mu_{n-1}^{(t)}] \qquad (n = 1, 2, 3, \ldots). \tag{7.4.23}$$

If the population starts from the gene frequency p, we get the following results for $n = 1$, 2, 3, and 4:

$$\mu_1^{(t)} = p, \tag{7.4.24}$$

$$\mu_2^{(t)} = p - p(1-p)e^{-\frac{1}{2N}t}, \tag{7.4.25}$$

$$V_x^{(t)} = p(1-p)\left(1 - e^{-\frac{1}{2N}t}\right), \tag{7.4.26}$$

$$\mu_3^{(t)} = p - \tfrac{3}{2}p(1-p)e^{-\frac{1}{2N}t} - \tfrac{1}{2}p(1-p)(2p-1)e^{-\frac{3}{2N}t}, \tag{7.4.27}$$

$$\begin{aligned}\mu_4^{(t)} = p &- \tfrac{9}{5}p(1-p)e^{-\frac{1}{2N}t} - p(1-p)(2p-1)e^{-\frac{3}{2N}t} \\ &- \tfrac{1}{5}p(1-p)(5p^2 - 5p + 1)e^{-\frac{6}{2N}t}.\end{aligned} \tag{7.4.28}$$

If we compare these formulae with 7.4.17, 7.4.18, 7.4.19, 7.4.20, and 7.4.21, it will be seen that even for a rather small N such as $N = 10$, they give very good approximations to the formulae based on the discrete model.

The above formulae may also be expressed in terms of the inbreeding coefficient, since

$$e^{-\frac{1}{2N}t} = 1 - f_t,$$

as shown by 7.3.18. Thus

$$\mu_1^{(t)} = p, \tag{7.4.29}$$

$$\mu_2^{(t)} = p - p(1-p)(1-f_t), \tag{7.4.30}$$

$$V_x^{(t)} = p(1-p)f_t, \tag{7.4.31}$$

$$\mu_3^{(t)} = p - \tfrac{3}{2}p(1-p)(1-f_t) - \tfrac{1}{2}p(1-p)(2p-1)(1-f_t)^3, \tag{7.4.32}$$

$$\mu_4^{(t)} = p - \tfrac{9}{5}p(1-p)(1-f_t) - p(1-p)(2p-1)(1-f_t)^3 - \tfrac{1}{5}p(1-p)[1-5p(1-p)](1-f_t)^6. \tag{7.4.33}$$

From the above process of calculation, it may be inferred that the solution of 7.4.23 for an arbitrary n, $(n \geq 2)$, with the initial condition $\mu_n^{(0)} = p^n$, must have the form

$$\mu_n^{(t)} = p + \sum_{i=1}^{n-1} C_n^{(i)} e^{-\lambda_i t}, \tag{7.4.34}$$

where

$$\lambda_i = \frac{i(i+1)}{4N} \tag{7.4.35}$$

and $C_n^{(i)}$'s are constants which do not depend on t. For a large value of t, only the first few of these constants are important and they are rather easily determined from 7.4.23. For example, substituting 7.4.34 into 7.4.23 and comparing the terms involving λ_1 in both sides of the resulting equation, we get

$$C_n^{(1)} = \frac{n(n-1)}{(n+1)(n-2)} C_{n-1}^{(1)},$$

from which we obtain

$$C_n^{(1)} = \frac{3(n-1)}{n+1} C_2^{(1)}, \tag{7.4.36}$$

where $C_2^{(1)} = -p(1-p)$ from 7.4.25. General terms are much more difficult to derive so we shall only give the result:

$$\mu_n^{(t)} = p + \sum_{i=1}^{n-1} (2i+1)p(1-p)(-1)^i F(1-i, i+2, 2, p) \times \frac{(n-1)(n-2)\cdots(n-i)}{(n+1)(n+2)\cdots(n+i)} e^{-\frac{i(i+1)}{4N}t}, \tag{7.4.37}$$

where F represents the hypergeometric function, that is,

$$F(1-i, i+2, 2, p) = 1 + \frac{(1-i)(i+2)}{1 \times 2} p + \frac{(1-i)(2-i) \times (i+2)(i+3)}{1 \times 2 \times 2 \times 3} p^2 + \cdots (i = 1, 2, 3, \ldots). \qquad 7.4.38$$

For $i = 1$, 2, and 3, the right-hand side of the above expression reduces to 1, $1 - 2p$, and $1 - 5p + 5p^2$, respectively.

As the process of random drift proceeds, the probability that the gene is either fixed in the population or lost increases gradually. It is possible to find this probability from the moment formula 7.4.37 by the following device: For the probability of fixation,

$$f(1, t) = \lim_{n \to \infty} \sum_{x=0}^{1} x^n f(x, t) = \lim_{n \to \infty} \mu_n^{(t)}. \qquad 7.4.39$$

This makes use of the fact that for $n = \infty$ and $0 \leqq x \leqq 1$, $x^n = 1$ only when $x = 1$; otherwise $x^n = 0$. The right side of 7.4.39 is readily evaluated from 7.4.37. Let us denote the above probability by $f(p, 1; t)$, meaning that this is the probability of the gene A_1 reaching fixation by the tth generation given that its initial frequency is p. Thus we have

$$f(p, 1; t) = p + \sum_{i=1}^{\infty} (2i+1)p(1-p)(-1)^i F(1-i, i+2, 2, p) e^{-\frac{i(i+1)}{4N}t}$$
$$= p - 3p(1-p)e^{-\frac{1}{2N}t} + 5p(1-p)(1-2p)e^{-\frac{3}{2N}t} - 7p(1-p)(1-5p+5p^2)e^{-\frac{6}{2N}t} + \cdots. \qquad 7.4.40$$

This can also be expressed as

$$f(p, 1; t) = p + \sum_{i=1}^{\infty} \frac{(-1)^i}{2} [P_{i-1}(r) - P_{i+1}(r)] e^{-\frac{i(i+1)}{4N}t}, \qquad 7.4.41$$

where $r = 1 - 2p$ and the $P_i(r)$'s represent the Legendre polynomials: $P_0(r) = 1$, $P_1(r) = r$, $P_2(r) = \frac{1}{2}(3r^2 - 1)$, $P_3(r) = \frac{1}{2}(5r^3 - 3r)$, etc.

The probability, denoted by $f(p, 0; t)$, of A_1 being lost or its allele A_2 being fixed by the tth generation is obtained simply by replacing p with $1 - p$ and r with $-r$ in the above expressions. Therefore the probability that both A_1 and A_2 coexist in the population at the tth generation is obtained from

$$\Omega_t = 1 - f(p, 1; t) - f(p, 0; t).$$

If we use the relation $P_i(-r) = (-1)^i P_i(r)$, we get

$$\Omega_t = \sum_{j=0}^{\infty} [P_{2j}(r) - P_{2j+2}(r)] e^{-\frac{(2j+1)(2j+2)t}{4N}}$$
$$= \tfrac{3}{2}(1 - r^2) e^{-\frac{1}{2N}t} + \cdots. \tag{7.4.42}$$

Thus for $t \to \infty$, we have the asymptotic formula

$$\Omega_t \sim 6p(1-p) e^{-\frac{1}{2N}t}. \tag{7.4.43}$$

Note that this is different from the formula for heterozygosity, i.e., 7.3.9.

A similar method for calculating moments can readily be extended to the case of three alleles A_1, A_2, and A_3, though the situation is more complicated (see Kimura, 1955*b*, 1956).

7.5 The Variance of a Quantitative Character Within and Between Subdivided Populations

When an infinitely large population is subdivided into isolated subgroups of finite size N, within each of which mating is at random, the process of random genetic drift will go on in each of them until the frequency of a particular allele ultimately becomes either 0 or 1. We will designate the subpopulations as lines.

First, consider a character determined by additive genes. Let α be the average effect of substituting A_1 for A_2, so that we can express the genotypic values of A_1A_1, A_1A_2, and A_2A_2 as 2α, α, and 0 respectively. If the frequency of A_1 in a particular line is x_t, the genotypic mean and the variance within the line are respectively $2\alpha x_t$ and $2\alpha^2 x_t(1 - x_t)$. Thus the genotypic variance within a line at generation t is

$$V_w^{(t)} = 2\alpha^2 E_\phi[x_t(1 - x_t)] = 2\alpha^2(\mu_1^{(t)} - \mu_2^{(t)}),$$

which reduces to

$$V_w^{(t)} = 2\alpha^2 p(1-p)(1-f_t) \tag{7.5.1}$$

if we apply 7.4.29 and 7.4.30. The variance between line means at the tth generation is

$$V_b^{(t)} = E_\phi[(2\alpha x_t)^2] - [E_\phi(2\alpha x_t)]^2$$
$$= 4\alpha^2[\mu_2^{(t)} - (\mu_1^{(t)})^2],$$

which similarly reduces to

$$V_b^{(t)} = 2\alpha^2 p(1-p)2f_t. \tag{7.5.2}$$

The total genotypic variance (see Chapter 4) is obtained from

$$V_h^{(t)} = E_\phi[(2\alpha)^2 x_t^2 + \alpha^2 2x_t(1 - x_t)] - [E_\phi(2\alpha x_t)]^2$$
$$= 2\alpha^2[\mu_1^{(t)} - 2(\mu_1^{(t)})^2 + \mu_2^{(t)}].$$

In terms of f_t, this becomes

$$V_h^{(t)} = 2\alpha^2 p(1 - p)(1 + f_t). \qquad 7.5.3$$

This total genotypic variance may also be derived by noting that the frequencies of A_1A_1, A_1A_2, and A_2A_2 in the total population with inbreeding coefficient f are

$$P_{11} = pf + p^2(1 - f),$$
$$2P_{12} = 2p(1 - p)(1 - f), \qquad 7.5.4$$
$$P_{22} = (1 - p)f + (1 - p)^2(1 - f),$$

and that the total genotypic variance for such population is given by

$$V_h = (2\alpha)^2 P_{11} + \alpha^2 2P_{12} - (2\alpha P_{11} + 2\alpha P_{12})^2,$$

and therefore

$$V_h = 2\alpha^2 p(1 - p)(1 + f),$$

as was discussed earlier (see 3.10.6).

The total genotypic variance (7.5.3) is of course equal to the sum of the preceding two variances $V_w^{(t)}$ and $V_b^{(t)}$. Since $V_0 = 2\alpha^2 p(1 - p)$ is the genotypic variance expected when the entire population is a panmictic unit, we may write

$$V_w^{(t)} = V_0(1 - f_t), \qquad 7.5.5$$
$$V_b^{(t)} = V_0 2f_t, \qquad 7.5.6$$
$$V_h^{(t)} = V_0(1 + f_t). \qquad 7.5.7$$

These results for additive genes were given by Wright (1951) and were shown to hold under quite general conditions (Wright, 1952). They hold not only when lines are completely isolated and drifting toward fixation but also for cases when a steady state has been reached under mutation, crossbreeding, and selection, as long as f_t represents the inbreeding coefficient of individuals relative to the total population.

Returning to the case of random genetic drift in completely isolated lines, if the lines are started as random samples from a very large parental stock for which $f_0 = 0$, we have approximately

$$V_w^{(t)} = V_0 e^{-\frac{t}{2N}} \qquad 7.5.8$$

and

$$V_b^{(t)} = V_0 2\left(1 - e^{-\frac{t}{2N}}\right). \qquad 7.5.9$$

To summarize, the variance within lines decreases at the rate of $1/2N$ per generation and finally becomes 0, while the variance between lines increases with time and finally becomes $2V_0$.

The situation becomes much more complicated if there is dominance. Let Y_{11}, Y_{12}, and Y_{22} be respectively the genotypic values of A_1A_1, A_1A_2, and A_2A_2. For a particular line in which the frequency of A_1 is x_t, the mean genotypic value is

$$M(x_t) = Y_{11}x_t^2 + 2Y_{12}x_t(1 - x_t) + Y_{22}(1 - x_t)^2$$

and the genotypic variance within the line is

$$V(x_t) = Y_{11}^2x_t^2 + 2Y_{12}^2x_t(1 - x_t) + Y_{22}^2(1 - x_t)^2 - M^2(x_t),$$

of which

$$2x_t(1 - x_t)[(Y_{11} - Y_{12})x_t + (Y_{12} - Y_{22})(1 - x_t)]^2$$

is the additive or genic component and

$$x_t^2(1 - x_t)^2(Y_{11} - 2Y_{12} + Y_{22})^2$$

is the dominance component (cf. Section 4.1).

To make the expressions simpler, we will choose a scale such that

$$Y_{11} = 1,$$
$$Y_{12} = h,$$
$$Y_{22} = 0.$$

The mean genotypic value averaged over the whole population in the tth generation is

$$M_h^{(t)} = E_\phi[M(x_t)] = E_\phi[2hx_t + (1 - 2h)x_t^2]$$

or

$$M_h^{(t)} = 2h\mu_1^{(t)} + (1 - 2h)\mu_2^{(t)}. \quad 7.5.10$$

In terms of the inbreeding coefficient, this becomes

$$M_h^{(t)} = p - (1 - 2h)p(1 - p)(1 - f_t). \quad 7.5.11$$

This result is also obtained from

$$M_h^{(t)} = P_{11} + h2P_{12},$$

by using 7.5.4.

The total genotypic variance is obtained from

$$V_h^{(t)} = E_\phi[x_t^2 + h^2 2x_t(1 - x_t)] - \{E_\phi[x_t^2 + h2x_t(1 - x_t)]\}^2,$$

and this reduces to

$$V_h^{(t)} = 2h^2\mu_1^{(t)} + (1 - 2h^2)\mu_2^{(t)} - [2h\mu_1^{(t)} + (1 - 2h)\mu_2^{(t)}]^2. \qquad 7.5.12$$

In terms of f_t, it is written as

$$\begin{aligned} V_h^{(t)} = p(1 - p)[1 &- (1 - 2p - 2h^2 + 4ph)(1 - f_t) \\ &- (1 - 2h)^2 p(1 - p)(1 - f_t)^2]. \end{aligned} \qquad 7.5.13$$

It may be noted that this is also obtained from

$$V_h^{(t)} = P_{11} + h^2 2P_{12} - [M_h^{(t)}]^2$$

by using 7.5.4 and 7.5.11.

If we regard the total genotypic mean and variance as functions of the inbreeding coefficient rather than functions of t, writing them as $M_h(f)$ and $f(V_h)$, we have

$$\begin{aligned} M_h(f) &= p - (1 - 2h)p(1 - p)(1 - f), \\ V_h(f) &= p(1 - p) - p(1 - p)(1 - 2p - 2h^2 + 4ph)(1 - f) \\ &\quad - p^2(1 - p)^2(1 - 2h)^2(1 - f)^2. \end{aligned} \qquad 7.5.14$$

Since these two relations can be derived directly from 7.5.4, it may be seen that they hold for any distribution as long as p represents the average gene frequency and f is defined such that 7.5.4 holds for the whole population. The above relations are also written

$$\begin{aligned} M_h(f) &= M_h(0)(1 - f) + M_h(1)f, \\ V_h(f) &= V_h(0)(1 - f) + V_h(1)f + [M_h(0) - M_h(1)]^2 f(1 - f), \end{aligned} \qquad 7.5.15$$

as was done earlier (3.10.1 and 3.10.3).

The genotypic variance within lines is

$$\begin{aligned} V_{w(h)}^{(t)} &= E_\phi[V(x_t)] \\ &= 2h^2\mu_1^{(t)} + (1 - 6h^2)\mu_2^{(t)} - 4h(1 - 2h)\mu_3^{(t)} \\ &\quad - (1 - 2h)^2\mu_4^{(t)}, \end{aligned} \qquad 7.5.16$$

of which the additive component or the genic variance within lines is

$$\begin{aligned} V_{w(g)}^{(t)} &= E_\phi[2x_t(1 - x_t)[(1 - h)x_t + h(1 - x_t)]^2] \\ &= 2h^2\mu_1^{(t)} + 2h(2 - 5h)\mu_2^{(t)} + 2(1 - 2h)(1 - 4h)\mu_3^{(t)} \\ &\quad - 2(1 - 2h)^2\mu_4^{(t)}, \end{aligned} \qquad 7.5.17$$

and the dominance component is

$$\begin{aligned} V_{w(d)}^{(t)} &= E_\phi[x_t^2(1-x_t)^2(1-2h)^2] \\ &= (1-2h)^2[\mu_2^{(t)} - 2\mu_3^{(t)} + \mu_4^{(t)}]. \end{aligned} \qquad 7.5.18$$

The variance between line means is obtained from

$$V_b^{(t)} = E_\phi[M^2(x_t)] - \{E_\phi[M(x_t)]\}^2,$$

which leads to

$$\begin{aligned} V_b^{(t)} = {} & 4h^2\mu_2^{(t)} + 4h(1-2h)\mu_3^{(t)} + (1-2h)^2\mu_4^{(t)} \\ & - [2h\mu_1^{(t)} + (1-2h)\mu_2^{(t)}]^2. \end{aligned} \qquad 7.5.19$$

If the random genetic drift proceeds within each of the completely isolated lines, the above results are expressed, by applying 7.4.29–7.4.33, in terms of the inbreeding coefficient as follows:

$$\begin{aligned} V_{w(h)}^{(t)} = {} & \tfrac{2}{5}(2 - 3h + 3h^2)p(1-p)(1-f_t) \\ & + (1-2h)p(1-p)(2p-1)(1-f_t)^3 \\ & + \tfrac{1}{5}(1-2h)^2p(1-p)[1-5p(1-p)](1-f_t)^6, \end{aligned} \qquad 7.5.20$$

$$\begin{aligned} V_{w(g)}^{(t)} = {} & \tfrac{1}{5}(3 - 2h + 2h^2)p(1-p)(1-f_t) \\ & + (1-2h)p(1-p)(2p-1)(1-f_t)^3 \\ & + \tfrac{2}{5}(1-2h)^2p(1-p)[1-5p(1-p)](1-f_t)^6, \end{aligned} \qquad 7.5.21$$

$$\begin{aligned} V_{w(d)}^{(t)} = {} & \tfrac{1}{5}(1-2h)^2p(1-p)(1-f_t) \\ & - \tfrac{1}{5}(1-2h)^2p(1-p)[1-5p(1-p)](1-f_t)^6, \end{aligned} \qquad 7.5.22$$

$$\begin{aligned} V_b^{(t)} = {} & p(1-p) + \tfrac{1}{5}(-9 + 6h + 4h^2 + 10p - 20hp)p(1-p)(1-f_t) \\ & - (1-2h)^2p^2(1-p)^2(1-f_t)^2 \\ & - (1-2h)p(1-p)(2p-1)(1-f_t)^3 \\ & - \tfrac{1}{5}(1-2h)^2p(1-p)[1-5p(1-p)](1-f_t)^6. \end{aligned} \qquad 7.5.23$$

These calculations are tedious but straightforward.

One of the most interesting situations arises when the gene A_1 is completely recessive and occurs in the original stock at a very low frequency. In this case, the frequency of homozygous-recessive individuals will increase with time and genotypic variance within lines increases up to a certain value of inbreeding coefficient, as shown by Robertson (1952).

For a completely recessive and rare gene ($h = 0, p \approx 0$), we have the following results:

$$V_{w(h)}^{(t)} = \tfrac{1}{5}p(1 - f_t)[4 - 5(1 - f_t)^2 + (1 - f_t)^5], \quad \textbf{7.5.24}$$

$$V_{w(g)}^{(t)} = \tfrac{1}{5}p(1 - f_t)[3 - 5(1 - f_t)^2 + 2(1 - f_t)^5], \quad \textbf{7.5.25}$$

$$V_{w(d)}^{(t)} = \tfrac{1}{5}p(1 - f_t)[1 - (1 - f_t)^5], \quad \textbf{7.5.26}$$

$$V_b^{(t)} = p[1 - \tfrac{9}{5}(1 - f_t) + (1 - f_t)^3 - \tfrac{1}{5}(1 - f_t)^6], \quad \textbf{7.5.27}$$

$$V_h^{(t)} = pf_t. \quad \textbf{7.5.28}$$

Figure 7.5.1 shows change of these variances in terms of change in the inbreeding coefficient. As will be seen in the figure, the genotypic variance within lines first increases with f and reaches its maximum value when f is roughly 1/2 and decreases thereafter to become 0 when f reaches unity. Its additive component behaves somewhat similarly. Generally, such a pattern of change in V_w, i.e., increase followed by decrease, occurs when $6p^2 < 1$ or $p < 0.41$. For a higher gene frequency, the variance within lines always decreases with increasing f, as in the case of no dominance. For details, see Robertson (1952).

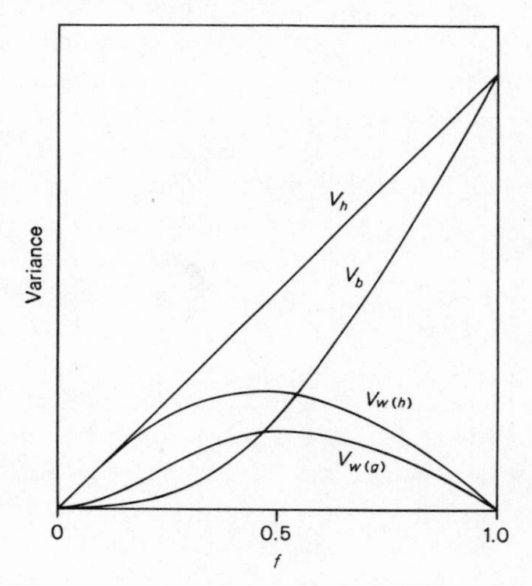

Figure 7.5.1. Change of total genotypic variance (V_h), variance between lines (V_b), variance within lines ($V_{w(h)}$), and its additive component ($V_{w(g)}$) of a character governed by very rare recessive genes (Robertson, 1952).

An essentially different case in which lines are partially isolated and random drift toward fixation is counterbalanced by occasional crossing between lines so that there is a steady-state gene frequency distribution has been worked out by Wright (1952). He has shown that the result for the variance within lines, assuming a completely recessive character, does not differ very much in terms of f and p from the above case of complete isolation.

7.6 Effective Population Number

7.6.1 Introduction

In the preceding section of this chapter, we assumed an ideal population of N breeding individuals which are produced each generation by random union of N male and N female gametes regarded as random samples from the population of the previous generation. We then studied the change of mean, variance, and higher moments in gene frequency due to this random sampling of gametes. We also discussed here and in Chapter 3 decrease in heterozygosity of an individual and increase of genetic homogeneity within a finite population. For a given initial gene frequency, these quantities are expressed solely in terms of population number N.

On the other hand, the breeding structure of an actual population is likely to be much more complicated and may differ from this ideal population in many respects. Thus it is desirable to have formulae through which such complicated situations are reduced to the equivalent ideal case which we understand and for which we have formulae. A few of the simpler formulae were given in Section 3.13 where the idea of an effective population number was first mentioned.

The very useful concept of effective population number was introduced by Wright (1931) to meet this need. In a finite population, as we have discussed, there is a decrease in homozygosity (inbreeding effect) and a random drift in gene frequencies because of sampling variance (variance effect). In simpler cases a population has the same effective number for either effect, and for this reason Wright used them more or less interchangeably. But for more complex situations it is necessary to make a distinction (Crow, 1954; Crow and Morton, 1955; Kimura and Crow, 1963). Our treatment in this section follows the latter paper.

7.6.2 Inbreeding Effective Number

We first consider a monoecious diploid population, mating at random and including the possibility of self-fertilization. We assume that each individual has an equal chance of contributing to the next generation. Then the number,

k, of gametes contributed by an individual to the next generation will follow a binomial distribution. A particular gamete has a probability $1/N$ of coming from a particular one of the N parents and there are $2N$ gametes in all. So the probability of k gametes coming from a particular parent is

$$\binom{2N}{k}\left(\frac{1}{N}\right)^k\left(1-\frac{1}{N}\right)^{2N-k}. \quad 7.6.2.1$$

In a population of stable size,

$$\bar{k} = 2 \quad 7.6.2.2$$

and the variance is

$$V_k = 2N\left(\frac{1}{N}\right)\left(1-\frac{1}{N}\right) = 2\left(1-\frac{1}{N}\right). \quad 7.6.2.3$$

More generally, if the average number of contributed gametes is $\bar{k}$, the variance of k will be

$$V_k = N\bar{k}\left(\frac{1}{N}\right)\left(1-\frac{1}{N}\right) = \bar{k}\left(1-\frac{1}{N}\right). \quad 7.6.2.4$$

Under this circumstance, each parent has the same expected number of offspring and the probability of two randomly chosen gametes coming from the same parent is $1/N$. For a large N the frequency distribution will approach the Poisson distribution for which

$$V_k = \bar{k}, \quad 7.6.2.5$$

as discussed in A.5.

When such an ideal situation is not realized, we define the effective population number by the reciprocal of the probability that two randomly chosen gametes come from the same parent. To see this point, let us consider first a population of monoecious diploids. We assume that mating is at random, though the expected number of offspring is not necessarily the same for each individual. Let P_t be the probability that two uniting gametes (or equivalently, under the assumption of random mating, two randomly chosen gametes) come from the same individual of the previous generation, $t-1$. Then, using the same reasoning as we have previously employed in Section 3.11, we obtain

$$f_t = P_t\left(\frac{1+f_{t-1}}{2}\right) + (1-P_t)f_{t-1}$$

or

$$f_t = \frac{P_t}{2} + \left(1-\frac{P_t}{2}\right)f_{t-1}, \quad 7.6.2.6$$

where f_t is the inbreeding coefficient in generation t. On the other hand, in our ideal population consisting of N_{t-1} individuals in generation $t-1$, the corresponding relation is

$$f_t = \frac{1}{2N_{t-1}} + \left(1 - \frac{1}{2N_{t-1}}\right) f_{t-1}. \qquad 7.6.2.7$$

The above two expressions agree with each other if

$$P_t = \frac{1}{N_{t-1}}.$$

The heterozygosity decreases at the rate

$$-\frac{\Delta H_{t-1}}{H_{t-1}} = \frac{\Delta f_{t-1}}{1 - f_{t-1}} = \frac{P_t}{2}$$

(cf. 7.3.16). Thus two monoecious populations with equal P_t are equivalent with respect to the inbreeding effect and we can define the *inbreeding effective number* by the relation:

$$N_{e(f)} = \frac{1}{P_t}. \qquad 7.6.2.8$$

Note that the effective number thus defined is determined by the number of individuals in the parental generation. From the above definition, we can derive a concrete formula for the effective number when the distribution of the number of contributed gametes (k) is known.

Let k_i be the number of successful gametes from the ith parent in generation $t-1$ ($i = 1, 2, \ldots, N_{t-1}$). The number of ways in which two gametes can be chosen out of the total number of $N_{t-1}\bar{k}$ gametes is

$$\binom{N_{t-1}\bar{k}}{2},$$

of which

$$\sum_i \binom{k_i}{2}$$

is the number of cases in which two gametes come from the same parent. Thus

$$P_t = \frac{\sum_i \binom{k_i}{2}}{\binom{N_{t-1}\bar{k}}{2}} = \frac{\sum_i k_i(k_i - 1)}{N_{t-1}\bar{k}(N_{t-1}\bar{k} - 1)}, \qquad 7.6.2.9$$

where $\bar{k}$ is the average number of contributed gametes per parental individual, and

$$\bar{k} = \frac{\sum_i k_i}{N_{t-1}}.$$

We will define V_k by

$$V_k = \frac{\sum_i (k_i - \bar{k})^2}{N_{t-1}} = \frac{\sum_i k_i^2}{N_{t-1}} - \bar{k}^2, \tag{7.6.2.10}$$

the variance of the number of gametes contributed per individual in the parent generation. With this definition,

$$\sum_i k_i^2 = N_{t-1}(V_k + \bar{k}^2),$$

and, if we note that

$$N_{t-1}\bar{k} = 2N_t, \tag{7.6.2.11}$$

the above probability may be expressed as follows:

$$P_t = \frac{V_k + \bar{k}(\bar{k} - 1)}{\bar{k}(2N_t - 1)}.$$

Thus the inbreeding effective number is given by

$$N_{e(f)} = \frac{2N_t - 1}{\bar{k} - 1 + V_k/\bar{k}}, \tag{7.6.2.12}$$

where N_t is the number of individuals in the tth generation.

For the ideal case in which the distribution of k is binomial with mean $\bar{k}$ and variance

$$V_k = N_{t-1}\bar{k}\left(\frac{1}{N_{t-1}}\right)\left(1 - \frac{1}{N_{t-1}}\right)$$
$$= \bar{k}\left(1 - \frac{1}{N_{t-1}}\right),$$

the inbreeding effective number reduces to the actual number in generation $t - 1$;

$$N_{e(f)} = \frac{N_{t-1}\bar{k} - 1}{\bar{k} - \dfrac{1}{N_{t-1}}} = N_{t-1},$$

as it should.

When k does not follow the binomial distribution but the population number, N, remains constant ($\bar{k} = 2$), the formula for the inbreeding effective number becomes

$$N_{e(f)} = \frac{4N - 2}{V_k + 2}, \tag{7.6.2.13}$$

as first shown by Wright (1938*b*). If all individuals contribute equally to the next generation ($V_k = 0$), the above formula reduces to

$$N_{e(f)} = 2N - 1; \tag{7.6.2.14}$$

the effective size is approximately twice as large as the actual size, as we mentioned in Section 3.13. On the other hand, if only one individual contributes the entire next generation,

$$V_k = \frac{(\bar{k}N_{t-1})^2}{N_{t-1}} - \bar{k}^2 = \bar{k}^2(N_{t-1} - 1),$$

and 7.6.2.12 reduces to

$$N_{e(f)} = \frac{N_{t-1}\bar{k} - 1}{\bar{k}(N_{t-1} - 1) + \bar{k} - 1} = 1,$$

as it should.

The assumption that all individuals have an equal expectation of offspring is unlikely to be met in nature and the effective number in natural conditions is usually considerably smaller than the actual number as pointed out repeatedly by Wright.

Next we will consider a population with separate sexes, still assuming random mating. Here the situation is somewhat complicated because we must consider three consecutive generations. We will define P_t as the probability that two homologous genes in two individuals in generation t came from the same individual in the previous generation, $t - 1$. Then we have

$$f_{t+1} = P_t\left(\frac{1 + f_{t-1}}{2}\right) + (1 - P_t)f_t$$

or

$$f_{t+1} = \frac{P_t}{2} + (1 - P_t)f_t + \frac{P_t}{2} f_{t-1},$$

as shown in Section 3.11.

For an ideal population consisting of N^*_{t-1} males and N^{**}_{t-1} females in which each individual has the same expected number of offspring as the others of the same sex, the probability P_t is given by

$$P_t = \frac{1}{4}\left(\frac{1}{N^*_{t-1}} + \frac{1}{N^{**}_{t-1}}\right) \tag{7.6.2.15}$$

(see 3.11.5).

If the numbers of males and females are equal,

$$N^*_{t-1} = N^{**}_{t-1} = \frac{N_{t-1}}{2},$$

and we get

$$P_t = \frac{1}{N_{t-1}}.$$

It has been shown already (see 3.11.9, 3.11.5) that, for a constant number of males and females, heterozygosity decreases at a rate of approximately $1/(2N_e + 1)$ per generation with N_e given by

$$N_e = \frac{4N^*N^{**}}{N^* + N^{**}}.$$

This is again the reciprocal of P_t, though P_t for this case is slightly different from the monoecious case. Thus any two populations (with separate sexes) having equal P_t are equivalent with respect to the inbreeding effect.

The formula corresponding to 7.6.2.12 may be derived as follows: Let k_i be the number of gametes contributed by the ith individual in generation $t - 1$. Then

$$P_t = \frac{\sum_i \binom{k_i}{2}}{\binom{N_{t-1}\bar{k}}{2} - N_t}. \tag{7.6.2.16}$$

The term $-N_t$ in the denominator, which did not appear in the case of a monoecious population, comes from the fact that only alleles that did not enter into the same individual in generation t can unite to form generation $t + 1$.

We now proceed as in 7.6.2.9–7.6.2.12.

Noting that $N_t = N_{t-1}\bar{k}/2$, the above expression is simplified and we get

$$P_t = \frac{V_k + \bar{k}^2 - \bar{k}}{\bar{k}(N_{t-1}\bar{k} - 2)}.$$

The effective number is then given by

$$N_{e(f)} = \frac{1}{P_t} = \frac{N_{t-1}\bar{k} - 2}{\bar{k} - 1 + \dfrac{V_k}{\bar{k}}}$$

or

$$N_{e(f)} = \frac{2N_t - 2}{\bar{k} - 1 + \dfrac{V_k}{\bar{k}}}. \qquad \mathbf{7.6.2.17}$$

Note that the numerator is now $2N_t - 2$ rather than $2N_t - 1$ as in the case of a monoecious population. The difference, however, is important only when N_t is very small.

When the numbers of males and females, N^*_{t-1} and N^{**}_{t-1}, are different it is sometimes convenient to calculate the mean and variance of k separately for each sex and then combine them to get $\bar{k}$ and V_k from

$$\bar{k} = m\bar{k}^* + (1 - m)\bar{k}^{**},$$
$$V_k = mV^*_k + (1 - m)V^{**}_k + m(1 - m)(\bar{k}^* - \bar{k}^{**})^2, \qquad \mathbf{7.6.2.18}$$

where k^* and k^{**} refer to the number of gametes from males and females and m is the proportion of males (cf. 3.10.3). Thus if k^*_i and k^{**}_j are the numbers of gametes from the ith male and jth female respectively, then

$$\bar{k}^* = \frac{\sum_i k^*_i}{N^*_{t-1}}, \qquad \bar{k}^{**} = \frac{\sum_j k^{**}_j}{N^{**}_{t-1}}$$

and

$$V^*_k = \frac{\sum_i (k^*_i - \bar{k}^*)^2}{N^*_{t-1}}, \qquad V^{**}_k = \frac{\sum_j (k^{**}_j - \bar{k}^{**})^2}{N^{**}_{t-1}}.$$

Also, since m and $1 - m$ represent the proportion of males and females in generation $t - 1$,

$$m = \frac{N^*_{t-1}}{N_{t-1}}, \qquad 1 - m = \frac{N^{**}_{t-1}}{N_{t-1}}.$$

If the number of gametes contributed per individual follows the binomial distribution,

$$V^*_k = \bar{k}^*\left(1 - \frac{1}{N^*_{t-1}}\right),$$

$$V^{**}_k = \bar{k}^{**}\left(1 - \frac{1}{N^{**}_{t-1}}\right),$$

and we obtain

$$N_{e(f)} = 4m(1 - m)N_{t-1}, \quad 7.6.2.19$$

or

$$N_{e(f)} = \frac{4N^*_{t-1}N^{**}_{t-1}}{N^*_{t-1} + N^{**}_{t-1}}. \quad 7.6.2.20$$

This agrees with our earlier result (3.11.5), first given by Wright (1931).

On the other hand, if the population consists of a single pair, a male and female,

$$N_{t-1} = 2, \; V_k = 0,$$

and 7.6.2.17 reduces to

$$N_{e(f)} = 2,$$

as it should.

Notice that the effective number related to an autozygosity increase in generation $t + 1$ is a function of the population number, N_{t-1}, two generations earlier. This is as expected, since with separate sexes two homologous genes could not come from a common ancestor more recent than a grandparent.

7.6.3 Variance Effective Number

We have just developed the concept of effective population number as this relates to the change in the probability of identity by descent. We now consider a definition of effective number that renders different populations comparable as regards the sampling variance of the gene frequency. This we call the variance effective number.

In many cases the two concepts lead to the same consequence, but not in general.

In an ideal population the sampling variance of the gene frequency drift from parent to offspring generation is $V_{\delta p} = p(1 - p)/2N$. So the natural definition of the variance effective number is obtained by setting the actual variance, $V_{\delta p}$, equal to $p(1 - p)/N_{e(v)}$, where $N_{e(v)}$ is the variance effective number.

Consider first a population of monoecious diploids and let N_{t-1} be the number of individuals in generation $t - 1$. As in the previous calculation, we will denote the number of gametes contributed per individual by k and define its mean and variance by

$$\bar{k} = \frac{\sum k_i}{N_{t-1}}, \qquad V_k = \frac{\sum k_i^2}{N_{t-1}} - \bar{k}^2, \quad 7.6.3.1$$

where k_i is the number of gametes contributed by the ith individual and the summation is over all the individuals in generation $t-1$ $(i = 1, 2, \ldots, N_{t-1})$.

In our retrospective approach of defining the effective number from the observed distribution of contributing gametes, the value of k for a specified individual is fixed and not a random variable. However, if we pick out, conceptually, an individual at random from the population, k is a random variable with mean $\bar{k}$ and variance V_k.

$$E(k) = \bar{k}, \quad E[(k-\bar{k})^2] = V_k. \tag{7.6.3.2}$$

Since each individual can contribute both male and female gametes, we will denote the number of the two types by k^* and k^{**} such that for the ith individual

$$k_i = k_i^* + k_i^{**}.$$

Their average is equal but they have their own variances:

$$\bar{k}^* = \frac{\sum k_i^*}{N_{t-1}} = \bar{k}^{**} = \frac{\sum \bar{k}_i^{**}}{N_{t-1}} = \frac{\bar{k}}{2}, \tag{7.6.3.3}$$

$$V_k^* = \frac{\sum k_i^{*2}}{N_{t-1}} - \bar{k}^{*2}, \qquad V_k^{**} = \frac{\sum k_i^{**2}}{N_{t-1}} - \bar{k}^{**2}. \tag{7.6.3.4}$$

We will assume that the population in generation $t-1$ contains a pair of alleles A_1 and A_2 with frequencies p and $1-p$. The number of individuals with genotypes A_1A_1, A_1A_2, and A_2A_2 within the population will be denoted by n_{11}, n_{12}, and n_{22} $(n_{11} + n_{12} + n_{22} = N_{t-1})$.

While all the gametes from an A_1A_1 individual contain allele A_1, half of those from a heterozygote (A_1A_2) do, so we will designate by l^* and l^{**} the number of A_1 gametes among male and female gametes produced by a heterozygote. For given values of k^* and k^{**}, l^* and l^{**} are random variables which follow the binomial distribution with means and variances given by

$$E(l^*) = \frac{k^*}{2}, \qquad E(l^{**}) = \frac{k^{**}}{2} \tag{7.6.3.5}$$

$$E\left[\left(l^* - \frac{k^*}{2}\right)^2\right] = \frac{k^*}{4}, \qquad E\left[\left(l^{**} - \frac{k^{**}}{2}\right)^2\right] = \frac{k^{**}}{4}. \tag{7.6.3.6}$$

Let us first consider a collection of male gametes which are produced by individuals of generation $t-1$ and which are destined to form generation t. The total number of A_1 genes contained in them may be expressed as

$$\Sigma_{11} k^* + \Sigma_{12} l^*,$$

where Σ_{11} and Σ_{12} denote respectively summation over A_1A_1 and A_1A_2 individuals in generation $t-1$.

Similarly, the corresponding quantity for female gametes is

$$\Sigma_{11}k^{**} + \Sigma_{12}l^{**}.$$

Thus the change of gene frequency between generations t and $t-1$ is given by

$$\delta p = \frac{1}{N_{t-1}\bar{k}}[(\Sigma_{11}k^* + \Sigma_{12}l^*) + (\Sigma_{11}k^{**} + \Sigma_{12}l^{**})] - p, \tag{7.6.3.7}$$

from which we have

$$N_{t-1}\bar{k}\delta p = (\Sigma_{11}k^* + \Sigma_{12}l^*) + (\Sigma_{11}k^{**} + \Sigma_{12}l^{**}) - N_{t-1}\bar{k}p. \tag{7.6.3.8}$$

Let us define random variables X^* and X^{**} by

$$X^* = \Sigma_{11}(k^* - \bar{k}^*) + \Sigma_{12}\frac{k^* - \bar{k}^*}{2} + \Sigma_{12}\left(l^* - \frac{k^*}{2}\right) \tag{7.6.3.9}$$

and

$$X^{**} = \Sigma_{11}(k^{**} - \bar{k}^{**}) + \Sigma_{12}\frac{k^{**} - \bar{k}^{**}}{2} + \Sigma_{12}\left(l^{**} - \frac{k^{**}}{2}\right). \tag{7.6.3.10}$$

Then 7.6.3.8 may be expressed as

$$N_{t-1}\bar{k}\delta p = X^* + X^{**}, \tag{7.6.3.11}$$

which may be derived by noting that

$$\left(\Sigma_{11}\bar{k}^* + \Sigma_{12}\frac{\bar{k}^*}{2}\right) + \left(\Sigma_{11}\bar{k}^{**} + \Sigma_{12}\frac{\bar{k}^{**}}{2}\right) = N_{t-1}p\bar{k}.$$

Now, from 7.6.3.2 and 7.6.3.5, we have

$$E(X^*) = E(X^{**}) = 0.$$

Furthermore, assuming independence of male and female gametes, we have

$$E(X^*X^{**}) = 0.$$

Thus, squaring both sides of 7.6.3.11 and taking expectations, we have

$$(N_{t-1}\bar{k})^2 V_{\delta p} = E(X^{*2}) + E(X^{**2}), \tag{7.6.3.12}$$

where

$$V_{\delta p} = E[(\delta p)^2].$$

In order to evaluate $E(X^{*2})$, we note that in the right side of 7.6.3.9 the first two terms are independent of the last term, because variation in the number of contributed gametes is independent of the variation in the number of A_1 genes within the gametes produced by heterozygotes.

Thus

$$E(X^{*2}) = E\left[\Sigma_{11}(k^* - \bar{k}^*) + \Sigma_{12}\frac{k^* - \bar{k}^*}{2}\right]^2 + E\left[\Sigma_{12}\left(l^* - \frac{k^*}{2}\right)\right]^2. \tag{7.6.3.13}$$

The first term in the right side of this equation may be evaluated by using the relation

$$C^*_{kk'} = -\frac{V^*_k}{N_{t-1} - 1}, \tag{7.6.3.14}$$

where $C^*_{kk'}$ is the covariance between the numbers of gametes contributed by two different males, that is,

$$C^*_{kk'} = \frac{\sum_{i \neq j}(k^*_i - \bar{k}^*)(k^*_j - \bar{k}^*)}{N_{t-1}(N_{t-1} - 1)}. \tag{7.6.3.15}$$

The relation 7.6.3.14 is a direct consequence of the identity

$$\sum(k^*_i - \bar{k}^*) = 0,$$

because, by squaring both sides, we have

$$[\sum(k^*_i - \bar{k}^*)]^2 = 0$$

or

$$N_{t-1}V^*_k + N_{t-1}(N_{t-1} - 1)C^*_{kk'} = 0,$$

which is equivalent to 7.6.3.14.

If we pick at random two individuals from the population, then the expected value of the cross product $(k^*_i - \bar{k}^*)(k^*_j - \bar{k}^*)$ is equal to $C^*_{kk'}$, i.e.,

$$E[(k^*_i - \bar{k}^*)(k^*_j - \bar{k}^*)] = C^*_{kk'} \qquad (i \neq j). \tag{7.6.3.16}$$

Noting the above and using 7.6.3.14, we obtain

$$\begin{aligned} &E\left[\Sigma_{11}(k^* - \bar{k}^*) + \Sigma_{12}\frac{k^* - \bar{k}^*}{2}\right]^2 \\ &= \left(n_{11} + \frac{n_{12}}{4}\right)V^*_k + \left[n_{11}(n_{11} - 1) + n_{11}n_{12} + \frac{n_{12}(n_{12} - 1)}{4}\right]C^*_{kk'} \\ &= \left(n_{11} + \frac{n_{12}}{4}\right)V^*_k + \left[\left(n_{11} + \frac{n_{12}}{2}\right)^2 - \left(n_{11} + \frac{n_{12}}{4}\right)\right]\left(-\frac{V^*_k}{N_{t-1} - 1}\right) \\ &= \frac{V^*_k}{N_{t-1} - 1}\left[N_{t-1}\left(n_{11} + \frac{n_{12}}{4}\right) - \left(n_{11} + \frac{n_{12}}{2}\right)^2\right]. \end{aligned} \tag{7.6.3.17}$$

The second term in the right side of 7.6.3.13 may easily be evaluated if we note that $l^* - (1/2)k^*$ follows the binomial distribution with mean 0 and variance $k^*/4$. The distributions are independent between two different individuals. Thus

$$E\left[\Sigma_{12}\left(l^* - \frac{k^*}{2}\right)\right]^2 = E\left[\Sigma_{12}\left(\frac{k^*}{4}\right)\right] = \frac{n_{12}}{4}\bar{k}^*. \tag{7.6.3.18}$$

Combining 7.6.3.17 and 7.6.3.18, we get

$$E(X^{*2}) = \frac{V_k^*}{N_{t-1} - 1}\left[N_{t-1}\left(n_{11} + \frac{n_{12}}{4}\right) - \left(n_{11} + \frac{n_{12}}{2}\right)^2\right] + \frac{n_{12}}{4}\bar{k}^*. \tag{7.6.3.19}$$

Similarly, for female gametes we have

$$E(X^{**2}) = \frac{V_k^{**}}{N_{t-1} - 1}\left[N_{t-1}\left(n_{11} + \frac{n_{12}}{4}\right) - \left(n_{11} + \frac{n_{12}}{2}\right)^2\right] + \frac{n_{12}}{4}\bar{k}^{**}. \tag{7.6.3.20}$$

Adding these two equations, we obtain

$$(N_{t-1}\bar{k})^2 V_{\delta p} = \frac{V_k}{N_{t-1} - 1}\left[N_{t-1}\left(n_{11} + \frac{n_{12}}{4}\right) - \left(n_{11} + \frac{n_{12}}{2}\right)^2\right] + \frac{n_{12}}{4}\bar{k}, \tag{7.6.3.21}$$

because

$$V_k = V_k^* + V_k^{**}, \qquad \bar{k} = \bar{k}^* + \bar{k}^{**}. \tag{7.6.3.22}$$

Let us define a coefficient α_{t-1} by the relation

$$\begin{aligned} \frac{n_{11}}{N_{t-1}} &= (1 - \alpha_{t-1})p^2 + \alpha_{t-1}p, \\ \frac{n_{12}}{N_{t-1}} &= (1 - \alpha_{t-1})2p(1 - p), \\ \frac{n_{22}}{N_{t-1}} &= (1 - \alpha_{t-1})(1 - p)^2 + \alpha_{t-1}(1 - p). \end{aligned} \tag{7.6.3.23}$$

The coefficient α_{t-1} is a measure of departure from Hardy–Weinberg proportions, whether because of inbreeding or other factors. Then the right side of 7.6.3.21 is much simplified, giving

$$(N_{t-1}\bar{k})^2 V_{\delta p} = \frac{N_{t-1}^2 V_k p(1 - p)(1 + \alpha_{t-1})}{2(N_{t-1} - 1)} + \frac{N_{t-1}\bar{k}p(1 - p)(1 - \alpha_{t-1})}{2}$$

or

$$\left[\frac{2N_{t-1}\bar{k}}{p(1 - p)}\right] V_{\delta p} = \frac{N_{t-1}}{N_{t-1} - 1}\frac{V_k}{\bar{k}}(1 + \alpha_{t-1}) + (1 - \alpha_{t-1}). \tag{7.6.3.24}$$

The variance effective number is defined by

$$N_{e(v)} = \frac{p(1-p)}{2V_{\delta p}}, \qquad \textbf{7.6.3.25}$$

since the sampling variance for the ideal situation of N monoecious individuals mating at random is

$$V_{\delta p} = \frac{p(1-p)}{2N}.$$

From 7.6.3.24 and 7.6.3.25, we obtain

$$N_{e(v)} = \frac{p(1-p)}{2V_{\delta p}} = \frac{N_{t-1}\bar{k}}{\frac{s_k^2}{\bar{k}}(1+\alpha_{t-1}) + (1-\alpha_{t-1})}$$

or

$$N_{e(v)} = \frac{2N_t}{\frac{s_k^2}{\bar{k}}(1+\alpha_{t-1}) + (1-\alpha_{t-1})}, \qquad \textbf{7.6.3.26}$$

which is the required formula for the variance effective number. In this formula

$$s_k^2 = \frac{\sum_i (k_i - \bar{k})^2}{N_{t-1} - 1} = \frac{N_{t-1}}{N_{t-1} - 1} V_k,$$

which includes the Gaussian correction, $N_{t-1}/(N_{t-1} - 1)$, and N_t is the number of individuals in generation t, $N_t = N_{t-1}\bar{k}/2$.

Notice that, whereas the inbreeding effective number is naturally related to the number in the parent (or with separate sexes, the grandparent) generation, the variance effective number is related to the number in the progeny generation. This is to be expected, since the probability of identity by descent depends on the number of ancestors whereas the sampling variance depends on the size of the sample, i.e., the number of offspring.

We now consider some special cases. When k follows the binomial distribution

$$V_k = \bar{k}\left(1 - \frac{1}{N_{t-1}}\right)$$

or

$$s_k^2 = \bar{k},$$

and 7.6.3.26 reduces to

$$N_{e(v)} = N_t$$

irrespective of the coefficient α_{t-1}.

When the population keeps a constant number N, $\bar{k} = 2$, and if $\alpha_{t-1} = 0$, 7.6.3.26 becomes

$$N_{e(v)} = \frac{4N}{s_k^2 + 2}. \qquad \textbf{7.6.3.27}$$

However, in a finite population under random mating, the expected value of α_{t-1} is not 0 but $-1/(2N_{t-1} - 1)$ (cf. 2.10.1). Thus if the parent generation were derived from random mating, but α_{t-1} is not known, we substitute the above expected value for α_{t-1} in formula 7.6.3.26. This gives

$$N_{e(v)} = \frac{(2N_{t-1} - 1)\bar{k}}{2\left(1 + \frac{V_k}{\bar{k}}\right)}. \qquad \textbf{7.6.3.28}$$

For example, with self-fertilization, $N_{t-1} = 1$, $\bar{k} = 2$, $V_k = 0$, and $N_{e(v)}$ becomes unity as expected.

If all the individuals contribute equally to the next generation ($V_k = 0$) and if $\bar{k} = 2$, we have

$$N_{e(v)} = 2N - 1, \qquad \textbf{7.6.3.29}$$

namely, the effective number is about twice the actual number.

When sexes are separate, the formula for the effective number may be derived from the following consideration: The gene frequency in the next generation is the average of gene frequencies of male and female gametes, i.e.,

$$p' = \frac{p^{*\prime} + p^{**\prime}}{2}.$$

Therefore

$$V_{\delta p} = \frac{1}{4}(V_{\delta p}^{*} + V_{\delta p}^{**})$$

$$= \frac{1}{4}\left[\frac{p(1-p)}{2N_e^{*}} + \frac{p(1-p)}{2N_e^{**}}\right].$$

Thus we obtain

$$N_{e(v)} = \frac{p(1-p)}{2V_{\delta p}} = \frac{1}{\frac{1}{4}\left(\frac{1}{N_e^{*}} + \frac{1}{N_e^{**}}\right)}$$

or

$$N_{e(v)} = \frac{4N_e^* N_e^{**}}{N_e^* + N_e^{**}}, \qquad 7.6.3.30$$

where

$$N_e^* = \frac{N_{t-1}\bar{k}^*}{(1 - \alpha_{t-1}^*) + (1 + \alpha_{t-1}^*)\dfrac{s_k^{*2}}{\bar{k}^*}}, \qquad 7.6.3.31$$

and the expression for N_e^{**} is similar.

If the males and females are equal in number and have equal progeny distributions and if α_{t-1} is not known, we obtain, noting that $\alpha_{t-1}^* = \alpha_{t-1}^{**} = -1/(N_{t-1} - 1)$,

$$N_e = \frac{(N_{t-1} - 1)\bar{k}}{1 + \dfrac{V_k}{\bar{k}}}.$$

In the special case of sib mating, $N_{t-1} = 2$, $\bar{k} = 2$, $V_k = 0$, and we get $N_e = 2$ as expected.

When the effective number changes from generation to generation, the representative effective number over T generations may also be obtained from the consideration of variance as we did for inbreeding (Section 3.13). Let $N_{e(v)}^{(t)}$ be the effective number in the tth generation. Then, as shown in 7.3.4, the gene frequency variances in two consecutive generations are related by

$$V_p^{(t)} = \left(1 - \frac{1}{2N_{e(v)}^{(t)}}\right)V_p^{(t-1)} + \frac{p(1-p)}{2N_{e(v)}^{(t)}}. \qquad 7.6.3.32$$

Starting with $V_p^{(0)} = 0$, the variance after T generations is

$$V_p^{(T)} = p(1-p)\left[1 - \prod_{t=1}^{T}\left(1 - \frac{1}{2N_e^{(t)}}\right)\right] \qquad (T = 1, 2, \ldots). \qquad 7.6.3.33$$

If $\bar{N}_{e(v)}$ is the representative effective size, then

$$\left(1 - \frac{1}{2\bar{N}_{e(v)}}\right)^T = \prod_{t=1}^{T}\left(1 - \frac{1}{2N_{e(v)}^{(t)}}\right)$$

or

$$T \log\left(1 - \frac{1}{2\bar{N}_{e(v)}}\right) = \sum_{t=1}^{T} \log\left(1 - \frac{1}{2N_{e(v)}^{(t)}}\right).$$

Thus for large $N_{e(v)}^{(t)}$'s and a relatively small T, we have

$$\frac{T}{\overline{N}_{e(v)}} \doteq \sum_{t=1}^{T} \frac{1}{N_{e(v)}^{(t)}}$$

or

$$\overline{N}_{e(v)} = \frac{T}{\left(\sum_{t=1}^{T} \frac{1}{N_{e(v)}^{(t)}}\right)}. \qquad \text{7.6.3.34}$$

Thus, as for the inbreeding effective number, the representative variance effective number is approximately the harmonic mean of the individual effective numbers over the whole time as pointed out by Wright (1938*a*). He gives an example in which the breeding population increases in five generations in geometric series from 10 to 10^6 and then returns to 10 and repeats the cycle. For this case, the representative effective size turns out to be 54, which is much nearer to the minimum number than to the maximum.

In many natural populations, the breeding number may stay fairly constant with small fluctuations around a certain mean. If it changes fortuitously from generation to generation with mean $\hat{N}$ and standard deviation σ_N, the effective number is given by

$$N_{e(v)} = \hat{N} - \frac{\sigma_N^2}{\hat{N}}, \qquad \text{7.6.3.35}$$

as long as σ_N is much smaller than $\hat{N}$. This is derived as follows: Let δN be the deviation of N from its mean $\hat{N}$, such that

$$N = \hat{N} + \delta N$$

and

$$E(\delta N) = 0, \qquad E(\delta N)^2 = \sigma_N^2.$$

Since

$$V_{\delta p} = \frac{p(1-p)}{2N} = \frac{p(1-p)}{2\hat{N}\left(1 + \frac{\delta N}{\hat{N}}\right)}$$

$$= \frac{p(1-p)}{2\hat{N}}\left[1 - \frac{\delta N}{\hat{N}} + \left(\frac{\delta N}{\hat{N}}\right)^2 - \cdots\right],$$

neglecting small terms, we obtain

$$E(V_{\delta p}) = \frac{p(1-p)}{2\hat{N}}\left[1 + \frac{\sigma_N^2}{\hat{N}^2}\right].$$

Thus

$$N_{e(v)} = \frac{p(1-p)}{2E(V_{\delta p})} = \frac{\hat{N}}{1 + \dfrac{\sigma_N^2}{\hat{N}^2}} \approx \hat{N}\left(1 - \frac{\sigma_N^2}{\hat{N}^2}\right) = \hat{N} - \frac{\sigma_N^2}{\hat{N}},$$

as was to be shown.

7.6.4 Comparison of the Two Effective Numbers

To make our comparison simpler, we will consider a population of monoecious organisms mating at random. From 7.6.2.12 and 7.6.3.28, the inbreeding and variance effective numbers are respectively

$$N_{e(f)} = \frac{N_{t-1}\bar{k} - 1}{\dfrac{V_k}{\bar{k}} + \bar{k} - 1} \qquad 7.6.4.1$$

and

$$N_{e(v)} = \frac{(2N_{t-1} - 1)\bar{k}}{2\left(1 + \dfrac{V_k}{\bar{k}}\right)}. \qquad 7.6.4.2$$

As noted already, the inbreeding effective number is more naturally related to the number of the parents, while the variance effective number is related to that of the offspring. The former is usually much smaller than the latter if a large number of offspring is produced out of a small number of parents. In an extreme case of $N_{t-1} = 1$ and $\bar{k} \to \infty$, the inbreeding effective number becomes unity while the variance effective number is infinity. On the other hand, if each parent produces just one offspring ($\bar{k} = 1$, $V_k = 0$) the inbreeding effective number becomes infinite but the variance effective number stays finite and equal to twice the number of offspring.

However, these are extreme examples. If the population size is constant ($N_{t-1} = N_t = N$, $\bar{k} = 2$) both formulae reduce to

$$N_{e(f)} = N_{e(v)} = \frac{4N - 2}{V_k + 2} \qquad 7.6.4.3$$

with a slight correction if there are separate sexes; the 2 in the numerator is replaced by 4.

Table 7.6.4.1 A comparison of the inbreeding and variance effective numbers for monoecious and bisexual populations. In the monoecious population self-fertilization is permitted. The values for separate sexes are for the case where both sexes have the same progeny distribution. N_t = number of individuals in generation t, $\bar{k}$ = mean number of progeny per parent, V_k = variance in number of progeny per parent, α = measure of departure from Hardy–Weinberg proportions; symbols with an asterisk refer to one sex only. (From Kimura and Crow, 1963*a*.)

	INBREEDING EFFECTIVE NUMBER		VARIANCE EFFECTIVE NUMBER	
	MONOECIOUS	SEPARATE SEXES	MONOECIOUS	SEPARATE SEXES
Equation number	7.6.2.12	7.6.2.17	7.6.3.26	7.6.3.31
General	$\dfrac{N_{t-1}\bar{k}-1}{\bar{k}-1+\dfrac{V_k}{\bar{k}}}$	$\dfrac{N_{t-2}\bar{k}-2}{\bar{k}-1+\dfrac{V_k^*}{\bar{k}}}$	$\dfrac{2N_t}{1-\alpha_{t-1}+\dfrac{(1+\alpha_{t-1})s_k^2}{\bar{k}}}$	$\dfrac{2N_t}{1-\alpha_{t-1}^*+\dfrac{(1+\alpha_{t-1}^*)s_k^2}{\bar{k}}}$
Ideal population $s_k^2=\bar{k}$, $V_k=\bar{k}(N-1)/N$, $V_k^*=\bar{k}(N-2)/N$	N_{t-1}	N_{t-2}	N_t	N_t
Constant population size Parents in random-mating proportions $\bar{k}=2$, $\alpha_{t-1}=-1/(2N_{t-1}-1)$, α_{t-1}^* $=-1/(N_{t-1}-1)$	$\dfrac{4N-2}{V_k+2}$	$\dfrac{4N-4}{V_k^*+2}$	$\dfrac{4N-2}{V_k+2}$	$\dfrac{4N-4}{V_k^*+2}$
Equal progeny numbers $V_k=s_k^2=0$	$\dfrac{N_{t-1}\bar{k}-1}{\bar{k}-1}$	$\dfrac{N_{t-2}\bar{k}-2}{\bar{k}-1}$	$\dfrac{2N_t}{1-\alpha_{t-1}}$	$\dfrac{2N_t}{1-\alpha_{t-1}^*}$
Equal progeny number Constant population size Parents in random-mating proportions	$2N-1$	$2N-2$	$2N-1$	$2N-2$
Decreasing population ($\bar{k}=1$) Equal progeny numbers Parents in random-mating proportions	∞	∞	$2N_t-1$	$2N_t-2$
Constant population size Homozygous parents ($\alpha=1$)	$2N-1$	$2N-2$	∞	∞
Self-fertilization	1		1	
Sib mating		2		2

If the population is growing, the inbreeding effective number, being related to the earlier generation, is usually less than the variance number. If the population is diminishing the reverse is true. Since in the long run a population cannot increase or decrease indefinitely (unless it is to become extinct), the periods of growth and decline must approximately cancel and the two effective numbers are roughly the same.

Table 7.6.4.1. summarizes the inbreeding and variance effective numbers and includes a number of special cases.

Crow and Morton (1955) have reported experiments where, by appropriate use of gene markers, the progeny from a large mixed group of Drosophila parents could be individually identified as to parentage. From such experiments the mean and variance of the number of progeny per parent could be computed. These data, plus others from various studies in the literature (originally done for other purposes), are given in Table 7.6.4.2.

Table 7.6.4.2. Relation of effective and actual population number in experimental and census data. (From Crow and Morton, 1955.)

ORGANISM	$\bar{k}$	$\frac{V_k}{\bar{k}}$	$\frac{V_k}{\bar{k}}$ ADJUSTED TO $\bar{k}=2$	$\frac{N_e}{N}$
Drosophila females				
Adult progeny	13.9	4.73	1.82	.71
Egg laying (5 day)	37.2	11.45	1.77	.72
Eggs, lifetime				
wild-type, 25°	714	130	1.36	.85
wild-type, 30°	430	154	1.71	.74
wild-type, 19°	941	111	1.23	.90
vestigial	432	102	1.46	.81
Lymnaea	391	134	1.68	.75
Women				
New South Wales	6.2	2.61	1.52	.79
England	3.5	2.54	1.88	.69
U.S., born 1839	5.5	1.28	1.10	.95
U.S., born 1866	3.0	1.93	1.63	.76
Drosophila males	17.9	11.35	3.18	.48

The variance-mean ratio $V_k/\bar{k}$ is given first for the actual data, then after adjustment to $\bar{k} = 2$. Clearly, any population that is not becoming extinct or growing explosively has a mean number of progeny per parent of approximately 2, and this adjustment was made by assuming that the deaths occur at random. This method for making such a correction, as well as a justification for it, is given in the paper by Crow and Morton referred to above.

It is quite striking that the figures for females of species as divergent as Drosophila, a snail, and *Homo sapiens* have values as similar as those in the last two columns. This must mean that the effective population number, in so far as this is determined by differential fertility, is not greatly different from the number of fertile adults.

The main factor not accounted for in these data is intrabrood correlation in fate. In insects where a number of eggs are laid in one spot, all or none may be destroyed, and in such a situation the effective number could be considerably less than the census number.

7.6.5 An A Priori Approach to Predicting Effective Number in Selection Programs

Under artificial selection, the effective population number may be smaller than the actual number, because the expected number of offspring is different in different individuals. This problem was taken up by Robertson (1961) who derived a different formula for effective number from the following considerations:

Suppose there are N males and N females which are sampled from a large gene pool. If the expected contribution of the ith pair to the next gene pool is K_i, the gene frequency of the newly formed gene pool is $\sum K_i p_i / \sum K_i$, where p_i is the gene frequency in the ith pair. Thus the sampling variance of gene frequency for one generation is

$$\frac{p(1-p)\sum K_i^2}{4(\sum K_i)^2},$$

because each pair is a sample of 4 genes. By equating this to the ideal case of N pairs with variance $p(1-p)/4N$, the effective number is given by

$$N_e = \frac{(\sum K_i)^2}{\sum K_i^2}. \qquad 7.6.5.1$$

If we assume that $K_1, K_2, \ldots, K_N$ form a sample from a distribution with mean μ_K and variance V_K and that each K_i is independent, then

$$E(\sum K_i)^2 = N^2\mu_K^2 + NV_K$$

and

$$E(\sum K_i^2) = N(\mu_K^2 + V_K).$$

Substituting these in 7.6.5.1 we obtain

$$N_e = \frac{N + C^2}{1 + C^2}, \qquad 7.6.5.2$$

where C is the coefficient of variation of K between mating pairs, i.e., σ_K/μ_K.

Exactly the same result may be obtained by considering N monoecious individuals, if K_i denotes the expected contribution of the ith individual.

In contrast to the previous retrospective approach of measuring effective number from the observed distribution of the number of offspring, this approach predicts the effective number from a knowledge in the variation of selective advantage between families. Actually C^2 in 7.6.5.2 may be equated to the variance of the relative selective advantages. With $C = 0$ the effective number reduces to the actual number, as it should.

Though his formula is based on the consideration of variance, we can give it an inbreeding interpretation, since

$$\frac{1}{N_e} = \frac{\sum K_i^2}{(\sum K_i)^2}$$

is the a priori probability that two gametes from same parents (in the case of separate sexes) or from a same parent (in the case of a monoecious organism) will enter into a specified pair of offspring.

Robertson's idea was used by Nei and Murata (1966) to derive a formula for effective population size when fertility is inherited. In a population of stable size and with random mating, their formula becomes

$$N_e = \frac{4N}{(1 + 3h^2)V_K + 2}, \qquad 7.6.5.3$$

where h^2 is the heritability of progeny number. For example, if $V_K = 3$ and $h^2 = 0.3$, $N_e = 0.52N$.

In general the effective size decreases as the heritability of fitness increases.

7.7 Problems

1. The effective number of alleles, as used in 7.2.5, is always less than or equal to the actual number. Suppose that the frequencies of alleles $A_1, A_2, A_3, A_4, \ldots$ are $\frac{1}{2}, \frac{1}{4}, \frac{1}{8}, \frac{1}{16}, \ldots$. In an infinite population the number of alleles is infinite, of course. What is the effective number? (Answer: 3)
2. Compare 7.3.9 and 7.4.43. Suggest an intuitive reason why these have the same final rate of decrease. A glance ahead to Figure 8.4.1 may be helpful.

3. In a group of subpopulations, each of effective size N and starting with the same initial frequency, what length of time is required for the gene frequency variance between populations to go half way to its equilibrium value (see 7.3.4)?
4. Prove equations 7.6.2.18.
5. Explain why α appears in the variance effective number formula (7.6.3.26) but not in the inbreeding effective number formula (7.6.2.17).
6. How would the smallness of a population affect the rate of approach to linkage equilibrium; i.e., would it enhance the rate, retard it, or have no systematic effect?
7. What is the effective number of a herd of cattle with one bull and 40 cows?
8. What is the effective number of a population of 100 males and 100 females and each parent has exactly two offspring?
9. An island population contains exactly 100 persons, equally divided into males and females. The initial frequency of a recessive gene is p. What is the probability of homozygosity for this gene 10 generations later? If in the 10th generation two cousins marry, what is the probability that their child will be homozygous for this gene?
10. Haldane (1936) showed that in a sib-mated line the equilibrium probability of a locus being heterozygous for a neutral gene is 12 times the mutation rate. Verify this, using the methods of equations 7.2.9 and following. Show also that for a self-fertilized population the value is 4 times the mutation rate.
11. Show that in a randomly mating population with each parent having an equal expectation of progeny about half of the random gene frequency drift is caused by Mendelian segregation from heterozygotes and half from unequal numbers of progeny.

8 STOCHASTIC PROCESSES IN THE CHANGE OF GENE FREQUENCIES

The various factors which change gene frequencies in natural populations may conveniently be classified into three groups.

The first consists of factors which cause directed changes, such as mutation and migration occurring at constant rates, and selection of a constant intensity. Wright (1949) called these systematic evolutionary pressures. The second group consists of factors which produce random fluctuations in gene frequencies, of which two different types may be recognized. One type is the random sampling of gametes in a finite population. This becomes important, as we have emphasized earlier, when the population is small. Even in an infinite population the fate of mutant genes depends very much on chance as long as such genes occur in small numbers. This will be discussed in more detail in Section 8.8. The other type is random fluctuation in systematic pressures, of which fluctuation in selection intensity is especially important.

There is a third group of factors consisting of events, such as chromosome rearrangements, duplication or deficiency of a nucleotide sequence, or polyploidy, that may occur only once in the history of a species. These are hardly amenable to any effective quantitative treatment.

The purpose of this chapter is to discuss the first two factors, systematic and random (or deterministic and stochastic), as they act to change the genetic makeup of the population.

8.1 The Rate of Evolution by Mutation and Random Drift

In Chapter 7, Section 7.2, we considered the amount of polymorphism and heterozygosity that is maintained in a population of finite size by recurrent mutation. It was assumed that the mutations were completely neutral selectively and that many allelic states are possible. We showed that if the mutation rate is larger than the reciprocal of the effective population number the population is polymorphic and the typical individual in the population will be heterozygous for this locus. More specifically, the majority of population will be homozygous for the locus if $4N_e u < 1$ and heterozygous if $4N_e u > 1$. The actual distribution of the number of alleles maintained in this way is more difficult and will be discussed in Chapter 9, along with the complicating effects of selection.

A similar situation arises when we look at the process in time. A newly arisen mutation is very likely to be lost from the population within a few generations because of accidents of the Mendelian process and variations in the number of progeny from different individuals in the population. On the other hand, if the population is finite, a minority of mutants may be lucky enough to persist in the population and ultimately become the prevailing type. To be sure, the likelihood of this is extremely small in a population of moderate size; but the probability is not zero and in the long time of evolutionary history events with small probabilities do occur.

We shall consider now the simple problem of how frequently such neutral-gene replacement is expected to occur. Later in the chapter (Section 8.8 and 8.9) we deal with the process in more detail, including the effects of selection.

Consider a population of size N (actual number, not effective number). If we look sufficiently long into the future the population of genes at a particular locus will all be descended from a single allele in the present generation. In the vocabulary of Chapter 3, they all will be identical by descent and the population will be autozygous for this locus. This is the result of the inexorable process of random gene frequency drift. If, in the present generation, an allele A_1 exists in frequency p, the probability is simply p that the lucky allele from which the whole population of genes is descended is A_1 rather than some other allele.

Now, if mutation occurs at a rate u per gene per generation, then the number of new mutants at this locus in the present generation is $2Nu$. The probability that a particular gene will eventually be fixed in the population is $1/2N$. So the probability of a mutant gene arising in this generation and eventually being incorporated into the population is $2Nu(1/2N) = u$. We have the remarkably simple result that

$$\text{Rate of neutral gene substitution} = u. \tag{8.1.1}$$

That is to say that, viewed over a long time period, the rate of evolution by fixation of neutral mutants is equal to the mutation rate. Stated in another way, the average interval between the occurrence of successful mutants is $1/u$.

The observed rate of evolution of amino acids in mammalian hemoglobins is about one replacement per codon per 10^9 years. This could be accounted for entirely by neutral substitutions if the mutation rate to such alleles were 10^{-9} per codon per year. This is not to assert that this is necessarily the major mechanism by which amino acids evolve, but we would not be surprised if it turns out that an appreciable fraction of nucleotide replacement in evolution is of this type. For a discussion, see Kimura (1968), King and Jukes (1969), and Crow (1969).

We have not considered the time required for a successful mutant to go from a single representative to complete fixation. Later, in Section 8.9, it will be shown that the average time required for those mutants which are successful to change frequency from $1/2N$ to 1 is

$$\bar{t} = 4N_e. \tag{8.1.2}$$

This does not depend on the mutation rate. This is as expected; since only one representative will be fixed, the time required does not depend on how many mutants there are. Note also that the relevent population number this time is the effective number, not the actual number.

These points are illustrated in Figure 8.1.1. The rate of gene substitution, when we consider a time period that is long with respect to the time required for a single substitution to occur, is u. This is given by the reciprocal of the time interval between the occurrence of successive successful mutants, as shown on the graph. The time required for a particular gene to be substituted is $\bar{t}$, also shown.

If $\bar{t}$ is small relative to $1/u$, that is, if $4N_e u \ll 1$, then the population is monomorphic most of the time, as illustrated by situation b in the figure. On the other hand, if $\bar{t}$ is comparable to $1/u$ as might be the case in a large population with a high rate of mutation to neutral alleles, the population

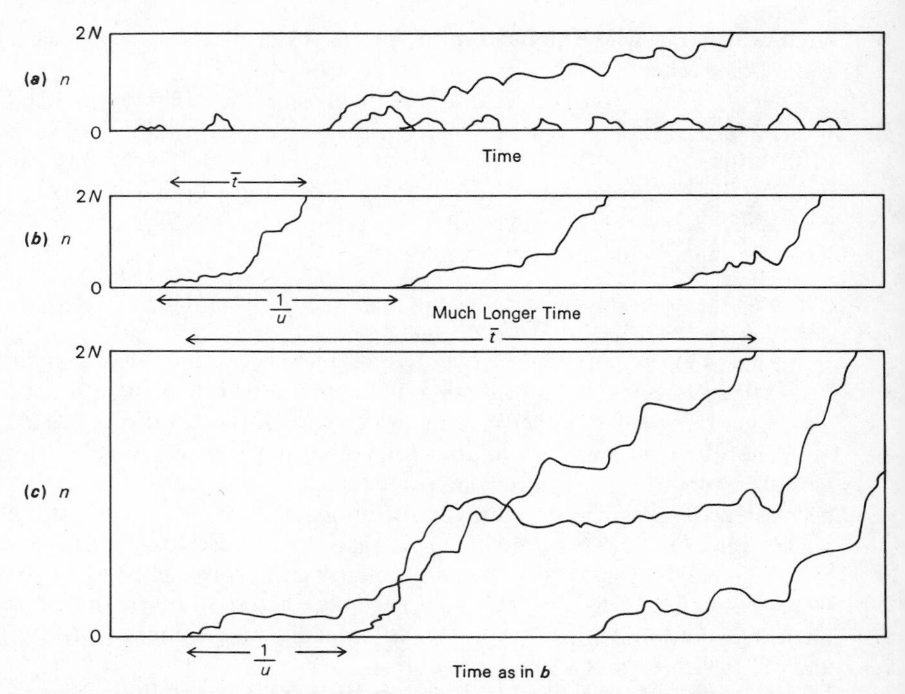

Figure 8.1.1. Gene substitution by random drift and mutation. The abscissa is time over a very long period; the ordinate is the number of mutant genes descended from a single mutant. The upper figure (*a*) is intended to illustrate that, while most mutants persist a few generations and then are lost, an occasional one increases to eventual fixation. The second drawing (*b*) shows only the mutants that eventually become incorporated into the population. The time scale is therefore much longer. The third (*c*) shows a situation in a larger population or one with a higher rate of occurrence of neutral mutants. In this case the time required for a replacement ($\bar{t}$) is comparable to that between such events ($1/u$), with the result that there is considerable transient polymorphism.

will have considerable transient polymorphism. At any one time it is likely to have more than one allele, although these will be different alleles at different times. This is illustrated by situation *c*.

Later in the chapter (Section 8.8), it will be shown that the probability of fixation of a single gene that is slightly favorable is approximately $2s$. So, even favorable mutants are lost most of the time. However, if they occur with any appreciable frequency they are substituted considerably more rapidly than neutral alleles, as expected.

8.2 Change of Gene Frequencies as a Stochastic Process

In many mathematical problems arising in population genetics, the process of change in gene frequency may be treated as deterministic, as we have done in Chapters 1 through 6. This approach was extensively developed by Haldane (1924 and later), especially for single-locus problems. It is still useful. There are many circumstances where such an approach is sufficiently realistic to yield interesting and reliable information, as we already illustrated.

Yet, when we consider that actual populations are all finite, that many mutant genes may be represented only once at the moment of their occurrence, and that organic evolution has proceeded over an enormous period of time in an ever-fluctuating environment, we realize the necessity of an approach that can take indeterminacy into account.

In this chapter we treat the process of change in gene frequencies as a stochastic process. By this we mean a mathematical formulation of chance events in a process that proceeds with time.

The pioneering work in this field has been done by Fisher (1922, 1930) and Wright (1931 and later). These authors have been mainly concerned with the state of statistical equilibrium that is reached when the form of the distribution becomes constant. The problem of constructing the entire history of change in gene frequencies starting from arbitrary initial frequencies is more complicated. Several practically important cases have been solved by one of the authors and he has reviewed the historical development of the subject elsewhere (Kimura, 1964).

A mathematical approach which has proven to be very powerful makes use of " diffusion " models, in which two diffusion equations, the Kolmogorov forward and backward equations, play a central role.

8.3 The Diffusion Equation Method

In population genetics the fundamental quantity used for describing the genetic composition of a Mendelian population is gene frequency rather than genotype frequency. The main reason for this, as was discussed in Section 2.1, is that each gene is a self-reproducing entity and its frequency changes almost continuously with time as long as the population is reasonably large. On the other hand, genotypes are produced anew in each generation by recombination of genes and therefore do not have the continuity that genes have.

Also, we note that in actual evolution the gene frequency changes are typically very slow. To be sure, there are some exceptions. One is the rapid increase in the melanic gene in some Lepidoptera in industrial areas. Another is the development of resistance to insecticides and antibiotics. But, as pointed out by Haldane (1949*d*), the typical rate of evolution shown by fossil records is of the order of one-tenth of a darwin, where one darwin represents a change by a factor e in a million years, or 10^{-6} per year. On the ordinary scales by which we consider time, this is exceedingly slow.

Therefore, in the following treatment we will regard the process of change of gene frequency as a continuous stochastic process. Roughly speaking, this means that as the time interval becomes smaller, the amount of change in gene frequency during that interval is expected to be smaller. More strictly, the process is called a continuous stochastic process if for any given positive value ε, however small, the probability that the change in gene frequency x during time interval $(t, t + \delta t)$ exceeds ε is $o(\delta t)$, that is, an infinitesimal of higher order than δt. Furthermore, we will assume that the process of change in gene frequencies is Markovian, that is, the probability distribution of gene frequencies at a given time t_1 depends on the gene frequencies at a preceding time t_0 but not on the previous history which has led to the gene frequencies at t_0, where $t_0 < t_1$.

The fundamental equations used to study this continuous Markov process are the Kolmogorov forward and backward equations (Kolmogorov, 1931). The forward equation is called the Fokker–Planck equation by physicists and it was introduced into the field of population genetics by Wright (1945), while the backward equation was first used in this field to study the problem of gene fixation by Kimura (1957; see also 1962).

Before going into the details of their derivation and use, we will present these two equations in the form that may be most easily applied to population genetics.

Let us assume that a pair of alleles A_1 and A_2 are segregating in a population. We will denote by $\phi(p, x; t)$ the probability density that the frequency of A_1 lies between x and $x + dx$ at time t, given that it is p at time $t = 0$. Then, as will be shown later, $\phi(p, x; t)$ satisfies the following Kolmogorov forward equation:

$$\frac{\partial \phi(p, x; t)}{\partial t} = \frac{1}{2} \frac{\partial^2}{\partial x^2} \{V_{\delta x} \phi(p, x; t)\} - \frac{\partial}{\partial x} \{M_{\delta x} \phi(p, x; t)\}. \tag{8.3.1}$$

The time parameter t in the above equation is a continuous variable and can be measured in any unit, but for practical purposes, it may most conveniently be measured here with the length of one generation as the unit. In this equation, gene frequency x at the tth generation is a random variable but the initial frequency p is fixed. That is to say, we consider the process of change in the forward direction. Also $M_{\delta x}$ and $V_{\delta x}$ are respectively the mean and the variance of δx, the amount of change in gene frequency per generation. Both $M_{\delta x}$ and $V_{\delta x}$ may depend on x and t.

Next, let us consider the process of change in the reverse direction and assume the gene frequency x in the tth generation is fixed and the initial frequency p is a random variable. That is, we consider the process retrospectively. Then, as will be shown later, ϕ satisfies the following Kolmogorov backward equation:

$$\frac{\partial\phi(p, x; t)}{\partial t} = \frac{1}{2} V_{\delta p} \frac{\partial^2}{\partial p^2} \phi(p, x; t) + M_{\delta p} \frac{\partial}{\partial p} \phi(p, x; t). \qquad 8.3.2$$

Actually, this is a restricted form of Kolmogorov backward equation that is applicable to the *time homogeneous case*. It is valid when both $M_{\delta x}$ and $V_{\delta x}$ do not depend on the time parameter t. In many problems arising in population genetics, both $M_{\delta x}$ and $V_{\delta x}$, though depending on the gene frequency x, do not depend on t, so 8.3.2 has a wide use.

Returning to the forward equation, we will investigate how this may be derived. In this equation, p is fixed, so we will omit this letter and write $\phi(x; t)$ for $\phi(p, x; t)$.

In a population of N diploid individuals, gene frequency changes from 0 to 1 in steps of size $1/2N$, but in the present formulation, x is regarded as a continuous variable that changes smoothly from 0 to 1. Thus, $\phi(x; t)$ gives the probability distribution at time t such that, when $1/2N$ is substituted for dx, $\phi(x, t)dx$ gives an approximation to the probability that the gene frequency is x at the tth generation. We can also interpret $\phi(x; t)dx$ as the relative frequency or proportion of populations with gene frequency x among a hypothetical aggregate consisting of an infinite number of populations satisfying the same condition. Thus, if we denote the frequency of such a class by $f(x, t)$, we have

$$f(x, t) = \phi(x; t)dx, \qquad 8.3.3$$

in which $1/2N$ may be substituted for dx. Here we should note that approximation 8.3.3 as well as equation 8.3.1 is valid for the frequencies of unfixed classes; that is, for $0 < x < 1$. As to frequencies of fixed classes, $f(0, t)$ and $f(1, t)$, separate treatments are required as will be discussed later.

Several methods of derivation of the forward equation, differing in mathematical rigor and sophistication, are available. We would like to present first a very elementary derivation based on the geometrical interpretation of the process of change in gene frequency (Kimura, 1955). Though unsatisfactory from the pure mathematical standpoint, it has the advantage of being quite elementary and at the same time helps us see the physical meaning of the terms involved.

Let $y = \phi(x, t)$ represent the curve for the probability distribution of the gene frequency at time t. As shown in Figure 8.3.1, the distribution is approxi-

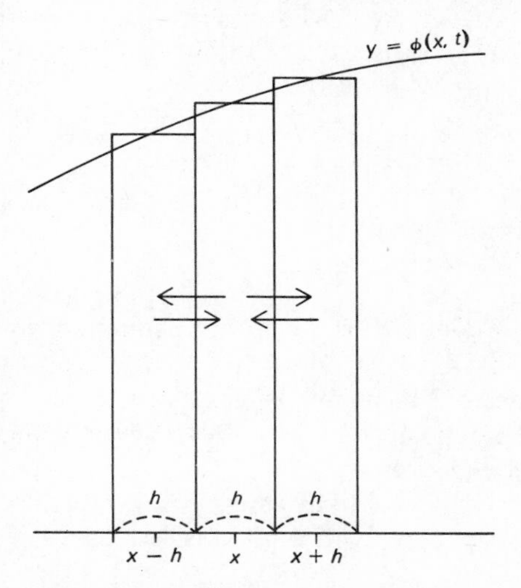

Figure 8.3.1. Diagram to show the meaning of terms in the Kolmogorov forward (Fokker–Planck) equation as applied to population genetics. (From Kimura, 1955).

mated by histograms, each column having width h. We represent the gene frequency of each class by its middle point. Consider the class with gene frequency x. For sufficiently small h, the area of the column $\phi(x, t)h$ gives the probability that the population has gene frequency in the interval $x - \frac{1}{2}h$ to $x + \frac{1}{2}h$. After a time interval of length δt, the population with gene frequency x will move to another class owing to systematic as well as random changes. However, because of the assumption of a continuous stochastic process, we make this movement sufficiently small by taking δt small that consideration of the two adjacent classes is sufficient.

Let $m(x, t)\delta t$ be the probability that the population moves to the higher class $(x + h)$ by systematic pressure. Let $v(x, t)\delta t$ be the probability that it moves outside the class by random fluctuation, half the time to the left class $(x - h)$ and the other half to the right class $(x + h)$. Any asymmetrical displacement may be included in the term $m(x, t)$. Also, greater displacements than mentioned above should have an infinitesimal probability of higher order ($o(\delta t)$) and can be neglected. Thus the probability that the population will have gene frequency in the range $x - \frac{1}{2}h$ to $x + \frac{1}{2}h$ after time δt is obtained by considering the exchange of gene frequencies between these adjacent classes.

$$\begin{aligned}\phi(x, t + \delta t)h = {} & \phi(x, t)h - \{v(x, t) + m(x, t)\}\, \delta t \phi(x, t)h \\ & + \tfrac{1}{2}v(x - h, t)\delta t \phi(x - h, t)h + \tfrac{1}{2}v(x + h, t)\delta t \phi(x + h, t)h \\ & + m(x - h, t)\delta t \phi(x - h, t)h. \end{aligned} \tag{8.3.4}$$

The second term in the right-hand side of the above equation is the amount of loss resulting from movement to other classes, the third term is the contribution from the left class, and the fourth from the right class, both due to random change; the final term is the contribution from the left class owing to systematic change.

If we denote by $V(x, t)\delta t$ the variance of the change in x per δt due to random change,

$$V(x, t)\delta t = h^2 \tfrac{1}{2}v(x, t)\delta t + (-h)^2 \tfrac{1}{2}v(x, t)\delta t$$

or

$$V(x, t) = h^2 v(x, t). \tag{8.3.5}$$

Similarly, if $M(x, t)\delta t$ is the mean change in x per δt,

$$M(x, t)\delta t = hm(x, t)\delta t$$

or

$$M(x, t) = hm(x, t). \tag{8.3.6}$$

Substituting 8.3.5 and 8.3.6 in the right side of 8.3.4 and dividing both sides by $h\delta t$, we have, after some rearrangement,

$$\begin{aligned}\frac{\phi(x, t + \delta t) - \phi(x, t)}{\delta t} & \\ = \frac{1}{2}\, & \frac{\dfrac{V\phi(x + h, t) - V\phi(x, t)}{h} - \dfrac{V\phi(x, t) - V\phi(x - h, t)}{h}}{h} \\ & - \frac{M\phi(x, t) - M\phi(x - h, t)}{h},\end{aligned}$$

where terms such as $V\phi(x, t)$ and $M\phi(x, t)$ are abbreviations of $V(x, t)\phi(x, t)$

and $M(x, t)\phi(x, t)$ respectively. Taking the limits $\delta t \to 0$, $h \to 0$, we obtain

$$\frac{\partial \phi(x, t)}{\partial t} = \frac{1}{2} \frac{\partial^2}{\partial x^2} \{V(x, t)\phi(x, t)\} - \frac{\partial}{\partial x} \{M(x, t)\phi(x, t)\}, \tag{8.3.7}$$

which is equivalent to 8.3.1, $V(x, t)$ and $M(x, t)$ in this equation corresponding respectively to $V_{\delta x}$ and $M_{\delta x}$ in equation 8.3.1. We may note here that these two sets of quantities are defined in a slightly different way. Namely, in the above derivation $V(x, t)$ represents the variance per infinitesimal time of the random component of change for which the mean is 0. Any systematic component of change is included in $M(x, t)$. On the other hand, $V_{\delta x}$ and $M_{\delta x}$ in equation 8.3.1 represent the variance and the mean of the change of gene frequency *per generation*. In practice, quantities such as mutation rates, rate of migration, intensity of selection, and effect of random sampling of gametes which determine the rate of change in gene frequency are all measured or expressed with one generation as the time unit. So, for practical purposes, expressions $V_{\delta x}$ and $M_{\delta x}$ might be more convenient than $V(x, t)$ and $M(x, t)$.

In the above derivation leading to 8.3.7, we assumed that the systematic pressure pushes the gene frequency toward the right, but no essential change is required for the argument if it pushes toward the left, in which case we simply use $m(x + h, t)\delta t\phi(x + h, t)h$ as the last term of 8.3.4. From the above derivation it is evident that the first and the second terms in the right-hand side of 8.3.7, and therefore 8.3.1, give respectively the rates of change due to random fluctuation and systematice pressures.

We will now attempt a mathematically more satisfactory but less intuitive derivation of the Kolmogorov forward equation (8.3.7).

Let us denote by $g(x, \xi; t, \delta t)$ the probability density that the gene frequency changes from x to $x + \xi$ during time interval $(t, t + \delta t)$. Then we have

$$\phi(x; t + \delta t) = \int \phi(x - \xi; t)g(x - \xi, \xi; t, \delta t)\, d\xi, \tag{8.3.8}$$

where the integral is over all possible values of ξ. The above relation follows directly from the assumption that the process is Markovian. That is, the probability that the gene frequency is x at time $t + \delta t$ is equal to the sum of probabilities of cases in which the gene frequency is $x - \xi$ at time t and then the gene frequency increases by ξ during the subsequent time interval of length δt. In this expression, if we take δt small, then, because of the assumption of a continuous stochastic process, the change ξ during this time interval is practically restricted to a very small value. Thus, expanding the integrand in terms of ξ, we have

$$\phi(x; t+\delta t) = \int \left\{ \phi g - \xi \frac{\partial(\phi g)}{\partial x} + \frac{\xi^2}{2!} \frac{\partial^2(\phi g)}{\partial x^2} - \frac{\xi^3}{3!} \frac{\partial^3(\phi g)}{\partial x^3} + \cdots \right\} d\xi, \tag{8.3.9}$$

where ϕ and g stand for $\phi(x; t)$ and $g(x, \xi; t, \delta t)$. In the following treatment we will assume that the order between summation, integration, and differentiation may be interchanged freely. From 8.3.9, neglecting terms involving ξ^3 and higher powers of ξ, we have

$$\phi(x; t+\delta t) = \phi \int g\, d\xi - \frac{\partial}{\partial x} \left\{ \phi \int \xi\, g d\xi \right\} + \frac{1}{2} \frac{\partial^2}{\partial x^2} \left\{ \phi \int \xi^2 g\, d\xi \right\}. \tag{8.3.10}$$

Noting that

$$\int g d\xi = 1,$$

transferring the first term in the right-hand side of 8.3.10 to the left, and then dividing both sides by δt, we get

$$\frac{\phi(x; t+\delta t) - \phi(x, t)}{\delta t} = -\frac{\partial}{\partial x} \left\{ \phi(x, t) \frac{1}{\delta t} \int \xi g(x, \xi; t, \delta t)\, d\xi \right\} + \frac{1}{2} \frac{\partial^2}{\partial x^2} \left\{ \phi(x, t) \frac{1}{\delta t} \int \xi^2 g(x, \xi; t, \delta t)\, d\xi \right\}. \tag{8.3.11}$$

At the limit as $\delta t \to 0$, if we define $M(x, t)$ and $V(x, t)$ by

$$M(x, t) = \lim_{\delta t \to 0} \frac{1}{\delta t} \int \xi g(x, \xi; t, \delta t)\, d\xi \tag{8.3.12}$$

and

$$V(x, t) = \lim_{\delta t \to 0} \frac{1}{\delta t} \int \xi^2 g(x, \xi; t, \delta t)\, d\xi, \tag{8.3.13}$$

equation 8.3.11 yields the Kolmogorov forward equation given before as 8.3.7. The above derivation may still be unsatisfactory from the standpoint of mathematical rigor. For more rigorous derivations, readers may refer to the mathematical literature, for example, Kolmogorov (1931).

Going back to equation 8.3.1 and substituting $\phi(x, t)$ for $\phi(p, x; t)$, the Kolmogorov forward equation may most conveniently be expressed for our purpose as

$$\frac{\partial \phi(x, t)}{\partial t} = \frac{1}{2} \frac{\partial^2}{\partial x^2} \{V_{\delta x} \phi(x, t)\} - \frac{\partial}{\partial x} \{M_{\delta x} \phi(x, t)\}. \quad 8.3.14$$

In applying this equation to population genetics, it is often very useful to keep in mind that the quantity

$$-\frac{1}{2} \frac{\partial}{\partial x} \{V_{\delta x} \phi(x, t)\} + M_{\delta x} \phi(x, t),$$

which we will denote by $P(x, t)$ and which enters the right-hand side of the above equation as $-\partial P(x, t)/\partial x$, represents the rate per generation of net flow of probability mass across the point x. With the help of Figure 8.3.1 we will again try to show this using a geometrical interpretation.

The net flow of probability mass across the point $x + h/2$ during the short time interval $(t, t + \delta t)$, which we denote by $P(x + \frac{1}{2}h, t)\delta t$, is given by

$$\begin{aligned} P(x + \tfrac{1}{2}h, t)\delta t = m(x, t)\delta t \phi(x, t)h \\ + \tfrac{1}{2}v(x, t)\delta t \phi(x, t)h \\ - \tfrac{1}{2}v(x + h, t)\delta t \phi(x + h, t)h. \end{aligned} \quad 8.3.15$$

Here we consider only the exchange of frequencies between the classes having gene frequency x and $x + h$. Substituting $m(x, t)h = M(x, t)$ and $v(x, t)h = V(x, t)/h$ in the above equation, we get

$$P(x + \tfrac{1}{2}h, t) = M(x, t)\phi(x, t) - \frac{1}{2} \frac{V(x + h, t)\phi(x + h, t) - V(x, t)\phi(x, t)}{h},$$

which gives

$$P(x, t) = M(x, t)\phi(x, t) - \frac{1}{2} \frac{\partial}{\partial x} \{V(x, t)\phi(x, t)\} \quad 8.3.16$$

at the limit of $h \to 0$. Again, for practical purposes, it is convenient to use the mean and the variance of the gene frequency change per generation, that is, $M_{\delta x}$ and $V_{\delta x}$ for $M(x, t)$ and $V(x, t)$, to give

$$P(x, t) = -\frac{1}{2}\frac{\partial}{\partial x}\{V_{\delta x}\phi(x, t)\} + M_{\delta x}\phi(x, t). \tag{8.3.17}$$

With this expression for the probability flux, the forward equation may be written as

$$\frac{\partial \phi(x, t)}{\partial t} = -\frac{\partial P(x, t)}{\partial x}. \tag{8.3.18}$$

As pointed out following equation 8.3.3, our fundamental equation 8.3.1 is valid only for gene frequencies in the interval $0 < x < 1$ (unfixed classes). Therefore, separate treatments are required to obtain probabilities for $x = 0$ and $x = 1$ (terminal classes). In order to obtain the rate of change in the frequencies of these terminal classes, we make use of the fact just established, that is, $P(x, t)$ in 8.3.17 gives the probability flux across the point x at time t. Here we will consider a special but important case in which the change of these frequencies is entirely due to inflow of probability mass from the unfixed classes. In the terminology of the mathematical theory of probability, the boundaries ($x = 0$ and $x = 1$) act as absorbing barriers. In such a case, we have

$$\frac{df(0, t)}{dt} = -P(0, t), \tag{8.3.19}$$

$$\frac{df(1, t)}{dt} = P(1, t), \tag{8.3.20}$$

where $f(0, t)$ and $f(1, t)$ are the frequencies of classes having gene frequency 0 and 1 at the tth generation.

In the particularly important case in which the random fluctuation is solely due to random sampling of gametes such that $V_{\delta x} = x(1 - x)/2N_e$, and the systematic pressure is solely due to selection such that $M_{\delta x} = x(1 - x)s(x, t)$, where $s(x, t)$ is the selection coefficient,

$$-P(0, t) = \lim_{x \to 0}\left[\frac{1}{2}\frac{\partial}{\partial x}\left\{\frac{x(1 - x)}{2N_e}\phi(x, t)\right\} - x(1 - x)s(x, t)\phi(x, t)\right]$$

$$= \frac{1}{4N_e}\phi(0, t).$$

Thus, from 8.3.19, we have

$$\frac{df(0, t)}{dt} = \frac{1}{2}\phi(0, t)\frac{1}{2N}\left(\frac{N}{N_e}\right). \tag{8.3.21}$$

In the right-hand side of the above equation, $\phi(0, t)$ is approximately equal to $\phi(1/2N, t)$. Since $\phi(1/2N, t)\,(1/2N)$ is our approximation for the frequency of subterminal class $f(1/2N, t)$, we have

$$\frac{df(0, t)}{dt} = \frac{1}{2} f\left(\frac{1}{2N}, t\right)\left(\frac{N}{N_e}\right). \tag{8.3.22}$$

That is, the contribution from the unfixed classes to the rate of change in the terminal class with $x = 0$ is half the frequency of the subterminal class with $x = 1/2N$ multiplied by the ratio N/N_e. In a special case in which the actual size of the population is equal to its effective size, this ratio reduces to unity.

In a like manner, we have

$$\frac{df(1, t)}{dt} = \frac{1}{2}\,\phi(1, t)\,\frac{1}{2N}\left(\frac{N}{N_e}\right) \approx \frac{1}{2} f\left(1 - \frac{1}{2N}, t\right)\left(\frac{N}{N_e}\right) \tag{8.3.23}$$

for the terminal class with $x = 1$ under a similar condition.

The diffusion equation method can readily be extended to treat the cases of two or more random variables. For example, for two *independently* segregating loci, each with a pair of alleles A_1 and A_2 in the first locus and B_1 and B_2 in the second, the corresponding Kolmogorov forward equation becomes

$$\frac{\partial \phi}{\partial t} = \frac{1}{2}\frac{\partial^2}{\partial x^2}(V_{\delta x}\phi) + \frac{1}{2}\frac{\partial^2}{\partial y^2}(V_{\delta y}\phi) - \frac{\partial}{\partial x}(M_{\delta x}\phi) - \frac{\partial}{\partial y}(M_{\delta y}\phi), \tag{8.3.24}$$

where $\phi \equiv \phi(p, q; x, y; t)$ stands for the probability density that the frequencies of A_1 and B_1 are x and y at the tth generation, given that their frequencies are p and q at $t = 0$.

Some of the applications of the forward equation to more concrete problems of population genetics, such as constructing the process of random genetic drift owing to small population number, will be presented in following sections.

Next, we will attempt to derive the Kolmogorov backward equation given as 8.3.2, in which we consider x fixed and p a variable. In this formulation we reverse the time sequence and look at the process retrospectively. Also, in order to make our treatment simpler, we will restrict our consideration to the cases in which the process is *time homogeneous*; that is, we consider only those cases in which if x_{t_1} and x_{t_2} are frequencies of a gene at times t_1 and t_2, then the probability distribution of x_{t_2}, given x_{t_1}, which in general is a function of t_1 and t_2, depends only on the time difference $t_2 - t_1$.

For such a time-homogeneous Markov process, we have

$$\phi(p, x; t + \delta t) = \int g(p, \xi; \delta t)\phi(p + \xi, x; t)\, d\xi. \tag{8.3.25}$$

The above relation, which is analogous to relation 8.3.8, may be derived by considerations similar to those by which 8.3.8 was derived. Note here, however, that g in the above relation depends on three variables, p, ξ, and δt, but not on t. This is due to the assumption of time homogeneity, that is, the probability that the gene frequency change from p to $p + \xi$ during time interval of length δt is the same for any t (generation).

Expanding $\phi(p + \xi, x; t)$ inside the integral in terms of ξ but neglecting terms involving ξ^3 and higher powers of ξ, we obtain, at the limit of $\delta t \to 0$, the following equation:

$$\frac{\partial \phi(p, x; t)}{\partial t} = \frac{V(p)}{2} \frac{\partial^2 \phi(p, x; t)}{\partial p^2} + M(p) \frac{\partial \phi(p, x; t)}{\partial p}, \tag{8.3.26}$$

where

$$M(p) = \lim_{\delta t \to 0} \frac{1}{\delta t} \int \xi g(p, \xi; \delta t)\, d\xi, \tag{8.3.27}$$

$$V(p) = \lim_{\delta t \to 0} \frac{1}{\delta t} \int \xi^2 g(p, \xi; \delta t)\, d\xi. \tag{8.3.28}$$

Thus, substituting the mean and the variance of the amount of change per generation, $M_{\delta p}$ and $V_{\delta p}$ for $M(p)$ and $V(p)$, in the above equation, we obtain 8.3.2, the Kolmogorov backward equation as applied to population genetics.

One of the very important uses of this equation is its application to the problem of gene fixation. If we take $x = 1$, ϕ in the backward equation may be interpreted as the probability that the gene becomes fixed (established) in the population by the tth generation, given that it is p at the start. We will denote this probability by $u(p, t)$, for which we have

$$\frac{\partial u(p, t)}{\partial t} = \frac{V_{\delta p}}{2} \frac{\partial^2 u(p, t)}{\partial p^2} + M_{\delta p} \frac{\partial u(p, t)}{\partial p}. \tag{8.3.29}$$

The probability of fixation will then be obtained by solving the above equation with the boundary conditions

$$u(0, t) = 0, \qquad u(1, t) = 1; \tag{8.3.30}$$

that is, the probability is 0 if $p = 0$ and is 1 if $p = 1$.

Of special interest in population genetics is the ultimate probability of gene fixation defined by

$$u(p) = \lim_{t \to \infty} u(p, t), \tag{8.3.31}$$

for which

$$\frac{\partial u}{\partial t} = 0.$$

Thus, $u(p)$ satisfies the ordinary differential equation

$$\frac{V_{\delta p}}{2} \frac{d^2 u(p)}{dp^2} + M_{\delta p} \frac{du(p)}{dp} = 0 \tag{8.3.32}$$

with boundary conditions

$$u(0) = 0, \qquad u(1) = 1. \tag{8.3.33}$$

The problem of gene fixation is important in the theory of evolution and also for the study of breeding. We will elaborate the application of equation 8.3.32 to this problem later in this chapter. In the next few sections, we will apply the Kolmogorov forward equation to solve some concrete problems arising in population genetics.

8.4 The Process of Random Genetic Drift Due to Random Sampling of Gametes

We will first consider the simplest situation in which a pair of alleles A_1 and A_2 are segregating with respective frequencies x and $1 - x$ in a random-mating population of N monoecious individuals and the only factor which causes gene frequency change is the random sampling of gametes in reproduction. As time goes on, the gene frequencies tend to deviate from their initial values and eventually one of the two alleles becomes fixed in the population. This is the simplest important stochastic process in the change of gene frequencies in a Mendelian population. Since Wright's work in 1931, the process has been known by the term 'drift' or more adequately 'random genetic drift'.

In the previous chapter, we studied the law of change in the mean, variance, and the higher moments. The mean and the variance of the change in gene frequency x per generation are, respectively,

$$M_{\delta x} = 0$$

and

$$V_{\delta x} = \frac{x(1-x)}{2N_e},$$

as shown in 7.3.1 and 7.3.2. For simplicity, we assume in this section that the effective population number N_e is equal to N. Thus the forward equation 8.3.1 becomes

$$\frac{\partial \phi}{\partial t} = \frac{1}{4N}\frac{\partial^2}{\partial x^2}\{x(1-x)\phi\}, \qquad (0 < x < 1), \tag{8.4.1}$$

where $\phi \equiv \phi(p, x; t)$ is the probability density that the gene frequency becomes x in the tth generation, given that it is p at $t = 0$. In terms of the Dirac delta function $\delta(\cdot)$, the initial condition may be expressed in the form

$$\phi(p, x; 0) = \delta(x - p). \tag{8.4.2}$$

The required solution of 8.4.1 that satisfies the initial condition 8.4.2 was first obtained by Kimura (1955*a*). It is expressed in terms of the hypergeometric function as follows:

$$\phi(p, x; t) = \sum_{i=1}^{\infty} p(1-p)i(i+1)(2i+1)F(1-i, i+2, 2, p) \times F(1-i, i+2, 2, x)e^{\frac{-i(i+1)t}{4N}}, \tag{8.4.3}$$

where $F(\cdot, \cdot, \cdot, \cdot)$ stands for the hypergeometric function so that

$$F(1-i, i+2, 2, x) = 1 + \frac{(1-i)(i+2)}{1 \times 2}x + \frac{(1-i)(2-i)(i+2)(i+3)}{1 \times 2 \times 2 \times 3}x^2 + \cdots. \tag{8.4.4}$$

The above solution, 8.4.3, may also be expressed in terms of the Gegenbauer polynomial (see Korn and Korn, 1968), defined by

$$T_{i-1}^1(z) = \frac{i(i+1)}{2}F\left(i+2, 1-i, 2; \frac{1-z}{2}\right) \tag{8.4.5}$$

as follows:

$$\phi(p, x; t) = \sum_{i=1}^{\infty} \frac{(2i+1)(1-r^2)}{i(i+1)} T_{i-1}^1(r)T_{i-1}^1(z)e^{\frac{-i(i+1)t}{4N}}, \tag{8.4.6}$$

where $r = 1 - 2p$, $z = 1 - 2x$, and $T_0^1(z) = 1$, $T_1^1(z) = 3z$, $T_2^1(z) = (3/2)(5z^2 - 1)$, $T_3^1(z) = (5/2)(7z^3 - 3z)$, $T_4^1(z) = (15/8)(21z^4 - 14z^2 + 1)$, etc.

The right-hand side of equation 8.4.3 or 8.4.6 is an infinite series, but for a large value of t, only the first few terms are of any significance in determining the actual form of the distribution. Thus, for a large t, we have

$$\phi(p, x; t) = 6p(1-p)e^{-\frac{1}{2N}t} + 30p(1-2p)(1-2x)e^{-\frac{3}{2N}t} + \cdots. \tag{8.4.7}$$

In particular, at the limit of $t \to \infty$, we obtain the asymptotic formula

$$\phi \sim Ce^{-\frac{1}{2N}t}, \qquad \textbf{8.4.8}$$

in which C is a constant. This means that after a large number of generations, the probability distribution for unfixed classes ($0 < x < 1$) becomes flat and decays at the rate of $1/(2N)$ per generation. This is called the state of steady decay, and $1/(2N)$ corresponds to the smallest eigenvalue of the partial differential equation 8.4.1. The relation 8.4.8 was first obtained by Wright (1931). Figure 8.4.1 illustrates such a state of steady decay, when fixation or loss of an allele occurs at a constant rate.

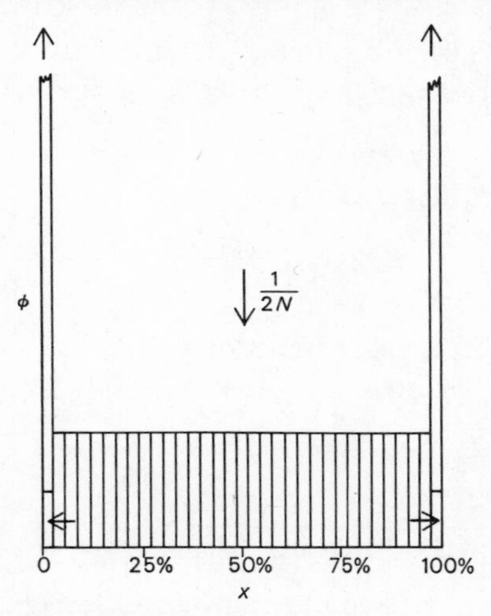

Figure 8.4.1 The distribution after many generations (roughly, any time after $t = 2N$) when the distribution is of steady form. All frequencies between 0 and 1 exclusive are equally probable and are decreasing at the same rate, $1/2N$. Fixation or loss of the allele proceeds at a constant rate, $1/4N$. (Adapted from Wright, 1931.)

The complete solution, 8.4.3, enables us to construct the more detailed process of change in the frequency distribution of unfixed classes as shown in Figure 8.4.2*a* and 8.4.2*b*. In Figure 8.4.2*a*, the initial gene frequency (p)

is 0.5. It may be seen from the figure that after $2N$ generations the distribution curve becomes almost flat and the genes are still unfixed in about 50% of the cases. In Figure 8.4.2*b*, the initial gene frequency is assumed to be 0.1 and it takes $4N$ or $5N$ generations before the distribution curve becomes practically flat. By that time, however, gene A_1 is either fixed in the population or lost from it in more than 90% of the cases. So the asymptotic formula

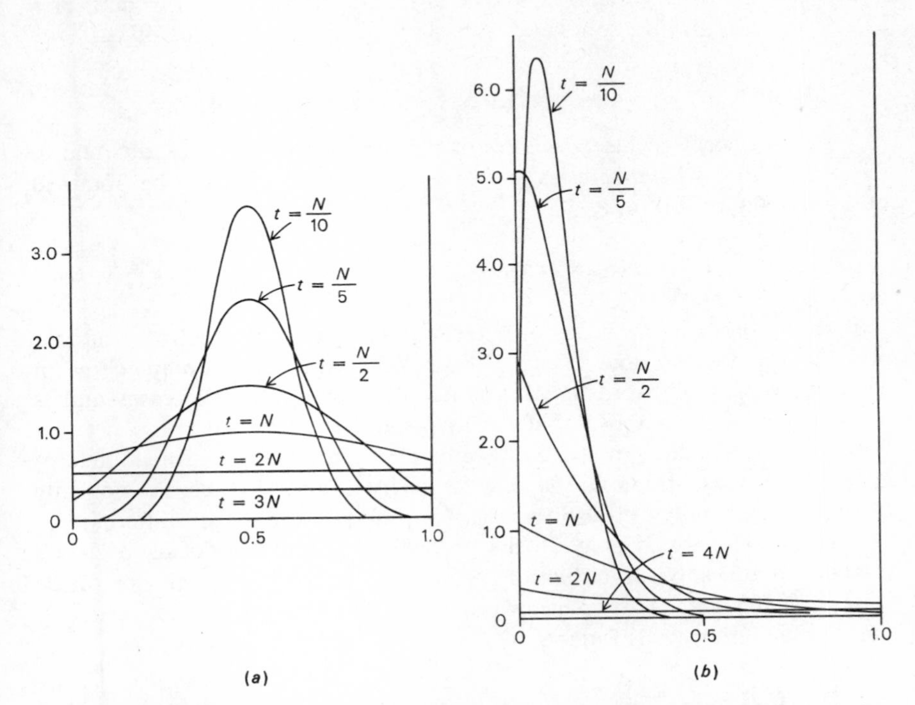

Figures 8.4.2*a,b*. The process of random genetic drift due to small population number, in which it is assumed that mutation, migration, and selection are absent, and the random change in the gene frequency from generation to generation is caused by random sampling of gametes. In Figure 8.4.2*a*, the initial frequency of A_1 is 0.5, while in Figure 8.4.2*b*, the initial frequency is 0.1. In both figures, t stands for time and N stands for the effective population number. The abscissa is the frequency of A_1 in the population and the ordinate is the corresponding probability density. (From Kimura, 1955*a*.)

8.4.8 may not be as useful for $p = 0.1$ as in the case $p = 0.5$. Actually, from 8.4.7, $C = 6p(1 - p)$ and this constant is small if p is near 0 or 1. Going back to the complete solution 8.4.6, the probability, Ω_t, of both A_1 and A_2

co-existing in the population in the tth generation may be obtained by integrating $\phi(p, x; t)$ with respect to x from 0 to 1. This gives

$$\Omega_t = \sum_{j=0}^{\infty} \{P_{2j}(r) - P_{2j+2}(r)\} e^{-\frac{(2j+1)(2j+2)t}{4N}}, \tag{8.4.9}$$

where $P(\cdot)$ represents the Legendre polynomials; $P_0(r) = 1$, $P_1(r) = r$ $P_2(r) = (1/2)(3r^2 - 1)$, etc. The above formula is an infinite series, but for a large t, we may use the following formula to compute the value of Ω_t:

$$\Omega_t = 6p(1-p)e^{-\frac{1}{2N}t} + 14p(1-p)(1-5p+5p^2)e^{-\frac{6}{2N}t} + \cdots. \tag{8.4.10}$$

The frequency of heterozygotes or the probability that an individual in the population is heterozygous at a given generation can also be obtained by using 8.4.6 as follows:

$$H_t = \int_0^1 2x(1-x)\phi(p, x; t)\, dx = 2p(1-p)e^{-\frac{t}{2N}}. \tag{8.4.11}$$

This shows that the frequency of heterozygotes decreases exactly at the rate of $1/(2N)$ per generation. It agrees with 7.3.7 which was obtained by an elementary method. Actually, this holds also for multiallelic cases and is independent of the number of alleles involved.

The above treatment should have made it clear that the genetic heterogeneity of a population and the heterozygosity of an individual are not only distinct conceptually but also their probabilities Ω_t and H_t are different.

The processes of change in the probability distribution of fixed classes may be obtained by using relations 8.3.21 and 8.3.23. The frequency of the class in which gene A_1 is fixed, or the probability that A_1 becomes fixed by the tth generation, is as follows:

$$\begin{aligned} f(p, 1; t) &= \frac{1}{4N}\int_0^t \phi(p, 1; \tau)\, d\tau \\ &= p + \frac{1}{2}\sum_{i=1}^{\infty} \frac{(2i+1)(1-r^2)}{i(i+1)} T_{i-1}^1(r)(-1)^i e^{-\frac{i(i+1)t}{4N}} \\ &= p + \sum_{i=1}^{\infty}(2i+1)p(1-p)F(i+2, 1-i, 2, p)(-1)^i e^{-\frac{i(i+1)t}{4N}}. \end{aligned} \tag{8.4.12}$$

Similarly, the frequency of the class in which A_1 is lost may be obtained by integrating $\phi(p, 0; \tau)/(4N)$, or more simply, by replacing p with $(1-p)$ in the above expression for $f(p, 1; t)$. With these expressions, it can be shown that

$$f(p, 0; t) + \Omega_t + f(p, 1; t) = 1$$

and, at the limit of $t = \infty$, we have

$$f(p, 0; \infty) = 1 - p, \qquad \Omega_\infty = 0, \qquad f(p, 1; \infty) = p. \tag{8.4.13}$$

We might very roughly characterize the above process of random genetic drift by the following example. If we start out with 1000 populations, each of size 100 individuals and containing 50% of gene A_1, then after about 200 generations A_1 is either fixed or lost in roughly 500 populations. In the remaining 500, the distribution of the frequency of A_1 is practically flat. When this state is reached, 1/200 of the unfixed populations become fixed for A_1 or its allele each generation from that time until eventually every population will be homogeneous for either A_1 or its allele.

In his first treatment of the process of random genetic drift, Fisher (1922) used the transformed gene frequency rather than the gene frequency itself. A main reason for this is that if the gene frequency is transformed from x to θ by the relation (angular transformation)

$$\theta = \cos^{-1}(1 - 2x), \tag{8.4.14}$$

the sampling variance of gene frequency per generation becomes roughly independent of the gene frequency, where θ in radians changes from 0 to π as x changes from 0 to 1. This may be seen as follows: From the above relation, we have

$$\delta\theta = \frac{1}{\sqrt{x(1-x)}}\,\delta x - \frac{1-2x}{4\{x(1-x)\}^{\frac{3}{2}}}(\delta x)^2 + \cdots, \tag{8.4.15}$$

where δx is the amount of change in x per generation and $\delta\theta$ is the corresponding change in θ. Then, if we note that

$$M_{\delta x} = E(\delta x) = 0, \qquad V_{\delta x} = E\{(\delta x)^2\} = \frac{x(1-x)}{2N}, \tag{8.4.16}$$

we obtain, after neglecting higher order terms,

$$\begin{cases} M_{\delta\theta} = -\dfrac{1}{4N}\cot\theta \\ V_{\delta\theta} = \dfrac{1}{2N} \end{cases} \tag{8.4.17}$$

It follows then that if a population starts from a fixed gene frequency p, the variance of θ after t generations is given approximately by

$$V_\theta(t) = \frac{t}{2N}. \tag{8.4.18}$$

We should note here that the above formula for V_θ is valid only when t is much smaller than N. Also, we should note that the expected value of $\delta\theta$ now depends on θ, since in calculating $E(\delta\theta)$ from 8.4.15, the second term in the right-hand side is nonzero and cannot be neglected even if the first term is zero. Fisher (1922) neglected $M_{\delta\theta}$ and wrote the differential equation for the probability distribution using only $V_{\delta\theta} = 1/2N$. This led him to the incorrect result of $1/4N$ as the rate of steady decay. Later (1930) he incorporated $M_{\delta\theta} = -(\cot\theta)/4N$ into the equation to obtain the correct result.

Nevertheless, this type of transformation which makes sampling variance nearly constant is rather convenient for treating data on random genetic drift over a relatively short period or if the gene frequency is restricted to a range not very far from 0.5.

So far we have considered a pair of alleles A_1 and A_2. With more than two alleles the situation is of course more complex, but the same principles apply. Figure 8.4.3 shows the steady-state situation for three alleles. At this

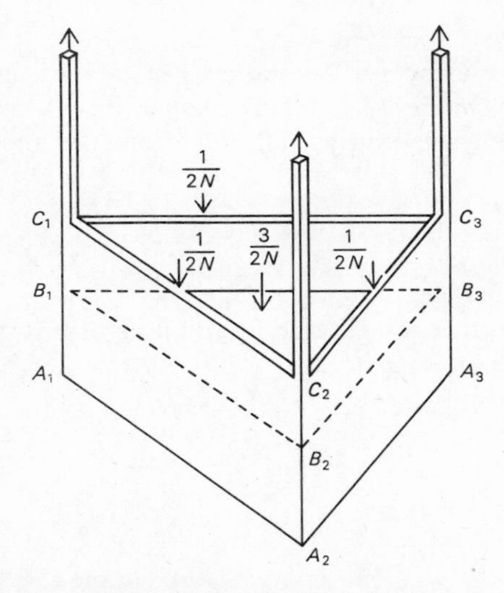

Figure 8.4.3. The distribution for three alleles at a steady state under random drift. (From Kimura, 1955*b*).

state the rate of change from populations with three alleles to populations with two alleles is $3/2N$ per generation. Then each of the populations with two alleles changes to a population with one allele at a rate of $1/2N$, as shown above. At the same time the rate of fixation is $1/2N$ for each of the three alleles, so that every generation $3/2N$ of the populations becomes fixed for one of the three alleles. The extension to more than three alleles follows naturally. The rate of change, when a steady state has been reached, from k alleles to $k-1$ alleles is $k(k-1)/4N$ per generation, where N is the effective population number. This result is from Kimura (1955*b*).

The problem becomes much more difficult if we consider two linked loci that are segregating simultaneously. Although a theory comparable to that of single locus has not been developed, the amount of linkage disequilibrium caused by random sampling of gametes in a finite population has been clarified by Hill and Robertson (1968) and also by Ohta and Kimura (1969). Let us assume that a pair of alleles A_1 and A_2 are segregating in the first locus, and B_1 and B_2 in the second locus. If we denote by X_1, X_2, X_3, and X_4 the respective frequencies of the 4 types of chromosomes A_1B_1, A_1B_2, A_2B_1, and A_2B_2, then $D = X_1X_4 - X_2X_3$ represents the amount of linkage disequilbrium. Hill and Robertson (1968) showed that in a small population $E(D^2)$, that is the mean square of D, may become large even if $E(D)$, the mean value of D, is 0. Using the method of moment generating matrix, they obtained analytical expressions for the 3 quantities, $E\{x(1-x)y(1-y)\}$, $E\{D(1-2x)(1-2y)\}$, and $E(D^2)$ in the case of no crossing-over, where x and y are respectively the frequencies of A_1 and B_1 in the first and second loci. Ohta and Kimura (1969) obtained more general expressions for an arbitrary recombination fraction c, based on the diffusion models. An interesting property first discovered by Hill and Robertson is that the expectation of $r^2 = D^2/\{x(1-x)y(1-y)\}$ settles down quickly in the process of random drift to a constant value which depends only on N_ec. Note that r is the correlation coefficient between gene frequencies at two segregating loci. They also inferred from simulation studies that $E(r^2)$ approaches $1/(4N_ec)$ as N_ec increases. Ohta and Kimura (1969) considered a quantity $\sigma_d^2 = E(D^2)/E\{x(1-x)y(1-y)\}$ and showed that it takes a value similar to $E(r^2)$. They obtained an analytical expression for σ_d^2 and showed that

$$\sigma_d^2 \approx 1/(4N_ec)$$

for a large N_ec.

It is interesting to note that a relation similar to this holds for the case in which a steady state is reached with recurrent mutations and random genetic drift as shown by Ohta and Kimura (1969*a*).

8.5 Change of Gene Frequency Under Linear Pressure and Random Sampling of Gametes

In the previous section, we have considered the process of random drift caused by random sampling of gametes alone. We will now investigate the process in which the effects of mutation and migration are also included.

Since the effects of mutation and migration on the rate of change in gene frequency are linear functions of the gene frequency, we may call them collectively linear pressure.

Let us consider a random-mating population of effective size N_e in which a pair of alleles A_1 and A_2 are segregating with respective frequencies x and $1 - x$. If we suppose that this population exchanges individuals with another population at the rate m per generation, then the rate of change in x due to this cause is

$$m(x_I - x)$$

per generation, where x_I is the frequency of A_1 in the immigrants (cf. 6.5.1). Here we will assume that x_I is a constant. This may be a good approximation if the immigrants represent a random sample from the entire species. If mutation rates are not negligible, we may replace m by $m + u + v$ and mx_I by $mx_I + v$, where u and v are respectively the mutation rates of A_1 to and from its allele A_2. Though the pressure of selection is intrinsically nonlinear, in certain cases, like selection acting at the neighborhood of the equilibrium gene frequency, it may be treated as if it were linear with good approximation. However, the range of applicability is quite restricted.

The change of mean, variance, and the higher moments of the distribution of gene frequency under linear pressure and random sampling of gametes can be worked out by applying the method we used in Chapter 7, Section 4 for the case of random sampling of gametes alone. In the present case we note that

$$E_\delta(\delta x_t) = m(x_I - x_t) \tag{8.5.1}$$

rather than $E_\delta(\delta x_t) = 0$ as in 7.4.5, where x_t is the value of x at the tth generation. For the mean, we obtain

$$\mu_1'^{(t+1)} = (1 - m)\mu_1'^{(t)} + mx_I,$$

which leads to

$$\frac{d\mu_1'^{(t)}}{dt} = -m(\mu_1'^{(t)} - x_I)$$

for the continuous model. The solution of the above equation gives the mean gene frequency at the tth generation:

$$\mu_1'^{(t)} = x_I + (p - x_I)e^{-mt}, \tag{8.5.2}$$

where p is the value of x at $t = 0$. Similarly, we can work out the second and the higher moments step by step. The general formula for the nth moment of the gene frequency around the origin is as follows (Crow and Kimura, 1956):

$$\mu_n'^{(t)} = \sum_{i=0}^{\infty} \binom{n}{i} \frac{\Gamma(B+n)\Gamma(A+2i)\Gamma(A-B+i)\Gamma(A+i-1)}{\Gamma(A+n+i)\Gamma(B+i)\Gamma(A-B)\Gamma(A+2i-1)}$$
$$\times F(A+i-1, -i, A-B, 1-p) \exp\left\{-i\left(m+\frac{i-1}{4N_e}\right)t\right\} \quad \text{8.5.3}$$

In the above equation, $A = 4N_e m$, $B = 4N_e m x_I$ $(1 > x_I > 0)$, p is the initial frequency of A_1, and $F(\cdot, \cdot, \cdot, \cdot)$ denotes the hypergeometric function.

Since in this case the mean and variance of the rate of change in gene frequency x are respectively

$$M_{\delta x} = m(x_I - x) \quad \text{8.5.4}$$

and

$$V_{\delta x} = \frac{x(1-x)}{2N_e}, \quad \text{8.5.5}$$

the forward equation 8.3.1 becomes

$$\frac{\partial \phi}{\partial t} = \frac{1}{4N_e}\frac{\partial^2}{\partial x^2}\{x(1-x)\phi\} - m\frac{\partial}{\partial x}\{(x_I - x)\phi\}, \quad \text{8.5.6}$$

where $\phi \equiv \phi(p, x; t)$ is the probability density that the frequency of A_1 becomes x at the tth generation, given that it is p at $t = 0$. The initial condition for the equation is

$$\phi(p, x; 0) = \delta(x - p). \quad \text{8.5.7}$$

The moment formula 8.5.3 suggests that the solution to the above equation must have the form

$$\phi(p, x; t) = \sum_{i=0}^{\infty} X_i(x) \exp\left\{-i\left(m + \frac{i-1}{4N_e}\right)t\right\}. \quad \text{8.5.8}$$

By comparing

$$\mu_n'^{(t)} = \int_0^1 x^n \phi(p, x; t)\, dx$$

with 8.5.3 we can get the appropriate expression for ϕ, which turns out to be the pertinent solution of 8.5.6. It is given by 8.5.8 in which

$$X_i(x) = x^{B-1}(1-x)^{(A-B)-1} F(A+i-1, -i, A-B, 1-x)$$
$$\times F(A+i-1, -i, A-B, 1-p)$$
$$\times \frac{\Gamma(A-B+i)\Gamma(A+2i)\Gamma(A+i-1)}{i!\,\Gamma^2(A-B)\Gamma(B+i)\Gamma(A+2i-1)} \quad \text{8.5.9}$$

At $t \to \infty$, our formula 8.5.8 converges to Wright's well-known formula for the steady-state gene frequency distribution under migration:

$$\phi(p, x; \infty) = \frac{\Gamma(4Nm)}{\Gamma(4Nmx_I)\Gamma(4Nm(1-x_I))} x^{4Nmx_I - 1}(1-x)^{4Nm(1-x_I)-1}. \tag{8.5.10}$$

Figures 8.5.1*a*, 8.5.1*b*, and 8.5.1*c* show the asymptotic behavior of the distribution curve for three different cases: $4N_e m = 0.2$, $4N_e m = 2$, and $4N_e m = 6$. In all the three cases illustrated the gene frequency, x_I, of the immigrants is 0.5 and the initial gene frequency, p, of the population is assumed to be 0.2. We will study the nature of steady-state distribution (8.5.10) in some detail in the next chapter. In the above treatment, it has been assumed that x_I is neither 0 nor 1. In terms of mutation pressure alone, this corresponds to the case of reversible mutation for which $u > 0$ and $v > 0$.

Next, we will investigate the case of irreversible mutation. Let us assume that A_2 mutates to A_1 at the rate v per generation ($v > 0$) but there is no mutation in the reverse direction ($u = 0$). If x_t is the frequency of A_1 at the tth generation, then the amount of change in x_t in one generation is

$$\delta x_t = v(1 - x_t) + \xi_t, \tag{8.5.11}$$

where ξ_t is the amount of change due to random sampling of gametes with mean and variance

$$E_\delta(\xi_t) = 0, \qquad E_\delta(\xi_t^2) = \frac{x_t(1-x_t)}{2N_e}. \tag{8.5.12}$$

Using the same procedure we used to derive 8.5.3, as well as 7.4.37 in the previous chapter, we obtain the following formula for the nth moment of the gene frequency distribution about origin in the tth generation:

$$\begin{aligned} \mu_n'^{(t)} = 1 - (1-p) \sum_{i=1}^{\infty} \binom{n}{i} (-1)^{i-1} \frac{\Gamma(c+n)\Gamma(c+i-1)(c+2i-1)}{\Gamma(c)\Gamma(c+n+i)} \\ \times F(1-i, i+c, c, p) \exp\left\{-i\left(v + \frac{i-1}{4N_e}\right)t\right\}. \end{aligned} \tag{8.5.13}$$

where $c = 4N_e v$, and $\exp\{\cdot\}$, $\Gamma(\cdot)$, and $F(\cdot, \cdot, \cdot, \cdot)$, respectively, denote the exponential, the gamma, and the hypergeometric functions.

In particular, the first two moments are

$$\mu_1'^{(t)} = 1 - (1-p)e^{-vt} \tag{8.5.14}$$

and

$$\mu_2'^{(t)} = 1 - (1-p)\frac{2c+2}{c+2} e^{-vt} - (1-p)\left(p - \frac{c}{c+2}\right) e^{-2\left(v + \frac{1}{4N_e}\right)t}. \tag{8.5.15}$$

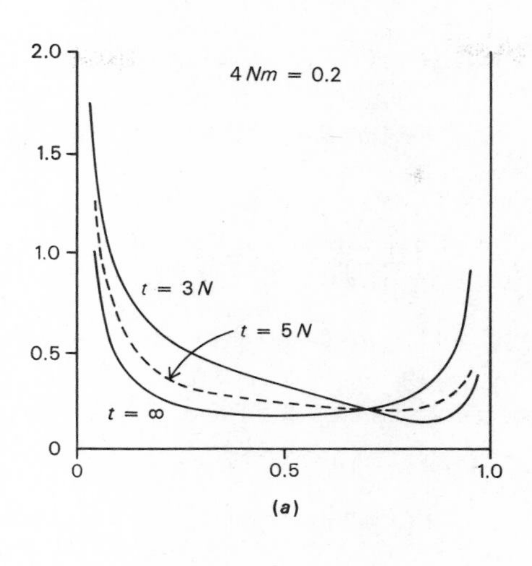

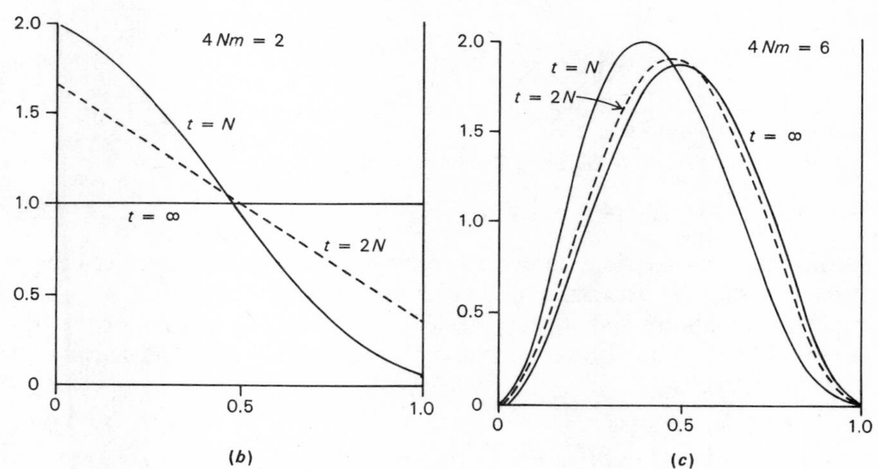

Figures 8.5.1*a,b,c*. Asymptotic behavior of distribution curves for a finite population with migration or other linear pressures. In all three drawings, the gene frequency of the immigrants is assumed to be 0.5, and the initial frequency in the population 0.2. The abscissa is the gene frequency x; the ordinate is the probability density ϕ. N: population number. m: rate of migration. (From Crow and Kimura, 1956.)

The gene frequency distribution $\phi(p, x; t)$ which satisfies the forward equation

$$\frac{\partial \phi}{\partial t} = \frac{1}{4N_e} \frac{\partial^2}{\partial x^2} \{x(1-x)\phi\} - v \frac{\partial}{\partial x} \{(1-x)\phi\} \tag{8.5.16}$$

can be worked out as in the case of reversible mutation. The pertinent solution is as follows.

$$\begin{aligned}\phi(p, x; t) &= \sum_{i=0}^{\infty} \frac{(c+1+2i)\Gamma(c+1+i)\Gamma(c+i)}{i!(i+1)![\Gamma(c)]^2} \\ &\quad \times (1-p)F(-i, i+c+1, c, p) \\ &\quad \times x^{c-1}F(-i, i+c+1, c, x)e^{-\frac{(i+1)(c+i)t}{4N_e}} \\ &= c(c+1)(1-p)x^{c-1}e^{-vt} \\ &\quad + \frac{c^2(c+1)(c+3)}{2}(1-p)\left(1 - \frac{c+2}{c}p\right)x^{c-1} \\ &\quad \times \left(1 - \frac{c+2}{c}x\right)e^{-2\left(v+\frac{1}{4N_e}\right)t} + \cdots,\end{aligned} \tag{8.5.17}$$

where $c = 4N_e v$ and $0 < x < 1$.

For a very large t, we have the asymptotic formula

$$\phi(p, x; t) \sim 4N_e v(4N_e v + 1)(1-p)x^{4N_e v - 1}e^{-vt}. \tag{8.5.18}$$

At this state, the distribution curve for intermediate gene frequencies decays at the rate of v per generation, as shown by Wright (1931).

The probability that A_1 becomes fixed by the tth generation or the frequency of the terminal class with $x = 1$ can be obtained by letting $n \to \infty$ in $\mu_n'^{(t)}$. This yields

$$\begin{aligned}f(1, t) &= 1 - (1-p)\sum_{i=1}^{\infty}(-1)^{i-1}\frac{\Gamma(i-1+c)(c+2i-1)}{\Gamma(c)i!} \\ &\quad \times F(1-i, i+c, c, p)\exp\left\{-i\left(v + \frac{i-1}{4N_e}\right)t\right\} \\ &= 1 - (c+1)(1-p)e^{-vt} - \frac{(c+2)(c+3)}{2}(1-p) \\ &\quad \times \left(p - \frac{c}{2+c}\right)e^{-2\left(v+\frac{1}{4N_e}\right)t} - \cdots.\end{aligned} \tag{8.5.19}$$

For a large t, we have

$$f(1, t) \sim 1 - (4N_e v + 1)(1 - p)e^{-vt} \tag{8.5.20}$$

showing that A_1 becomes irreversibly fixed in the population. Also, from 8.5.17 and 8.5.19, we note that

$$f(1, t) + \int_0^1 \phi(p, x; t)\, dx = 1.$$

Figure 8.5.2 illustrates the probability distribution of intermediate gene frequencies for $c = 4N_e v = 0.5$ at $t = 12N_e$, assuming $p = 0$.

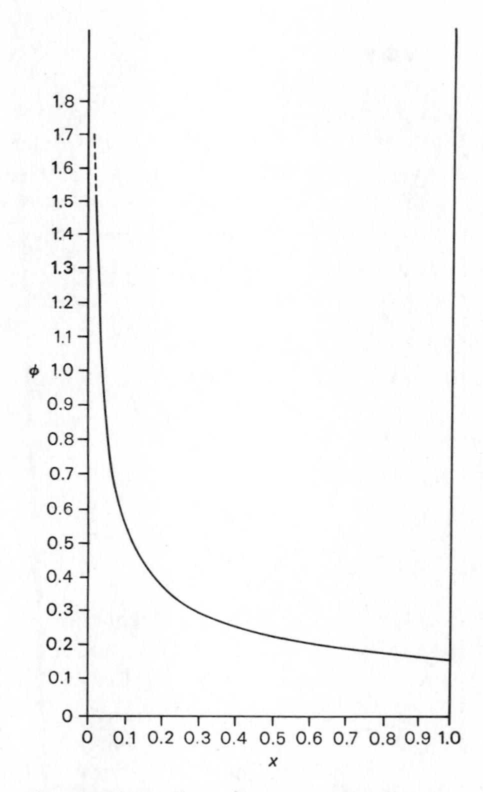

Figure 8.5.2. Gene frequency distribution under irreversible mutation for the case of $c = 4N_e v = 0.5$, $t = 12N_e$, and $p = 0$.

8.6 Change of Gene Frequency Under Selection and Random Sampling of Gametes

For our understanding of the process of evolution and breeding, it is quite important to clarify the process by which gene frequencies change in a finite population under selection. However, because of the mathematical difficulties involved, so far only the simplest case has been solved completely. The main difficulty stems from the fact that under selection the rate of change in gene frequency is a nonlinear function of the gene frequency. In the present section we will summarize the main results obtained by Wright, Kimura, and Robertson.

8.6.1 Genic Selection

Let us consider the simplest case of no dominance, where the heterozygote has fitness midway between the fitnesses of the two homozygotes. We will assume that a pair of alleles A_1 and A_2 are segregating with respective frequencies x and $1 - x$ in a finite population of effective size N_e. If we denote by s the selective advantage of A_1 over A_2 or, more precisely, if we assume that the selective advantage of A_1 over A_2 for a short time interval of length δt is $s\delta t$, then the mean rate of change in x per generation is given by

$$M_{\delta x} = sx(1 - x). \qquad \textbf{8.6.1.1}$$

The variance of the rate of change in x is

$$V_{\delta x} = \frac{x(1 - x)}{2N_e} \qquad \textbf{8.6.1.2}$$

Thus, the forward equation 8.3.1 becomes

$$\frac{\partial \phi}{\partial t} = \frac{1}{4N_e} \frac{\partial^2}{\partial x^2} \{x(1 - x)\phi\} - s \frac{\partial}{\partial x} \{x(1 - x)\phi\} \qquad (0 < x < 1), \qquad \textbf{8.6.1.3}$$

where $\phi \equiv \phi(p, x; t)$ is the probability density that the frequency of A_1 becomes x at the tth generation, given that it is p at $t = 0$. This equation has been used to analyze the gene frequency change in very small experimental populations of *Drosophila melanogaster* (Wright and Kerr, 1954). In that paper, Wright devised an ingenious method to analyze the process of steady decay in which the distribution curve, while keeping a constancy in form, decreases in height at a constant rate determined by the smallest eigenvalue (λ_0) of equation 8.6.1.3. The complete solution of the problem has been given by Kimura (Kimura, 1955; Crow and Kimura, 1956).

In order to solve equation 8.6.1.3, let us put

$$\phi \propto e^{2cx} V(x) e^{-\lambda t},$$

where $c = N_e s$ and $V(x)$ is a function of x but not of t. If we substitute this in 8.6.1.3, we have

$$x(1-x)\frac{d^2V}{dx^2} + 2(1-2x)\frac{dV}{dx} - \{2 + 4c^2x(1-x) - 4N_e\lambda\}V = 0. \quad \text{8.6.1.4}$$

Then by the substitution

$$x = \frac{(1-z)}{2}, \quad \text{8.6.1.5}$$

equation 8.6.1.4 becomes

$$(1-z^2)\frac{d^2V}{dz^2} - 4z\frac{dV}{dz} + \{(4N_e\lambda - 2 - c^2) + c^2z^2\}V = 0, \quad \text{8.6.1.6}$$

where $z = 1 - 2x$ $(-1 < z < 1)$. This type of differential equation is known as the oblate spheroidal equation. Actually, 8.6.1.6 is a special case of the oblate spheroidal wave equation,

$$(1-z^2)V'' - 2(m+1)zV' + (b + c^2z^2)V = 0, \quad \text{8.6.1.7}$$

with $m = 1$ and $b = 4N_e\lambda - 2 - c^2$. We want here the solutions which are finite at the singularities, $z = \pm 1$, and reduce to the Gegenbauer polynomial if there is no selection, that is, if $s = 0$ and therefore $c = 0$. This is because if there is no selection the process reduces to that of random drift which was discussed in 8.4. Such a solution has been studied by Stratton *et al.* (1941) and is expressed in the form

$$V_{1k}^{(1)}(z) = \sum_{n=0,\,1}{}' f_n^k T_n^1(z). \quad \text{8.6.1.8}$$

where $k = 0, 1, 2, \ldots$ (corresponding to l in the notation of Stratton *et al.*). In the above expression, f_n^k's are constants and $T_n^1(z)$'s are the Gegenbauer polynomials defined in 8.4.5. The primed summation is over even values of n if k is even, odd values of n if k is odd.

The desired solution of 8.6.1.3 is given by summing $V_{1k}^{(1)}(z)$ for all possible values of k, after having multiplied through by $e^{-2cx-\lambda_k t}$, where λ_k is the kth eigenvalue.

$$\phi(p, x; t) = \sum_{k=0}^{\infty} C_k e^{-\lambda_k t + 2cx} V_{1k}^{(1)}(z) \quad \text{8.6.1.9}$$

The coefficients, C_k's, can be obtained from the initial condition

$$\phi(p, x; 0) = \delta(x - p), \qquad \text{8.6.1.10}$$

by using the orthogonal relations of the eigenfunctions;

$$\int_{-1}^{1} (1 - z^2) V_{1k}^{(1)}(z) V_{1l}^{(1)}(z)\, dz = \delta_{kl} 2 \sum_{n=0,1}' (f_n^k)^2 \frac{(n+2)!}{n!(2n+3)}.$$

They are given by

$$C_k = \frac{(1 - r^2) e^{-c(1-r)} V_{1k}^{(1)}(r)}{\sum_{n=0,1}' \dfrac{(n+1)(n+2)}{(2n+3)} (f_n^k)^2}, \qquad \text{8.6.1.11}$$

where $r = 1 - 2p$, $c = N_e s$, and the primed summation is over even values of n if k is even, over odd values of n if k is odd.

The solution 8.6.1.9 with coefficients given by 8.6.1.11 gives the probability distribution of unfixed classes in the tth generation. As t increases, the exponential terms in 8.6.1.9 decrease in absolute value very rapidly, and for large t only the first few terms are important. The numerical values of the first few eigenvalues λ_0, λ_1, and λ_2 can be obtained from the tables of the separation constants ($B_{1,k}$) which are listed in the book of Stratton *et al.* (1941). This is done by using the relation

$$4N\lambda_k = c^2 - B_{1,k}.$$

The boundaries $x = 0$ and $x = 1$ act as absorbing barriers and the smallest eigenvalue λ_0 gives the ultimate rate of decay of the frequency-distribution curve for unfixed classes; that is,

$$\lim_{t \to \infty} \frac{1}{\phi} \frac{d\phi}{dt} = -\lambda_0$$

and has special importance. It is equivalent to K in Wright and Kerr (1954). For small values of c, the eigenvalues can be expanded into a power series in c and we obtain the following formula for λ_0 (Kimura, 1955):

$$2N\lambda_0 = 1 + \frac{2}{5} c^2 - \frac{2}{5^3 \times 7} c^4 - \frac{2^2}{3 \times 5^5 \times 7} c^6 - \frac{2 \times 31}{5^6 \times 7^3 \times 11} c^8 - \frac{17507389}{2^2 \times 3^4 \times 5^9 \times 7^4 \times 11 \times 13} c^{10} - \cdots, \qquad \text{8.6.1.12}$$

or in terms of $2c = 2N_e s$, it is also given by

$$2N_e \lambda_0 = 1 + \frac{(2N_e s)^2}{10} - \frac{(2N_e s)^4}{7{,}000} - \frac{(2N_e s)^6}{1{,}050{,}000} - 4.108 \times 10^{-9}(2N_e s)^8 - 7.869 \times 10^{-11}(2N_e s)^{10} \cdots \qquad 8.6.1.13$$

It is impressive to see that Wright's empirical formula (Wright and Kerr, 1954, p. 236) is exact up to the coefficients of the order of 10^{-9}. For the purpose of numerical calculation the above power series is not suitable except for small values of c. According to Stratton *et al.* (1941), convergence of the power series is usually slow and it proves unsatisfactory for values of $c^2 > 1$. However, in the present case ($m = 1, l = 0$ in the spheroidal function), the above series gives the right answer to three significant figures even for $c = 3$. This agrees with the statement of Wright and Kerr (1954). Figure 8.6.1.1 gives the relation between c and $2N\lambda_0$. More exact values are listed

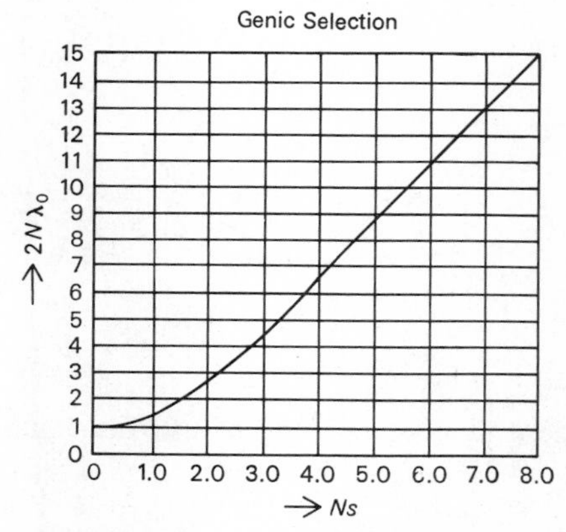

Figure 8.6.1.1. Relations between rate of steady decay (λ_0) and intensity of selection (s) in the process of genic selection in finite populations. N is effective size of population. (Kimura, 1955).

in Table 8.6.1.1. From Figure 8.6.1.1 it looks as if $2N\lambda_0$ increases linearly with c for large value of c, though having no proof we are not certain about it. The eigenfunctions $V_{1k}^{(1)}(z)$ corresponding to λ_k's are given by 8.6.1.8. The coefficients f_n^k corresponding to the first three eigenvalues are found in the tables of Stratton *et al.* (1941, pp. 116, 118, and 120). It will be noted here that for $c = 0$, all the formulae given above reduce to the ones for the case of random drift studied in 8.4.

Table 8.6.1.1. Relation between $c(=N_e s)$ and $2N\lambda_0$. (From Kimura, 1955.)

c	$2N_e\lambda_0$	c	$2N_e\lambda_0$
0.0	1.00000	3.5	5.43183
0.5	1.09985	4.0	6.54540
1.0	1.39765	4.5	7.66121
1.5	1.88771	5.0	8.75330
2.0	2.55927	6.0	10.85728
2.5	3.39445	7.0	12.89983
3.0	4.36529	8.0	14.91989

Note: In the new table of spheroidal wave functions by Stratton *et al.* (1956), t is tabulated for c (denoted as g in the table) up to $c = 8.0$ (pp. 506–508), from which $2N\lambda_0$ can be calculated by the relation

$$2N\lambda_0 = 1 + \left(\frac{2}{5} - \frac{t}{2}\right)c^2$$

The eigenfunction $V_{10}^{(1)}(z)$ corresponding to the smallest eigenvalue λ_0 is of special significance, since it gives the frequency distribution of unfixed classes at the state of steady decay, when it is multiplied by $e^{c(1-z)}$. It is expressed by

$$V_{10}^{(1)}(z) = f_0^0 T_0^1(z) + f_2^0 T_2^1(z) + f_4^0 T_4^1(z) + \cdots. \tag{8.6.1.14}$$

The coefficients f_0^0, f_2^0, f_4^0, etc., depend on c. When $c = 1.0$, for example, $f_0^0 = 1.0208, f_2^0 = 0.013980, f_4^0 = 0.000096$, etc.

Figure 8.6.1.2 illustrates the distribution at steady decay of unfixed classes for some values of c. The area under each curve is adjusted so that it is unity. The case $c = 1.7$ corresponds to the case experimentally studied by Wright and Kerr (1954) and the present result agrees quite well with theirs (including the rate of decay). The rate of fixation of A_1 may be calculated from

$$\frac{df(p, 1; t)}{dt} = \frac{\phi(p, 1; t)}{4N_e}, \tag{8.6.1.15}$$

as shown in 8.3.23. Here $f(p, 1; t)$ stands for the probability that A_1 has become fixed by the tth generation. Similarly, for the rate of loss (or fixation of A_2), we have

$$\frac{df(p, 0; t)}{dt} = \frac{\phi(p, 0; t)}{4N_e}. \tag{8.6.1.16}$$

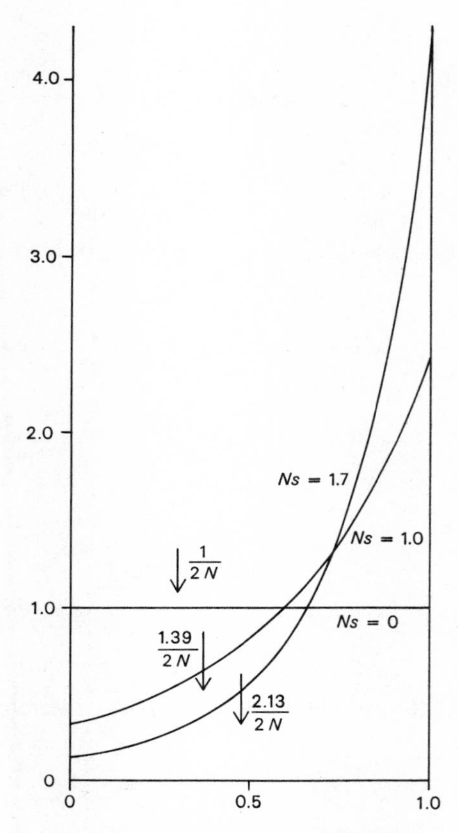

Figure 8.6.1.2. Frequency distribution of unfixed classes at the state of steady decay for various values of Ns. The area under each curve is adjusted so that it is unity. Numerals beside the arrows indicate rates of steady decay. N is effective size of population; s is selection coefficient. (From Kimura, 1955.)

8.6.2 Case of Complete Dominance

Let us suppose that A_1 is completely dominant over A_2 and the dominant genotypes A_1A_1 and A_1A_2 have selective advantage s over the homozygous recessive A_2A_2. If x is the frequency of A_1, then the mean and the variance of the rate of change in x are

$$M_{\delta x} = sx(1-x)^2 \qquad \text{8.6.2.1}$$

from 5.3.9 and

$$V_{\delta x} = \frac{x(1-x)}{2N_e}, \qquad \text{8.6.2.2}$$

respectively, so that the probability density $\phi(p, x; t)$ satisfies the differential equation (cf. 8.3.1);

$$\frac{\partial \phi}{\partial t} = \frac{1}{4N_e} \frac{\partial^2}{\partial x^2} \{x(1-x)\phi\} - s \frac{\partial}{\partial x} \{x(1-x)^2 \phi\} \qquad \text{8.6.2.3}$$

with the initial condition

$$\phi(p, x; t) = \delta(x-p). \qquad \text{8.6.2.4}$$

If we apply the transformation

$$\left.\begin{aligned} \phi &= \exp\left\{-\lambda t + 2cx\left(1 - \frac{x}{2}\right)\right\} w \\ x &= \frac{1-z}{2} \end{aligned}\right. \qquad \text{8.6.2.5}$$

to the differential equation 8.6.2.3, we obtain the ordinary differential equation

$$(1-z^2)w'' - 4zw' + \left\{\Lambda - 2 - \frac{c}{2}(z^2-1) + \frac{c^2}{4}(z^2-1)(1+z)^2\right\} w = 0, \qquad \text{8.6.2.6}$$

where $\Lambda = 4N_e \lambda$ and $c = N_e s$.

As in the previously treated case of genic selection (sec. 8.6.1), we try to expand the solution into a series of Gegenbauer polynomials. Let

$$w = \sum_{n=0}^{\infty} d_n T_n^1(z), \qquad \text{8.6.2.7}$$

where d_n's are constants. If we substitute this into 8.6.2.6 and repeatedly use the recurrence relation

$$zT_n^1(z) = \frac{n+2}{2n+3} T_{n-1}^1(z) + \frac{n+1}{2n+3} T_{n+1}^1(z),$$

in which we set

$$T_{-1}^1(z) \equiv 0,$$

we obtain a nine-term recursion formula for the d_n's. Now we expand Λ and the d_n's into power series of c:

$$\begin{aligned}
\Lambda &= k_0 + k_1 c + k_2 c^2 + k_3 c^3 + \cdots, \\
d_1 &= (\alpha_1^1 c + \alpha_2^1 c^2 + \alpha_3^1 c^3 + \cdots)\, d_0, \\
d_2 &= (\alpha_1^2 c + \alpha_2^2 c^2 + \alpha_3^2 c^3 + \cdots)\, d_0, \\
d_3 &= (\alpha_2^3 c^2 + \alpha_3^3 c^3 + \alpha_4^3 c^4 + \cdots)\, d_0, \\
d_4 &= (\alpha_2^4 c^2 + \alpha_3^4 c^3 + \cdots)\, d_0, \\
d_5 &= (\alpha_3^5 c^3 + \cdots)\, d_0, \text{ etc.,}
\end{aligned}$$

and substitute these into the recursion formula. By picking out coefficients of equal power of c, we can determine the k's and α's, by means of which the eigenvalue λ (or Λ) and the eigenfunction w are expressed. The most important information is the smallest eigenvalue (λ_0), which gives the rate of steady decay, and the corresponding eigenfunction. To get λ_0, we set $k_0 = 2$, since for $c = 0$, $\Lambda(=4N_e\lambda)$ should be 2, as shown in the treatment of pure random drift in which the final rate of decay is $1/2N_e$ (see 8.4).

Though the calculation involved is quite tedious, we can obtain the desired coefficients step by step. For the smallest eigenvalue λ_0 we get (Kimura, 1957),

$$\begin{aligned}
2N_e\lambda_0 &= 1 - \frac{1}{5}c + \frac{199}{2 \times 5^3 \times 7}c^2 + \frac{17}{2 \times 5^5 \times 7}c^3 \\
&\quad - \frac{23 \times 41 \times 29599}{2^3 \times 3^3 \times 5^6 \times 7^3 \times 11}c^4 - \cdots \qquad \textbf{8.6.2.8} \\
&= 1 - (0.2)c + (0.113714\cdots)c^2 \\
&\quad + (0.000388\cdots)c^3 - (0.002191\cdots)c^4 - \cdots.
\end{aligned}$$

The coefficients of the eigenfunction are as follows:

$$\left.\begin{aligned}
&\alpha_1^1 = 0,\ \alpha_2^1 = -\frac{1}{2 \times 3 \times 7},\ \alpha_3^1 = \frac{11}{3^4 \times 5^2 \times 7}, \\
&\alpha_1^2 = -\frac{1}{2 \times 3 \times 5^2},\ \alpha_2^2 = \frac{13}{3^2 \times 5^4},\ \alpha_3^2 \approx -7.31 \times 10^{-5}, \\
&\alpha_2^3 = \frac{1}{2 \times 3^2 \times 5 \times 7}, \ldots, \\
&\alpha_2^4 \approx 2.49 \times 10^{-4}, \text{ etc.}
\end{aligned}\right\} \qquad \textbf{8.6.2.9}$$

The same method may be applied to get similar expansions for other eigenvalues and eigenfunctions.

The gene frequency distribution at the state of steady decay is given by

$$\phi(x) = \exp\left\{2cx\left(1 - \frac{x}{2}\right)\right\}w_0. \tag{8.6.2.10}$$

It will be convenient to adjust d_0 so that

$$\int_0^1 \phi(x)\,dx = 1,$$

from which the fixed classes are excluded. The rate of fixation and loss of the gene A_1 per generation at this state is then given by $\phi(0)/4N_e$ and $\phi(1)/4N_e$ and therefore

$$4N_e\lambda_0 = \phi(0) + \phi(1). \tag{8.6.2.11}$$

This follows from 8.3.21 and 8.3.23.

A numerical example will be given here. For weak selection favoring the dominant, $N_e s = c = 1/2$; we get from 8.6.2.8,

$$2N_e\lambda_0 = 0.928$$

and from 8.6.2.9,

$$w_0 = d_0\{T_0^1(z) - 0.0058T_1^1(z) - 0.0028T_2^1(z) + 0.0004T_3^1(z) + 0.00006T_4^1(z) + \cdots\},$$

in which $T_0^1(z) = 1, T_1^1(z) = 3z, T_2^1(z) = (3/2)(5z^2 - 1), T_3^1(z) = (5/2)(7z^3 - 3z)$, etc. Values of $\phi(x)$ at 0, 0.1, 0.2, ..., and 1.0 are listed in Table 8.6.2.1 and the curve is illustrated in Figure 8.6.2.1. They are adjusted by Simpson's rule

Table 8.6.2.1. Frequency distributions at steady decay with $2N_e s = 1$ and $2N_e s = -1$. (From Kimura, 1957.)

x	$2N_e s = 1$	$2N_e s = -1$
0.0	0.688	1.389
0.1	0.764	1.251
0.2	0.838	1.142
0.3	0.910	1.056
0.4	0.977	0.990
0.5	1.037	0.940
0.6	1.088	0.903
0.7	1.128	0.879
0.8	1.155	0.865
0.9	1.168	0.860
1.0	1.166	0.866

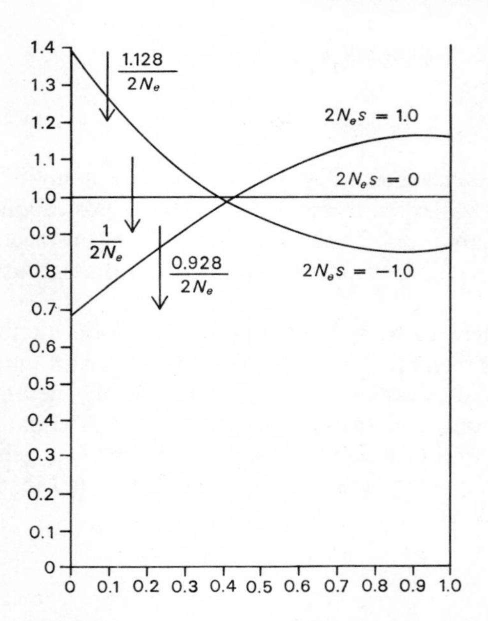

Figure 8.6.2.1. Curves giving the frequency distribution of unfixed classes at the state of steady decay for $2N_e s = 1.0$, 0, and -1.0, where s is the selective advantage of dominants (A_1A_1 and A_1A_2) over the recessive (A_2A_2). N_e is the effective size of the population. Rates of steady decay are indicated beside the arrows.

so that the area under the curve is unity. $\phi(1) + \phi(0)$ comes out 1.855, while $4N_e\lambda_0$ is 1.856. The agreement is satisfactory for this level of approximation. As a second example, we assume weak selection against the dominants: $N_e s = c = -1/2$. $2N_e\lambda_0$ is 1.128 and values of $\phi(x)$ are given in Table 8.6.2.1 and are also illustrated in Figure 8.6.2.1. In this case $\phi(1) + \phi(0)$ comes out 2.254 while $4N_e\lambda_0$ is 2.256. Again the agreement is satisfactory. A most remarkable fact shown by the above examples is that, compared with the case of pure random drift, selection favoring the dominant allele ($s > 0$) decreases the final rate of decay, while selection against the dominant allele ($s < 0$) increases it. That this is always true for weak selection follows from 8.6.2.8, since the most influential term $-1/5c$ is negative if $c(=N_e s)$ is positive, and positive if c is negative. Formula 8.6.2.8 also suggests that $2N_e\lambda_0$ has a minimum at about $c = 0.9$, when $c > 0$, after which $2N_e\lambda_0$ starts to increase with increasing c.

Miller (1962) has shown that, when c is very large,

$$N_e \lambda_0 \approx 0.718 \left(\frac{c}{2}\right)^{\frac{1}{2}}. \qquad 8.6.2.12$$

Furthermore, he worked out exact values of $N_e \lambda_0$ by numerical analysis to cover the range of $0 \leqq c \leqq 60$ (Miller's α in his Tables 1 and 2 corresponds to $c/2$ in the present notation). Table 8.6.2.2 and Figure 8.6.2.2 are constructed based on his results. (In Figure 8.6.2.2 the curve for small c is drawn based on the expansion 8.6.2.8.)

For the continuous treatment to be applicable, the population number should be fairly large so that its reciprocal is negligible compared with unity. If the population is extremely small, we must treat the problem by the methods of finite Markov chains. Assuming that the effective number N_e is equal to the actual number N, the transition probability that the number of A_1 genes in the population becomes j in the next generation, given that it is i in the present generation, will be given by

$$P_{j/i} = \binom{2N}{j} x'^j (1 - x')^{2N-j} \qquad (i, j = 0, 1, \ldots, 2N), \qquad 8.6.2.13$$

where $x' = x + \delta x$, in which $x = i/2N$ and δx is the change of gene frequency by selection per generation and is $sx(1 - x)^2$ if s is small. The rate of decay

Table 8.6.2.2. Relation between $c(=N_e s)$ and $2N_e \lambda_0$ in the case of complete dominance, where s is the selection coefficient for the dominant. This table is constructed based on the values given by Miller (1962) in his Tables 1 and 2.

c	$2N_e \lambda_0$	c	$2N_e \lambda_0$
0	1.0000	20	4.3886
1	0.9120	22	4.6332
2	1.0244	24	4.8668
3	1.2688	28	5.3058
4	1.5586	32	5.7142
6	2.0900	36	6.0972
8	2.5300	40	6.4594
10	2.9118	44	6.8038
12	3.2542	48	7.1326
14	3.5676	52	7.4480
16	3.8586	56	7.7512
18	4.1312	60	8.0438

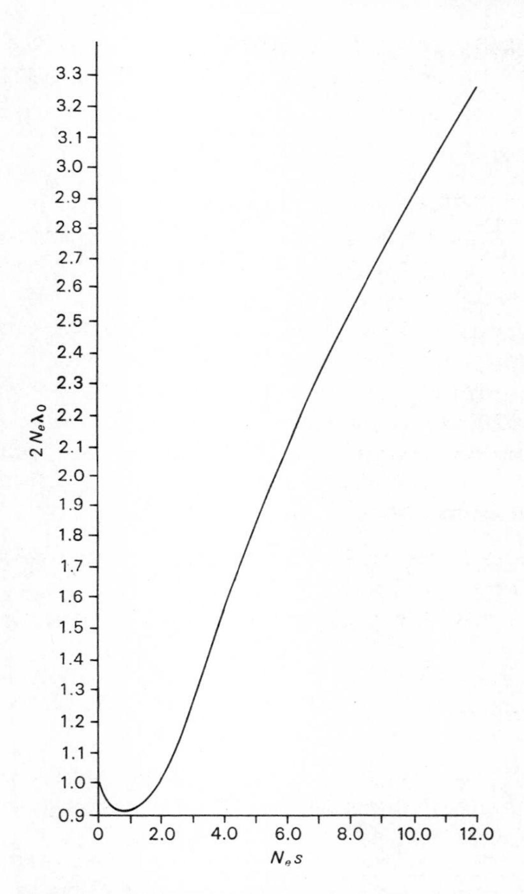

Figure 8.6.2.2. Relation between the rate of steady decay (λ_0) and selection coefficient (s) in the case of complete dominance. N_e is effective population size.

of the unfixed classes and their limiting distribution can be obtained by iteration. For example, if $N = 4$ and $2Ns = 1$, the limiting form of the distribution (fixed classes excluded) is shown in Table 8.6.2.3, with rate of decay (λ_0) 11.875%, giving $2N\lambda_0 = 0.9500$. If there is no selection ($s = 0$), it turns out that the rate of decay becomes $1/2N = 0.125$ or $2N\lambda_0 = 1.0$. It may be interesting to note that with this magnitude of selection for dominants, the rate of decay is smaller than in the case of no selection. This agrees with the result of continuous treatment.

Table 8.6.2.3. Frequency distribution of unfixed classes at steady decay with $N = 4$, $2Ns = 1$, where s is the selection coefficient for dominants. (From Kimura, 1957.)

CLASSES	FREQUENCIES %
$7A_1 + 1A_2$	14.80
$6A_1 + 2A_2$	16.48
$5A_1 + 3A_2$	16.32
$4A_1 + 4A_2$	15.51
$3A_1 + 5A_2$	14.28
$2A_1 + 6A_2$	12.64
$1A_1 + 7A_2$	9.97
TOTAL	100.00

8.6.3 Arbitrary Degree of Dominance

We will consider a more general case in which the selective advantages of A_1A_1 and A_1A_2 over A_2A_2 are s and sh respectively.

The partial differential equation corresponding to 8.6.2.3 in the foregoing section is now

$$\frac{\partial \phi}{\partial t} = \frac{1}{4N_e} \frac{\partial^2}{\partial x^2} \{x(1-x)\phi\} - s \frac{\partial}{\partial x} \{(h + (1-2h)x)x(1-x)\phi\}. \quad 8.6.3.1$$

As in the previous cases, the rate of steady decay is given by the smallest eigenvalue λ_0, which can be expanded into a power series of c if c is small (Kimura, 1957):

$$2N_e \lambda_0 = 1 + K_1 c + K_2 c^2 + K_3 c^3 + K_4 c^4 + \cdots, \quad 8.6.3.2$$

where

$$K_1 = -\frac{1}{5}D, \; K_2 = \frac{1}{2 \times 5} + \frac{2^2 \times 3}{5^3 \times 7} D^2,$$

$$K_3 = \frac{1}{2 \times 5^3 \times 7} D - \frac{2^2}{5^5 \times 7} D^3,$$

$$K_4 = -\frac{1}{2^3 \times 5^3 \times 7} - \frac{7^3}{2 \times 3^3 \times 5^5} D^2 - \frac{2^2 \times 3^5}{5^6 \times 7^3 \times 11} D^4, \text{ etc.},$$

$$c = N_e s, \qquad \text{and} \qquad D = 2h - 1.$$

It may be easily seen that the above formula reduces to 8.6.1.12 if there is no dominance ($h = 1/2$), and therefore $D = 0$, provided that $2s$ is used instead of s to express the selective advantage of A_1A_1. For the case of complete dominance, $D = 1$ or -1 according to whether the mutant gene is dominant or recessive. In the former case of $D = 1$, 8.6.3.2 agrees with 8.6.2.8.

The above notation for expressing the selective advantage of A_1A_1 and A_1A_2 over A_2A_2 as s and sh may not be very convenient for overdominant genes, in which case it is preferable to take the heterozygote as standard and denote the selective advantages of two homozygotes, A_1A_1 and A_2A_2 as $-s_1$ and $-s_2$. The latter notation is related to the former by

$$\left.\begin{array}{l} s_1 = s(h-1) \\ s_2 = sh, \end{array}\right\} \tag{8.6.3.3}$$

$$\left.\begin{array}{l} s = s_2 - s_1 \\ D = 2h - 1 = \dfrac{s_1 + s_2}{s_2 - s_1} \end{array}\right\}. \tag{8.6.3.4}$$

Thus in the symmetric case of $s_1 = s_2 \equiv \bar{s}$, $c = N_e s = 0$ and $Dc = N_e sD = N_e(s_1 + s_2) = 2N_e\bar{s}$, and therefore 8.6.3.2 reduces to

$$\begin{aligned} 2N_e\lambda_0 &= 1 - \frac{2}{5}(N_e\bar{s}) + \frac{2^4 \times 3}{5^3 \times 7}(N_e\bar{s})^2 \\ &\quad - \frac{2^5}{5^5 \times 7}(N_e\bar{s})^3 - \frac{2^6 \times 3^5}{5^6 \times 7^3 \times 11}(N_e\bar{s})^4 - \cdots \\ &= 1 - 0.4(N_e\bar{s}) + (0.05485\cdots)(N_e\bar{s})^2 \\ &\quad - (0.001462\cdots)(N_e s)^3 - (0.000263\cdots)(N_e s)^4 - \cdots \end{aligned} \tag{8.6.3.5}$$

The power series expansions 8.6.3.2 and 8.6.3.5 are only valid for small value of $N_e(s_1 + s_2)$. On the other hand, for the case of overdominant genes, knowledge of the rate of decay for fairly large values of $N_e(s_1 + s_2)$ is required, and an important contribution has been made by Miller (1962) in evaluating the smallest eigenvalues for such a case. His work together with that of Robertson (1962) will be presented briefly in the following material.

8.6.4 Overdominant Case

Consider a pair of overdominant alleles A_1 and A_2; if we designate by s_1 and s_2 the selection coefficients against the homozygotes A_1A_1 and A_2A_2 in such a way that the rate of change of the frequency of A_1 by selection is given by

$$\{s_2 - (s_1 + s_2)x\}x(1 - x),$$

then the equation corresponding to 8.6.3.1 in the previous section becomes

$$\frac{\partial \phi}{\partial t} = \frac{1}{4N_e} \frac{\partial^2}{\partial x^2} \{x(1-x)\phi\} - \frac{\partial}{\partial x} \{[s_2 - (s_1 + s_2)x]x(1-x)\phi\}, \quad 8.6.4.1$$

where x is the frequency of A_1.

Let $\bar{s}$ be the average of the two selection coefficients,

$$\frac{s_1 + s_2}{2} = \bar{s}, \quad 8.6.4.2$$

and let $\hat{x}$ be the equilibrium frequency of A_1 in an infinite population,

$$\frac{s_2}{s_1 + s_2} = \hat{x}. \quad 8.6.4.3$$

Then equation 8.6.4.1 may be expressed as

$$4N_e \frac{\partial \phi}{\partial t} = \frac{\partial^2}{\partial x^2} \{x(1-x)\phi\} - 4N_e \bar{s} \frac{\partial}{\partial x} \{2(\hat{x} - x)x(1-x)\phi\} \quad 8.6.4.4$$

or, denoting $2\hat{x} - 1$ by $\hat{z}$, i.e.,

$$\hat{x} = (\hat{z} + 1)/2, \quad 8.6.4.5$$

we have

$$4N_e \frac{\partial \phi}{\partial t} = \frac{\partial^2}{\partial x^2} \{x(1-x)\phi\} + 4N_e \bar{s} \frac{\partial}{\partial x} \{(2x - 1 - \hat{z})x(1-x)\phi\}. \quad 8.6.4.6$$

The smallest eigenvalue, λ_0, of the above equation has been worked out by Miller (1962). Because of the symmetry involved between a pair of alleles, it is sufficient to study the case of

$$\hat{x} \geqq 0.5$$

or

$$\hat{z} \geqq 0.$$

For a large value of $c \equiv N_e \bar{s}$, and for the range of $1 > \hat{z} \geqq 0$, Miller obtained the asymptotic expansion

$$2N_e \lambda_0 = \frac{c}{T(c, \hat{z})} \left\{ \frac{(1-\hat{z})e^{-c(1-\hat{z})^2}}{S\{c(1-\hat{z})^2\}} + \frac{(1+\hat{z})e^{-c(1+\hat{z})^2}}{S\{c(1+\hat{z})^2\}} \right\}, \quad 8.6.4.7$$

where

$$S(X) = 1 + \frac{1}{2}\frac{1}{X} + \frac{1 \times 3}{2^2}\frac{1}{X^2} + \cdots \quad 8.6.4.8$$

and

$$T(c, \hat{z}) = \sum_{i=0}^{\infty} \frac{C_{2i}\Gamma(i + \frac{1}{2})}{c^{i+\frac{1}{2}}} = \frac{C_0}{\sqrt{c}}\sqrt{\pi} + \cdots \quad 8.6.4.9$$

in which

$$c = N_e \bar{s} = \frac{N_e(s_1 + s_2)}{2} \quad 8.6.4.10$$

and C_i's are given by the recurrence relation

$$(1 - \hat{z}^2)C_{i+1} = 2\hat{z}C_i + C_{i-1}, \quad 8.6.4.11$$

starting from

$$C_0 = \frac{1}{1 - \hat{z}^2}, \quad C_1 = \frac{2\hat{z}}{(1 - \hat{z}^2)^2}. \quad 8.6.4.12$$

In the special case of $\hat{z} = 0$ (i.e., $s_1 = s_2$), we get

$$2N_e \lambda_0 = \frac{2c^{\frac{3}{2}}e^{-c}}{\sqrt{\pi}\{S(c)\}^2}. \quad 8.6.4.13$$

Table 8.6.4.1. Values of $2N_e\lambda_0$ as a function of $c(=N_e(s_1 + s_2)/2)$ and $\hat{x}$. This table is based on the values given by Miller (1962) in his Table 1.

c \ $\hat{x}$	0.5	0.6	0.7	0.75	0.8	0.9
0.0	1.0000	1.0000	1.0000	1.0000	1.0000	1.0000
0.5	0.8136	0.8174	0.8292	0.8382	0.8490	0.8766
1.0	0.6532	0.6682	0.7130	0.7466	0.7876	0.8916
1.5	0.5172	0.5482	0.6408	0.7096	0.7932	1.0038
2.0	0.4038	0.4528	0.5982	0.7060	0.8360	1.1590
3.0	0.2358	0.3154	0.5516	0.7258	0.9352	1.4532
4.0	0.12984	0.2230	0.5082	0.7256	0.9930	1.6730
5.0	0.06770	0.15656	0.4550	0.6990	1.0110	1.8374
6.0	0.03362	0.10780	0.3964	0.6558	1.0032	1.9648
7.0	0.016004	0.07242	0.3384	0.6044	0.9790	2.0656
8.0	0.007362	0.04750	0.2840	0.5498	0.9444	2.1460
9.0	0.003294	0.03046	0.2352	0.4948	0.9030	2.2106
10.0	0.0014410	0.01918	0.19230	0.4414	0.8578	2.2622
11.0	0.0006192	0.01186	0.15562	0.3910	0.8100	2.3032
12.0	0.0002622	0.00722	0.12472	0.3440	0.7614	2.3352

Also, if $2N_e(s_1 + s_2)(1 - \hat{x})^2$ is large, formula 8.6.4.7 can be reduced to a simpler form:

$$2N_e\lambda_0 = \sqrt{\frac{2N_e^3(s_1 + s_2)^3}{\pi}} \{2\hat{x}(1 - \hat{x})^2 e^{-4c(1-\hat{x})^2} + 2\hat{x}^2(1 - \hat{x})e^{-4c\hat{x}^2}\}. \tag{8.6.4.14}$$

Miller also worked out by numerical analysis the eigenvalue for various values of c ($\geqq 0$) up to $c = 12$. In Table 8.6.4.1, $2N_e\lambda_0$ is tabulated as a function of c and $\hat{x}$ based on the results obtained by Miller. Figure 8.6.4.1 is

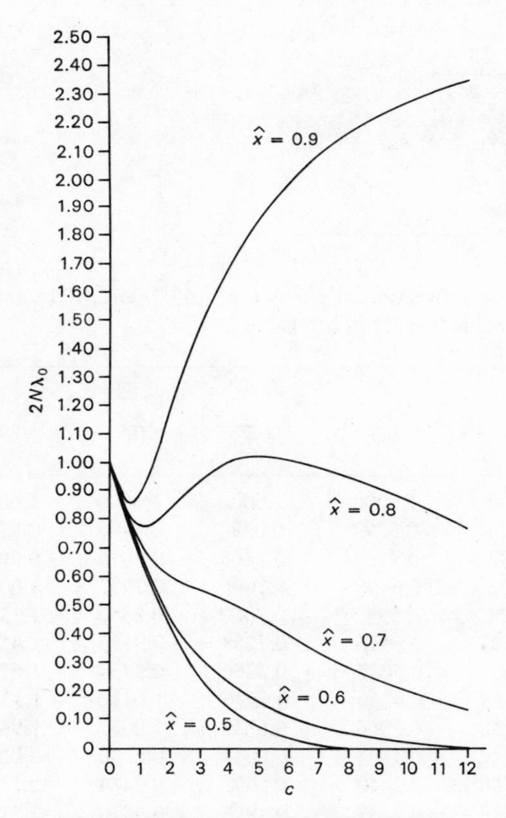

Figure 8.6.4.1. $2N_e\lambda_0$ for overdominant genes is plotted against $c(=N_e(s_1 + s_2)/2)$ for various values of equilibrium gene frequency $\hat{x}$.

drawn from the values in Table 8.6.4.1 for the cases of $\hat{x} = 0.5, 0.6, 0.7, 0.8$, and 0.9. One of the most remarkable features disclosed from the figure is that if s_1 and s_2 differ to such an extent that the equilibrium frequency $\hat{x}$ is higher than 0.8 (or, because of symmetry, less than 0.2), overdominance accelerates rather than retards fixation, as compared with the neutral case. This was first discovered by Robertson (1962), who presented this fact in the form of a graph shown in Figure 8.6.4.2, where the term retardation factor is defined as the reciprocal of $2N_e \lambda_0$. According to him, selection for the heterozygote is a factor retarding fixation only if the equilibrium frequency $\hat{x}$ lies inside the range 0.2–0.8. For equilibrium gene frequencies outside this range, there is a range of values of $N_e(s_1 + s_2)$ for which heterozygote advantage accelerates fixation and the more extreme the equilibrium frequency, the wider this range. However, for all values of $\hat{x}$ except 0 or 1, an increase in the values of $N_e(s_1 + s_2)$ eventually leads to a retardation. Table 8.6.4.2 shows values of $N_e(s_1 + s_2)$ which give retardation by a factor of 100 for various equilibrium gene frequencies (Robertson, 1962).

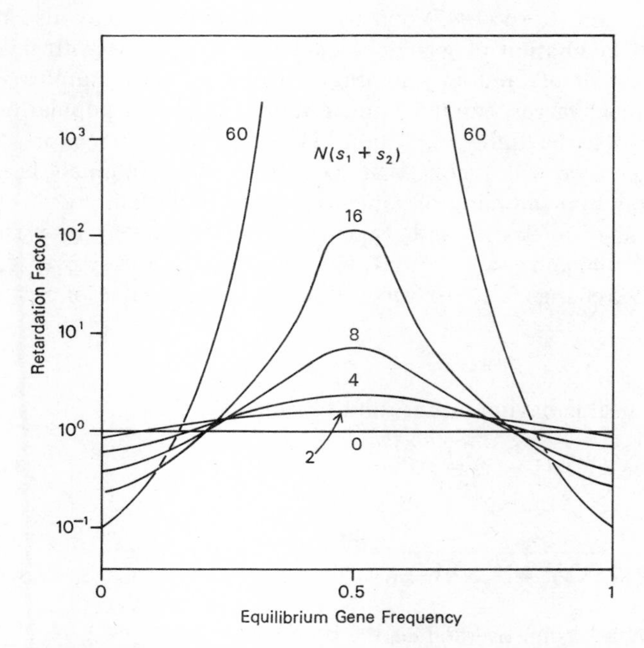

Figure 8.6.4.2. Graphs showing retardation factor as a function of equilibrium gene frequency for various values of $N_e(s_1 + s_2)$. (From Robertson, 1962.)

Table 8.6.4.2. Values of $N_e(s_1 + s_2)$ which give $2N_e\lambda_0 = 0.01$ for various equilibrium gene frequencies. (From Robertson, 1962.)

$\hat{x}$	0.05	0.1	0.2	0.3	0.4	0.5
$N_e(s_1 + s_2)$	2100	490	110	45	22	15

8.7 Change of Gene Frequency Due to Random Fluctuation of Selection Intensities

So far we have considered stochastic processes of gene frequency change due to random sampling of gametes. In the present section, we will investigate a similar process due to random fluctuation of selection intensities.

Such a fluctuation of selection intensities may be caused by random change in environment affecting large and small populations equally, as suggested by Fisher and Ford (1947) and by Wright (1948). It may also be caused by random fluctuation of genetic background associated with individual genes as a result of random sampling of gametes. If the number of segregating loci is much larger than the number of individuals in a population, the second cause may be quite important. However, in this section, to simplify the problem, we will assume that the population is infinitely large and the effect of random sampling of gametes may be neglected.

Consider a pair of alleles A_1 and A_2 with respective frequencies x and $1 - x$ in the population, and assume that A_1 has selective advantage s over A_2. Then the amount of change in x per generation for a given value of s is

$$\delta x = sx(1 - x).$$

If s is a random variable having mean $\bar{s}$ and variance V_s, then

$$M_{\delta x} = E\{\delta x\} = E(s)x(1 - x) = \bar{s}x(1 - x) \tag{8.7.1}$$

and

$$V_{\delta x} = \text{var}(s)x^2(1 - x)^2 = V_s x^2(1 - x)^2. \tag{8.7.2}$$

With these expressions for $M_{\delta x}$ and $V_{\delta x}$, the forward equation 8.3.1 becomes

$$\frac{\partial \phi}{\partial t} = \frac{V_s}{2}\frac{\partial^2}{\partial x^2}\{x^2(1 - x)^2\phi\} - \bar{s}\frac{\partial}{\partial x}\{x(1 - x)\phi\} \qquad (0 < x < 1). \tag{8.7.3}$$

The solution of the above equation has not been obtained, but in the special case of $\bar{s} = 0$, namely, when gene A_1 is selectively neutral on the long-term average, so that

$$\frac{\partial \phi}{\partial t} = \frac{V_s}{2} \frac{\partial^2}{\partial x^2} \{x^2(1-x)^2 \phi\} \qquad (0 < x < 1), \tag{8.7.4}$$

the solution can be obtained as follows (Kimura, 1954).

Letting

$$u = \tfrac{1}{2} \exp\left\{\frac{V_s}{8} t\right\} x^{\frac{3}{2}}(1-x)^{\frac{3}{2}} \phi \tag{8.7.5}$$

and

$$\xi = \log_e \left(\frac{x}{1-x}\right) \qquad (-\infty < \xi < \infty), \tag{8.7.6}$$

and substituting these in 8.7.4, we obtain the heat conduction equation,

$$\frac{\partial u}{\partial t} = \frac{V_s}{2} \frac{\partial^2 u}{\partial \xi^2}. \tag{8.7.7}$$

It is known that this equation has a unique solution which is continuous over $-\infty$ to $+\infty$ when $t \geqq 0$ and which reduces to $u(\xi, 0)$ when $t = 0$. The solution is given by

$$u(\xi, t) = \frac{1}{\sqrt{2\pi V_s t}} \int_{-\infty}^{\infty} e^{-\frac{(\xi-\eta)^2}{2V_s t}} u(\eta, 0)\, d\eta. \tag{8.7.8}$$

Therefore the required solution of 8.7.4, when the initial distribution of gene frequency is $\phi(x, 0)$, is given by

$$\phi(x, t) = \frac{1}{\sqrt{2\pi V_s t}} \frac{\exp\left\{\frac{-V_s t}{8}\right\}}{\{x(1-x)\}^{\frac{3}{2}}} \times \int_0^1 \exp\left\{-\frac{\left(\log \frac{x(1-y)}{(1-x)y}\right)^2}{2V_s t}\right\} \sqrt{y(1-y)}\,\phi(y, 0)\, dy, \tag{8.7.9}$$

where $\exp\{\cdot\}$ denotes the exponential function.

If the initial condition is not a continuous distribution but is a fixed gene frequency p, such that $\phi(x, 0) = \delta(x - p)$, the above formula becomes

$$\phi(p, x; t) = \frac{1}{\sqrt{2\pi V_s t}} \exp\left\{ -\frac{V_s}{8} t - \frac{\left[\log \frac{x(1-p)}{(1-x)p}\right]^2}{2V_s t} \right\} \times \frac{[p(1-p)]^{\frac{1}{2}}}{[x(1-x)]^{\frac{3}{2}}}, \qquad 8.7.10$$

giving the probability density that the frequency of A_1 is x at the tth generation, given that it is p at $t = 0$.

The change in the distribution curve with time is illustrated in Figure 8.7.1, assuming $p = 0.5$ and $V_s = 0.0483$. As the figure shows, the distribution curve is unimodal for a considerable number of generations (in this case, up to the 27th generation), after which it becomes bimodal. Generally, with an arbitrary positive value of V_s, the distribution curve is unimodal if the number of generations is less than $4/(3V_s)$ but becomes bimodal if it exceeds this value (still assuming $p = 0.5$). In the 100th generation, gene frequencies in our example which give maximum probability (corresponding to peaks in the curve) are approximately 0.0007 and 0.9993, where the height of the curve is about 11.37. As time goes on, the distribution curve appears more and more U-shaped, though it is never truly U-shaped, since its value at either terminal is always 0. This means that as time elapses the gene frequency shifts toward either terminal of the distribution ($x = 0$ or 1) indefinitely and accumulates in the neighborhood just short of fixation or loss but never goes to fixation or loss completely (at least theoretically). To distinguish this from the fixation or loss in the case of random drift in small populations, the terms "quasi-fixation" and "quasi-loss" have been proposed (Kimura, 1954).

The same problem may be treated using a discrete model as follows: Let $x^{(t)}$ be the frequency of A_1 in the tth generation and let s be the selection coefficient against its allele A_2 such that the selective values of A_1A_1, A_1A_2, and A_2A_2 are respectively 1, $1 - s$, and $(1 - s)^2$. Then we have

$$x^{(t+1)} = \frac{x^{(t)}}{1 - s(1 - x^{(t)})} \qquad 8.7.11$$

or

$$\frac{x^{(t+1)}}{1 - x^{(t+1)}} = \frac{x^{(t)}}{1 - x^{(t)}} \times \frac{1}{1 - s}. \qquad 8.7.12$$

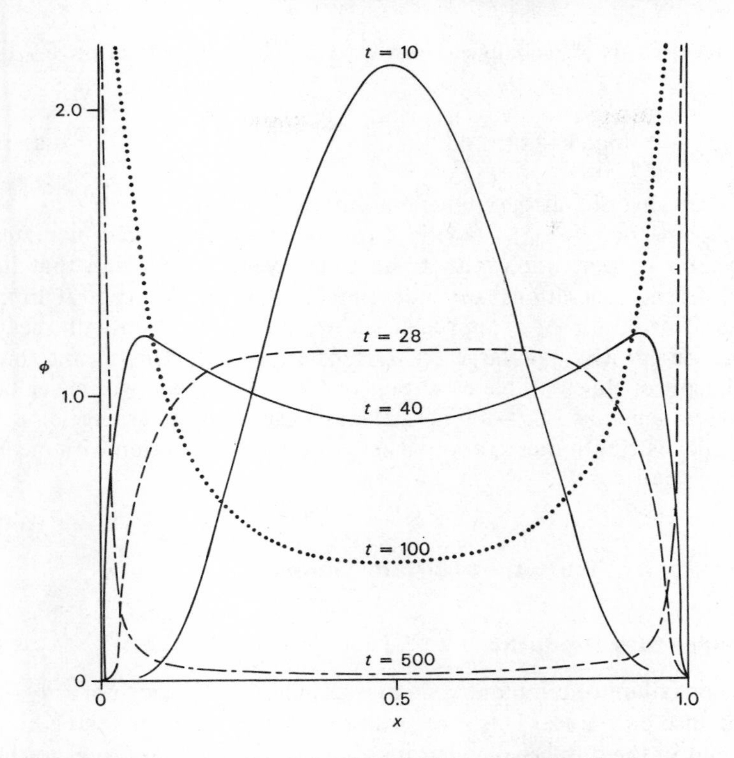

Figure 8.7.1. The change in the gene frequency distribution under random fluctuation of selection intensities. In this illustration it is assumed that the gene is selectively neutral when averaged over a very long period, that there is no dominance, and $p = 0.5$, $V_s = 0.0483$. Abscissa: gene frequency x; Ordinate: probability density ϕ; t is time in generations. (From Kimura, 1954.)

Taking the logarithm of both sides and putting

$$z_t = \log_e\left(\frac{x^{(t)}}{1 - x^{(t)}}\right), \tag{8.7.13}$$

equation 8.7.12 becomes

$$z_{t+1} = z_t - \log_e(1 - s), \tag{8.7.14}$$

where z_t, as given by 8.7.13, is the logit of $x^{(t)}$ and changes continuously from $-\infty$ to $+\infty$ as $x^{(t)}$ changes from 0 to 1. The above relation 8.7.14 leads to

$$z_{t+1} = z_0 - \sum_{i=0}^{t} \log_e(1 - s_i), \tag{8.7.15}$$

where s_i is the value of s at the ith generation.

If we assume that $-\log_e(1 - s)$ is a random variable which fluctuates from generation to generation with mean 0 and variance σ^2, and that its values in different generations are independent, then by the central limit theorem, the distribution of z_t approaches a normal distribution with mean z_0 and variance $\sigma^2 t$ as t gets large (cf. Dempster, 1955*a*). This means that in a finite length of time complete fixation or loss of the gene will never be realized. More generally, if $-\log_e(1 - s)$ has mean $\overline{m}$ and variance σ^2 per generation, the distribution of z_t approaches a normal distribution with mean $z_0 + \overline{m}t$ and variance $\sigma^2 t$.

8.8 Probability of Fixation of Mutant Genes

8.8.1 Introductory Remarks

The success or failure of a mutant gene in a population depends not only on selection but also on chance. This is because chance is involved in segregation at meiosis and in the number of offspring that each individual leaves to the next generation. Even for an advantageous gene, natural selection cannot insure its future survival, until individuals carrying that gene become sufficiently numerous in the population.

It is important to know the probability of the ultimate success (i.e., fixation) of mutant genes, because the fixation of advantageous genes in the population is a key factor in the evolution of the species.

The first significant contribution to this problem was made by Haldane (1927), based on a method suggested by Fisher (1922). This was followed by more detailed study by Fisher (1930, 1930*e*). They made use of the method which is now standard in the treatment of a stochastic process called the branching process. Equivalent results have been obtained by Wright (1931) from the study of distribution curves. The probability was also estimated for a recessive gene by Haldane (1927) and Wright (1942). Subsequently, more general results based on the continuous model (diffusion model) and including the case of any arbitrary degree of dominance were obtained by Kimura (1957). The probability of eventual fixation, $u(p)$, was expressed in terms of the initial frequency p, the selection coefficients, and the effective population number. This function was used by Robertson (1960*a*) in his theory of selection limits in plant and animal breeding. A still more general, but quite simple, expression for $u(p)$ was obtained by Kimura (1962) in terms of the initial frequency and the mean and variance of the rate of change of gene frequency per generation. These results were obtained by using the Kolmogorov backward equation.

Before discussing these more recent studies, we will summarize the branching-process method together with some of the main results obtained by Fisher and Haldane.

8.8.2 Discrete Treatment Based on the Branching-process Method

In this treatment, we assume a population consisting of an infinite but "countable" number of individuals and consider the absolute number of mutant genes in the population, rather than their relative frequency. Though the applicability of this method is restricted to simpler genetic situations (mainly genic selection and low initial frequency), it has the advantage of being somewhat more exact than the continuous treatment in the region where the frequency of mutant genes is very low. In a sense, this branching-process treatment is complementary to the continuous treatment which we will present in the next subsection in order to treat much more general cases. However, as we will see later, the results obtained by the former method for simple cases suggest that the continuous treatment may be quite satisfactory down to the extreme situation where a single mutant gene appears in a population.

In the present section, the classical results obtained by Fisher and Haldane will be summarized. Suppose a mutant gene has appeared in a population. Let $p_0, p_1, p_2, \ldots$ be the probabilities that the mutant gene will become 0, 1, 2, ... in number in the next generation, i.e., it will be represented by 0, 1, 2, ... descendants ($\sum_{k=0}^{\infty} p_k = 1$). In particular, p_0 is the probability that it will be lost by the next generation. Assume that $0 < p_0 < 1$.

Instead of working directly on the probabilities, we will make use of the probability-generating function defined by

$$f(x) = p_0 + p_1 x + p_2 x^2 + \cdots = \sum_{k=0}^{\infty} p_k x^k, \qquad 8.8.2.1$$

where the probability of leaving k mutant genes for the next generation is given as the coefficient of x^k.

The mean of this distribution is given by

$$M_1 = f'(1) = \sum_{k=1}^{\infty} k p_k. \qquad 8.8.2.2$$

Likewise the variance is given by

$$V_1 = f''(1) + f'(1) - \{f'(1)\}^2. \qquad 8.8.2.3$$

In the next generation, if individuals having the mutant genes reproduce independently so that each mutant gene again leaves 0, 1, 2, ... descendant genes with probabilities $p_0, p_1, p_2, \ldots$, as in the previous generation, then it is not difficult to show that the probability-generating function for the number of mutant genes after two generations is given by

$$f(f(x)) = \sum p_k(f(x))^k. \qquad 8.8.2.4$$

This is readily verified by expanding the expression and collecting the terms that are coefficients of the same power of x. The coefficient of x^0 is the probability of no descendant now of the mutant gene two generations back, the coefficient of x^1 the probability of one, the coefficient of x^2 the probability of two, and so on.

If this pattern of reproduction continues, the probability-generating function for the number of mutant genes after t generations ($t = 1, 2, \ldots$) is given by the tth iteration of the generating function $f(x)$, namely,

$$\underbrace{f(f(f \cdots f}_{t \text{ times}}(x) \cdots)),$$

which may be denoted by $f_t(x)$. In particular, $f_1(x) = f(x)$.

Thus

$$f_t(x) = f(f_{t-1}(x)) = f_{t-1}(f(x)), \qquad 8.8.2.5$$

($t \geqq 2$). The mean number of the mutant genes after t generations may be calculated from

$$M_t = f_t'(1) = f_{t-1}'(f(1))f'(1) = M_{t-1} \times M_1,$$

noting $f(1) = 1$. Thus

$$M_t = M_1^t, \qquad 8.8.2.6$$

as might be expected.

In connection with the fate of mutant genes, the important quantity is the probability that the mutant gene will be lost by the tth generation. This is given by $f_t(0)$, which we will denote by v_t.

From 8.8.2.5, we have

$$f(v_t) = v_{t+1}. \qquad 8.8.2.7$$

Thus it may be seen that the probability of ultimate loss which is denoted by v and which is equal to v_∞ satisfies the equation

$$f(v) = v. \qquad 8.8.2.8$$

Furthermore, it can be shown (Feller, 1950; Harris, 1963) that the probability v is the smallest positive root of the above equation. From this, it is possible to prove that if

$$M_1 = f'(1) \leqq 1,$$

then $v_\infty = 1$. That is, unless each mutant gene leaves on the average more than one descendant per generation, it is certain that the gene will ultimately be lost from the population.

In order to get more detailed information on the process of elimination as well as the probability of survival for advantageous mutant genes, it is necessary to assume a particular form of distribution for the number of descendant genes. A simple but realistic assumption is that the values p_0, $p_1, p_2, \ldots$ follow Poisson distribution such that

$$p_k = \frac{c^k}{k!} e^{-c}, \qquad 8.8.2.9$$

where c is the average number of descendant genes which a mutant gene leaves in the next generation. The probability-generating function 8.8.2.1 for this distribution is

$$f(x) = e^{c(x-1)}. \qquad 8.8.2.10$$

Let us put

$$s = \log_e c;$$

then s represents the selective advantage of the mutant gene if the population number is constant: The gene is disadvantageous if $s < 0$, neutral if $s = 0$, and advantageous if $s > 0$. For small absolute value of s, we have approximately

$$s = c - 1.$$

In Table 8.8.2.1, the process of elimination is tabulated for some cases of disadvantageous genes, $c = 0.90, 0.95, 0.97, 0.99$, and also for a neutral gene, $c = 1.00$. In the latter case, the probability of survival is roughly equal to twice the reciprocal of the number of generations elapsed. It will be noted from the table that the process of elimination is quite rapid for $c = 0.90$.

For advantageous genes ($c > 1$), the probability of ultimate survival, u, may be obtained by solving the equation

$$1 - u = e^{-cu}, \qquad 8.8.2.11$$

which is derived from 8.8.2.7, 8.8.2.10, and the definition

$$u = 1 - v = 1 - v_\infty. \qquad 8.8.2.12$$

Table 8.8.2.1. Probability of survival for deleterious and neutral genes, calculated from $u_t = 1 - e^{-cu_{t-1}}$, $u_0 = 1.0000$.

GENERATION	$c = 0.90$	$c = 0.95$	$c = 0.97$	$c = 0.99$	$c = 1.00$
0	1.0000	1.0000	1.0000	1.0000	1.0000
1	0.5934	0.6132	0.6209	0.6284	0.6321
5	0.1972	0.2319	0.2462	0.2607	0.2680
10	0.0849	0.1186	0.1338	0.1499	0.1582
20	0.0236	0.0487	0.0625	0.0787	0.0875
50	0.0009	0.0076	0.0155	0.0287	0.0376
100	—	—	0.0028	0.0111	0.0193
200	—	—	—	0.0030	0.0098
300	—	—	—	0.0010	0.0065
400	—	—	—	—	0.0049
500	—	—	—	—	0.0039
∞	0.0000	0.0000	0.0000	0.0000	0.0000

From 8.8.2.11, we obtain

$$u = 2s - \frac{5}{3}s^2 + \frac{7}{9}s^3 - \frac{131}{540}s^4 + \cdots. \qquad 8.8.2.13$$

Thus, if s is small ($s > 0$), we have approximately

$$u = 2s. \qquad 8.8.2.14$$

That is, the probability of ultimate survival of an individual mutant gene is approximately equal to twice its selective advantage, if the advantage is small (Haldane, 1927). Table 8.8.2.2 shows for a few examples ($c = 1.01$, 1.03, 1.05, 1.10, and 1.50) how the probability of survival changes with time. It may be seen from the table that for $c = 1.50$ the final value is approached very rapidly. If the gene survives the first five generations it is here to stay.

The above treatment based on the branching-process method, though simple and intuitively appealing, has the drawback that it cannot be used to calculate the probability of fixation in a small population for genes with an arbitrary degree of dominance. The diffusion-equation method, as will be discussed below, can overcome this difficulty and enables us to obtain general but simple answers to many important problems in gene fixation, sometimes with remarkable accuracy despite the fact that it is essentially an approximation. It makes use of the Kolmogorov backward equation, first used in this context by Kimura (1957, 1962).

Table 8.8.2.2. Probability of survival for advantageous mutant genes, as calculated from $u_t = 1 - e^{-cu_{t-1}}$, starting with $u_0 = 1.00000$.

GENERATION	$c = 1.01$	$c = 1.03$	$c = 1.05$	$c = 1.10$	$c = 1.50$
0	1.00000	1.00000	1.00000	1.00000	1.00000
1	0.63578	0.64299	0.65006	0.66712	0.77686
5	0.27541	0.29015	0.30497	0.34212	0.60502
10	0.16674	0.18424	0.20232	0.24928	0.58486
20	0.09694	0.11707	0.13879	0.19763	0.58283
50	0.04800	0.07266	0.10107	0.17713	0.58281
100	0.03069	0.06047	0.09427	0.17613	0.58281
200	0.02272	0.05781	0.09370	0.17613	0.58281
300	0.02074	0.05768	0.09370	0.17613	0.58281
400	0.02009	0.05768	0.09370	0.17613	0.58281
500	0.01986	0.05768	0.09370	0.17613	0.58281
∞	0.01973	0.05768	0.09370	0.17613	0.58281

8.8.3 Continuous Treatment Based on the Kolmogorov Backward Equation

In this section, we will treat the problem with much wider scope. Instead of restricting our attention to the probability of fixation of individual mutant genes, we will consider more generally the probability, $u(p, t)$, that the mutant gene becomes fixed in the population by the tth generation, given that its frequency is p at $t = 0$. Here it will be assumed that the population number is large enough so that the process of change in gene frequency is treated as a continuous stochastic process with good approximation.

Then, as shown in Section 8.3, the Kolmogorov backward equation is applicable to the problem and we have (see 8.3.29)

$$\frac{\partial u(p, t)}{\partial t} = \frac{V_{\delta p}}{2}\frac{\partial^2 u(p, t)}{\partial p^2} + M_{\delta p}\frac{\partial u(p, t)}{\partial p}, \tag{8.8.3.1}$$

where $M_{\delta p}$ and $V_{\delta p}$ are the mean and the variance in the change of gene frequency per generation, the gene frequency p being considered as variable.

The required probability, $u(p, t)$, may be obtained by solving the above partial differential equation with boundary conditions

$$u(0, t) = 0, \qquad u(1, t) = 1. \tag{8.8.3.2}$$

In the simplest case of random drift in a finite population of effective size N_e with no mutation and selection involved, we have

$$V_{\delta p} = \frac{p(1-p)}{2N_e}, \qquad M_{\delta p} = 0,$$

and 8.8.3.1 reduces to

$$\frac{\partial u(p, t)}{\partial t} = \frac{p(1-p)}{4N_e} \frac{\partial^2 u(p, t)}{\partial p^2}. \tag{8.8.3.3}$$

The pertinent solution of this equation is

$$u(p, t) = p + \sum_{i=1}^{\infty} (2i+1)p(1-p)(-1)^i \times F(1-i, i+2, 2, p)e^{-\frac{i(i+1)t}{4N_e}}, \tag{8.8.3.4}$$

which agrees exactly with the result obtained as 8.4.12 by a different method.

Let us now consider the ultimate probability of fixation defined by

$$u(p) = \lim_{t \to \infty} u(p, t), \tag{8.8.3.5}$$

for which $\partial u/\partial t = 0$ and which therefore satisfies the equation

$$\frac{V_{\delta p}}{2} \frac{d^2 u(p)}{dp^2} + M_{\delta p} \frac{du(p)}{dp} = 0, \tag{8.8.3.6}$$

with boundary conditions

$$u(0) = 0, \qquad u(1) = 1. \tag{8.8.3.7}$$

The above differential equation may be written in the form

$$\frac{d}{dp}\left(\log \frac{du(p)}{dp}\right) = -\frac{2M_{\delta p}}{V_{\delta p}},$$

and it may be seen that the solution which satisfies the boundary conditions 8.8.3.7 is given by

$$u(p) = \frac{\int_0^p G(x)\, dx}{\int_0^1 G(x)\, dx}, \tag{8.8.3.8}$$

where

$$G(x) = e^{-\int \frac{2M_{\delta x}}{V_{\delta x}} dx}, \tag{8.8.3.9}$$

in which $M_{\delta x}$ and $V_{\delta x}$ are the mean and the variance of the amount of change in gene frequency x per generation (Kimura, 1962). Formula 8.8.3.8 gives the (ultimate) probability of fixation of a mutant gene whose initial frequency is p. It has a pleasing simplicity and generality comparable to Wright's formula for the probability distribution of gene frequency at equilibrium (see Section 9.1 in the next chapter). His formula may be expressed in the present notation as

$$\phi(x) = \frac{C}{V_{\delta x} G(x)}, \qquad 8.8.3.10$$

where C is a constant determined such that

$$\int_0^1 \phi(x)\, dx = 1.$$

It may be remarked that the above formula for $u(p)$ is the steady-state solution of the Kolmogorov backward equation and is the counterpart of Wright's formula for $\phi(x)$, which is the steady-state solution of the forward equation.

In a population consisting of N individuals in each generation (N not necessarily equal to N_e), the chance of ultimate fixation of an individual mutant gene is given by $u(p)$ with $p = 1/2N$; i.e.,

$$u = u(1/2N). \qquad 8.8.3.11$$

However, if the population consists of different numbers of males (N^*) and females (N^{**}), a slight modification taking $p = 1/(4N^*)$ or $p = 1/(4N^{**})$ may be required, depending on whether the mutant gene occurred in a male or in a female, as pointed out by Moran (1961) and Watterson (1962). Subsequently we will investigate several more special cases by applying these results. We will consider a population of sexually reproducing diploid individuals and denote the mutant gene by A, assuming its initial frequency p.

Let us start from the simplest case of genic selection in which A has a constant selective advantage s over its alleles in a population of effective size N_e. If the frequency of A is x, then the mean and the variance in the rate of change in x per generation are

$$\left.\begin{aligned} M_{\delta x} &= sx(1-x) \\ V_{\delta x} &= \frac{x(1-x)}{2N_e}, \end{aligned}\right\} \qquad 8.8.3.12$$

so that $2M_{\delta x}/V_{\delta x} = 4N_e s$, $G(x) = e^{-4N_e sx}$, and we obtain from 8.8.3.8

$$u(p) = \frac{1 - e^{-4N_e sp}}{1 - e^{-4N_e s}}. \qquad 8.8.3.13$$

This formula is a special case of formula 8.8.3.21 which we will derive later. A corresponding result has been obtained by Moran (1962, p. 118) for the haploid case. For $|2N_e s| < \pi$, the right side of the above equation may be expanded in terms of $4N_e s$ as follows:

$$u(p) = \sum_{i=1}^{\infty} \frac{\phi_i(p)(-1)^{i-1}}{i!} (4N_e s)^{i-1}$$
$$= p + 2N_e sp(1-p) + \frac{(2N_e s)^2}{3} p(p-1)(2p-1) + \cdots, \quad \textbf{8.8.3.14}$$

where $\phi_i(p)$'s are Bernoulli polynomials.

The probability of fixation of an individual mutant gene is obtained from 8.8.3.13 by putting $p = 1/(2N)$.

$$u = \left(1 - e^{-\frac{2N_e s}{N}}\right) \Big/ (1 - e^{-4N_e s}) \quad \textbf{8.8.3.15}$$

For a random-mating population with Poisson distribution of progeny number, $N_e = N$ and the above formula reduces to

$$u = \frac{1 - e^{-2s}}{1 - e^{-4Ns}}. \quad \textbf{8.8.3.16}$$

If $|s|$ is small, we obtain

$$u = \frac{2s}{1 - e^{-4Ns}} \quad \textbf{8.8.3.17}$$

as a good approximation. This agrees with the results obtained by Fisher (1930) and Wright (1931). This formula is good even for negative s, though u for such a case is very small unless $|Ns|$ is not large. For a positive s and very large N we obtain the result, already worked out in the previous section, that the probability of ultimate survival of an advantageous mutant gene is approximately twice the selection coefficient (see 8.8.2.14), that is,

$$u \approx 2s. \quad \textbf{8.8.3.18}$$

If N_e differs from N, the corresponding formula derived from 8.8.3.15 is

$$u \approx 2s\left(\frac{N_e}{N}\right). \quad \textbf{8.8.3.19}$$

In other words, the classical result $u \approx 2s$ should be modified by a factor of N_e/N (Kimura, 1964). According to Crow and Morton (1955), estimated values of N_e/N are $0.71 \sim 0.9$ for Drosophila, 0.75 for Lymnaea, and $0.69 \sim 0.95$ for man. In natural as well as controlled populations, the effective size

N_e will usually be different from the actual size N. Therefore, formula 8.8.3.19 should be more useful than 8.8.3.18.

On the other hand, if we let $s \to 0$ in 8.8.3.15, we obtain $u = 1/(2N)$ irrespective of the effective size N_e. This is the special case of $u = p$, the result known for a neutral gene.

We will now consider a more general case of zygotic selection. Let s and sh be the selective advantage of the mutant homozygote and the heterozygote respectively. Then

$$\left. \begin{aligned} M_{\delta x} &= sx(1-x)[h + (1-2h)x] \\ V_{\delta x} &= \frac{x(1-x)}{2N_e}, \end{aligned} \right\} \tag{8.8.3.20}$$

and therefore

$$G(x) = e^{-2cDx(1-x)-2cx},$$

where $c = N_e s$ and $D = 2h - 1$. Thus we obtain

$$u(p) = \frac{\int_0^p e^{-2cDx(1-x)-2cx}\, dx}{\int_0^1 e^{-2cDx(1-x)-2cx}\, dx} \tag{8.8.3.21}$$

(Kimura, 1957). The rate of approach to the ultimate state of complete fixation or loss is given by the smallest eigenvalue λ_0 of the equation

$$\frac{\partial u}{\partial t} = \frac{p(1-p)}{4N_e}\frac{\partial^2 u}{\partial p^2} + sp(1-p)\{h + (1-2h)p\}\frac{\partial u}{\partial p}. \tag{8.8.3.22}$$

In this case λ_0 is equal to the smallest eigenvalue of the adjoint form 8.6.3.1 and is given by 8.6.3.2 with $c = N_e s$ and $D = 2h - 1$.

For a completely recessive gene $h = 0$ or $D = -1$, and we have

$$u(p) = \frac{\int_0^p e^{-2cx^2}\, dx}{\int_0^1 e^{-2cx^2}\, dx}. \tag{8.8.3.23}$$

If s is positive and small but $N_e s$ is large, the above formula leads approximately to

$$u = \sqrt{\frac{2s}{\pi N}} \tag{8.8.3.24}$$

for $N_e = N$ and $p = 1/(2N)$. This gives the probability of fixation of an individual mutant gene which is advantageous but completely recessive in a randomly mating population consisting of N individuals each generation and distribution of progeny number following the Poisson distribution. It is interesting to note that the more exact value given here in 8.8.3.24 lies between $\sqrt{s/N}$, the value estimated by Haldane (1927) in his treatment of this as a branching process, and $\sqrt{s/2N}$, obtained by Wright (1942) with his method of integral equations. Also, Wright's numerical approximation $1.1\sqrt{s/2N}$ is very close to the present value.

From 8.8.3.21, it is also possible to calculate the chance of fixation of a nearly recessive gene with a selective advantage in the homozygous state (Kimura, 1957). For example, consider a case with $N_e = N = 10^3$ and $s = 10^{-1}$. If the mutant gene is completely recessive ($h = 0$), $u \approx 0.8 \times 10^{-2}$. With a slight phenotypic effect of $h = 0.01$ in the heterozygote, $u \approx 0.9 \times 10^{-2}$, while with $h = 0.1$, $u \approx 2.3 \times 10^{-2}$.

The function $u(p)$ defined in 8.8.3.5 plays a very important role in the theory of limits in artificial selection put forward by Robertson (1960*a*), who refers to $u(p)$ as "the expected limit," since if we assign value 1 to the phenotype of the mutant homozygote, $u(p)$ is the expected value with respect to this locus at the limit. Thus the total advance by selection is $u(p) - p$. If $N_e s$ is small in absolute value, we can expand $u(p)$ in terms of $N_e s$. For the case of no dominance as shown in 8.8.3.14, we obtain

$$u(p) - p = 2N_e sp(1 - p) + \cdots .$$

We note here that $sp(1 - p)$ is the change of p in one generation by selection. This means that, for a character governed by additive genes, the total advance is $2N_e$ times the change in the first generation, provided that $N_e s$ per locus is small. Similarly it has been shown by Robertson (1960*a*) that, in the case of recessive character, the ratio of expected total response to initial change in the first generation is $2N_e$ times $(1 + p)/3p$.

In most of animal- and plant-breeding theory, such as we have discussed earlier in the book, the major aim has been to maximize the rate of progress under selection. The selection limit theory inquires into the other aspect of the problem and asks such questions as "Under what selection program is the highest ultimate performance attained?" In general the two questions do not have the same answer.

The Kolmogorov backward equation can be extended to treat the probability of joint fixation of mutant genes involving two or more loci. As an example, we will consider a case of two independent loci each with a pair of alleles, A_1 and A_2, in the first locus, and B_1 and B_2 in the second. Let $u(p, q; t)$ be the probability that both A_1 and B_1 become fixed in the population by the tth generation, given that their initial frequencies (at $t = 0$) are

p and q respectively. Then $u(p, q; t)$ satisfies the following partial differential equation:

$$\frac{\partial u(p, q; t)}{\partial t} = \frac{1}{2} V_{\delta p} \frac{\partial^2 u(p, q; t)}{\partial p^2} + \frac{1}{2} V_{\delta q} \frac{\partial^2 u(p, q; t)}{\partial q^2} + M_{\delta p} \frac{\partial u(p, q; t)}{\partial p} + M_{\delta q} \frac{\partial u(p, q; t)}{\partial q}. \quad 8.8.3.25$$

The above equation is an extension of equation 8.8.3.1 (still assuming that the process is time homogeneous) and is the counterpart of equation 8.3.24.

The probability of ultimate joint fixation defined by

$$u(p, q) = \lim_{t \to \infty} u(p, q; t)$$

satisfies the partial differential equation

$$\frac{1}{2} V_{\delta p} \frac{\partial^2 u(p, q)}{\partial p^2} + \frac{1}{2} V_{\delta q} \frac{\partial^2 u(p, q)}{\partial q^2} + M_{\delta p} \frac{\partial u(p, q)}{\partial p} + M_{\delta q} \frac{\partial u(p, q)}{\partial q} = 0 \quad 8.8.3.26$$

and the boundary conditions that both $u(p, 1)$ and $u(1, q)$ reduce to the appropriate formulae for the probability of fixation at a single locus.

In the simplest case of a haploid population with effective size N_e, if the fitnesses of four genotypes are given as in Table 8.8.3.1, then equation 8.8.3.26 becomes

$$\frac{p(1-p)}{2N_e} \frac{\partial^2 u}{\partial p^2} + \frac{q(1-q)}{2N_e} \frac{\partial^2 u}{\partial q^2} + p(1-p)(s_1 + \varepsilon q) \frac{\partial u}{\partial p} + q(1-q)(s_2 + \varepsilon p) \frac{\partial u}{\partial q} = 0. \quad 8.8.3.27$$

Table 8.8.3.1. Table of fitnesses of four genotypes measured in Malthusian parameters, taking the fitness of $A_2 B_2$ as standard.

GENOTYPE	FITNESS (MALTHUSIAN PARAMETERS)	FREQUENCY
$A_1 B_1$	$s_1 + s_2 + \varepsilon$	pq
$A_1 B_2$	s_1	$p(1-q)$
$A_2 B_1$	s_2	$(1-p)q$
$A_2 B_2$	0	$(1-p)(1-q)$

The general solution of the above equation has not been obtained; equation 10.17 given in Kimura (1964, p. 228) is unfortunately erroneous. However, for the special case of $s_1 = s_2 = 0$, the pertinent solution

$$u(p, q) = \frac{1 - e^{-2N_e \varepsilon pq}}{1 - e^{-2N_e \varepsilon}} \qquad 8.8.3.28$$

was obtained by Kimura and it was checked through simulation experiments by Ohta (1968). Also, if $|N_e s_1|$, $|N_e s_2|$, and $|N_e \varepsilon|$ are all small, the approximate solution of 8.8.3.27 is

$$u(p, q) = pq\{1 + N_e s_1(1 - p) + N_e s_2(1 - q) + N_e \varepsilon(1 - pq) + \cdots\}. \qquad 8.8.3.29$$

Equation 8.8.3.28 may be used to study the effect of epistasis on the probability of joint fixation when mutant genes are individually neutral but become advantageous in combination.

8.9 The Average Number of Generations Until Fixation of a Mutant Gene in a Finite Population

Once the probability of fixation is known, it is natural to ask how long it takes for a mutant gene to reach fixation, excluding the cases in which it is eventually lost from the population. The basic theory to answer this question has been worked out by Kimura and Ohta (1969). They have shown that a single mutant gene, if it is selectively neutral, takes about $4N_e$ generations on the average to reach fixation in a population of effective size N_e. More generally, if p is the initial frequency of the mutant allele, the average number of generations until fixation (excluding the cases of loss) is given by

$$\bar{t}_1(p) = \int_p^1 \psi(\xi)u(\xi)\{1 - u(\xi)\}\, d\xi + \frac{1 - u(p)}{u(p)} \int_0^p \psi(\xi)u^2(\xi)\, d\xi, \qquad 8.9.1$$

where $u(p)$ is the probability of ultimate fixation (see equation 8.8.3.8) and

$$\psi(x) = 2\,\frac{\int_0^1 G(x)\, dx}{V_{\delta x}\, G(x)},$$

in which $G(x)$ is given by 8.8.3.9 in the previous section. Similarly, the average number of generations until loss (excluding the cases of ultimate fixation) is

$$\bar{t}_0(p) = \frac{u(p)}{1 - u(p)} \int_p^1 \psi(\xi)\{1 - u(\xi)\}^2\, d\xi + \int_0^p \psi(\xi)\{1 - u(\xi)\}u(\xi)\, d\xi. \qquad 8.9.2$$

In the simplest case of neutral mutations, formulae 8.9.1 and 8.9.2 reduce respectively to

$$\bar{t}_1(p) = -\frac{1}{p}\{4N_e(1-p)\log_e(1-p)\} \tag{8.9.3}$$

and

$$\bar{t}_0(p) = -4N_e\left(\frac{p}{1-p}\right)\log_e p. \tag{8.9.4}$$

If the mutant allele is represented only once at the moment of its occurrence in a population of actual size N and effective size N_e, we may set $p = 1/(2N)$

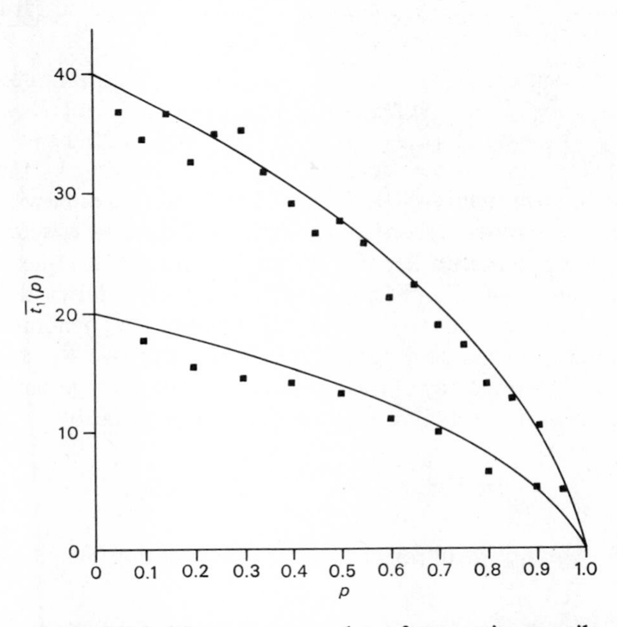

Figure 8.9.1. The average number of generations until fixation, $\bar{t}_1(p)$, as a function of initial gene frequency, p. It is assumed that the gene is selectively neutral. In this figure, the theoretical values based on formula 8.9.3 are represented by curves and those of Monte Carlo experiments by square dots. $2N_e = 20$ in the upper curve and $2N_e = 10$ in the lower one. (From Kimura and Ohta, 1969.)

in the above formulae. Thus, for a neutral mutation, we have

$$\bar{t}_1(1/2N) = -8NN_e\left(1 - \frac{1}{2N}\right)\log_e\left(1 - \frac{1}{2N}\right) \approx 4N_e \qquad \textbf{8.9.5}$$

and

$$\bar{t}_0(1/2N) = \frac{4N_e}{2N-1}\log_e(2N) \approx 2\left(\frac{N_e}{N}\right)\log_e(2N). \qquad \textbf{8.9.6}$$

For example, if $N = 10^4$ and $N_e = 0.8 \times 10^4$, $\bar{t}_1 = 32000$ and $\bar{t}_0 = 15.9$. This shows that in general a great majority, $1 - 1/(2N)$, of neutral mutant genes which appeared in a finite population are lost from the population within a small number of generations, while the remaining minority, $1/(2N)$, spread over the entire population to reach fixation, taking a very large number of generations. For details, readers may refer to Kimura and Ohta (1969). Figure 8.9.1 shows some results of Monte Carlo experiments to test equation 8.9.3.

We may add here that for the special case of selectively neutral mutants, not only the mean value of the length of time until fixation, but also its standard deviation has been obtained. It was shown by Narain (1969) and also by Kimura and Ohta (1969*a*) that the standard deviation of the length of time until fixation of a neutral mutant is about $(2.15)N_e$ generations. Furthermore, for this case, the entire probability distribution of the length of time until fixation has been worked out (Kimura, 1969*a*).

9 DISTRIBUTION OF GENE FREQUENCIES IN POPULATIONS

In earlier chapters we have emphasized, as Haldane first did, that selection and evolution have both static and dynamic aspects. Chapter 5 treated those factors, principally selection, that change the population to produce evolutionary transformations—the *dynamics* of evolution. In Chapter 6 we considered the approximate balance between various opposing forces that keeps the gene frequencies somewhere near an equilibrium—the *statics* of evolution.

Chapters 8 and 9 are analogous, but with the added complication that random factors are taken into account. Chapter 8 has been a stochastic treatment of the dynamics of evolution. This final chapter considers the interaction of the various factors, both deterministic and random, that lead to equilibria. However, we should expect to find instead of a single point of stable equilibrium a probability distribution. We will investigate the equilibrium distribution of gene frequencies that is attained under the joint action of systematic factors that tend to bring the population to a fixed value and random factors that cause random scatter therefrom.

The first attempt to deduce the gene frequency distribution was by Fisher (1922). This was clarified and developed in his 1930 book. In his 1922 paper, Fisher compared such an investigation to the analytical treatment of the Theory of Gases. Starting in 1931, Wright has published a large series of important papers in which the steady-state distribution has been worked out under more and more general assumptions. Much of our knowledge of this subject, a grossly disproportionate share, is based on the work of these two pioneers.

9.1 Wright's Formula for the Gene Frequency Distribution

In Chapter 8, we studied the change of gene frequencies as a stochastic process. For a pair of alleles at a single locus we designated by $\phi(p, x; t)$ the density of the transition probability that the gene frequency becomes x at the tth generation, given that it is p at $t = 0$. In the following material we will designate by $\phi(x)$ the density function of the steady-state distribution such that

$$\lim_{t \to \infty} \phi(p, x; t) = \phi(x). \qquad 9.1.1$$

That is, the distribution is independent of the initial frequency p. For example, we have shown (8.5.10) that under linear pressure and random sampling of gametes we get

$$\phi(x) = \phi(p, x; \infty) = \frac{\Gamma(4Nm)}{\Gamma(4Nmx_I)\Gamma(4Nm(1 - x_I))} \times x^{4Nmx_I - 1}(1 - x)^{4Nm(1 - x_I) - 1}, \qquad 9.1.2$$

which is independent of p. This independence may not be difficult to understand if we note that formula 9.1.2 represents the distribution at the state of balance between the loss of genes by random drift and their new production or introduction by mutation or migration.

We will sometimes speak of $\phi(x)$ as the frequency distribution since $\phi(x)\,dx$ gives the relative number of populations with gene frequency in the range of $x \sim x + dx$ among the hypothetical aggregate of an infinite number of populations satisfying the same conditions. In the case of a single locus with a pair of alleles, a quite general formula for $\phi(x)$ has been obtained by Wright (1938a). His formula may be expressed as follows:

$$\phi(x) = \frac{C}{V_{\delta x}} \exp\left(2 \int \frac{M_{\delta x}}{V_{\delta x}}\,dx\right), \qquad 9.1.3$$

where $M_{\delta x}$ and $V_{\delta x}$ are respectively the mean and variance of the rate of change in x per generation. C is a constant which is usually adjusted so that

$$\int_0^1 \phi(x)\,dx = 1; \qquad \textbf{9.1.4}$$

that is, the area under the curve is unity. Wright's formula (9.1.3) may also be expressed as

$$\phi(x) = \frac{C}{V_{\delta x}G(x)}, \qquad \textbf{9.1.5}$$

where

$$G(x) = e^{-\int \frac{2M_{\delta x}}{V_{\delta x}}dx}. \qquad \textbf{9.1.6}$$

It is remarkable that Wright (1938) derived the formula from a simple consideration that at equilibrium the mean and variance are unchanged between successive generations. Later he showed that all the moments are indeed unchanged for this formula (Wright, 1952*a*).

From our standpoint, however, it is more natural to derive this formula from 8.3.17 by imposing the condition $P(x, t) = 0$, that is, at equilibrium the net probability flux is 0 at every point in the interval (0, 1). This leads to

$$\frac{1}{2}\frac{d}{dx}(V_{\delta x}\phi) - M_{\delta x}\phi = 0. \qquad \textbf{9.1.7}$$

Let $V_{\delta x}\phi = f$; then

$$\frac{1}{f}\frac{df}{dx} = \frac{2M_{\delta x}}{V_{\delta x}}$$

or

$$\frac{d}{dx}\log f = \frac{2M_{\delta x}}{V_{\delta x}}.$$

Integrating both sides of the last expression, we get

$$\log f = \text{const} + \int \frac{2M_{\delta x}}{V_{\delta x}}\,dx$$

or

$$f = C\exp\left(2\int \frac{M_{\delta x}}{V_{\delta x}}\,dx\right), \qquad \textbf{9.1.8}$$

which is equivalent to 9.1.3 if we note that $\phi = f/V_{\delta x}$. One of the tacit assumptions involved here is that the process is time homogeneous so that $M_{\delta x}$ and $V_{\delta x}$ are independent of t.

The equilibrium distribution discussed above may be called the flux-zero distribution, because the net probability flux is 0, i.e., $P(x, t) = 0$.

On the other hand, there is an equilibrium distribution among mutant genes at different loci resulting from steady flux of mutations. In this case, a steady state is reached with respect to distribution of intermediate gene frequencies ($0 < x < 1$) but there is a constant probability flux from one terminal class to the other, i.e., $P(x, t) =$ constant. We will treat some problems relating to such a distribution later in this chapter (see Section 9.8).

In the next few sections we will investigate, using Wright's formula (9.1.3), some concrete forms of the equilibrium distribution, assuming various conditions of mutation, migration, selection, and random sampling of gametes.

9.2 Distribution of Gene Frequencies Among Subgroups Under Linear Pressure

Consider a species subdivided into partially isolated groups, each of which exchanges individuals with the general population at a constant rate. Let x be the frequency of allele A_1 in a subgroup which exchanges individuals at the rate m per generation. If x_I is the frequency of A_1 among the immigrants, the rate of change of x per generation is

$$\delta x = m(x_I - x). \tag{9.2.1}$$

We will assume as we did in Chapter 6 that the immigrants represent a random sample from the entire species so that x_I is constant and equal to the general mean $\bar{x}$ (see 9.2.6 below). Such a model was called by Wright (1951) the island model. We disregard selection, but the effect of mutation can be included in the parameter m (see 8.5). From 9.2.1, the mean of δx is

$$M_{\delta x} = m(x_I - x),$$

which may also be expressed as

$$M_{\delta x} = -m(1 - x_I)x + mx_I(1 - x). \tag{9.2.2}$$

If N_e is the effective size of the subgroup, the variance in δx due to random sampling of gametes is

$$V_{\delta x} = \frac{x(1 - x)}{2N_e}. \tag{9.2.3}$$

Thus

$$2\int \frac{M_{\delta x}}{V_{\delta x}}\,dx = -4N_e m(1-x_I)\int \frac{dx}{1-x} + 4N_e m x_I \int \frac{dx}{x}$$
$$= 4N_e m\{(1-x_I)\log(1-x) + x_I \log x\} + \text{const},$$

and from 9.1.3 we obtain

$$\phi(x) = Cx^{4N_e m x_I - 1}(1-x)^{4N_e m(1-x_I)-1}, \qquad 9.2.4$$

where the constant C is determined by the condition 9.1.4:

$$\int_0^1 \phi(x)\,dx = C\int_0^1 x^{4N_e m x_I - 1}(1-x)^{4Nm_e(1-x_I)-1}\,dx$$
$$= C \times B(4N_e m x_I, 4N_e m(1-x_I)) = 1,$$

in which $B(\cdot\,,\cdot)$ represents the beta function. Thus we get

$$C = \frac{1}{B(4N_e m x_I, 4N_e m(1-x_I))} = \frac{\Gamma(4N_e m)}{\Gamma(4N_e m x_I)\Gamma(4N_e m(1-x_I))}$$

and 9.2.4 becomes

$$\phi(x) = \frac{\Gamma(4N_e m)}{\Gamma(4N_e m x_I)\Gamma(4N_e m(1-x_I))} \qquad 9.2.5$$
$$\times\, x^{4N_e m x_I - 1}(1-x)^{4N_e m(1-x_I)-1},$$

where $\Gamma(\cdot)$ represents the gamma function. Note that the above formula agrees with 9.1.2. Figure 9.2.1 illustrates the distribution curve $y = \phi(x)$ for various values of $N_e m$, assuming $x_I = 0.5$. In this figure, the abscissa represents the frequency of A_1 in a subgroup and the ordinate represents a corresponding probability density. As is shown in the figure, the distribution curve changes from U-shaped to bell-shaped as $2N_e m$ changes from smaller ($2N_e m < 1$) to larger ($2N_e m > 1$) values, passing through a flat shape at the intermediate value of $2N_e m = 1$. If $2N_e m$ is less than unity, the curve is U-shaped and there is a considerable tendency for gene A_1 to be either fixed ($x = 1$) or lost ($x = 0$) within a subgroup. This tendency is larger, the smaller the value of $2N_e m$. On the other hand, if $2N_e m$ is larger than unity, the curve is bell-shaped and the gene frequency within a subgroup tends to approach the general mean (x_I), the tendency naturally being stronger the larger the value of $2N_e m$. Since $2N_e m = 1$ means an exchange of one individual in every two generations, this means that the isolation has to be fairly strong between subgroups for the tendency of random fixation or loss to be predominant.

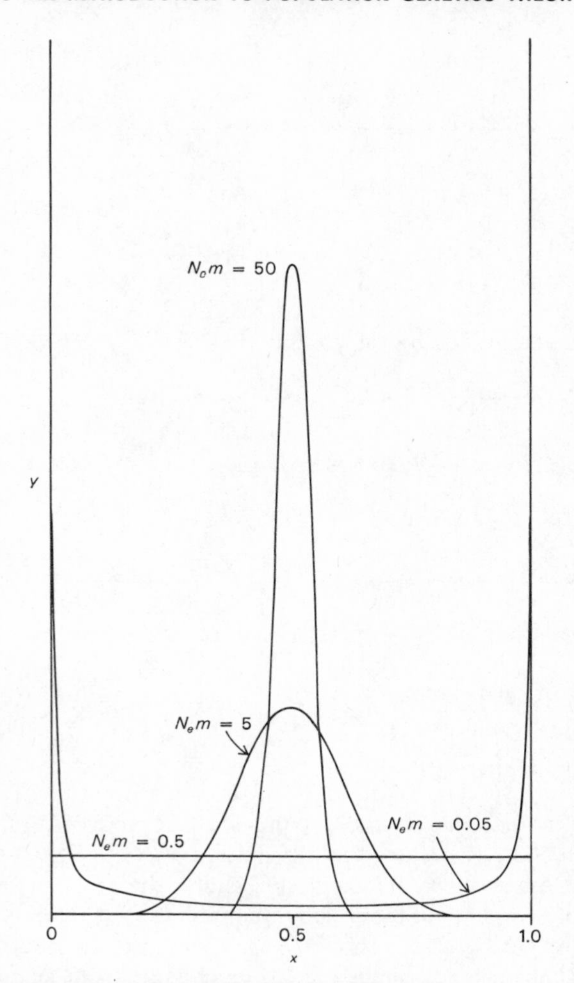

Figure 9.2.1. Distribution of gene frequencies among subgroups for various values of $N_e m$. Here the gene frequency in immigrants is assumed to be 0.5 ($x_I = 0.5$). Abscissa: gene frequency. Ordinate: probability density. (From Wright, 1940.)

However, in an actual situation, exchange of individuals may be restricted mainly to adjacent subgroups so that m in 9.2.5 should be replaced by $m(1 - r)$, where r is the correlation coefficient of gene frequencies between adjacent subgroups. Since r can be very near to unity in such a circumstance, a considerable amount of exchange would still lead to appreciable fixation or loss of an allele. Actually, a model of population structure in which the entire population is subdivided into colonies and migration in each generation is restricted to adjacent colonies ("stepping-stone model") was worked out by Kimura and Weiss (1964) and Weiss and Kimura (1965). They determined the correlation coefficient of gene frequencies between two arbitrary colonies for one-, two-, three-, and higher dimensional stepping-stone models (see Section 9.9).

For the distribution given in 9.2.5, the mean gene frequency over the whole species is

$$\bar{x} = \int_0^1 x\phi(x)\,dx = x_I. \tag{9.2.6}$$

This agrees with the assumption that in the island model migrants represent a random sample from the whole species. The variance of gene frequency among subgroups is

$$\sigma_x^2 = \int_0^1 (x - \bar{x})^2\phi(x)\,dx = \frac{x_I(1 - x_I)}{4N_e m + 1}, \tag{9.2.7}$$

as first given by Wright (1931). This was derived earlier by elementary methods (see 6.6.6).

This can also be derived from the following procedure: Let x_t be the frequency of A_1 in the tth generation; then

$$x_{t+1} = x_t + \delta x_t, \tag{9.2.8}$$

where δx_t has mean and variance given by 9.2.2 and 9.2.3. Using the same procedure we used in the study of pure random drift (compare Section 7.4), we get

$$\begin{aligned}\mu_1'^{(t+1)} &\equiv E(x_{t+1}) = E(x_t + \delta x_t)\\ &= \mu_1'^{(t)} + m(\bar{x} - \mu_1'^{(t)}),\end{aligned}$$

so that $\mu_1'^{(\infty)} = \bar{x}$. In addition, we get

$$\begin{aligned}\mu_2'^{(t+1)} &\equiv E(x_{t+1}^2) = E\{(x_t + \delta x_t)^2\}\\ &= \mu_2'^{(t)} + 2E_\phi\{x_t m(\bar{x} - x_t)\} + E_\phi\left\{\frac{x_t(1 - x_t)}{2N_e}\right\}\end{aligned} \tag{9.2.9}$$

or

$$\Delta\mu_2'^{(t)} = 2m(\mu_1'^{(t)}\bar{x} - \mu_2'^{(t)}) + \frac{1}{2N_e}(\mu_1'^{(t)} - \mu_2'^{(t)}).$$

For $t \to \infty$, $\Delta\mu_2'^{(\infty)} = 0$, $\mu_1'^{(\infty)} = \bar{x}$, and we obtain

$$\left(2m\bar{x}^2 + \frac{\bar{x}}{2N_e}\right) - \left(2m + \frac{1}{2N_e}\right)\mu_2'^{(\infty)} = 0.$$

Thus

$$\sigma_x^2 = \mu_2'^{(\infty)} - (\mu_1'^{(\infty)})^2 = \frac{4N_e m\bar{x}^2 + \bar{x}}{4N_e m + 1} - \bar{x}^2 = \frac{\bar{x}(1-\bar{x})}{4N_e m + 1}, \qquad \text{9.2.10}$$

which agrees with Wright's formula (9.2.7).

To derive the above formula for variance, we have assumed that the change in gene frequency by migration is small so that in calculating the sampling variance $x(1-x)/2N_e$, the gene frequency before migration may be used for x. On the other hand, if the amount of migration per generation is very large and if the sampling of gametes occurs after migration, it is more accurate to use the frequency after migration for x in calculating the variance. Thus, denoting by Δx the amount of deterministic change by migration and by ξ the amount of stochastic change by random sampling of gametes, we have $E(\xi_t) = 0$ and

$$E(\xi_t^2) = \frac{X_t(1-X_t)}{2N_e}, \qquad \text{9.2.11}$$

where $X_t = x_t + \Delta x_t = x_t + m(\bar{x} - x_t) = m\bar{x} + (1-m)x_t$ is the frequency after migration but before the sampling. Since $x_{t+1} = X_t + \xi_t$, formula 9.2.9 for μ_2' is modified as

$$\begin{aligned}\mu_2'^{(t+1)} &= E_\phi(X_t^2) + E_\phi\left\{\frac{X_t(1-X_t)}{2N_e}\right\}\\ &= \left(1 - \frac{1}{2N_e}\right)E(X_t^2) + \frac{1}{2N_e}E(X_t),\end{aligned} \qquad \text{9.2.12}$$

where $E(X_t) = m\bar{x} + (1-m)\mu_1'^{(t)}$ and

$$E(X_t^2) = (m\bar{x})^2 + 2m\bar{x}(1-m)\mu_1'^{(t)} + (1-m)^2\mu_2'^{(t)}.$$

At equilibrium in which $\mu_1'^{(t)} = \bar{x}$ and $\mu_2'^{(t+1)} = \mu_2'^{(t)} \equiv \mu_2'^{(\infty)} = \sigma_x^2 + \bar{x}^2$, we have $E(X_t) = \bar{x}$ and $E(X_t^2) = \bar{x}^2 + (1-m)^2\sigma_x^2$, and therefore 9.2.12 yields

$$\sigma_x^2 + \bar{x}^2 = \left(1 - \frac{1}{2N_e}\right)[\bar{x}^2 + (1-m)^2\sigma_x^2] + \frac{\bar{x}}{2N_e}$$

or

$$\sigma_x^2 = \frac{1}{2N_e}\left\{\frac{\bar{x}(1-\bar{x})}{1 - (1-m)^2\left(1 - \dfrac{1}{2N_e}\right)}\right\}. \qquad \text{9.2.13}$$

Unless an unusually large value is assigned to m, the difference between this improved formula, 9.2.13, and the approximate formula, 9.2.10, is negligible.

Assuming migration only, we may immediately adapt the above treatment to include reversible mutations. If there is mutation from A_1 to its allele A_2 at the rate of u per generation and also reverse mutation from A_2 to A_1 at the rate v, then the rate of change in x due to these mutations is $-ux + v(1 - x)$. Thus 9.2.2 becomes

$$M_{\delta x} = -[m(1 - x_I) + u]x + (mx_I + v)(1 - x). \qquad 9.2.14$$

This means that, in order to include mutations, we simply replace m by $m + u + v$ and mx_I by $mx_I + v$ in the various formulae.

Going back to distribution 9.2.5, let us now investigate the frequencies of fixed classes or the probabilities that allele A_1 is temporarily fixed or lost in the subgroup.

For this, we note first that although the constant C in 9.2.4 was determined by the condition

$$\int_0^1 \phi(x)\,dx = 1, \qquad 9.2.15$$

from which 9.2.5 was derived, the study of the stochastic process involved (see formula 8.5.9) shows that C is intrinsically determined by the process and it is not a constant determined by the arbitrary statistical procedure of making its integral equal to unity. That is to say, condition 9.2.15 follows directly from the nature of the stochastic process.

Thus it looks as if in the present treatment no probability mass is left to represent the fixed classes. It turns out, however (cf. Kimura, 1968), that the frequency $f(0)$ that A_1 is temporarily lost from the population is given by

$$f(0) = \int_0^{\frac{1}{2N}} \phi(x)\,dx. \qquad 9.2.16$$

Thus,

$$f(0) \approx \frac{\Gamma(4N_e m)}{\Gamma(4N_e mx_I + 1)\Gamma(4N_e m(1 - x_I))} \left(\frac{1}{2N}\right)^{4N_e mx_I} \qquad 9.2.17$$

approximately. This may also be derived by considering the balance between mutation and random extinction at the subterminal class, as suggested by Wright (1931).

Similarly, the frequency $f(1)$ that A_1 is temporarily fixed in the population (A_2 temporarily lost) is

$$f(1) = \int_{1-\frac{1}{2N}}^{1} \phi(x)\,dx$$

$$\approx \frac{\Gamma(4N_e m)}{\Gamma(4N_e mx_I)\Gamma(4N_e m(1 - x_I) + 1)} \left(\frac{1}{2N}\right)^{4N_e m(1 - x_I)}. \qquad 9.2.18$$

9.3 Distribution of Gene Frequencies Under Selection and Reversible Mutation

Let us assume a pair of alleles A_1 and A_2 with respective frequencies x and $1-x$ in a random-mating population of effective size N_e. Let u be the mutation rate from A_1 to A_2 and let v be the mutation rate in the reverse direction. Then the rate of frequency change of A_1 by mutation is

$$-ux + v(1-x).$$

If $\bar{a}$ is the average fitness of the population measured in Malthusian parameters, then the rate of change in gene frequency owing to natural selection is given by

$$\frac{x(1-x)}{2}\frac{d\bar{a}}{dx}.$$

Note that if the fitness is measured in selective values (W), the corresponding rate of change is given by Wright's formula

$$\frac{x(1-x)}{2\bar{W}}\frac{d\bar{W}}{dx},$$

where $\bar{W}$ is the mean selective value of the population (cf. 5.2.12). Combining the expressions for mutation and selection, the mean rate of change in x per generation is

$$M_{\delta x} = -ux + v(1-x) + \frac{x(1-x)}{2}\frac{d\bar{a}}{dx}. \qquad \mathbf{9.3.1}$$

For a population of effective size N_e, the variance in δx due to random sampling of gametes is

$$V_{\delta x} = \frac{x(1-x)}{2N_e}.$$

Therefore

$$2\int \frac{M_{\delta x}}{V_{\delta x}}\,dx = 4N_e u \log(1-x) + 4N_e v \log x + 2N_e \bar{a} + \text{const},$$

and we obtain from 9.1.3 the distribution formula for the frequency of A_1:

$$\phi(x) = Ce^{2N_e\bar{a}}x^{4N_e v-1}(1-x)^{4N_e u-1}. \qquad \mathbf{9.3.2}$$

In the simplest case of genic selection in which A_1 has selective advantage s over A_2, we have

$$\bar{a} = 2sx^2 + s2x(1-x) = 2sx,$$

if we assign relative fitnesses $2s$, s, and 0 to A_1A_1, A_1A_2, and A_2A_2 respectively. Then 9.3.2 reduces to

$$\phi(x) = Ce^{4N_e sx}x^{4N_e v-1}(1-x)^{4N_e u-1}. \qquad \textbf{9.3.3}$$

In Figures 9.3.1*a*, *b*, and *c* the frequency distribution of a deleterious gene

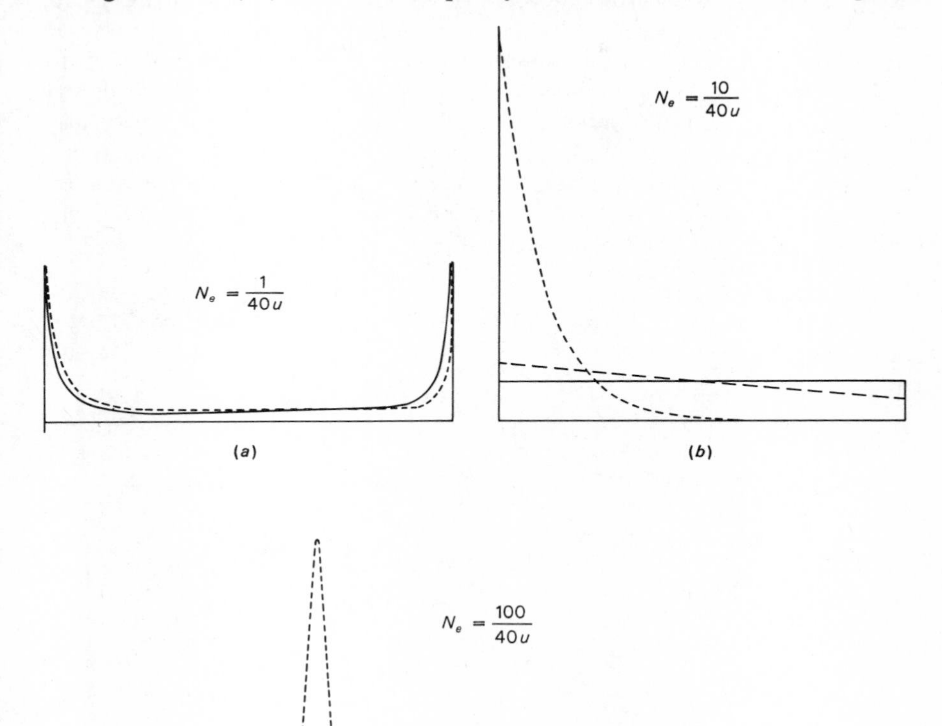

Figures 9.3.1*a,b,c*. Graphs showing the frequency distribution of a deleterious gene ($s < 0$) in a small (*a*), an intermediate (*b*), and a large (*c*) population, assuming equal mutation rates in both directions ($u = v$). In each figure, curves with solid, broken, and dotted lines represent cases with $s = -u/10$, $-u$, and $-10u$, respectively. (From Wright, 1937.)

($s < 0$) in a small population (a), i.e., $N_e = 1/40u$, an intermediate population (b), i.e., $N_e = 10/40u$, and a large population (c), i.e., $N_e = 100/40u$, is illustrated for three levels of selection intensities, assuming $u = v$. In each figure, curves with solid, broken, and dotted lines represent cases with $s = -u/10$, $-u$, and $-10u$, respectively. As will be seen from Figure 9.3.1a, fixation or loss of alleles predominates in a small population and selection is not very effective in determining the gene frequency distribution. On the other hand, in a large population (c) the gene frequency tends to cluster around the equilibrium value and small change in selection intensity may lead to a marked change in the distribution.

In the more general case of zygotic selection, if the relative fitnesses of three genotypes A_1A_1, A_1A_2, and A_2A_2 are s, sh, and 0, respectively, measured in Malthusian parameters, then

$$\bar{a} = sx^2 + 2shx(1 - x)$$

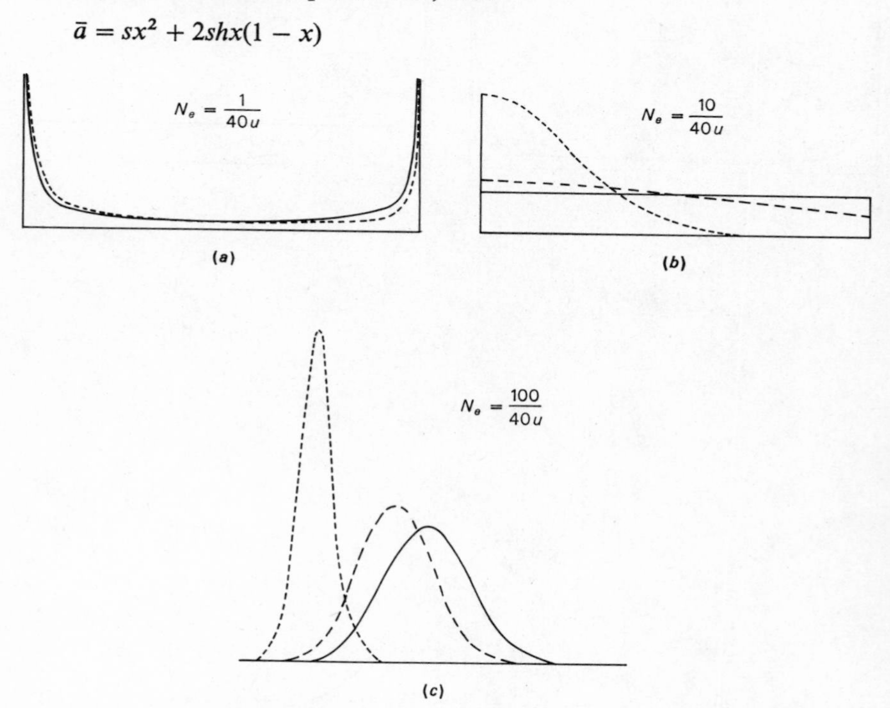

Figures 9.3.2*a,b,c*. Graphs showing the distribution of a completely recessive deleterious gene ($s < 0$, $h = 0$, $u = v$). Effective size $N_e = 1/40u$, and $10/40u$, and $100/40u$ in a, b, and c, respectively. In these figures the solid line represents the least selection ($s = -u/5$), the broken line selection 10 times as intense, and the dotted line selection 100 times as intense. (From Wright, 1937.)

and the distribution of x is given by

$$\phi(x) = Ce^{2N_e sx^2 + 4N_e shx(1-x)} x^{4N_e v - 1}(1 - x)^{4N_e u - 1}. \tag{9.3.4}$$

Figures 9.3.2*a*, *b*, and *c* illustrate the distribution of completely recessive deleterious genes ($s < 0$, $h = 0$) for three different population sizes and for three levels of selection intensity. If there is overdominance between alleles, it is more convenient to express the relative fitnesses of the three genotypes A_1A_1, A_1A_2, and A_2A_2 as $-s_1$, 0, and $-s_2$, so that

$$\bar{a} = -s_1x^2 - s_2(1 - x)^2.$$

Then the distribution formula for x becomes

$$\phi(x) = Ce^{-2N_e s_1 x^2 - 2N_e s_2(1-x)^2} x^{4N_e v - 1}(1 - x)^{4N_e u - 1}. \tag{9.3.5}$$

Using a distribution formula equivalent to 9.3.4 or 9.3.5, Nei and Imaizumi (1966*a*) investigated the amount of genetic variation expected in a finite population at equilibrium.

9.4 Distribution of Lethal Genes

If gene A_1 is lethal, its distribution will be restricted to the range of very low frequencies and mutation from the lethal gene A_1 to its normal allele A_2 may be neglected. Let x be the frequency of the lethal gene A_1. Suppose that A_1 is completely recessive to its allele (A_2) so that the selective values of three genotypes are $w_{11} = 0$ and $w_{12} = w_{22} = 1$. Then the rate of change in x by selection is

$$-\frac{x^2}{1 + x}.$$

The rate of change by mutation from A_2 to A_1 is

$$v(1 - x).$$

Thus, we can take

$$M_{\delta x} = v(1 - x) - \frac{x^2}{1 + x}, \tag{9.4.1}$$

as the mean of δx per generation. Using the binomial variance as before,

$$V_{\delta x} = \frac{x(1 - x)}{2N_e}, \tag{9.4.2}$$

9.1.3 yields

$$\phi(x) = C(1 - x^2)^{2N_e} x^{4N_e v - 1}(1 - x)^{-1}, \tag{9.4.3}$$

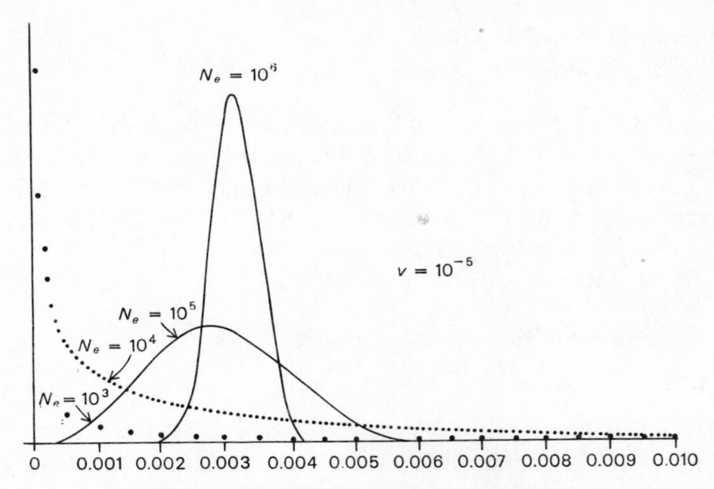

Figure 9.4.1. Probability distribution of the frequency of a lethal gene for various population sizes. The mutation rate (v) is assumed to be 10^{-5} per generation. Abscissa: Frequency of the lethal gene in the population. Ordinate: Probability density. (From Wright 1937.)

where constant C is determined from 9.1.4 as

$$C = \frac{2}{B(2N_e, 2N_e v) + B(2N_e, 2N_e v + \frac{1}{2})}, \tag{9.4.4}$$

where $B(\cdot, \cdot)$ designates the beta function. Figure 9.4.1 shows the probability distribution of x among populations containing at least one lethal gene for various effective population sizes, N_e. Since x is practically restricted to a very small positive value, we may substitute

$$\phi(x) = Ce^{-2N_e x^2} x^{4N_e v - 1} \tag{9.4.5}$$

for 9.4.3. This formula can also be obtained by using

$$M_{\delta x} = v - x^2 \tag{9.4.6}$$

instead of formula 9.4.1. The constant C is then

$$C = \frac{2(2N_e)^{2N_e v}}{\Gamma(2N_e v)}, \tag{9.4.7}$$

approximately. The mean of the distribution is easily found and we obtain

$$\bar{x} = \frac{\Gamma(2N_e v + \frac{1}{2})}{\sqrt{2N_e}\,\Gamma(2N_e v)}. \tag{9.4.8}$$

At the limit of $N_e \to \infty$, $\bar{x}$ becomes $\sqrt{v}$. This is the value expected in an infinitely large random-mating population and it can also be obtained from

$$M_{\delta x} = v - x^2 = 0.$$

On the other hand, if $N_e v$ tends to 0,

$$\bar{x} \approx v\sqrt{2\pi N_e}. \tag{9.4.9}$$

It may be seen easily that the expected frequency of lethal genes is much lower in a small than in a large population. In Figure 9.4.2, mean lethal frequency in a population is plotted as a function of the effective population number, assuming $v = 10^{-5}$. The figure shows clearly that for the well-known relation $\bar{x} = \sqrt{v}$ to be valid, the population number has to be hundreds of thousands, as pointed out by Robertson (1962).

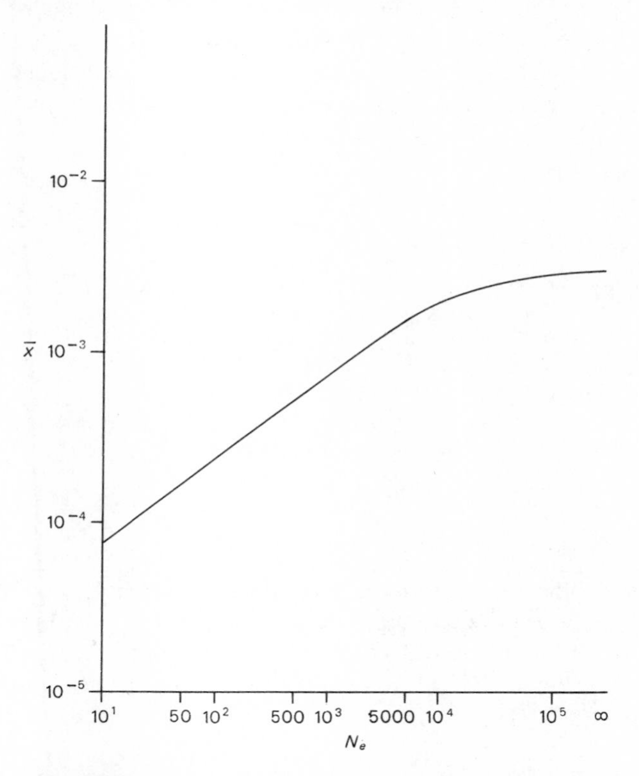

Figure 9.4.2. Average frequency of a completely recessive lethal gene in populations of various sizes where the mutation rate, v, is 10^{-5}.

If the lethal gene (A_1) is not completely recessive in fitness but is slightly deleterious in the heterozygote, as is the case for the majority of lethal genes in Drosophila, we should take

$$M_{\delta x} = v - hx(1-f) - \{x^2(1-f) + xf\} \approx v - hx - x^2 - xf \qquad \textbf{9.4.10}$$

as the mean change of the lethal frequency per generation, still assuming that x is small.

Here f denotes the inbreeding coefficient and we assume that it is small and constant. Using the approximate expression $V_{\delta x} = x/2N_e$ for variance, the formula for the frequency distribution becomes

$$\phi(x) = Ce^{-4N_e(h+f)x - 2N_e x^2} x^{4N_e v - 1}. \qquad \textbf{9.4.11}$$

In this case, if the mutation rate and the heterozygous disadvantage are around $v = 10^{-5}$ and $h = 1/40$, respectively, as observed for the majority of lethal genes in Drosophila, the mean gene frequency is little influenced by the effective population number and is roughly equal to $v/(h+f)$.

More generally, if $(h+f) \gg \sqrt{v}$, the selection against lethal genes is primarily exerted in the heterozygous state so that 9.4.11 may be replaced by a simpler formula,

$$\phi(x) = Ce^{-4N_e(h+f)x} x^{4N_e v - 1}, \qquad \textbf{9.4.12}$$

where

$$C = \frac{[4N_e(h+f)]^{4N_e v}}{\Gamma(4N_e v)};$$

that is, the distribution is approximated by the gamma distribution with mean and variance

$$\bar{x} = \frac{v}{h+f}$$

and

$$\sigma_x^2 = \frac{v}{4N_e(h+f)^2}.$$

In experimental studies of lethal genes, the directly observable quantity is the frequency of lethal-bearing chromosomes rather than individual lethal genes. So, let X_1 be the proportion of chromosomes carrying one or more lethal genes, and let $X_0 = 1 - X_1$. Then, assuming independent distribution of lethal genes at different loci,

$$X_0 = e^{-\Sigma_i x_i},$$

where x_i is the frequency of the lethal gene at the ith locus. Thus,

$$Q_1 \equiv -\log_e X_0 = \sum_i x_i$$

has a mean and variance

$$\bar{Q}_1 = \sum \bar{x}_i = \frac{U}{h+f}$$

and

$$\sigma^2_{Q_1} = \sum_i \sigma^2_{x_i} = \frac{U}{4N_e(h+f)^2},$$

if we assume that $(h+f)$ is the same for all relevant loci. In the above formulae, $U = \sum_i v_i$ is the lethal mutation rate per chromosome.

As pointed out by Nei (1968*a*), since the frequency of lethal genes at each locus follows a gamma distribution, their sum Q_1 is again distributed as a gamma distribution, so that we have

$$\phi(Q_1) = Ce^{-4N_e(h+f)Q_1}Q_1^{4NU-1},$$

where

$$C = \frac{\{4N_e(h+f)\}^{4N_eU}}{\Gamma(4N_eU)}.$$

Figure 9.4.3 illustrates this frequency distribution for various population sizes, assuming $U = 0.005$, $h = 0.025$, and $f = 0$.

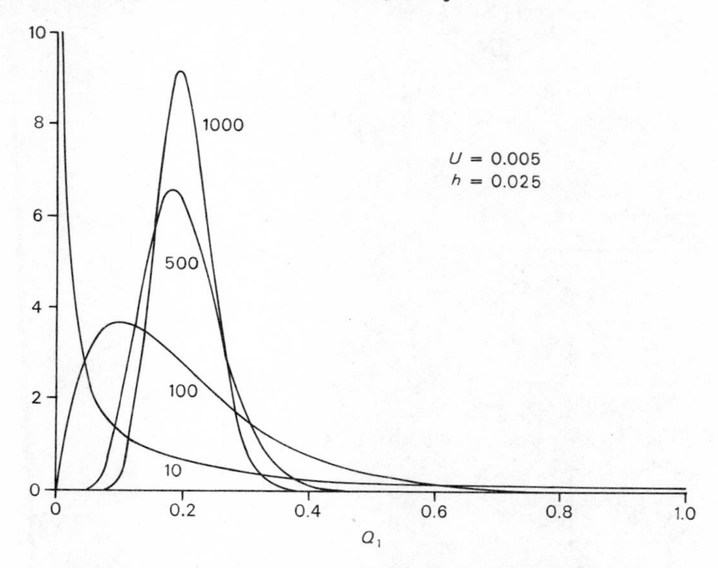

Figure 9.4.3. Frequency distribution of lethal chromosomes (Q_1). Here it is assumed that each lethal gene has a heterozygous selective disadvantage of 2.5% and that the lethal mutation rate per chromosome is 0.5%. The number beside each curve represents the effective population size. (From Nei, 1968.)

For more details on the distribution of lethal genes, readers may refer to Nei (1968*a*).

9.5 Effect of Random Fluctuation in Selection Intensity on the Distribution of Gene Frequencies

In the preceding sections, we have considered only those cases in which the sole factor producing random fluctuation of gene frequencies is the random sampling of gametes. However, as pointed out in Section 8.7, it is probable that random fluctuation in selection intensity is also at work in determining the gene frequency distribution in natural populations. In this section, we will investigate this effect for the case of no dominance, that is, the heterozygote having fitness midway between the two homozygotes. For a more detailed treatment of the subject, see Kimura (1955).

We will express the selective advantage of A_1 over A_2 by s, and assume that s is a random variable with mean $\bar{s}$ and variance V_s. Let u and v be the mutation rates as defined in 9.3. Then the mean and variance of the rate of change in gene frequency x are given by

$$M_{\delta x} = \bar{s}x(1-x) - ux + v(1-x)$$

$$V_{\delta x} = V_s x^2(1-x)^2 + \frac{x(1-x)}{2N_e}. \qquad 9.5.1$$

Substituting these in 9.1.3, we get

$$\phi(x) = Cx^{4N_e v-1}(1-x)^{4N_e u-1}(\lambda_1 - x)^{4N_e A-1}(x-\lambda_2)^{4N_e B-1}, \qquad 9.5.2$$

where

$$A = -\frac{\dfrac{\bar{s}}{2N_e V_s} + u\lambda_1 - v\lambda_2}{\lambda_1 - \lambda_2}$$

and

$$B = \frac{\dfrac{\bar{s}}{2N_e V_s} - v\lambda_1 + u\lambda_2}{\lambda_1 - \lambda_2}$$

(Kimura, 1955). Here λ_1 and λ_2 are given by

$$\lambda_1 = \frac{1 + \sqrt{1 + \dfrac{2}{N_e V_s}}}{2} \qquad (>1)$$

and

$$\lambda_2 = \frac{1 - \sqrt{1 + \dfrac{2}{N_e V_s}}}{2}, \qquad (<0).$$

In order to compare the relative effect resulting from the two different random factors, i.e., the random sampling of gametes and random fluctuation of selection intensities, we will study in some detail the symmetrical case where $\bar{s} = 0$ and $u = v$. With these assumptions, 9.5.2 reduces to

$$\phi(x) = C\{x(1-x)\}^{4N_e u - 1}\{(\lambda_1 - x)(x - \lambda_2)\}^{4N_e u - 1}. \qquad 9.5.3$$

If the population size is small such that $4N_e u$ is much less than 1, the genes are fixed most of the time and the distribution curves for unfixed classes are U-shaped, as shown in Figure 9.5.1*a*, where $N_e = 10^3$. If the variance of s is $0.1/2N_e$ or 5×10^{-5} the effect of the fluctuation is so small that, when drawn on the graph, the curve is indistinguishable from the one with no fluctuation. With $V_s = 10/2N_e$ or 5×10^{-3} the effect is still not striking; the frequencies of the subterminal classes merely rise by about 33% compared with the case of $V_s = 0$. With $V_s = 100/2N_e$ or 0.05, the subterminal classes become about 2.2 times as high as in the case where $V_s = 0$. $V_s = 0.05$ is a high fluctuation since $\sigma_s = \sqrt{V_s} = 0.2236$. Thus for small populations the effect of random fluctuation of selection is rather unimportant.

Figure 9.5.1*b* shows the distribution of gene frequencies in intermediate populations such that $4N_e u = 1$. If $N_e = N$, this means that one mutation appears every two generations on the average. For example, if we assume a mutation rate of one in one hundred thousand (10^{-5}), N_e should be 25,000, that is, 25 times as large as before. In such populations, if there is no fluctuation, there will be a flat distribution. That is, all the heterallelic classes are equally probable. Here the effect of the fluctuation may not be negligible. With $V_s = 10/2N_e$ or 2×10^{-4}, the distribution curve becomes U-shaped and the heights of the subterminal classes are about 4.9 times the heights of those where $V_s = 0$. With $V_s = 100/2N_e$ or 0.002, the distribution resembles that of a very small population ($4N_e u \approx 0$), the frequency of the subterminal classes rising about 48 times as high as when $V_s = 0$.

In Figure 9.5.1*c* the population size is assumed to be twice as large as before ($4N_e u = 2$), so that, if $N = N_e$, one new mutation is expected per generation. With a mutation rate of $u = 10^{-5}$, N_e should be 50,000. In this case it can be shown that if V_s does not exceed $1/N_e$ or 2×10^{-5}, the distribution curve is unimodal. On the other hand, if V_s exceeds this value, the curve becomes bimodal. As is shown in the figures, if $V_s = 10/2N$ or 10^{-4}, the gene frequencies giving the two modes of the distribution are about 0.053 and 0.947, respectively. With $V_s = 100/2N_e$ or 10^{-3}, the modal gene frequencies are about 0.5% and 99.5%, and the modal classes are about 104 times as frequent as the class with 50% gene frequency. Therefore in the last case the distribution curve looks as if it were U-shaped on the figure. However, the class frequencies fall sharply outside these modes and the frequencies of terminal classes are 0.

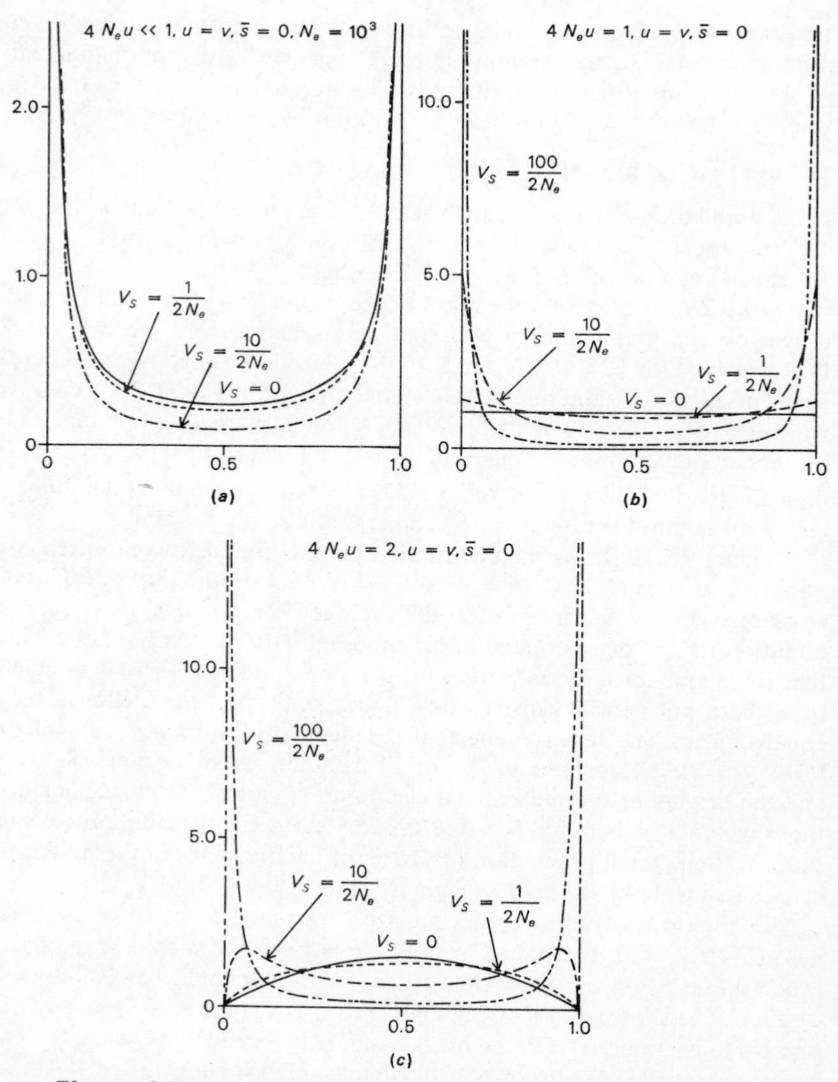

Figures 9.5.1*a,b,c*. Effects of random fluctuation of selection intensity on the probability distribution of gene frequencies in a small (*a*), an intermediate (*b*), and a large (*c*) population. In these figures it is assumed that there is no dominance between a pair of alleles which are on the average neutral ($\bar{s} = 0$). Also mutation rates in both directions are equal ($u = v$). N_e stands for effective size of population and V_s is the variance of the selection coefficient s. Abscissa: gene frequency (x). Ordinate: probability density. (From Kimura, 1955.)

9.6 Number of Neutral Alleles Maintained in a Finite Population

In Section 7.2 we discussed the amount of heterozygosity and the effective number of neutral alleles maintained in a population of effective size N_e. If the number of neutral alleles is potentially large enough that each mutant is a type not currently represented in the population, the probability of homozygosity at equilibrium was shown to be

$$f \approx \frac{1}{4N_e u + 1}, \qquad \textbf{9.6.1}$$

approximately, whether the population is monoecious or has separate sexes. N_e is the inbreeding effective population number as defined in Section 7.6.2.

The effective number of alleles maintained is $1/f$, or

$$n_e = 4N_e u + 1. \qquad \textbf{9.6.2}$$

If there are K alleles so that the chance of mutating to a particular one of them is $u/(K - 1)$, then

$$f = \frac{4N_e u\left(\dfrac{1}{K-1}\right) + 1}{4N_e u\left(\dfrac{K}{K-1}\right) + 1}. \qquad \textbf{9.6.3}$$

As we said, in this treatment N_e is the inbreeding effective number. Essentially the same formulae can be derived, using the variance effective number (see Kimura, 1968), which we now show.

Assume that there are K possible allelic states $A_1, A_2, \ldots, A_K$ and that each allele mutates with a rate $u/(K - 1)$ to one of the remaining $K - 1$ alleles. The unit as far as the definition of state is concerned may be a single nucleotide, a codon, a cistron, or any other entity that can be regarded as a stable unit. Let x be the frequency of one of the alleles, say A_1, and let X be its frequency after mutation. Then

$$X = x + u_1(1 - Kx),$$

where $u_1 = u/(K - 1)$. If we assume that random sampling of gametes is carried out after the deterministic change of gene frequency, then the change of gene frequency due to random sampling, which we denote by ξ, has mean and variance

$$E(\xi) = 0,$$

$$E(\xi) = \frac{X(1 - X)}{2N_e}.$$

In the above formula, N_e is the "variance" effective number of the population. Since the frequency of A_1 in the next generation is

$$x' = X + \xi = u_1 + (1 - u_1K)x + \xi,$$

if we denote by μ_1' and μ_2' the first and second moments of the frequency distribution of A_1 at equilibrium, we have

$$\mu_1' = E(x') = u_1 + (1 - u_1K)\mu_1', \tag{9.6.4}$$

$$\begin{aligned}\mu_2' &= E(x'^2) = E\{(X + \xi)^2\} = E\left\{X^2 + \frac{X(1-X)}{2N_e}\right\} \\ &= \frac{1}{2N_e}\{u_1 + (1 - u_1K)\mu_1'\} + \left(1 - \frac{1}{2N_e}\right)\{u_1^2 + 2u_1(1 - u_1K)\mu_1' \\ &\quad + (1 - u_1K)^2\mu_2'\}.\end{aligned} \tag{9.6.5}$$

Thus, from 9.6.4, we obtain

$$\mu_1' = \frac{1}{K}. \tag{9.6.6}$$

Substituting $1/K$ for μ_1' in 9.6.5, we obtain

$$\mu_2' = \frac{1 + (2N_e - 1)(2u_1 - u_1^2K)}{K\{2N_e - (2N_e - 1)(1 - u_1K)^2\}}, \tag{9.6.7}$$

where $u_1 = u/(K - 1)$. The average homozygosity or the sum of squares of the allelic frequencies may be expressed in terms of μ_2' as follows:

$$\overline{H}_0 = E\left(\sum_1^K x_i^2\right) = K\mu_2', \tag{9.6.8}$$

in agreement with 9.6.3 when very small quantities are ignored. The effective number of alleles n_e is equal to the reciprocal of $\overline{H}_0$ and therefore $n_e = 1/(K\mu_2')$.

If the number of allelic states is indefinitely large, $K = \infty$ so that $u_1 = 0$, $u_1K = u$, and $u_1^2K = 0$ in the above formulae and we obtain

$$n_e = 2N_e - (2N_e - 1)(1 - u)^2. \tag{9.6.9}$$

Note that N_e here represents the variance effective number of the population rather than the inbreeding effective number that appears in 9.6.2. Since the mutation rate is usually very much smaller than unity ($u \ll 1$), 9.6.9 reduces to

$$n_e = 4N_eu + 1, \tag{9.6.10}$$

in good agreement with 9.6.2.

The problem of finding the distribution of the allelic frequencies is more difficult but was solved by Kimura and Crow (1964). For neutral alleles, the steady-state distribution has a rather simple form:

$$\Phi(x) = 4M(1 - x)^{4M-1}x^{-1} \qquad \textbf{9.6.11}$$

(Kimura and Crow, 1964), where $M = N_e u$. N_e stands for the effective population number.

In Chapter 7 (cf. Table 7.6.5.1) we showed that the two definitions of effective population number are equivalent if the population is neither increasing nor decreasing. In much of the discussion to follow, this assumption is made, so it is not necessary to distinguish between the two effective numbers.

In this distribution $\Phi(x)\,dx$ gives an approximation to the expected number of alleles whose frequencies in the population lie within the range x to $x + dx$ $(0 < x < 1)$. It is important to note here that we are considering a frequency distribution, within a single population, of various alleles having a different number of representatives, but we are not considering a probability distribution for any particular allele.

Using the above distribution, f may be obtained by computing the sum of squares of allelic frequencies

$$f = \int_0^1 x^2\Phi(x)\,dx = \frac{1}{4N_e u + 1},$$

which agrees with the result obtained in 9.6.1. Thus

$$n_e = \frac{1}{f} = 4N_e u + 1\,. \qquad \textbf{9.6.12}$$

This shows that the effective number of alleles is determined solely by the effective population number and is independent of the actual population number.

The average (actual) number of alleles (n_a) can be obtained by summing the expected number of alleles in each frequency class from $1/2N$ to 1. Thus

$$n_a = \int_{\frac{1}{2N}}^1 \Phi(x)\,dx = 4N_e u \int_{\frac{1}{2N}}^1 (1 - x)^{4N_e u - 1} x^{-1}\,dx. \qquad \textbf{9.6.13}$$

This shows that the average number of alleles within a population depends both on the effective and the actual population numbers. Therefore n_a may be much more difficult to estimate from a sample than n_e, and less informative as to genetic variability.

Figure 9.6.1 illustrates the result of a Monte Carlo experiment on the number of neutral isoalleles in a small population consisting of 50 males and

50 females ($N = 100$), of which only 25 males and 25 females actually participate in breeding ($N_e = 50$). The simulation experiment was carried out using the IBM 7090 computer by generating pseudo-random numbers. In each generation, 100 male and 100 female gametes were randomly chosen from 25 breeding males and 25 breeding females to form the next generation.

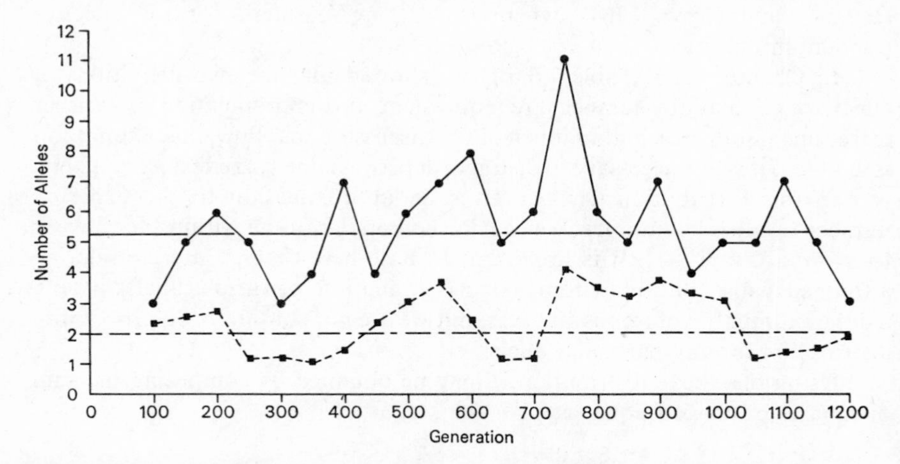

Figure 9.6.1. A result of Monte Carlo experiments on the number of neutral isoalleles. In this experiment actual population number $N = 100$, effective population number $N_e = 50$, and mutation rate $u = 0.005$. The average number of alleles (n_a) and the effective number of alleles (n_e) are plotted by round and square dots respectively. Horizontal lines, solid and broken, represent corresponding theoretical values. For details, see text.

Mutation to a new (not preexisting) allele was induced in each gamete with probability 0.005 prior to the formation of zygotes ($u = 0.005$). The initial population was set up such that it contained 200 different alleles. Outputs of both average and effective numbers of alleles were given at 50-generation intervals starting with generation 100, and the experiment was continued until generation 1200. The balance between mutation and random extinction of alleles was reached well before generation 100. The averages of 23 outputs are $n_a = 5.522$ and $\bar{f} = 0.520$ or $n_e = 1.923$. These should be compared with the theoretical values obtained from equations 9.6.13 and 9.6.12, setting $N = 100$, $N_e = 50$, and $u = 0.005$, that is,

$$n_a = \int_{\frac{1}{200}}^{1} x^{-1}\, dx = \log_e 200 \approx 5.298$$

and

$$n_e = 4N_e u + 1 = 2.$$

Thus, the agreement between the results of a Monte Carlo experiment and the theoretical predictions based on the diffusion approximation is satisfactory. For more extensive simulation experiments, readers may refer to Kimura (1968).

9.7 The Number of Overdominant Alleles in a Finite Population

Since Fisher's paper of 1922 it has been known that in an infinite population, overdominance leads unconditionally to stable polymorphism for a pair of alleles. For more than two alleles, however, a more delicate condition has to be satisfied at stable equilibrium (see 6.8). The complexity of the condition, however, does not vitiate the general conclusion that overdominance is a potent factor for maintaining selective polymorphism in an infinite population. As pointed out by Robertson (1962), selection for the heterozygote is a factor retarding fixation if the equilibrium gene frequency lies inside the range of 0.2–0.8, when a pair of alleles are involved (see 8.6.4).

On the other hand, for an equilibrium gene frequency outside this range, there is a range of values of $N_e(s_1 + s_2)$ for which heterozygote advantage in fact accelerates fixation. N_e is the effective population number and s_1 and s_2 are selection coefficients against both homozygotes. This might suggest that if there are a large number of overdominant alleles in a population, they will be lost by random drift as readily as neutral alleles, and overdominance is rather ineffective in keeping a large number of them in a finite population.

In this section we will investigate quantitatively the maximum number of overdominant alleles that can be maintained in a finite population. For this purpose, we will consider an ideal situation in which every mutation produces a novel allele which is different from the preexisting ones and every allele is heterotic with any other allele. Furthermore, we assume that all heterozygotes have equal fitness and all homozygotes also have equal fitness which is lower by s compared with that of the heterozygotes. Any asymmetry with respect to fitness within homozygotes or heterozygotes will reduce the number of alleles. Consider a random-mating population of effective size N_e. We will designate by x the relative frequency of an allele in the population and let $\Phi(x)\,dx$ be the expected number of alleles whose frequency is in the range $x \sim x + dx$. The relative frequency of each allele may change from generation to generation by mutation, selection, and random drift, but at statistical equilibrium there will be a stable distribution in x given by Wright's formula (cf. 9.1.3),

$$\Phi(x) = \frac{C}{V_{\delta x}} e^{2\int \frac{M_{\delta x}}{V_{\delta x}} dx}, \qquad 9.7.1$$

where C is a constant and $M_{\delta x}$ and $V_{\delta x}$ are the mean and variance of the rate of change in x.

Let u be the mutation rate per gene per generation and denote by f the sum of squares of allelic frequencies that are contained in the population

$$f = \sum_i x_i^2, \tag{9.7.2}$$

where x_i is the frequency of the ith allele A_i.

As in the previous section, we will denote by $n_e = 1/f$ the effective number of alleles maintained in the population. Since the rate of change of the frequency of a particular allele by mutation is $-ux$ and that by selection is

$$-sx(x - f), \tag{9.7.3}$$

we have

$$M_{\delta x} = -ux - sx(x - f). \tag{9.7.4}$$

In the above expression, $s(>0)$ is the selective disadvantage of a homozygote compared with a heterozygote measured in Malthusian parameters. For a discrete model, in which the selective values of a homozygote and a heterozygote are $1 - s$ and 1, respectively, the amount of change in x per generation is given by

$$-\frac{sx(x - f)}{1 - sf}, \tag{9.7.5}$$

but since we are concerned with cases in which sf is much smaller than unity ($0 < sf \ll 1$), this also leads to the same expression as 9.7.3 if we neglect this small quantity in the denominator. The variance in the rate of change in x is given by

$$V_{\delta x} = \frac{x(1 - x)}{2N_e}. \tag{9.7.6}$$

Since we are considering a situation in which a large number of alleles are maintained in a population and the effective range of x is essentially restricted to a very low value ($x \ll 1$), we will use the approximate expression

$$V_{\delta x} = \frac{x}{2N_e}. \tag{9.7.7}$$

Using 9.7.4 and 9.7.7, we obtain from 9.7.1 the distribution formula

$$\Phi(x) = Ce^{-2S(x-f)^2 - 4Mx}\, x^{-1}, \tag{9.7.8}$$

where

$$S = N_e s, \qquad M = N_e u. \tag{9.7.9}$$

In deriving this distribution formula, f is assumed to be a constant which is interpreted as the expected value of the sum of squares of allelic frequencies or, more simply, as the probability of allelism, the reciprocal of which is the effective number of alleles in a population.

Such a treatment regarding f as constant (for a given S and M) is again an approximation, but as will be shown later, this turns out to be satisfactory for our purpose.

Constant C in 9.7.8 is determined by the condition that the frequencies of all the alleles in the population add up to unity,

$$\sum_i x_i = 1,$$

or

$$\int_0^1 x\Phi(x)\, dx = 1. \qquad \textbf{9.7.10}$$

Note that usually ϕ in Wright's formula (9.1.3) represents probability density of gene frequencies rather than the expected number of alleles as we use in this section. Thus, usually, C is determined by condition 9.1.4 rather than 9.7.10.

From 9.7.8 and 9.7.10, we obtain

$$\frac{1}{C} = \int_0^1 e^{-2S(x-f)^2-4Mx}\, dx. \qquad \textbf{9.7.11}$$

This can also be expressed as

$$\frac{1}{C} = \frac{e^{-4Mf+\frac{2M^2}{S}}}{2\sqrt{S}} \int_{-X}^{2\sqrt{S}-X} e^{-\frac{\lambda^2}{2}}\, d\lambda, \qquad \textbf{9.7.12}$$

where

$$X = 2\sqrt{S}\left(f - \frac{M}{S}\right). \qquad \textbf{9.7.13}$$

At the equilibrium state in which random extinction is exactly balanced by mutational production of new alleles, we have the following condition at the subterminal class:

$$\underbrace{2Nu}_{\text{number of new mutants}} = \underbrace{\frac{1}{2}\Phi\left(\frac{1}{2N}\right)\frac{1}{2N}\left(\frac{N}{N_e}\right)}_{\text{number of extinctions}} \qquad \textbf{9.7.14}$$

(cf. 8.3.22). This gives

$$4M = Ce^{-2S\left(\frac{1}{2N}-f\right)^2-\frac{4M}{2N}}. \qquad \textbf{9.7.15}$$

Since u is very small compared with unity, the term $(4M/2N)$ may be neglected in the actual calculation of the exponential term in the above formula. Furthermore, we assume that the effective number of alleles is much smaller than the total number of genes in a population, i.e., $n_e = 1/f \ll 2N$. Then the above relation 9.7.15 may be reduced to a simpler form,

$$4M = Ce^{-2Sf^2}. \tag{9.7.16}$$

Combining 9.7.12 with 9.7.16 to eliminate C, we obtain

$$e^{-\frac{X^2}{2}} = r\int_{-X}^{2\sqrt{S}-X} e^{-\frac{\lambda^2}{2}}\,d\lambda, \tag{9.7.17}$$

where

$$r = \frac{2M}{\sqrt{S}} \tag{9.7.18}$$

(Kimura and Crow, 1964). Thus, for given values of $M(=N_e u)$ and $S(=N_e s)$, the value of f may be obtained by first solving 9.7.17 for X and then obtaining f from 9.7.13. Then the effective number of alleles (n_e) is obtained by $1/f$.

Unless S is small, $2\sqrt{S} - X$ in the integral of 9.7.17 may be replaced with good approximation by ∞ to give

$$r = \frac{e^{-\frac{X^2}{2}}}{\int_{-X}^{\infty} e^{-\frac{\lambda^2}{2}}\,d\lambda}. \tag{9.7.19}$$

In deriving the above relations, one of the essential assumptions we have made is that f in equation 9.7.8 is a constant which represents the expected value of the sum of squares of allelic frequencies, i.e.,

$$f = E\left(\sum_i x_i^2\right). \tag{9.7.20}$$

So, let us examine the corresponding value evaluated from the integral:

$$J = \int_0^1 x^2\Phi(x)\,dx = C\int_0^1 xe^{-2S(x-f)^2-4Mx}dx. \tag{9.7.21}$$

Noting 9.7.12 and 9.7.16, it can be shown that

$$J = f - \frac{M}{S}\,e^{-2S\left(1-2f+\frac{2M}{S}\right)}. \tag{9.7.22}$$

Thus, if

$$f \gg \frac{M}{S}\, e^{-2S\left(1-2f+\frac{2M}{S}\right)}, \tag{9.7.23}$$

or equivalently, if

$$u \ll sfe^{2S(1-2f)+4M}, \tag{9.7.24}$$

we have

$$\int_0^1 x^2\Phi(x)\,dx = f \tag{9.7.25}$$

with sufficient approximation, and f in 9.7.8 indeed represents the expected value of

$$\sum x_i^2 .$$

Therefore, under this condition, the present formulation is satisfactory for estimating this expected value by solving for f using 9.7.17 or 9.7.19.

In the above treatments we have studied the effective number of alleles, n_e, defined by

$$n_e = \frac{1}{f}. \tag{9.7.26}$$

This is different from the actual number of alleles, the expectation of which may be obtained from

$$n_a = \int_{\frac{1}{2N}}^{1-\frac{1}{2N}} \Phi(x)\,dx. \tag{9.7.27}$$

In the latter, the rarest allele is counted as equivalent to a more common type when the allele number is counted. Therefore, $n_e \leqq n_a$. For a discussion of actual and effective number of alleles, see Section 7.1.

The segregational load due to overdominant alleles is given by

$$L_s = sf = \frac{s}{n_e}. \tag{9.7.28}$$

In Table 9.7.1, the relation between $r(= 2M/\sqrt{S})$ and $f\sqrt{S}$ is tabulated for X (considered as a parameter) between 3.0 and -3.0, using the relation 9.7.19 and

$$f\sqrt{S} = \frac{r + X}{2}, \tag{9.7.29}$$

Table 9.7.1. Quantities needed for computing the effective number of alleles.

X	$r\left(=\frac{2M}{\sqrt{S}}\right)$	$\frac{n_e}{\sqrt{S}}$	$f\sqrt{S}$
3.0	0.00444	0.666	1.50
2.7	0.0105	0.738	1.36
2.5	0.0176	0.794	1.26
2.2	0.0360	0.894	1.12
2.0	0.0553	0.973	1.03
1.7	0.0984	1.11	0.899
1.5	0.139	1.22	0.819
1.25	0.204	1.38	0.727
1.0	0.288	1.55	0.644
0.75	0.389	1.76	0.570
0.5	0.509	1.98	0.505
0.25	0.646	2.23	0.448
0	0.798	2.51	0.399
−0.25	0.964	2.80	0.357
−0.5	1.14	3.12	0.321
−0.75	1.33	3.46	0.290
−1.0	1.53	3.81	0.263
−1.25	1.73	4.18	0.239
−1.5	1.94	4.56	0.219
−1.7	2.11	4.88	0.205
−2.0	2.37	5.36	0.187
−2.2	2.55	5.69	0.176
−2.5	2.82	6.19	0.161
−2.7	3.00	6.60	0.151
−3.0	3.28	7.11	0.141

which is derived from 9.7.13. Here the numerical calculation is facilitated by the observation that r is equal to the ratio between the height of the normal curve (with mean 0 and unit variance) at X and the area under the curve from $-X$ to $+\infty$. In Figure 9.7.1, the relation between $2M/\sqrt{S}$ and $n_e/\sqrt{S}$ (i.e., the reciprocal of $f\sqrt{S}$) is shown graphically. For example with $N_e = 10^5$, $s = 10^{-3}$, and $u = 10^{-5}$, we have $S = 100$ and $M = 1$ so that $r = 2M/\sqrt{S} = 0.2$ and the graph gives $n_e/\sqrt{S} \approx 1.35$ or $n_e = 13.5$.

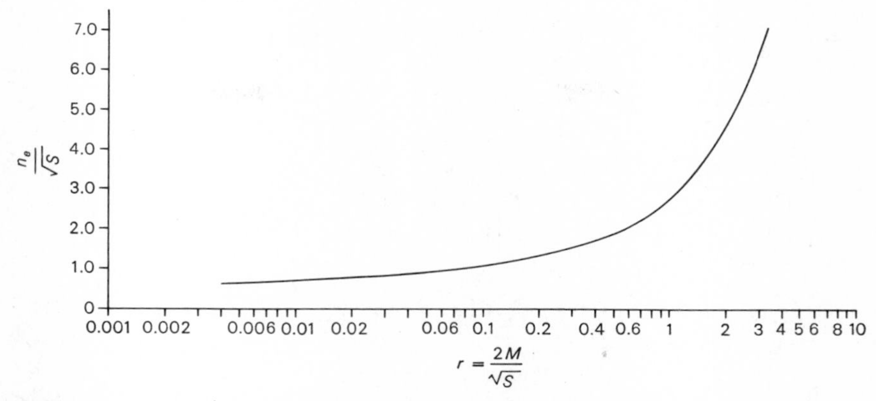

Figure 9.7.1. Graph showing the relationship between $r = 2M/\sqrt{S}$ and $n_e/\sqrt{S}$, where n_e is the effective number of alleles.

For values of r outside the range tabulated, the following approximations will be found useful:

(1) For a small value of r, use

$$f\sqrt{S} = \frac{1}{2}\sqrt{4.6 \log_{10}\left(\frac{0.4}{r}\right)}. \tag{9.7.30}$$

This is derived from the two approximate relations,

$$r = \frac{1}{\sqrt{2\pi}} e^{-\frac{X^2}{2}}$$

and

$$f\sqrt{S} = \frac{X}{2},$$

which hold when X is large and r is small. For example, with $N_e = 10^5$, $s = 0.1$, and $u = 0.5 \times 10^{-5}$, we have $\sqrt{S} = 100$, $2M = 1$, so that $r = 0.01$. Then formula 9.7.30 gives $f\sqrt{S} = 1.36$. We get practically the same value at this level of approximation from interpolation in Table 9.7.1. The effective number of alleles in this case is approximately 73.6.

(2) For a large value of r, use

$$f\sqrt{S} = \frac{1}{2r}\left(1 - \frac{1}{r^2}\right). \tag{9.7.31}$$

This is derived from 9.7.29 and the relation

$$r = |X| + \frac{1}{|X|} - \frac{2}{|X|^3} + \cdots,$$

which holds for a large negative value of X, for which we have

$$\int_{|X|}^{\infty} e^{-\frac{\lambda^2}{2}}\, d\lambda = \frac{1}{|X|}\, e^{-\frac{|X|^2}{2}} \left\{1 - \frac{1}{|X|^2} + \frac{3}{|X|^4} - \frac{1 \times 3 \times 5}{|X|^6} + \cdots\right\}.$$

Already at $r = 3$, 9.7.31 gives $f\sqrt{S} = 0.148$, a quantity very near to 0.151 which we get from Table 9.7.1.

Formula 9.7.31 may also be expressed in the form

$$n_e = 4M\left(1 + \frac{1}{r^2}\right). \qquad \textbf{9.7.32}$$

This can include the limiting case of $s = 0$ and therefore $r = \infty$ (i.e., neutral mutation). In this case, we get

$$n_e = 4M. \qquad \textbf{9.7.33}$$

Actually in this special case of neutral mutation, it was shown in the previous section that

$$n_e = 4M + 1. \qquad \textbf{9.7.34}$$

Since we are only concerned with the case of large n_e (or small f), formula 9.7.33 is sufficient for our purpose. Formula 9.7.32 is interesting in that it suggests that if r is large, or $2M$ is much larger than $\sqrt{S}$, selection plays only a minor role in maintaining a given number of alleles. In such a case, overdominance increases the number of alleles only about $100/r^2$ %. For example, in a population of $N_e = 10^5$, roughly 400 alleles may be maintained with the unusually high mutation rate of $u = 10^{-3}$, if the alleles are neutral. With the development of 1% overdominance ($s = 10^{-2}$), only 10 more alleles will be added. With overdominance of 10% ($s = 0.1$), 70 more alleles are added (cf. Table 9.7.1). Even with such a high overdominance, $s = 0.5$, the number of alleles added is roughly 200. No overdominance but a 50% increase in mutation rate can do the same. Thus, overdominance is extremely inefficient as a factor maintaining a large number of alleles in a population.

Figure 9.7.2 illustrates a Monte Carlo experiment performed to test the theoretical treatment of the number of overdominant alleles. The experiment was performed by using the IBM 7090 computer to simulate a population consisting of 50 breeding individuals in which one mutation occurred in each generation ($u = 0.01$, $N_e = N = 50$). The selection coefficient against homozygotes was $s = 0.1$ and the population contained 100 different alleles at the start (0th generation).

In the figure, experimental outcome on the effective number of alleles is shown from generation 60 through generation 400. Averaged over 18 outputs, the observed n_e is about 4.7, while the theoretical value for n_e (horizontal line in the figure) which is obtained from 9.7.17 is $n_e = 4.19$. So the two values for n_e agree fairly well.

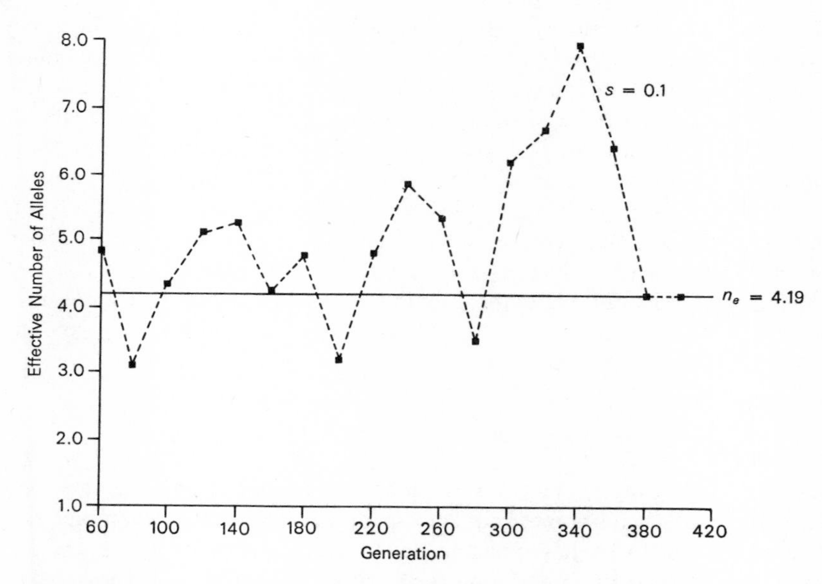

Figure 9.7.2. Monte Carlo experiment on the number of overdominant alleles, $s = 0.1$, $N = N_e = 50$, $u = 0.01$.

One of the most interesting applications of formula 9.7.17 is reported by Kerr (1967), who studied the number of sex-determining alleles in bee populations in Brazil. These alleles are haplo-viable but homozygous lethal so that among females they are overdominant with $s = 1$. In the population of Piracicaba, having an independently estimated effective population number (N_e) of 428, the observed number of alleles is $11.0 \sim 12.4$, while the corresponding theoretical value derived from the formula of Kimura and Crow (1964) is 11.2, if we take the mutation rate $u = 10^{-5}$, and it is 10.9 if $u = 10^{-6}$. So the number of alleles actually observed and the corresponding theoretical value are in good agreement. For the details, see the original paper by Kerr.

9.8 The Number of Heterozygous Nucleotide Sites Per Individual Maintained in a Finite Population by Mutation

In the previous section, we investigated the number of alleles maintained at a single locus, assuming that the number of possible allelic states is so large that each new mutant is a state not preexisting in the population.

In the present section, we will consider a different model and will investigate the number of heterozygous nucleotide sites per individual. We will assume that the total number of nucleotide sites per individual is so large, while the mutation rate per site is so low, that whenever a mutant appears in a population it represents a mutation at a site in which a previous mutant is not segregating. Such a model is also appropriate if we regard as a site a small group of nucleotides such as a codon rather than a single nucleotide. To treat this model, we will use the Kolomogorov backward equation as explained in Chapter 8, Section 3.

Let us assume that on the average in each generation mutant forms appear in the population in v_m sites. Consider a particular site in which a mutant has appeared. Let $\phi(p, x; t)$ be the probability density that the frequency of the mutant in the population becomes x after t generations, given that it is p at the start ($t = 0$). Then ϕ satisfies the Kolmogorov backward equation (see equation 8.3.2),

$$\frac{\partial\phi(p, x; t)}{\partial t} = \frac{1}{2} V_{\delta p} \frac{\partial^2\phi(p, x; t)}{\partial p^2} + M_{\delta p} \frac{\partial\phi(p, x;t)}{\partial p}. \qquad 9.8.1$$

In the above equation, $M_{\delta p}$ and $V_{\delta p}$ stand for the mean and the variance of the change of the mutant frequency per generation. For example, if the mutant in each locus has selective advantage s in homozygotes and sh in heterozygotes over the preexisting form, we have

$$M_{\delta p} = sp(1 - p)\{h + (1 - 2h)p\}, \qquad 9.8.2$$

and, if random fluctuation in mutant frequency is solely due to random sampling of gametes,

$$V_{\delta p} = \frac{p(1 - p)}{2N_e}, \qquad 9.8.3$$

where N_e is the variance effective number of the population.

In the following treatment, we will assume that $M_{\delta p}$ and $V_{\delta p}$ are the same for all the sites. Since a mutant form which appears in a finite population is either lost from the population or fixed in it within a finite length of time, under continued production of new mutations over many generations, there will be a stable distribution of mutant frequencies in different segregating sites. We will denote such a stable distribution by $\Phi(p, x)$, which is expressed in terms of $\phi(p, x; t)$ as

$$\Phi(p, x) = v_m \int_0^\infty \phi(p, x; t)\, dt, \qquad (0 < x < 1). \tag{9.8.4}$$

This distribution is interpreted to mean that $\Phi(p, x)\, dx$ stands for the *expected number* of sites having mutants in the frequency range x to $x + dx$. This may be seen as follows. Since new mutant forms appear in each generation in v_m sites, $v_m \phi(p, x; t)\, dx$ represents the contribution made by the mutants which appeared t generations earlier with initial frequency p to the present frequency class $x \sim x + dx$. Thus, adding up all the contributions made by the mutations in the past, the *expected number* of sites having mutants in the frequency range $x \sim x + dx$ at the present generation is

$$\left\{v_m \int_0^\infty \phi(p, x; t)\, dt\right\} dx.$$

Let us now consider the number of heterozygous sites per individual. Assuming random mating, the proportion of heterozygotes in a site having the mutant with frequency x is $2x(1 - x)$. So, the expected number of heterozygous sites per individual is

$$\begin{aligned} H(p) &= \int_0^1 2x(1 - x)\Phi(p, x)\, dx \\ &= v_m \int_0^1 2x(1 - x)\, dx \int_0^\infty \phi(p, x; t)\, dt. \end{aligned} \tag{9.8.5}$$

In order to obtain an equation for $H(p)$, we multiply each term of the backward equation 9.8.1 by $v_m 2x(1 - x)$ and then integrate each of the resulting terms first with respect to x over the open interval $(0, 1)$ and then with respect to t from $t = 0$ to $t = \infty$. This leads to

$$\int_0^\infty \frac{\partial}{\partial t}\left\{v_m \int_0^1 2x(1 - x)\phi(p, x; t)\right\} dt = \frac{1}{2} V_{\delta p} \frac{\partial^2}{\partial p^2} H(p) + M_{\delta p} \frac{\partial}{\partial p} H(p). \tag{9.8.6}$$

The left-hand side of this equation becomes

$$v_m\left\{\int_0^1 2x(1 - x)\phi(p, x; \infty)\, dx - \int_0^1 2x(1 - x)\phi(p, x; 0)\, dx\right\},$$

which is further reduced to $-2v_m p(1 - p)$ if we note that

$$\phi(p, x; \infty) = 0 \tag{9.8.7}$$

for $0 < x < 1$, and

$$\phi(p, x; 0) = \delta(x - p), \tag{9.8.8}$$

where $\delta(\cdot)$ is the Dirac delta function.

Of the above two conditions, the first, i.e., 9.8.7, follows from the fact that the mutant form is either fixed or lost in a finite time, while the second, i.e., 9.8.8, is based on the assumption that the initial frequency of the mutant is p. Thus, we obtain the ordinary differential equation for $H(p)$,

$$\tfrac{1}{2}V_{\delta p}H''(p) + M_{\delta p}H'(p) + 2v_m p(1-p) = 0. \tag{9.8.9}$$

Furthermore, since "mutations" at $p = 0$ and $p = 1$ do not contribute to the segregating sites, we have the boundary conditions

$$H(0) = H(1) = 0. \tag{9.8.10}$$

The solution of 9.8.9 which satisfies the boundary conditions 9.8.10 is

$$H(p) = \{1 - u(p)\}\int_0^p \psi_H(\xi)u(\xi)\,d\xi + u(p)\int_p^1 \psi_H(\xi)\{1 - u(\xi)\}\,d\xi, \tag{9.8.11}$$

where

$$\psi_H(\xi) = 4v_m\xi(1-\xi)\frac{\int_0^1 G(x)\,dx}{V_{\delta\xi}\,G(\xi)}, \tag{9.8.12}$$

and

$$u(p) = \frac{\int_0^p G(x)\,dx}{\int_0^1 G(x)\,dx} \tag{9.8.13}$$

(see 8.8.3.8) is the probability of ultimate fixation in which

$$G(x) = e^{-2\int_0^x \left(\frac{M_{\delta\xi}}{V_{\delta\xi}}\right)d\xi}.$$

In the special case of no dominance in which the mutant has selective advantage s in homozygotes and $s/2$ in heterozygotes over the preexisting form, 9.8.2 becomes

$$M_{\delta p} = \frac{s}{2}p(1-p)$$

and, if we combine this with 9.8.3, i.e., $V_{\delta p} = p(1-p)/(2N_e)$, we have $G(x) = e^{-2Sx}$, $u(p) = (1 - e^{-2Sp})/(1 - e^{-2S})$, and $\psi_H(\xi) = 4N_e v_m e^{2S\xi} \times (1 - e^{-2S})/S$, so that we obtain

$$H(p) = \frac{4N_e v_m}{S}\left(\frac{1 - e^{-2Sp}}{1 - e^{-2S}} - p\right), \tag{9.8.14}$$

where $S = N_e s$.

In a population consisting of N individuals, if the mutant in each site is represented only once at the moment of its occurrence, $p = 1/(2N)$ in the above formulae. Thus, for the case of no dominance, we have

$$H(1/2N) = \frac{4N_e v_m}{S}\left(\frac{1 - e^{-\frac{S}{N}}}{1 - e^{-2S}} - \frac{1}{2N}\right). \qquad 9.8.15$$

This reduces to

$$H(1/2N) \approx 4v_m \frac{N_e}{N} \qquad 9.8.16$$

if the mutant form is advantageous such that $2S = 2N_e s \gg 1$. On the other hand, if the mutant is deleterious such that $2N_e s' \gg 1$, in which $s' = -s$, 9.8.15 reduces to

$$H(1/2N) \approx \frac{2v_m}{Ns'}. \qquad 9.8.17$$

Finally, if the mutant is almost neutral such that $|2N_e s| \ll 1$, we have

$$H(1/2N) \approx 2v_m \frac{N_e}{N}. \qquad 9.8.18$$

These results suggest that mutant genes with definite advantage or disadvantage cannot contribute greatly to the heterozygosity of an individual because advantageous mutations occur rarely (as shown by small v_m in 9.8.16) and deleterious genes are eliminated rapidly (as shown by large Ns' in 9.8.17).

9.9 Decrease of Genetic Correlation with Distance in the Stepping-stone Model of Population Structure

Usually, a total population forming a species is not a random-mating unit, because the distance of individual migration is usually much smaller than the entire distribution range of the species. This phenomenon, which Wright called "isolation by distance" (Wright, 1943), will lead to local differentiation of gene frequencies due to random genetic drift. Wright considered a model in which a population is distributed uniformly over a large territory, but the parents of any given individual are drawn at random from a small surrounding region. The size of the neighborhood, that is, the population number in such a surrounding region, plays a fundamental role in his analysis. He studied, by his method of path coefficients, the pattern of change in the inbreeding coefficient of subgroups relative to a larger population in which they are contained (Wright, 1940, 1943, 1946, 1951). He has shown that the tendency toward local differentiation is much stronger in a linear than in a two-dimensional habitat.

The problem of local differentiation may also be studied in terms of change in correlation with distance, as considered by Malécot (1955), since individuals living nearby tend to be more alike than those living far apart.

In natural populations, individuals often are distributed more or less discontinuously to form numerous colonies, and individuals are exchanged mainly between adjacent or nearby colonies. To analyze such a situation, Kimura (1953) proposed a model of population structure which he termed the "stepping-stone model." A successful mathematical treatment of the model for the case of an infinite number of colonies was presented by Kimura and Weiss (1964) up to three-dimensional cases. A more general treatment that can cover any number of dimensions was also presented by Weiss and Kimura (1965). Earlier, Malécot (1959) studied a similar model and obtained the asymptotic solutions for one- and two-dimensional cases.

In the present section, we will consider mainly a one-dimensional stepping-stone model with an infinite array of colonies (see Figure 9.9.1).

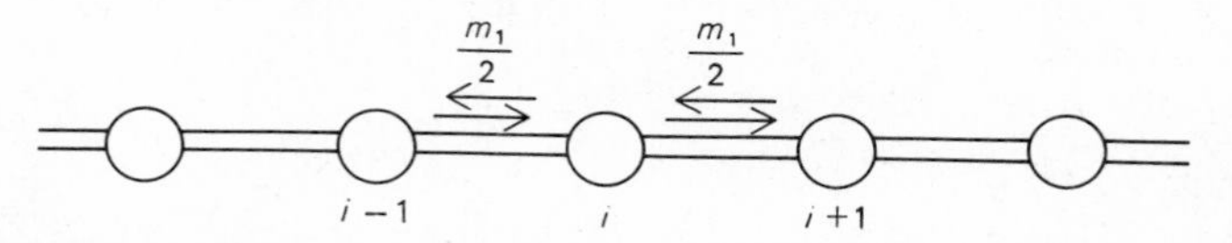

Figure 9.9.1. One-dimensional stepping-stone model.

Let us consider a pair of alleles A_1 and A_2 and let p_i be the frequency of A_1 in the ith colony in the present generation before migration. We assume that to go to the next generation, long-range migration, short-range migration, and random sampling of gametes occur in this order.

In long-range migration, we assume that each colony exchanges individuals with a random sample taken from the general gene pool with constant gene frequency p_I at the rate $\overline{m}_\infty$ per generation. The long-range migration is formally equivalent to mutation or, more generally, any linear systematic evolutionary pressure. For example, if there is mutation between A_1 and A_2 in addition to long-range migration, $\overline{m}_\infty$ should be replaced by $u + v + \overline{m}_\infty$ and $\overline{m}_\infty p_I$ by $v + \overline{m}_\infty p_I$ in the following treatment, where u is the mutation rate from A_1 to A_2 and v is the mutation rate in the reverse direction. Thus, by long-range migration, the gene frequency changes from p_i to

$$(1 - \overline{m}_\infty)p_i + \overline{m}_\infty p_I. \tag{9.9.1}$$

Next, we assume that short-range migration takes place. If such a migration is restricted to two neighboring colonies, each one step apart, and if m_1 is the rate of such short-range migration per generation so that $m_1/2$ is the proportion of individuals exchanged each generation between a pair of adjacent colonies (Figure 9.9.1), then the gene frequency becomes

$$P_i = (1 - m_1)\{(1 - \bar{m}_\infty)p_i + \bar{m}_\infty p_I\} + \frac{m_1}{2}\{(1 - \bar{m}_\infty)p_{i-1} + \bar{m}_\infty p_I\}$$

$$+ \frac{m_1}{2}\{(1 - \bar{m}_\infty)p_{i+1} + \bar{m}_\infty p_I\},$$

or

$$P_i = (1 - m_1)(1 - \bar{m}_\infty)p_i + \frac{m_1(1 - \bar{m}_\infty)}{2}(p_{i-1} + p_{i+1}) + \bar{m}_\infty p_I, \qquad 9.9.2$$

where P_i denotes the frequency of A_1 in the ith colony after long-range and short-range migrations.

Finally, we assume that random sampling of gametes takes place to form the next generation. Thus, if we denote by ξ_i the amount of change in the gene frequency resulting from random sampling of gametes in the ith colony, the frequency of A_1 in the next generation is

$$p_i' = P_i + \xi_i. \qquad 9.9.3$$

Since ξ_i represents the change due to random sampling, we have

$$E_\delta(\xi_i) = 0 \qquad 9.9.4$$

and

$$E_\delta(\xi_i^2) = \frac{P_i(1 - P_i)}{2N_e}, \qquad 9.9.5$$

where N_e is the effective population number of each colony and E_δ designates the operator for taking the expectation with respect to random sampling. In the following calculation we will denote by E_ϕ an operator for taking the expectation with respect to the existing gene frequency distribution, so that the operator E for taking the overall expectation is given by $E = E_\phi E_\delta$.

Let us first consider the mean $\bar{p}$ of the gene frequency distribution at equilibrium, i.e., $E(p_i) = \bar{p}$. Taking expectations of both sides of 9.9.3 and noting 9.9.2 and 9.9.4, we obtain

$$E(p_i') = (1 - \bar{m}_\infty)\bar{p} + \bar{m}_\infty p_I.$$

Since $E(p_i') = \bar{p}$ at equilibrium, we have

$$\bar{p} = p_I.$$

So, we will substitute $\bar{p}$ for p_I in the following treatment. In order to calculate the variance and the correlation coefficients of gene frequencies at equilibrium, we will let

$$\tilde{p}_i = p_i - \bar{p}$$

and

$$\tilde{P}_i = P_i - \bar{p}.$$

Then 9.9.2 becomes

$$\tilde{P}_i = \alpha\tilde{p}_i + \beta(\tilde{p}_{i-1} + \tilde{p}_{i+1}), \tag{9.9.6}$$

where $\alpha = (1 - m_1)(1 - \bar{m}_\infty)$ and $\beta = m_1(1 - \bar{m}_\infty)/2$. Also, 9.9.3 becomes

$$\tilde{p}'_i = \tilde{P}_i + \xi_i, \tag{9.9.7}$$

where $\tilde{p}'_i = p'_i - \bar{p}$.

We will denote by V_p the variance of the gene frequency distribution among colonies. Squaring both sides of 9.9.7 and taking expectations, the variance in the next generation is

$$V'_p \equiv E(\tilde{p}_i'^2) = E_\phi\{\tilde{P}_i^2 + 2\tilde{P}_i E_\delta(\xi_i) + E_\delta(\xi_i^2)\}.$$

If we apply 9.9.4 and 9.9.5 to the right-hand side of the above equation and note that $P_i = \tilde{P}_i + \bar{p}$, we obtain

$$V'_p = \left(1 - \frac{1}{2N_e}\right)E_\phi(\tilde{P}_i^2) + \frac{\bar{p}(1-\bar{p})}{2N_e}. \tag{9.9.8}$$

Now, from 9.9.6

$$E_\phi(\tilde{P}_i^2) = \alpha^2 V_p + 4\alpha\beta V_p r_1 + 2\beta^2 V_p(1 + r_2), \tag{9.9.9}$$

where r_j is the correlation coefficient of gene frequencies between two colonies which are j steps apart, i.e.,

$$r_j \equiv \frac{E_\phi(\tilde{p}_i\tilde{p}_{i+j})}{V_p}.$$

In particular, $r_0 = 1$, and also we assume $r_{-1} = r_1$. Substituting 9.9.9 for $E_\phi(\tilde{P}_i^2)$ in the right-hand side of 9.9.8, we have

$$V'_p = \left(1 - \frac{1}{2N_e}\right)\{\alpha^2 + 4\alpha\beta r_1 + 2\beta^2(1 + r_2)\}V_p + \frac{\bar{p}(1-\bar{p})}{2N_e}. \tag{9.9.10}$$

Thus at equilibrium in which $V_p' = V_p$, we obtain

$$V_p = \frac{1}{2N_e}\left\{\frac{\bar{p}(1-\bar{p})}{1-\left(1-\frac{1}{2N_e}\right)(\alpha^2+2\beta^2+4\alpha\beta r_1+2\beta^2 r_2)}\right\}, \tag{9.9.11}$$

where r_1 and r_2 are correlation coefficients of gene frequencies between colonies one and two steps apart.

In order to obtain the correlation coefficients $r_j (j = 1, 2, \ldots)$, let us consider the covariance (C) in the next generation,

$$C_j' \equiv E(\tilde{p}_i' \tilde{p}_{i+j}'),$$

in which $\tilde{p}_i' = \tilde{P}_i + \xi_i$ and $\tilde{p}_{i+j}' = \tilde{P}_{i+j} + \xi_{i+j}$. Since for $j \geqq 1$, ξ_i and ξ_{i+j} are mutually independent random variables, each with mean 0, we have

$$C_j' = E\{(\tilde{P}_i + \xi_i)(\tilde{P}_{i+j} + \xi_{i+j})\}$$

or

$$C_j' = E_\phi(\tilde{P}_i \tilde{P}_{i+j}).$$

Then, if we use relation 9.9.6, we get

$$\begin{aligned} C_j' &= E_\phi\{[\alpha\tilde{p}_i + \beta(\tilde{p}_{i-1} + \tilde{p}_{i+1})][\alpha\tilde{p}_{i+j} + \beta(\tilde{p}_{i+j-1} + \tilde{p}_{i+j+1})]\} \\ &= \alpha^2 C_j + 2\alpha\beta(C_{j+1} + C_{j-1}) + \beta^2(C_{j+2} + 2C_j + C_{j-2}). \end{aligned}$$

At equilibrium in which $C_j' = C_j$, noting $r_j = C_j/V_p$, we obtain

$$(\alpha^2 + 2\beta^2 - 1)r_j + 2\alpha\beta(r_{j+1} + r_{j-1}) + \beta^2(r_{j+2} + r_{j-2}) = 0, \qquad (j \geqq 1). \tag{9.9.12}$$

Equation 9.9.12 holds for $j \geqq 1$. However, for $j = 1$, r_{-1} should be replaced by r_1 to give

$$(\alpha^2 + 2\beta^2 - 1)r_1 + 2\alpha\beta(r_2 + 1) + \beta^2(r_3 + r_1) = 0. \tag{9.9.13}$$

The essential part of the mathematical treatment of the stepping-stone model is to find the solution (r_j) for the system of equations 9.9.12 that satisfies the boundary conditions, $r_0 = 1$ and $r_\infty = 0$. This was done by Kimura and Weiss (1964) and also by Weiss and Kimura (1965). Here we will present an elementary treatment which was given in Kimura and Weiss (1964). Let $r_j = \lambda^j$ and substitute in 9.9.12; then we have

$$(\alpha^2 + 2\beta^2 - 1) + 2\alpha\beta(\lambda + \lambda^{-1}) + \beta^2(\lambda^2 + \lambda^{-2}) = 0,$$

or

$$\{\alpha + \beta(\lambda + \lambda^{-1})\}^2 - 1 = 0,$$

from which we obtain the four roots,

$$\lambda_1 = \frac{(1-\alpha) + \sqrt{(1-\alpha)^2 - (2\beta)^2}}{2\beta},$$

$$\lambda_2 = \frac{(1-\alpha) - \sqrt{(1-\alpha)^2 - (2\beta)^2}}{2\beta},$$

$$\lambda_3 = -\frac{(1+\alpha) + \sqrt{(1+\alpha)^2 - (2\beta)^2}}{2\beta}, \tag{9.9.14}$$

$$\lambda_4 = -\frac{(1+\alpha) - \sqrt{(1+\alpha)^2 - (2\beta)^2}}{2\beta}.$$

Then the required solution may be expressed in the form

$$r_j = \sum_{i=1}^{4} K_i \lambda_i^j, \tag{9.9.15}$$

where K_i's are constants. However, since $\lambda_1 > 1$, $1 > \lambda_2 > 0$, $\lambda_3 < -1$, and $-1 < \lambda_4 < 0$, we must have $K_1 = K_3 = 0$ in order that r_j vanishes at $j = \infty$. Furthermore, in order that $r_0 = 1$, we must have $K_2 + K_4 = 1$. Thus, writing K for K_2, 9.9.15 becomes

$$r_j = K\lambda_2^j + (1-K)\lambda_4^j. \tag{9.9.16}$$

In order to determine K, we substitute $r_1 = K\lambda_2 + (1-K)\lambda_4$, $r_2 = K\lambda_2^2 + (1-K)\lambda_4^2$, and $r_3 = K\lambda_2^3 + (1-K)\lambda_4^3$ in 9.9.13. This yields

$$\begin{aligned} 2\alpha\beta + K\{(\alpha^2 + 3\beta^2 - 1)\lambda_2 + 2\alpha\beta\lambda_2^2 + \beta^2\lambda_2^3\} \\ + (1-K)\{(\alpha^2 + 3\beta^2 - 1)\lambda_4 + 2\alpha\beta\lambda_4^2 + \beta^2\lambda_4^3\} = 0. \end{aligned} \tag{9.9.17}$$

Then, if we use the relationships

$$\lambda_2^2 = \frac{1-\alpha}{\beta}\lambda_2 - 1, \qquad \lambda_2^3 = \left\{\frac{(1-\alpha)^2}{\beta^2} - 1\right\}\lambda_2 - \frac{1-\alpha}{\beta},$$

and

$$\lambda_4^2 = -\frac{1+\alpha}{\beta}\lambda_4 - 1, \qquad \lambda_4^3 = \left\{\frac{(1+\alpha)^2}{\beta^2} - 1\right\}\lambda_4 + \frac{1+\alpha}{\beta},$$

9.9.17 is reduced to

$$2\alpha\beta + K\{2\beta^2\lambda_2 - \beta(\alpha+1)\} + (1-K)\{2\beta^2\lambda_4 + \beta(1-\alpha)\} = 0.$$

This is further reduced, if we substitute 9.9.14 for λ_2 and λ_4, to give

$$R_1 - (R_1 + R_2)K = 0,$$

or

$$K = \frac{R_1}{R_1 + R_2}, \tag{9.9.18}$$

where

$$R_1 = \sqrt{(1+\alpha)^2 - (2\beta)^2}$$

and

$$R_2 = \sqrt{(1-\alpha)^2 - (2\beta)^2},$$

in which $\alpha = (1 - m_1)(1 - \overline{m}_\infty)$ and $\beta = m_1(1 - \overline{m}_\infty)/2$.

Therefore, the correlation of gene frequencies between two colonies which are j steps apart is given by 9.9.16 with λ_2 and λ_4 given by 9.9.14 and K given by 9.9.18. Then, applying this formula to calculate r_1 and r_2 in 9.9.11, the formula for variance becomes

$$V_p = \frac{\bar{p}(1-\bar{p})}{2N_e\left\{1 - \left(1 - \dfrac{1}{2N_e}\right)\left(1 - \dfrac{2R_1R_2}{R_1 + R_2}\right)\right\}}. \tag{9.9.19}$$

In the special case of no migration between adjacent colonies (island model), $m_1 = 0$ and therefore $R_1 = 1 + \alpha$, $R_2 = 1 - \alpha$. Thus, 9.9.19 agrees with 9.2.13 except that $\overline{m}_\infty$ rather than m is used to represent long-range migration.

Weiss and Kimura (1965) developed a more sophisticated method which can also treat higher dimensional cases. Using this method, the correlation in the one-dimensional model is expressed as

$$r_j = \frac{\dfrac{1}{2\pi}\displaystyle\int_0^{2\pi} \frac{\cos j\theta \, d\theta}{1 - H^2(\cos\theta)}}{\dfrac{1}{2\pi}\displaystyle\int_0^{2\pi} \frac{d\theta}{1 - H^2(\cos\theta)}} \qquad (j \geqq 0), \tag{9.9.20}$$

where

$$H(\cos\theta) = \alpha + 2\beta\cos\theta, \tag{9.9.21}$$

in which $\alpha = (1 - m_1)(1 - \overline{m}_\infty) = 1 - \overline{m}_\infty - m_1(1 - \overline{m}_\infty)$ and $2\beta = m_1 \times (1 - \overline{m}_\infty)$. Formula 9.9.20 may also be expressed in the form

$$r_j = \frac{A_1(j) + A_2(j)}{A_1(0) + A_2(0)}, \tag{9.9.22}$$

where

$$A_1(j) = \frac{1}{4\pi}\int_0^{2\pi} \frac{\cos j\theta \, d\theta}{1 - H(\cos\theta)} = \frac{1}{4\pi}\int_0^{2\pi} \frac{\cos j\theta \, d\theta}{(1-\alpha) - 2\beta\cos\theta} \tag{9.9.23}$$

and

$$A_2(j) = \frac{1}{2\pi}\int_0^{2\pi} \frac{\cos j\theta \, d\theta}{1 + H(\cos\theta)} = \frac{1}{4\pi}\int_0^{2\pi} \frac{\cos j\theta \, d\theta}{(1+\alpha) + 2\beta\cos\theta}. \tag{9.9.24}$$

To evaluate the above integrals the following formula will be found useful:

$$\frac{1}{2\pi}\int_0^{2\pi} \frac{\cos n\theta \, d\theta}{x + \cos\theta} = \begin{cases} \dfrac{1}{\sqrt{x^2 - 1}}(\sqrt{x^2 - 1} - x)^n, & x > 1 \\ \dfrac{(-1)^{n+1}}{\sqrt{x^2 - 1}}(x + \sqrt{x^2 - 1})^n, & x < -1 \end{cases} \tag{9.9.25}$$

Then,

$$A_1(j) = \frac{1}{4\beta}\frac{1}{\sqrt{a_1^2 - 1}}(a_1 - \sqrt{a_1^2 - 1})^j \tag{9.9.26}$$

and

$$A_2(j) = \frac{1}{4\beta}\frac{1}{\sqrt{a_2^2 - 1}}(\sqrt{a_2^2 - 1} - a_2)^j, \tag{9.9.27}$$

where

$$a_1 = \frac{1-\alpha}{2\beta} = 1 + \frac{\overline{m}_\infty}{m_1(1 - \overline{m}_\infty)}$$

and

$$a_2 = \frac{1+\alpha}{2\beta} = \frac{2 - \overline{m}_\infty}{m_1(1 - \overline{m}_\infty)} - 1.$$

Thus 9.9.20 agrees with 9.9.16.

Though the above solution for r_j given in 9.9.16 is perfectly general, a simple approximation formula is available for an interesting and important special case in which $\bar{m}_\infty \ll m_1$. In this case, $A_2(j)$ is small in comparison to $A_1(j)$ so that

$$r_j = e^{-\left(\sqrt{\frac{2\bar{m}_\infty}{m_1}}\right)j}, \qquad 9.9.28$$

approximately. If in addition m_1 is small so that $1 \gg m_1 \gg \bar{m}_\infty$, we have, from 9.9.19.

$$V_p = \frac{\bar{p}(1-\bar{p})}{1 + 4N_e\sqrt{2m_1\bar{m}_\infty}}, \qquad 9.9.29$$

approximately.

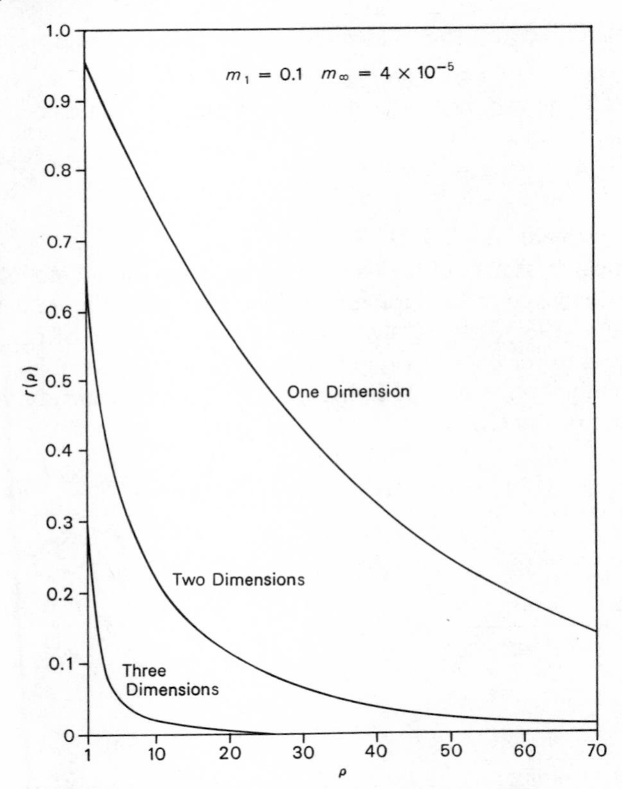

Figure 9.9.2. Decrease of genetic correlation with distance when $m_1 = 0.1$ and $\bar{m}_\infty = 4 \times 10^{-5}$. (From Kimura and Weiss, 1964.)

In Section 9.2 we derived formula 9.2.5, giving the gene frequency distribution among subgroups in the "island model." In this model, if the entire species is subdivided into colonies, each with effective size N_e, and each colony exchanges individuals at the rate m with a random sample taken from the entire species, the frequency distribution is given by

$$\phi(p) = Cp^{4N_e m\bar{p}-1}(1-p)^{4N_e m(1-\bar{p})-1}, \tag{9.9.30}$$

where C is a constant.

In the stepping-stone model which we are considering in this section, there is a correlation in gene frequency between immigrants and the receiving colony so that $m_1(1-r_1)$ gives the effective rate of exchange comparable to m in the island model. So, the gene frequency distribution among colonies in the stepping-stone model may be approximated by 9.9.30 if we substitute

$$m = \bar{m}_\infty + m_1(1-r_1). \tag{9.9.31}$$

When $\bar{m}_\infty \ll m_1$, we have $m \approx \sqrt{2m_1\bar{m}_\infty}$ since $r_1 \approx 1 - \sqrt{2\bar{m}_\infty/m_1}$ from 9.9.28. So, from 9.2.7, the variance of gene frequency among colonies is

$$\sigma_p^2 = \frac{\bar{p}(1-\bar{p})}{4N_e m + 1} \approx \frac{\bar{p}(1-\bar{p})}{4N_e\sqrt{2m_1\bar{m}_\infty} + 1},$$

which is in good agreement with 9.9.29.

The mathematical treatment of the two- and three-dimensional models is more difficult, but it is given in Kimura and Weiss (1964) and in more detail in Weiss and Kimura (1965). Figure 9.9.2 illustrates the decrease of genetic correlation with distance for one-, two-, and three-dimensional cases, assuming $\bar{m}_\infty = 4 \times 10^{-5}$ and $m_1 = 0.1$. The figure shows that it depends very much on the number of dimensions.

APPENDIX

SOME STATISTICAL AND MATHEMATICAL METHODS FREQUENTLY USED IN POPULATION GENETICS

The purpose of this appendix is to supply some of the methods that the reader may not have encountered or has forgotten. There is no attempt at mathematical rigor, although we have tried to supply enough of the background to make clear the general nature and limitations of the methods. This is written for biologists, not mathematicians, and if one of the latter chances to be reading this he is invited to look the other way or at least be tolerant of some of the intuitive arguments and cookbook attitudes.

No knowledge of statistics, matrices, or higher mathematics is assumed, but the reader is expected to know the elementary theory of probability and to be familiar with the differential and integral calculus.

This appendix is intended to be sufficient for any procedures used in the first six chapters, and most of Chapter 7. However, the last two chapters involve subjects of considerably greater mathematical difficulty and it seems to be impractical to include all the procedures here. The reader will either have to accept some results without derivation or look up the methods elsewhere. We have tried to provide references.

Among the various books that might be mentioned as sources of additional information we call particular attention to two that are by pioneers in the field of population genetics. One is the first volume of Sewall Wright's *Evolution and the Genetics of Populations*, which is devoted to the biometrical and statistical foundations of population genetics. The other is R. A. Fisher's *Statistical Methods for Research Workers*.

There is one method of great importance in population and biometrical genetics which we have omitted. This is Sewall Wright's method of path coefficients. Although we have used alternative methods in this book, the reader should realize that many of the results were first obtained by this method. The procedures are derived and explained in Wright's first volume (1968). A clear elementary exposition is given by Li (1955*a*).

A.1 Various Kinds of Averages

If there are a large number of observations or measurements it is usually difficult, if not impossible, to make much sense of them by examination of the individual values. We therefore make use of various derived quantities that extract from the data the information in which we are especially interested. We usually desire some measure of the central or typical value and some measure of the amount of variability. We might wish, in addition, to know other things, such as whether the values are symmetrically distributed about the central value. Usually, we are especially interested in two quantities, the mean and the variance.

The ordinary average, or *arithmetic mean*, is represented by M_X or $\overline{X}$ and is defined by

$$M_X = \overline{X} = \frac{1}{N}(X_1 + X_2 + X_3 + \cdots + X_N) = \frac{1}{N}\sum_{i=1}^{N} X_i, \qquad \text{A.1.1}$$

where $X_1, X_2, \ldots, X_N$ are the successive measurements, N is the number of measurements, and the Greek $\sum$ means to add all the X's from X_1 to X_N. The convention of indicating the mean by a superior bar is widely used.

The *geometric mean* is the *Nth* root of the product of the N values. It is frequently useful for data that are not symmetrically distributed. In population genetics the geometric mean is useful in measuring average population growth over a period of generations, because of the geometric nature of population increase. The geometric mean is defined symbolically by

$$G_X = (X_1 X_2 \cdots X_N)^{\frac{1}{N}} = \left(\prod_{1}^{N} X_i\right)^{\frac{1}{N}}. \qquad \text{A.1.2}$$

For computation, it is more conveniently defined as the antilog of

$$\bar{x} = \frac{1}{N} \sum x_i, \qquad \text{A.1.3}$$

where $x_i = \log X_i$. In other words, the geometric mean is the antilog of the arithmetic mean of the logarithms of the values.

Another mean that is used in population genetics, for example in the study of effects of chance in small populations, is the *harmonic mean.* The harmonic mean is the reciprocal of the mean of the reciprocals, or

$$\frac{1}{H_X} = \frac{1}{N} \sum \left(\frac{1}{X_i}\right). \qquad \text{A.1.4}$$

Table A.1. Some hypothetical data to illustrate computational methods.

i	X_i	$\frac{1}{X_i}$	$\log_{10} X_i$	$X_i - \bar{X}$	$(X_i - \bar{X})^2$
1	12	.083	1.079	2	4
2	9	.111	.954	−1	1
3	13	.077	1.114	3	9
4	7	.143	.845	−3	9
5	11	.091	1.041	1	1
6	11	.091	1.041	1	1
7	9	.111	.954	−1	1
8	11	.091	1.041	1	1
9	7	.143	.845	−3	9
Sum	90	.941	8.914	0	36

$$N = 9$$

$$M_X = \bar{X} = \frac{90}{9} = 10$$

$$H_X = \frac{1}{\left(\frac{.941}{9}\right)} = 9.56$$

$$G_X = \text{antilog} \frac{8.914}{9} = \text{antilog}\ .990 = 9.77$$

$$\text{Median} = 11$$

$$V_X = \frac{36}{9} = 4$$

For some purposes, the most appropriate summarizing value is not any of the mean values, but the *median*. The median is the middle value; that is, the value chosen so that there are an equal number of observations above and below it. If the total number of observations is even, the median is defined as the mean of the two central values. A familiar example of the use of the median is in the half life of a radioactive element. An exactly analogous problem arises in measuring the number of generations through with a harmful mutant gene persists in the population; it is sometimes convenient to measure its half life, or median persistence, rather than its mean persistence.

These various averages are illustrated with a simple numerical example in Table A.1.

A.2 Measures of Variability: The Variance and Standard Deviation

There are also several measures of variability, but we shall consider only two, the variance and the standard deviation. The *variance*, V, usually reflects the properties of greatest genetic interest. It is defined as the mean of the squares of the deviations of the individual items from their mean. Symbolically,

$$V_X = \frac{1}{N} \sum [(X_i - \overline{X})^2]. \qquad \text{A.2.1}$$

This is illustrated with a numerical example in Table A.1.

For computation, it is more convenient to write the variance formula as

$$V_X = \overline{X^2} - \overline{X}^2; \qquad \text{A.2.2}$$

that is, the mean of the squared value minus the square of the mean value. This can be derived as follows. From A.2.1,

$$V_X = \frac{1}{N} \sum (X_i^2 - 2\overline{X}X_i + \overline{X}^2)$$

$$= \frac{1}{N} [\sum X_i^2 - 2\overline{X} \sum X_i + N\overline{X}^2] \qquad (\text{since} \sum \overline{X}^2 = N\overline{X}^2)$$

$$= \frac{1}{N} [\sum X_i^2 - N\overline{X}^2] \qquad (\text{since} \sum X_i = N\overline{X})$$

$$= \overline{X^2} - \overline{X}^2.$$

Data are often grouped into classes with the same or similar measurements. In this case the mean and variance are computed by weighting each measurement by the number of individuals with that measurement. If there are n_1 individuals with measurement X_1, n_2 with measurement X_2, and so on, then

$$\overline{X} = \frac{n_1 X_1 + n_2 X_2 + \cdots + n_k X_k}{n_1 + n_2 + \cdots + n_k} = \frac{\sum n_i X_i}{\sum n_i} \tag{A.2.3}$$

and

$$V_X = \frac{\sum n_i (X_i - \overline{X})^2}{\sum n_i} = \frac{\sum n_i X_i^2 - N\overline{X}^2}{N}, \tag{A.2.4}$$

where $N = \sum n_i$.

Notice that, as before, the variance formula can be written

$$V_X = \overline{X^2} - \overline{X}^2,$$

where the superior bar now indicates the weighted mean.

In population genetics the frequencies of individuals with each measurement or attribute are usually expressed as a proportion of the total.

If $p_i = n_i/N$, then $\sum p_i = 1$,

$$\overline{X} = \sum p_i X_i, \tag{A.2.5}$$

and

$$V_X = \sum p_i (X_i - \overline{X})^2 = \sum p_i X_i^2 - \overline{X}^2. \tag{A.2.6}$$

The variance is always measured in units that are the square of the units of the original measurements. This is sometimes inconvenient, as when the variance in height of a group of persons is given as square inches, or variance in the yield of wheat is in square bushels. In order to return to the original dimensions, it is customary to take the square root of the variance, and call the resulting quantity the *standard deviation*.

Thus, the standard deviation, σ, is defined by

$$\sigma = \sqrt{V}. \tag{A.2.7}$$

The population variance is often denoted by σ^2.

Despite these dimensional difficulties, the variance is almost always the more useful quantity in population genetics. There are two principal reasons. One is that the variance has properties of additivity and subdivisibility whereas the standard deviation does not. This means that if a compound measure is the sum of two independent measures, the variance of the compound is the sum of the variances of the two parts. For example, if we can think of the yield of maize, Y, as being the sum of a genetic component G, and an independent environmental component, E, then $V_Y = V_G + V_E$.

The second property of the variance that makes it especially useful in population genetics is that the rate of evolutionary change is more closely related to the variance than to other measures of population variability.

The *variance of the sum* of two measurements may be derived as follows. Let X and Y be the two measurements.

$$V_{X+Y} = \frac{1}{N} \sum (X_i + Y_i - \overline{X + Y})^2$$

$$= \frac{1}{N} \sum (X_i - \overline{X} + Y_i - \overline{Y})^2 \qquad (\text{since } \overline{X + Y} = \overline{X} + \overline{Y})$$

$$= \frac{1}{N} [\sum (X_i - \overline{X})^2 + \sum (Y_i - \overline{Y})^2 + 2 \sum (X_i - \overline{X})(Y_i - \overline{Y})]$$

$$= V_X + V_Y + 2 \text{ cov}_{XY}, \qquad \text{A.2.8}$$

where

$$\text{cov}_{XY} = \frac{1}{N} \sum (X_i - \overline{X})(Y_i - \overline{Y}). \qquad \text{A.2.9}$$

Cov_{XY} is called the *covariance* of X and Y.

If the measurements are of several quantities, then the variance of the sum is

$$V_{\Sigma X} = \sum V_X + 2 \sum \text{cov}_{XX'}, \qquad \text{A.2.10}$$

where the last sum is the sum of all possible pairs of covariances. If there are measurements on n objects then there are $n(n - 1)/2$ covariances and n variances.

For example, if we had measurements on hand length, forearm length, and upper arm length for each of N persons, the variance of the total arm length would be given by the sum of the three variances and six covariances.

If the quantities X and Y are independent of each other they will tend to deviate from their means in opposite directions just as often as in the same direction. In the former case, the value of the product will be negative, in the latter, positive; so, on the average, $\sum (X_i - \overline{X})(Y_i - \overline{Y})$ will be 0.

Therefore, if the quantities X and Y are independent

$$V_{X+Y} = V_X + V_Y, \qquad \text{A.2.11}$$

or, if there are more than two quantities, and they are independent,

$$V_{\Sigma X} = \sum V_X. \qquad \text{A.2.12}$$

If there are n equally variable and independent quantities, then the variance of their sum is

$$V_{\Sigma X} = nV_X. \qquad \text{A.2.13}$$

If K is a constant

$$V_{KX} = \frac{1}{N}\sum(KX_i - \overline{KX})^2$$

$$= \frac{1}{N}\sum[K(X_i - \overline{X})]^2 \qquad (\text{since } \overline{KX} = K\overline{X}) \qquad \text{A.2.14}$$

$$= K^2\frac{1}{N}\sum(X_i - \overline{X})^2$$

$$= K^2V_X.$$

From the foregoing, we derive a very important formula—the variance of a mean. If $\overline{X}$ is the mean of N independent observations, using A.2.12 we obtain

$$V_{\overline{X}} = \left(\frac{1}{N}\right)^2 V_{\Sigma X} \qquad \left(\text{since } \overline{X} = \frac{1}{N}\sum X_i\right)$$

$$= \frac{1}{N}V_X \qquad (\text{from } A.2.13). \qquad \text{A.2.15}$$

A.3 Population Values and Sample Values

In the practical use of statistics we are often interested in drawing inferences about some population on the basis of a sample of observations from the population. In such circumstances we have little interest in the sample values themselves except in so far as they provide information about the population. If the sample is representative of the population, we can use it as the basis for estimates of the unknown population values.

If we wish to estimate the mean of a population, the mean of the sample is on the average a correct estimate. It is unbiased, in the sense that the expected value of $\overline{X}$ is μ, the population value.

On the other hand, the variance of a sample is a biased estimate of the variance of the population. However, there is a simple correction that removes the bias; simply divide by $N - 1$ instead of N.

We estimate σ^2, the population variance, by measuring the squared deviations from the true mean, μ, rather than the sample mean $\bar{X}$. Using E to stand for the expected or mean value, we have

$$V = E\left\{\frac{1}{N}\sum (X_i - \mu)^2\right\}$$

$$= E\left\{\frac{1}{N}\sum [(X_i - \bar{X}) + (\bar{X} - \mu)]^2\right\}$$

$$= E\left\{\frac{1}{N}\left[\sum (X_i - \bar{X})^2 + 2(\bar{X} - \mu)\sum (X_i - \bar{X}) + \sum (\bar{X} - \mu)^2\right]\right\}.$$

But, $\sum (X_i - \bar{X}) = 0$ and $E\{(1/N)\sum (\bar{X} - \mu)^2\}$ is the variance of the mean of X, which is V/N. Hence

$$V = E\left\{\frac{1}{N}\sum (X_i - \bar{X})^2\right\} + \frac{1}{N}V = \frac{1}{N-1}E\{\sum (X_i - \bar{X})^2\}$$

and an unbiased estimate of the population variance is

$$V = \frac{\sum (X_i - \bar{X})^2}{N-1}. \quad \text{A.3.1}$$

Most of the time in this book we are dealing with theoretical populations and their variances, so formulae A.2.1 and A.2.2 are appropriate; but whenever one is dealing with actual data and wants to estimate the population values from these data, A.3.1 should be used.

The same correction applies to the covariance if the purpose is to estimate the population parameter from sample measurements. Thus,

$$\text{cov}_{XY} = \frac{\sum (X_i - \bar{X})(Y_i - \bar{Y})}{N-1}. \quad \text{A.3.2}$$

The variance of the mean, as estimated from measurements on a sample of N observations, is

$$V_{\bar{X}} = \frac{\sum (X_i - \bar{X})^2}{N(N-1)}. \quad \text{A.3.3}$$

A.4 Correlation and Regression

The covariance is measured in units which are the squares of the original measurement units. So is the variance. It is frequently desirable to have a measurement of association that is dimensionless. For this purpose the conventional measurement is the *Coefficient of Correlation.* It is the covariance divided by the geometric mean of the two variances. Thus the correlation coefficient, r, is

$$r_{XY} = \frac{\text{cov}_{XY}}{\sqrt{V_X V_Y}} = \frac{1}{N} \sum \frac{X - \bar{X}}{\sigma_X} \frac{Y - \bar{Y}}{\sigma_Y}. \qquad \text{A.4.1}$$

For computation, the formula is conveniently written as

$$r_{XY} = \frac{\sum (X - \bar{X})(Y - \bar{Y})}{\sqrt{\sum (X - \bar{X})^2 \sum (Y - \bar{Y})^2}} \qquad \text{A.4.2}$$

$$= \frac{\overline{XY} - \bar{X}\bar{Y}}{\sqrt{(\overline{X^2} - \bar{X}^2)(\overline{Y^2} - \bar{Y}^2)}}. \qquad \text{A.4.3}$$

The value of the correlation coefficient ranges from -1 to $+1$. If there is no association between the variables (if X and Y are independent) the value is 0. With perfect linear association of X and Y, that is to say, if every increase in X is associated with a proportional increase in Y, the correlation is 1. Perfect negative association gives a correlation of -1.

The coefficient of correlation is closely related to the *regression* coefficient. The traditional way to express a dependent variable Y in terms of an independent variable X is to determine the curve of relationship by the method of least squares. By this procedure the curve is chosen so that the sum of the squares of the vertical deviations of the actual points from the curve is minimized. If the relationship is assumed to be linear, the regression coefficient is the slope of the line.

Suppose that there are a series of points, (X_i, Y_i), to which a straight line is to be fitted, as shown in Figure A.4.1. We wish to find the line for which the sum of the squared deviations, $\sum D_i^2$, is a minimum. Let the equation of the line be

$$Y' = f(X) = a + bX,$$

where a and b are to be determined. For any value X, the corresponding value on the line is $Y' = a + bX$, so we minimize the quantity

$$Q = \sum D_i^2 = \sum (Y - Y')^2 = \sum (Y - a - bX)^2. \qquad \text{A.4.4}$$

To do this, we follow the usual procedure and differentiate Q with respect to a and to b and set the resulting quantities equal to 0. This gives

$$\frac{\partial Q}{\partial a} = -2 \sum (Y - a - bX) = 0,$$

$$\frac{\partial Q}{\partial b} = -2 \sum X(Y - a - bX) = 0,$$

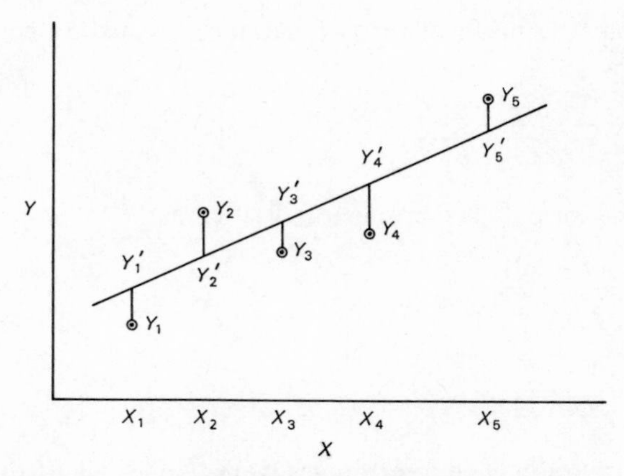

Figure A.4.1. The procedure for fitting a straight line by the least-squares method. X is the independent variable and Y the dependent. The line is determined so that the sum of the squares of the deviations from the line, $\sum(Y_i - Y_i')^2$, is minimized.

from which we have the two equations,

$$aN + b\sum X - \sum Y = 0,$$

$$a\sum X + b\sum X^2 - \sum XY = 0,$$

where N is the number of pairs of measurements and $\sum a = Na$. The solutions are

$$a = \overline{Y} - b\overline{X}, \qquad \text{A.4.5}$$

where

$$\overline{X} = \frac{\sum X}{N},$$

and

$$\overline{Y} = \frac{\sum Y}{N},$$

and

$$b = \frac{\sum XY - N\overline{X}\,\overline{Y}}{\sum X^2 - N\overline{X}^2} = \frac{\mathrm{cov}_{XY}}{V_X}. \qquad \text{A.4.6}$$

The line of best fit (by the least-squares criterion) is

$$\begin{aligned} Y' &= a + bX \\ &= \overline{Y} + b(X - \overline{X}), \end{aligned} \qquad \text{A.4.7}$$

where b (usually written as b_{YX}) is given by A.4.6. The slope, b_{YX}, is called the regression of Y on X. It gives the amount by which Y changes for a unit change in X. Notice that if the two variables are measured as deviations from their means, say $y = Y' - \overline{Y}$ and $x = X - \overline{X}$, the equation takes the simple form

$$y = bx. \qquad \text{A.4.7a}$$

This also shows that the regression line passes through the two means, $\overline{X}$ and $\overline{Y}$.

Another way of writing the regression, one that shows its relation to the correlation coefficient, is

$$b_{YX} = r_{XY}\frac{\sigma_Y}{\sigma_X}. \qquad \text{A.4.8}$$

The regression coefficient is concrete and is expressed in the appropriate units of measurement for the dependent and independent variables. For example, if X is the pounds of fertilizer applied per acre and Y is the yield of corn in bushels per acre, b_{YX} is expressed as bushels/pound. On the other hand, the correlation coefficient is dimensionless.

The correlation coefficient has two properties that are especially useful in interpreting genetic data. By substituting A.4.5 and A.4.8 into A.4.4, we see (after some algebraic rearrangement) that

$$\sum (Y - Y')^2 = (1 - r^2) \sum (Y - \overline{Y})^2. \qquad \text{A.4.9}$$

This tells us that r^2 is the fraction by which the squared deviations from the regression line are less than the squared deviations from the general mean, $\overline{Y}$. In this sense, a fraction r^2 of the variance of Y is associated with (or explained by) a linear change in X; the remainder, $1 - r^2$, represents the fraction of the variance due to random deviations from the linear association with X. In Chapter 4 we use r^2, where r is the correlation between genotypic and phenotypic measurements, as a measurement of *heritability*, that is to say, the extent to which phenotypic variance is accounted for by genotypic differences.

The other interpretation is this: If the quantitative trait or measurement can be regarded as the sum of a large number of components of which a fraction, r, are common to the measurements while the remaining fraction, $1 - r$, are independent, then the expected correlation between such measurements is r. For example, a mother and daughter are identical for half their genes, the other half being independent (assuming that the father and mother are unrelated). Thus the expected correlation between mother and child for a trait determined by a large number of additively acting, independent genes would be 1/2.

The correlation coefficient is useful in another, somewhat related way. Suppose that a quantitative trait or measurement, y_{ij}, of the jth member of the ith group is made up of three independent and additive components:

(1) an overall mean, μ,
(2) a component common to the group, g_i, and
(3) a component special to the individual, s_{ij}.

Then

$$y_{ij} = \mu + g_i + s_{ij}, \qquad \text{A.4.10}$$

and since they are deviations from the mean,

$$\sum_i g_i = 0, \qquad \sum_i \sum_j s_{ij} = 0. \qquad \text{A.4.11}$$

For example, the groups might be families. Then g_i is the component, genetic or environmental or both, common to all members of the ith family and s_{ij} is the additional component special to the jth member of the ith family.

Since the group and special factors are independent, the variance is

$$V_y = V(y_{ij}) = V_g + V_s. \qquad \text{A.4.12}$$

The covariance between the measurements, y_{ij} and y_{ik}, of two members of the same group is

$$\begin{aligned} \text{cov}(y_{ij}, y_{ik}) &= E\{(g_i + s_{ij})(g_i + s_{ik})\} \\ &= E(g_i^2) + E(g_i s_{ij}) + E(g_i s_{ik}) + E(s_{ij} s_{ik}). \end{aligned} \qquad \text{A.4.13}$$

The last three terms are all 0, because s_{ij} and s_{ik} are independent of g_i and of each other. Since $E(g_i^2) = V_g$,

$$\text{cov}(y_{ij}, y_{ik}) = V_g \qquad \text{A.4.14}$$

and the correlation between two members of a group is

$$r = \frac{V_g}{V_g + V_s} = \frac{V_g}{V_y}. \qquad \text{A.4.15}$$

For example, since the correlation between mother and daughter for independent and additive genes is 1/2, the variance within mother-daughter groups, V_s, is 1/2 the population variance, V_y.

If s_{ij} and s_{ik} are not independent, then A.4.14 is not correct. An example would be competition between litter mates, pre- or postnatally. The death of one may enhance the probability of survival of others; or if one is stunted and eats less, the others may have more. In extreme cases this might create negative values of the last term in A.4.13 large enough to offset the first and make the covariance negative. For many purposes, however, the assumption that the s_{ij}'s are independent is a reasonable one.

A.5 Binomial, Poisson, and Normal Distributions

If the probability of an event is p, the probability that in N independent trials the event will occur exactly n times is given by

$$\text{prob}(n) = \frac{N!}{n!(N-n)!}\, p^n(1-p)^{N-n}. \qquad \text{A.5.1}$$

Because this is the formula for a term in the expansion of $[p + (1-p)]^N$ this is called the *binomial disbribution.*

The extension to more than two kinds of events is straightforward and is given by the multinomial distribution. If p_1 is the probability of event 1, p_2 the probability of event 2, p_3 the probability of event 3, and so on, then the probability that in N trials event 1 will occur n_1 times, event 2 n_2 times, event 3 n_3 times, and so on, is

$$\text{prob}(n_1, n_2, \ldots) = \frac{N!}{n_1!\,n_2!\,n_3!\cdots}\, p_1^{n_1} p_2^{n_2} p_3^{n_3} \cdots, \qquad \text{A.5.2}$$

$$\sum n_i = N, \qquad \sum p_i = 1.$$

The theoretical mean and variance of the binomial distribution are easily obtained. If p is the probability of an event, then the expected number of occurrences in N trials is Np.

In a single trial there are two possible numbers of occurrences of the event, 0 or 1, with probability $(1-p)$ and p. The mean number of occurrences is, of course, p. Thus, by A.2.6 the variance is

$$\begin{aligned} V &= p(1-p)^2 + (1-p)(0-p)^2 \\ &= p(1-p). \end{aligned} \qquad \text{A.5.3}$$

Therefore, in N trials the variance of the number of occurrences is (by A.2.13) N times as large, or

$$V_N = Np(1-p). \qquad \text{A.5.4}$$

Likewise, from A.2.15, the variance of the proportion of occurrences in N trials of an event with probability p is

$$V_p = \frac{p(1-p)}{N}. \qquad \text{A.5.5}$$

A case of special interest arises when p is allowed to approach 0 at the same time that N becomes indefinitely large in such a way that the product Np remains of moderate value. The limiting form approached in this manner is the *Poisson distribution.* The probability of exactly n occurrences of the event when the mean number is $\mu(=Np)$ is given by

$$\text{prob}\ (n) = \frac{e^{-\mu}\mu^n}{n!}. \qquad \text{A.5.6}$$

In particular, the probability of no occurrence is $e^{-\mu}$.

The variance formula is easily obtained from formula A.5.4. As p approaches 0 while $Np = \mu$ remains finite, the variance approaches

$$V_n = \mu. \qquad \text{A.5.7}$$

That is to say, the variance is the same as the mean for the Poisson distribution.

The distribution of measurements of many biological materials, and of a great many other things, is often approximated by a symmetrical, bell-shaped curve such as is shown in Figure A.5.1. This is the curve of the *normal distribution* and has the equation

$$Y = \frac{1}{\sigma\sqrt{2\pi}} e^{-\frac{(X-\mu)^2}{2\sigma^2}}, \qquad \text{A.5.8}$$

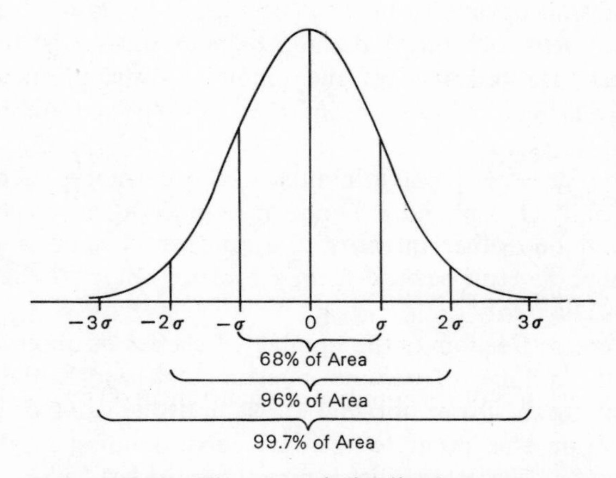

Figure A.5.1. The normal distribution curve.

where μ is the mean and σ^2 is the variance. The normal distribution is also the limit of the binomial distribution as N gets large while p remains finite.

As can be seen from the figure, approximately 68% of the observations are within one standard deviation of the mean and about 96% are within two standard deviations. The proportion of observations lying between any two multiples of the standard deviation may be obtained by numerical integration of A.5.8 between the appropriate limits.

If the observations are not normally distributed, it is often possible to transform them to yield a distribution that is approximately normal. For example, data that are skewed are often rendered roughly normal by replacing the numbers with their logarithms or square roots.

A convenient property of the normal distribution is that with large samples the mean tends to be normally distributed even when the parent distribution is not normal. Therefore, if the standard deviation of the mean (often called standard error) is known, it is possible to use the normal distribution to give an idea of the precision of the estimate. For example, with an appropriately drawn sample, the mean of the sample is expected to lie within two standard errors of the population mean about 96% of the time.

A.6 Significance Tests and Confidence Limits

1. Significance Tests for Enumeration Data: The Chi-square Test

We frequently want to compare observed results with those values that are predicted on the basis of some hypothesis. How far from the expectation must the observations be before they can be regarded as real differences and not simply statistical accidents? One widely used approach is to ask the question: What is the probability that, if the expectations are correct, results would be obtained that deviate from the expectations by as much as or more than those which were observed? An approximate answer to this question is given by the Chi-square test.

The value of χ^2 is given by

$$\chi^2 = \sum \left[\frac{(\text{observed} - \text{expected})^2}{\text{expected}}\right]. \qquad \text{A.6.1}$$

From the value of χ^2 and one other quantity, the number of degrees of freedom, the desired probability can be read from a table or chart. A chart is given at the end of this chapter.

The number of degrees of freedom is the number of classes of observations minus one, minus the number of parameters estimated from the data. To read the chart, look along the lower horizontal axis until the value of χ^2 is reached; then go up from this point to the line corresponding to the number of degrees of freedom. The probability is directly to the left.

As an example, suppose a plant breeder who expected a 9 : 3 : 3 : 1 ratio in an F_2 cross observed 98, 22, 25, and 15 in the four categories. Since the total number is 160, the expected numbers on the 9 : 3 : 3 : 1 hypothesis are 90, 30, 30, and 10. He would compute χ^2 by

$$\chi_3^2 = \frac{(98-90)^2}{90} + \frac{(22-30)^2}{30} + \frac{(25-30)^2}{30} + \frac{(15-10)^2}{10}$$

$$= 6.18, \qquad \text{prob} = .11.$$

There are three degrees of freedom, indicated in the subscript to χ^2. There are four categories of plants. In this case no parameters were estimated from the data, so the number of degrees of freedom is $4 - 1 = 3$.

The probability, 0.11, is obtained from the chart as the probability associated with a χ^2 of 6.17 and 3 degrees of freedom. The interpretation is that, if one were to repeat this experiment 100 times with 9 : 3 : 3 : 1 expectations, 11 times he would expect to get results that deviate from the expected 90, 30, 30, and 10 by as much as or more than the observed set of results did.

Stated another way, and somewhat more precisely, the probability of obtaining a value of χ^2 equal to or greater than that observed, if the hypothesis is correct, is .11. This is useful practically because the probability given by the χ^2 procedure is a good approximation to that obtained by an exact calculation based on the multinomial distribution.

As a second example, consider the data in Table A.6.1. These are for a double backcross in mink where the theoretical expectation with equal viabilities and no linkage is 1 : 1 : 1 : 1. The data are from R. M. Shackleford.

Again, there are 3 degrees of freedom, but this time the value of χ^2 is much larger. From the chart it can be seen that the probability is less than the smallest value on the chart, .0001. With probabilities as small as this we know that (1) either the hypothesis was wrong, or (2) a very improbable event has happened. In this case, we conclude that something is wrong with the hypothesis; to a geneticist it is obvious that what is wrong is that the genes are linked.

By convention, if the probability given by the chart is less than .05 we say that the difference between the observations and the expectations is *significant*. If the probability is less than .01 we say that the difference is *highly significant*.

The additivity property of χ^2 is illustrated by the further analysis in Table A.6.1. The comparison number 2 tests the segregation ratio for the *E, e* locus. Here, $\chi^2 = 6.05$ for 1 degree of freedom. This corresponds to a probability a little larger than 0.01. This is significant, but not highly significant; we conclude that the hypothesis of equality is doubtful, probably attributable to viability differences. The comparison number 3 tests the segregation ratio of the *B*, *b* locus. Here $\chi^2 = 0.80$, corresponding to a probability of about 0.4; there is no reason to doubt the correctness of this expectation. Comparison number 4 tests the hypothesis of nonlinkage.

Table A.6.1. An example of the calculation of χ^2.

	GENOTYPE	PHENOTYPE	OBSERVED NUMBER (O)	EXPECTED NUMBER (E)	$\frac{(O-E)^2}{E}$
1	*Ee Bb*	Ebony	22	20	.200
	Ee bb	Palomino	7	20	8.450
	ee Bb	Dark	14	20	1.800
	ee bb	Pastel	37	20	14.450
	TOTAL		80	80	$24.900 = \chi_3^2$
2	Ebony + Palomino		29	40	3.025
	Dark + Pastel		51	40	3.025
	TOTAL		80	80	$6.050 = \chi_1^2$
3	Ebony + Dark		36	40	.400
	Palomino + Pastel		44	40	.400
	TOTAL		80	80	$.800 = \chi_1^2$
4	Ebony + Pastel		59	40	9.025
	Palomino + Dark		21	40	9.025
	TOTAL		80	80	$18.050 = \chi_1^2$
	TOTAL for 2, 3, and 4				$24.900 = \chi_3^2$

Here there is a large χ^2, corresponding to a probability of less than .0001. So we conclude that definitely the two loci are not independent.

By this analysis the total χ^2 for 3 degrees of freedom has been broken up in 3 components, each with 1 degree of freedom and each testing a separate aspect of the observations. The additivity principle is illustrated by the fact that these three values, each of 1 degree of freedom, add up to the original χ^2 with 3 degrees of freedom.

As a final example, consider the data in Table 2.2.2 in Chapter 2. In this case we have

	MM	*MN*	*NN*	TOTAL
OBSERVED	363	634	282	1279
EXPECTED	361.5	636.9	280.5	1278.9

$$\chi_1^2 = 0.027, \text{ probability} = 0.87$$

There is no reason to question the assumptions that led to these expectations —namely, correct estimation of the gene frequencies, random mating, equal viabilities, no segregation bias, etc. This differs from the earlier examples in one regard, however. The number of degrees of freedom is not 2, as might have been expected, but 1. This is because the data themselves were used to estimate the gene frequency on which the expectations are based. Therefore one more degree of freedom has been removed; the number of degrees of freedom, then, is $3 - 1 - 1 = 1$. The manner in which the allele frequency was computed is given in Table 2.1.1.

There are two cautions that should be mentioned. The first is that one must use numbers, not proportions or percentages, in the calculations. The second is that the method is only approximate, and the degree of approximation gets worse as the numbers become small. A conservative rule is not to use the χ^2 method if any expected number is less than 5.

2. Confidence Limits for Enumeration Data A somewhat similar problem arises when statistical procedures are used for estimation of some population parameter, rather than to test a hypothesis. This is done by determining fiducial or confidence limits. (Actually there are subtle differences in these two concepts, but they do not enter into the computations for the kinds of examples that we are discussing.)

If the sample is large enough that the normal distribution can be invoked, we can assign confidence limits as follows. If p is the estimate and s is the standard deviation of the estimate, then the approximate upper and lower confidence limits are

$$\begin{aligned} \text{upper limit} &= L_u = p + ts, \\ \text{lower limit} &= L_l = p - ts, \end{aligned} \qquad \text{A.6.2}$$

where t is chosen according to the probability level desired. The value of t is obtained from the t chart at the end of this appendix, using infinite degrees of freedom and a probability corresponding to the complement of the confidence level desired. For 95% "confidence," we choose t corresponding to a probability of .05.

For convenience, here are t values corresponding to several frequently used confidence levels.

CONFIDENCE LEVEL	t
.50	0.67
.90	1.64
.95	1.96
.98	2.33
.99	2.58

Here is a simple numerical illustration. Suppose in a count of 400 persons, 160 are found to have blue eyes. These are a random sample of a large population (theoretically infinite) within which we should like to estimate the true frequency of blue eyes. On the basis of the sample, the estimate is $p = 160/400 = .40$. The variance of this proportion, from A.5.5, is $(.4)(1 - .4)/400 = .0006$. The standard deviation, s, is the square root of this, or .0245. If we desire a 98% confidence statement we choose $t = 2.33$. Thus the upper limit is approximately

$$L_u = .40 + (2.33)(.0245) = .457$$

and the lower limit is

$$L_l = .40 - (2.33)(.0245) = .343.$$

We therefore have 98% "confidence" that the value lies between .343 and .457.

We can get an improvement in precision, especially when p is far from 0.5, by asking the following questions: What value of the true proportion would lead to a probability α of getting the observed number or something larger and what value would lead to a probability α of getting the observed number or something less? These two values would then be the limits for the confidence value of $1 - 2\alpha$.

Suppose that p is the observed proportion and π is the true value. The probability of equalling or exceeding p in a sample of N is gotten by computing $t = (p - \pi)/s$, where $s = \sqrt{\pi(1 - \pi)/N}$ and looking up the probability in a table or chart of the normal integral. The upper limit is gotten the same way. Considering both upper and lower limits, we solve the quadratic,

$$t^2 = \frac{N(p - \pi)^2}{\pi(1 - \pi)},$$

or

$$(N + t^2)\pi^2 - (2Np + t^2)\pi + Np^2 = 0 \qquad \text{A.6.3}$$

and the two solutions for π are the upper and lower limits.

Thus

$$L_u = \frac{(2Np + t^2) + \sqrt{(2Np + t^2)^2 - 4Np^2(N + t^2)}}{2(N + t^2)}$$

$$L_l = \frac{(2Np + t^2) - \sqrt{(2Np + t^2)^2 - 4Np^2(N + t^2)}}{2(N + t^2)} \qquad \text{A.6.4}$$

For example, suppose we observe 8 successes in 40 observations. Then $p = 8/40 = 0.2$ and $N = 40$. If we desire 99% confidence limits, we choose $t = 2.58$, leading to 0.086 and 0.400 as the lower and upper 99% limits. If we had used equations A.6.2 we would have gotten the less accurate values 0.037 and 0.363.

This procedure is only as accurate as the normal approximation to the binomial. The exact answers may be obtained by computing the appropriate binomial values. Charts giving binomial limits are available in standard handbooks (Beyer, 1966; Pearson and Hartley, 1958). From these charts the exact limits to the problem above are 0.07 and 0.41, so A.6.4 gave a very good approximation.

The exact meaning of fiducial and confidence limits has been a matter of a great deal of controversy. Some have questioned whether it is proper to speak of the probability that the "true" value lies between certain limits. For an interesting and characteristically polemic discussion of the deeper meanings, see Fisher (1956).

3. Significance Tests for Measurement Data If the data are from measurements rather than counts, the appropriate test of significance is based on the t distribution. If we have measurements on two groups and would like to know if their means are significantly different we compute t as follows:

$$t = \frac{\bar{X}_a - \bar{X}_b}{\sqrt{\dfrac{(N_a + N_b)[(N_a - 1)V_a + (N_b - 1)V_b]}{N_a N_b (N_a + N_b - 2)}}}, \qquad \text{A.6.5}$$

where $\bar{X}_a$ and $\bar{X}_b$ are the means of the two samples and V_a and V_b the two variances, computed from A.3.1,

$$V = \frac{\sum (X_i - \bar{X})^2}{N - 1} = \frac{\sum X_i^2 - N\bar{X}^2}{N - 1}.$$

N_a and N_b are the number of measurements in the two samples.

This value of t is introduced into the t chart exactly as χ^2 was in the χ^2 chart. The number of degrees of freedom is $N_a + N_b - 2$. A probability less than .05 is interpreted as a significant difference and less than .01 is conventionally regarded as highly significant.

If the two sets of measurements are paired in some way—for example if they are measures of blood pressure of the same person before and after administration of a drug—it is usually more appropriate to test the hypothesis that the two means do not differ by treating each of the differences as a variable. If $d_1 = X_{a1} - X_{b1}$, the difference between the first measurements in each group, and so on,

$$t = \frac{\bar{d}}{\sqrt{V_d}}, \qquad \text{A.6.6}$$

where

$$V_{\bar{d}} = \frac{\sum (d_i - \bar{d})^2}{N(N-1)} = \frac{\sum d_i^2 - N\bar{d}^2}{N(N-1)}.$$

This value of t is entered into the chart with $N - 1$ degrees of freedom.

The probability is that of obtaining an absolute value of $\bar{d}$ as large as or larger than the observed value if the population true difference is 0.

4. The Significance of a Correlation Coefficient When a correlation coefficient has been measured and we desire to know whether it is significantly different from 0, the t test is again appropriate. In this case

$$t = \frac{r}{\sqrt{\dfrac{(1-r)^2}{(N-2)}}}, \qquad \text{A.6.7}$$

where r is the observed correlation coefficient and N is the number of pairs of measurements. The appropriate number of degrees of freedom is $N - 2$.

5. Confidence Limits for the Mean with Measurement Data The confidence limits for the mean of a series of measurements are the same as A.6.2. The upper and lower limits are

$$L_u = \bar{X} + ts_{\bar{X}}, \qquad \text{A.6.8}$$
$$L_l = \bar{X} - ts_{\bar{X}},$$

where $\bar{X}$ is the observed mean, $s_{\bar{X}}$ is the standard deviation of the mean (the square root of the variance, as given below, and t is chosen to correspond to the probability desired with degrees of freedom equal to $N - 1$.

For example, if the mean of 11 sample items is 25 and the standard deviation of the mean is computed to be 4, we might desire the 95% confidence limits. Looking for the value of t corresponding to $P = .05$ and 10 degrees of freedom, we find from the χ^2 and t chart that the value is between 2.2 and 2.3 (actually the value is 2.23). Thus the 95% confidence limits for the population mean are $25 \pm (2.23)(4)$, or 16.1 and 33.9.

The t values in the χ^2 and t chart and in the table in Section A.6.2 are for two-tailed tests. That is, they give combined probabilities for deviations in both directions. If deviations in only one direction are considered, then the probability of a deviation as large or larger than that observed is only half as large. If confidence intervals are desired where a deviation in only one direction makes sense, then choose a t value corresponding to a probability twice as large as that given in the chart, e.g., choose t corresponding to 0.02 instead of 0.01.

A.7 Matrices and Determinants

A matrix, **A**, is a set of quantities arranged in rectangular form, such as

$$\begin{pmatrix} a_{11} & a_{12} & a_{13} & a_{14} \\ a_{21} & a_{22} & a_{23} & a_{24} \\ a_{31} & a_{32} & a_{33} & a_{34} \\ a_{41} & a_{42} & a_{43} & a_{44} \end{pmatrix} = \mathbf{A}, \qquad \text{A.7.1}$$

where the first subscript indicates the row and the second the column. The element in row i and column j is thus designated by a_{ij}. The dimensions of a matrix are specified by the numbers of rows and columns; a matrix with m rows and n columns is called an $m \times n$ matrix.

If two matrices have the same dimensions, the processes of addition and subtraction are direct. The corresponding elements are simply added (or subtracted). If matrices **A** and **B** are added to produce matrix **C**, then $c_{ij} = a_{ij} + b_{ij}$. For example,

$$\begin{pmatrix} 3 & 6 & 1 \\ 2 & 1 & 0 \end{pmatrix} + \begin{pmatrix} 2 & -4 & 1 \\ 3 & 5 & -7 \end{pmatrix} = \begin{pmatrix} 5 & 2 & 2 \\ 5 & 6 & -7 \end{pmatrix}.$$

The rule for the multiplication of two matrices is more complicated. The element in row i and column j of the product matrix is the sum of the products of the elements of row i of the first matrix and column j of the second. That is

$$c_{ij} = \sum_k a_{ik} b_{kj}. \qquad \text{A.7.2}$$

This makes it necessary that the first matrix have the same number of columns as the second has rows. Here is an example:

$$\begin{pmatrix} 3 & -2 & 6 \\ 1 & 2 & 0 \end{pmatrix} \begin{pmatrix} 2 & 1 \\ 3 & 1 \\ 1 & 4 \end{pmatrix} = \begin{pmatrix} 6 & 25 \\ 8 & 3 \end{pmatrix}.$$

The element in the first row and second column of the product matrix is $(3 \times 1) + (-2 \times 1) + (6 \times 4) = 25$, for example.

The product of an $m \times k$ and a $k \times n$ matrix is of dimension $m \times n$. Notice that with matrices **AB** is not necessarily equal to **BA**. In fact, unless special restrictions are placed on the dimensions of **A** and **B** one of the products may not even be defined. For example if **A** is $m \times k$ and **B** is $k \times n$, $\mathbf{A} \times \mathbf{B}$ is possible, but $\mathbf{B} \times \mathbf{A}$ is not.

A matrix with only one row is called a row vector; a matrix with only one column is called a column vector. In population genetics, the most usual multiplication is $(m \times m)(m \times 1)$; that is, a square matrix by a column vector. For example,

$$\begin{pmatrix} 3 & 1 & 0 \\ 1 & 2 & 1 \\ 1 & 0 & 3 \end{pmatrix} \begin{pmatrix} 2 \\ 3 \\ 1 \end{pmatrix} = \begin{pmatrix} 9 \\ 9 \\ 5 \end{pmatrix}.$$

Multiplication of a matrix by a constant yields a matrix in which every element is multiplied by that number: e.g.

$$kA = \begin{pmatrix} ka_{11} & ka_{12} & ka_{13} \\ ka_{21} & ka_{22} & ka_{23} \\ ka_{31} & ka_{32} & ka_{33} \end{pmatrix}. \qquad \text{A.7.3}$$

Note that this is equivalent to multiplication by the matrix

$$\begin{pmatrix} k & 0 & 0 \\ 0 & k & 0 \\ 0 & 0 & k \end{pmatrix}.$$

When $k = 1$ we have the unit or identity matrix, usually designated by **I**:

$$\begin{pmatrix} 1 & 0 & 0 & 0 \\ 0 & 1 & 0 & 0 \\ 0 & 0 & 1 & 0 \\ 0 & 0 & 0 & 1 \end{pmatrix} = \mathbf{I}. \qquad \text{A.7.4}$$

The unit matrix plays the same role in the algebra of matrices as the number 1 does in ordinary algebra. Notice that multiplication of a square matrix by the unit matrix of the same dimensions leads to the original matrix.

A square matrix has a corresponding *determinant*, written as

$$|A| = \begin{vmatrix} a_{11} & a_{12} & a_{13} & a_{14} \\ a_{21} & a_{22} & a_{23} & a_{24} \\ a_{31} & a_{32} & a_{33} & a_{34} \\ a_{41} & a_{42} & a_{43} & a_{44} \end{vmatrix}. \qquad \text{A.7.5}$$

The *minor* of an element in a determinant is the determinant that remains when the row and column of the element are deleted. The *cofactor* of an element is the minor, prefixed by a sign which is determined by the position of the element. If the sum of the subscripts is even the sign is positive; if the sum is odd the sign is negative. We shall designate the cofactor of the element a_{ij} by A_{ij}. The value of a determinant is given by taking any row or column and summing the product of each element in the row or column by its cofactor. Thus, in the determinant above the cofactor of a_{32} is

$$A_{32} = (-1)^{3+2} \begin{vmatrix} a_{11} & a_{13} & a_{14} \\ a_{21} & a_{23} & a_{24} \\ a_{41} & a_{43} & a_{44} \end{vmatrix}, \qquad \text{A.7.6}$$

and the value of the determinant is

$$|A| = a_{11}A_{11} + a_{21}A_{21} + a_{31}A_{31} + a_{41}A_{41}. \qquad \text{A.7.7}$$

In this case the first column was used as the basis for expansion, but any other row or column could have been; for example

$$|A| = a_{31}A_{31} + a_{32}A_{32} + a_{33}A_{33} + a_{34}A_{34}.$$

Each cofactor may itself be evaluated as a determinant, and so on until the cofactors are single elements.

The *inverse* of a square matrix is the analog of the reciprocal of a number. The inverse of a square matrix $\mathbf{A}$ is denoted by $\mathbf{A}^{-1}$ and satisfies the relation $\mathbf{AA}^{-1} = \mathbf{A}^{-1}\mathbf{A} = \mathbf{I}$. The process of obtaining the inverse is rather troublesome. To obtain an element, which is designated a^{ij}, we consider the corresponding element of the original matrix with rows and columns transposed, a_{ji}. We take the cofactor of this element and divide by the determinant of the entire original matrix. In symbols, the element a^{ij} in the inverse matrix $\mathbf{A}^{-1}$ is given by $A_{ji}/|A|$ in the original matrix. If the value of the determinant is 0, the matrix has no inverse.

As an example, consider the square matrix,

$$\mathbf{A} = \begin{pmatrix} 3 & 1 & 1 \\ 2 & 2 & 1 \\ 2 & 2 & 3 \end{pmatrix}. \qquad \text{A.7.8}$$

The determinant of this matrix is

$$|A| = 3\begin{vmatrix} 2 & 1 \\ 2 & 3 \end{vmatrix} - 2\begin{vmatrix} 1 & 1 \\ 2 & 3 \end{vmatrix} + 2\begin{vmatrix} 1 & 1 \\ 2 & 1 \end{vmatrix}$$
$$= 3 \times 4 - 2 \times 1 + 2(-1) = 8.$$

The cofactors are

$$A_{11} = (-1)^2 \begin{vmatrix} 2 & 1 \\ 2 & 3 \end{vmatrix} = 4,$$

$$A_{12} = (-1)^3 \begin{vmatrix} 2 & 1 \\ 2 & 3 \end{vmatrix} = -4,$$

and so on, so the matrix of cofactors is

$$\begin{pmatrix} 4 & -4 & 0 \\ -1 & 7 & -4 \\ -1 & -1 & 4 \end{pmatrix}.$$

The inverse matrix is obtained by dividing by $|A|$ and interchanging rows and columns,

$$\mathbf{A}^{-1} = \begin{pmatrix} \frac{4}{8} & -\frac{1}{8} & -\frac{1}{8} \\ -\frac{4}{8} & \frac{7}{8} & -\frac{1}{8} \\ 0 & -\frac{4}{8} & \frac{4}{8} \end{pmatrix}. \qquad \text{A.7.8a}$$

It can easily be verified in this example that the product of A and its inverse is the identity matrix; $\mathbf{AA}^{-1} = \mathbf{I}$.

Although the procedure given here always works, it is not the best for computational efficiency. Much more rapid procedures for large matrices are given in textbooks on numerical methods (e.g., Faddeev and Faddeeva, 1963).

As an example of the use of matrices in population genetics, consider successive generations of self-fertilization. We let D, H, and R stand for the proportions of dominant homozygotes, heterozygotes, and recessive homozygotes. Using the subscript t to indicate time measured in generations we can write the frequencies in generation t in terms of the corresponding frequencies in the previous generation. With self-fertilization (see Table 3.1.1),

$$\begin{aligned} D_t &= D_{t-1} + \tfrac{1}{4}H_{t-1}, \\ H_t &= \tfrac{1}{2}H_{t-1}, \\ R_t &= \tfrac{1}{4}H_{t-1} + R_{t-1}. \end{aligned} \qquad \text{A.7.9}$$

This can be written in matrix form

$$\begin{pmatrix} D_t \\ H_t \\ R_t \end{pmatrix} = \begin{pmatrix} 1 & \frac{1}{4} & 0 \\ 0 & \frac{1}{2} & 0 \\ 0 & \frac{1}{4} & 1 \end{pmatrix} \begin{pmatrix} D_{t-1} \\ H_{t-1} \\ R_{t-1} \end{pmatrix}. \qquad \text{A.7.10}$$

Since there is the same relation between the genotype frequencies at times $t-1$ and $t-2$ as between those at times t and $t-1$, we can write

$$\begin{pmatrix} D_t \\ H_t \\ R_t \end{pmatrix} = \begin{pmatrix} 1 & \frac{1}{4} & 0 \\ 0 & \frac{1}{2} & 0 \\ 0 & \frac{1}{4} & 1 \end{pmatrix}^2 \begin{pmatrix} D_{t-2} \\ H_{t-2} \\ R_{t-2} \end{pmatrix}$$

$$= \begin{pmatrix} 1 & \frac{1}{4} & 0 \\ 0 & \frac{1}{2} & 0 \\ 0 & \frac{1}{4} & 1 \end{pmatrix}^t \begin{pmatrix} D_0 \\ H_0 \\ R_0 \end{pmatrix}. \qquad \text{A.7.11}$$

If **A** is the matrix in A.7.11, then by the rule for matrix multiplication

$$\mathbf{A}^2 = \begin{pmatrix} 1 & \frac{3}{8} & 0 \\ 0 & \frac{1}{4} & 0 \\ 0 & \frac{3}{8} & 1 \end{pmatrix}, \quad \mathbf{A}^3 = \begin{pmatrix} 1 & \frac{7}{16} & 0 \\ 0 & \frac{1}{8} & 0 \\ 0 & \frac{7}{16} & 1 \end{pmatrix}.$$

As a numerical example, consider a population in which the initial frequencies of the three types are $D_0 = 1/4$, $H_0 = 1/2$, and $R_0 = 1/4$. What are the frequencies after three generations of self-fertilization? This is given by

$$\begin{pmatrix} D_3 \\ H_3 \\ R_3 \end{pmatrix} = \begin{pmatrix} 1 & \frac{7}{16} & 0 \\ 0 & \frac{1}{8} & 0 \\ 0 & \frac{7}{16} & 1 \end{pmatrix} \begin{pmatrix} \frac{1}{4} \\ \frac{1}{2} \\ \frac{1}{4} \end{pmatrix}, \tag{A.7.12}$$

from which we can write

$$\begin{aligned} D_3 &= 1 \times \frac{1}{4} + \frac{7}{16} \times \frac{1}{2} = \frac{15}{32}, \\ H_3 &= \frac{1}{8} \times \frac{1}{2} = \frac{1}{16}, \\ R_3 &= \frac{7}{16} \times \frac{1}{2} + 1 \times \frac{1}{4} = \frac{15}{32}. \end{aligned} \tag{A.7.13}$$

These are the same values obtained by another method in Table 3.1.1.

A.8 Eigenvalues and Eigenvectors

We would like to have a way to obtain quantities after several generations without the necessity for repeated matrix multiplication. This can often be done. Consider the recurrence relations

$$\begin{aligned} x_t &= a_{11}x_{t-1} + a_{12}\,y_{t-1} + a_{13}\,z_{t-1}, \\ y_t &= a_{21}x_{t-1} + a_{22}\,y_{t-1} + a_{23}\,z_{t-1}, \\ z_t &= a_{31}x_{t-1} + a_{32}\,y_{t-1} + a_{33}\,z_{t-1}, \end{aligned} \tag{A.8.1}$$

where, as before, the subscript t measures time in generations. In matrix form

$$\begin{pmatrix} x_t \\ y_t \\ z_t \end{pmatrix} = \begin{pmatrix} a_{11} & a_{12} & a_{13} \\ a_{21} & a_{22} & a_{23} \\ a_{31} & a_{32} & a_{33} \end{pmatrix} \begin{pmatrix} x_{t-1} \\ y_{t-1} \\ z_{t-1} \end{pmatrix}. \tag{A.8.2}$$

We consider the related quantities l, m, and n, chosen so that

$$\begin{pmatrix} a_{11} & a_{12} & a_{13} \\ a_{21} & a_{22} & a_{23} \\ a_{31} & a_{32} & a_{33} \end{pmatrix} \begin{pmatrix} l \\ m \\ n \end{pmatrix} = \lambda \begin{pmatrix} l \\ m \\ n \end{pmatrix}. \tag{A.8.3}$$

The column vector is called an eigenvector. Equations A.8.3 may be written

$$\begin{aligned} a_{11}l + a_{12}m + a_{13}n &= \lambda l, \\ a_{21}l + a_{22}m + a_{23}n &= \lambda m, \\ a_{31}l + a_{32}m + a_{33}n &= \lambda n, \end{aligned} \tag{A.8.4}$$

which is the same as

$$\begin{aligned} (a_{11} - \lambda)l + a_{12}m + a_{13}n &= 0 \\ a_{21}l + (a_{22} - \lambda)m + a_{23}n &= 0 \\ a_{31}l + a_{32}m + (a_{33} - \lambda)n &= 0. \end{aligned} \tag{A.8.4a}$$

From the rules of algebra, we know that this system of homogeneous equations has a nontrivial solution for l, m, and n if and only if

$$\begin{vmatrix} a_{11} - \lambda & a_{12} & a_{13} \\ a_{21} & a_{22} - \lambda & a_{23} \\ a_{31} & a_{32} & a_{33} - \lambda \end{vmatrix} = 0, \tag{A.8.5}$$

or, in abbreviated form,

$$|\mathbf{A} - \lambda \mathbf{I}| = 0, \tag{A.8.5a}$$

where **I** is the identity matrix.

A.8.5*a* is the *characteristic equation* of the matrix **A** and has three roots. We assume that the three roots, λ_1, λ_2, and λ_3, are distinct; that is, that no two have the same value. These roots are called *eigenvalues* (also characteristic roots, latent roots, characteristic values, and proper values). When λ_1 is substituted into A.8.4*a*, the equations can be solved for the ratios l_1, m_1, and n_1. Likewise, l_2, m_2, and n_2 are obtained by substituting λ_2; and so on for l_3, m_3, and n_3.

We note here that l, m, and n are not uniquely determined, since (for example) we could divide each equation by n and have three equations in two variables, l/n and m/n. However, the ratio $l : m : n$ can be uniquely determined if condition A.8.5 is met; so one of the three can be chosen arbitrarily.

The relation between x_{t-1} and x_{t-2} is the same as that between x_t and x_{t-1}. Therefore, from A.8.2, we can write

$$\begin{pmatrix} x_t \\ y_t \\ z_t \end{pmatrix} = \begin{pmatrix} a_{11} & a_{12} & a_{13} \\ a_{21} & a_{22} & a_{23} \\ a_{31} & a_{32} & a_{33} \end{pmatrix}^2 \begin{pmatrix} x_{t-2} \\ y_{t-2} \\ z_{t-2} \end{pmatrix} \qquad \text{A.8.6}$$

$$= \begin{pmatrix} a_{11} & a_{12} & a_{13} \\ a_{21} & a_{22} & a_{23} \\ a_{31} & a_{32} & a_{33} \end{pmatrix}^t \begin{pmatrix} x_0 \\ y_0 \\ z_0 \end{pmatrix}.$$

The initial values x_0, y_0, and z_0 are known, and are treated as constants. They can be expressed in terms of the l, m, and n's by

$$\begin{aligned} x_0 &= c_1 l_1 + c_2 l_2 + c_3 l_3, \\ y_0 &= c_1 m_1 + c_2 m_2 + c_3 m_3, \\ z_0 &= c_1 n_1 + c_2 n_2 + c_3 n_3. \end{aligned} \qquad \text{A.8.7}$$

If the three eigenvalues are distinct, these equations can be solved and the c's thereby determined. Then a general solution is possible. Substituting A.8.7 into A.8.6 gives

$$\begin{pmatrix} x_t \\ y_t \\ z_t \end{pmatrix} = \begin{pmatrix} a_{11} & a_{12} & a_{13} \\ a_{21} & a_{22} & a_{23} \\ a_{31} & a_{32} & a_{33} \end{pmatrix}^t \begin{pmatrix} c_1 l_1 + c_2 l_2 + c_3 l_3 \\ c_1 m_1 + c_2 m_2 + c_3 m_3 \\ c_1 n_1 + c_2 n_2 + c_3 n_3 \end{pmatrix} \qquad \text{A.8.8}$$

$$= c_1 \mathbf{A}^t \begin{pmatrix} l_1 \\ m_1 \\ n_1 \end{pmatrix} + c_2 \mathbf{A}^t \begin{pmatrix} l_2 \\ m_2 \\ n_2 \end{pmatrix} + c_3 \mathbf{A}^t \begin{pmatrix} l_3 \\ m_3 \\ n_3 \end{pmatrix}.$$

From A.8.3

$$\mathbf{A} \begin{pmatrix} l_i \\ m_i \\ n_i \end{pmatrix} = \lambda_i \begin{pmatrix} l_i \\ m_i \\ n_i \end{pmatrix}.$$

Notice that, by the rules of matrix multiplication, if $\mathbf{x}$ is a column vector

$$\mathbf{A}^2\mathbf{x} = \mathbf{A}(\mathbf{A}\mathbf{x}) = \mathbf{A}(\lambda\mathbf{x}) = \lambda(\lambda\mathbf{x}) = \lambda^2\mathbf{x}.$$

Continuing this,

$$\mathbf{A}^t \begin{pmatrix} l_i \\ m_i \\ n_i \end{pmatrix} = \lambda_i^t \begin{pmatrix} l_i \\ m_i \\ n_i \end{pmatrix}.$$

Substituting in A.8.8, we obtain

$$\begin{pmatrix} x_t \\ y_t \\ z_t \end{pmatrix} = c_1\lambda_1^t \begin{pmatrix} l_1 \\ m_1 \\ n_1 \end{pmatrix} + c_2\lambda_2^t \begin{pmatrix} l_2 \\ m_2 \\ n_2 \end{pmatrix} + c_3\lambda_3^t \begin{pmatrix} l_3 \\ m_3 \\ n_3 \end{pmatrix},$$

which may be written as

$$\begin{aligned} x_t &= c_1 l_1 \lambda_1^t + c_2 l_2 \lambda_2^t + c_3 l_3 \lambda_3^t, \\ y_t &= c_1 m_1 \lambda_1^t + c_2 m_2 \lambda_2^t + c_3 m_3 \lambda_3^t, \\ z_t &= c_1 n_1 \lambda_1^t + c_2 n_2 \lambda_2^t + c_3 n_3 \lambda_3^t, \end{aligned} \qquad \text{A.8.9}$$

where the c's, l's, m's, n's, and λ's are all constants which have been determined or can be determined.

This gives us an explicit expression for x, y, and z in any generation without requiring a series of matrix multiplications.

One of the eigenvalues, say λ_1, will be the largest and, as t gets large, λ_1^t is much larger than λ_2^t and λ_3^t. Thus after a few generations the structure of the population will be determined entirely by the largest or dominant eigenvalue. Hence, for many purposes, only the largest root need be found.

In this book we are mainly concerned with the cases in which the elements of a matrix are probabilities, which are necessarily non-negative. Therefore the largest eigenvalue, λ_1, is positive. Furthermore, the absolute value of any eigenvalue is not greater than one.

As the other roots become small, we see from A.8.9 that

$$x_t : y_t : z_t \to l_1 : m_1 : n_1 \qquad \text{A.8.10}$$

as $t \to \infty$. This tells us that the frequencies of the three types represented by x, y, and z eventually reach a constant ratio measured by the values of the largest eigenvector corresponding to the largest eigenvalue.

If some of the roots are repeated there may not be a corresponding number of independent eigenvectors, but this situation does not arise in the examples that we are considering.

As a numerical example, consider the recurrence relations for repeated sib mating as given in equations 3.8.6.

$$\begin{aligned} h_t &= \qquad\quad k_{t-1}, \\ k_t &= \tfrac{1}{4}h_{t-1} + \tfrac{1}{2}k_{t-1}, \end{aligned} \qquad \text{A.8.11}$$

where the quantities are defined in Section 3.8. The quantity of greatest interest is h_t, the relative heterozygosity at generation t.

The characteristic equation is

$$\begin{vmatrix} -\lambda & 1 \\ \frac{1}{4} & \frac{1}{2} - \lambda \end{vmatrix} = 0. \qquad \text{A.8.12}$$

On expansion of the determinant, we have

$$\lambda^2 - \tfrac{1}{2}\lambda - \tfrac{1}{4} = 0,$$

giving the roots

$$\lambda_1 = \frac{1 + \sqrt{5}}{4} = .809,$$

$$\lambda_2 = \frac{1 - \sqrt{5}}{4} = -.309. \quad \text{A.8.13}$$

The general solution for h_t is given by

$$h_t = a\lambda_1^t + b\lambda_2^t, \quad \text{A.8.14}$$

where a and b can be determined by the value of h for the first two generations. Putting $t = 0$ and $t = 1$, we obtain respectively

$$h_0 = a + b,$$
$$h_1 = a\lambda_1 + b\lambda_2.$$

These can be solved for a and b, giving

$$a = \frac{h_1 - \lambda_2 h_0}{\lambda_1 - \lambda_2},$$

$$b = \frac{h_1 - \lambda_1 h_0}{\lambda_2 - \lambda_1}. \quad \text{A.8.15}$$

From the definition, h_t is the amount of heterozygosity expressed as a fraction of the original amount. If we start with a randomly mating population, $h_0 = h_1 = 1$. For example, $h_3 = 5/8$.

Notice that when t becomes large

$$h_t \approx \lambda_1 h_{t-1} = 0.809 h_{t-1}. \quad \text{A.8.16}$$

Thus, after a few generations the heterozygosity decreases by a fraction 0.191 each generation.

A.9 The Method of Maximum Likelihood

The principle of this method is to choose as the estimate of the parameter the value that maximizes the probability of the observed results. In addition to its intuitive appeal, this method has been shown by Fisher to be an optimum procedure by several criteria (Fisher 1958).

As an illustration consider a problem where the answer is already known from other considerations, the estimation of gene frequency in a 2-allele locus with the heterozygote recognizable. We assume random mating.

GENOTYPE	AA	Aa	aa	TOTAL
EXPECTED PROPORTION	p^2	$2p(1-p)$	$(1-p)^2$	1
OBSERVED NUMBER	A	B	C	N

The probability of the observed numbers, A, B, and C, as a function of the unknown parameter, p, is

$$\text{prob} = \frac{N!}{A!\,B!\,C!}\,p^{2A}(2p(1-p))^B(1-p)^{2C}$$

$$= \frac{N!}{A!\,B!\,C!}\,2^B p^{2A+B}(1-p)^{B+2C}. \qquad \text{A.9.1}$$

The same value of p that maximizes the probability will also maximize the logarithm of the probability and the algebra is thereby greatly simplified.

$$L = \log \text{prob} = (2A + B)\log p + (B + 2C)\log(1 - p) + K, \qquad \text{A.9.2}$$

where K is a constant.

To find the value of p that maximizes L, we equate the derivative of L with respect to p to 0.

$$\frac{dL}{dp} = \frac{2A + B}{p} - \frac{B + 2C}{1 - p} = 0,$$

which has the solution

$$p = \frac{2A + B}{2N}, \qquad \text{A.9.3}$$

as expected.

The theoretical variance of a maximum-likelihood estimate appropriate when the sample is large is given by the estimated value of the negative reciprocal of the second derivative

$$\frac{1}{V_p} = -E\left(\frac{d^2L}{dp^2}\right). \qquad \text{A.9.4}$$

The estimated value symbol "E" is taken to mean "replace the observed quantities with their maximum-likelihood estimates."

$$\frac{d^2L}{dp^2} = -\frac{2A + B}{p^2} - \frac{B + 2C}{(1 - p)^2}.$$

Replacing $(2A + B)$ by $2Np$ and $(B + 2C)$ by $2N(1 - p)$, we obtain

$$V_p = \frac{p(1-p)}{2N}, \tag{A.9.5}$$

which is as expected, for this is equivalent to a binomial sample of $2N$ genes.

When more than one parameter is to be estimated, as with multiple alleles, the procedure is a straightforward extension, though naturally more complicated. Consider the 3-allele model with dominance in Table A.9.1.

Table A.9.1. A 3-allele model with dominance. A_1 is dominant to A_2 and A_3, while A_2 is dominant to A_3.

PHENOTYPE	A_1	A_2	A_3	TOTAL
GENOTYPES	A_1A_1, A_1A_2, A_1A_3	A_2A_2, A_2A_3	A_3A_3	
OBSERVED NUMBERS	A	B	C	N
EXPECTED PROPORTIONS	$p_1^2 + 2p_1p_2 + 2p_1p_3$	$p_2^2 + 2p_2p_3$	p_3^2	1

To simplify the calculations we make use of a very convenient property of maximum-likelihood estimates, that of *functional invariance*. This means that any function of a maximum-likelihood estimate is the maximum-likelihood estimate of that function. Therefore, in this example, we let

$$y = p_3^2, \qquad x = (p_2 + p_3)^2. \tag{A.9.6}$$

Inversely,

$$p_3 = \sqrt{y}, \qquad p_2 = \sqrt{x} - \sqrt{y}, \qquad p_1 = 1 - \sqrt{x}. \tag{A.9.6a}$$

Then the expected proportions for the phenotypes A_1, A_2, and A_3 are $1 - x$, $x - y$, and y. The probability of the observed numbers in terms of x and y is

$$\text{prob} = \frac{N!}{A!\,B!\,C!}(1-x)^A(x-y)^B y^C \tag{A.9.7}$$

and, taking logarithms as before,

$$L = A\log(1-x) + B\log(x-y) + C\log y + K, \tag{A.9.8}$$

where K is a constant.

$$\frac{\partial L}{\partial x} = \frac{-A}{1-x} + \frac{B}{x-y} = 0, \tag{A.9.9}$$

$$\frac{\partial L}{\partial y} = \frac{-B}{x-y} + \frac{C}{y} = 0. \tag{A.9.9a}$$

Solving for x and y, we obtain

$$y = \frac{C}{N}, \qquad x = \frac{B+C}{N}, \tag{A.9.10}$$

and, by the principle of functional invariance (see A.9.6*a*),

$$p_3 = \sqrt{\frac{C}{N}}, \qquad p_2 = \sqrt{\frac{B+C}{N}} - \sqrt{\frac{C}{N}}, \qquad p_1 = 1 - \sqrt{\frac{B+C}{N}}. \tag{A.9.11}$$

The same principle applies to more involved cases. The difficulty is that the likelihood equations are often impossible to solve except by successive approximations. The formulae for the variances are also more involved, but straightforward. We shall give the arithmetic procedures later in this section.

One more very useful property of maximum-likelihood estimates (and many other large sample estimates as well) is that if X is a quantity whose variance is known and Y is a function of X, the variance of Y is given by

$$V_Y = V_X \left(\frac{dY}{dX}\right)^2. \tag{A.9.12}$$

For example, if there are n people with genotype *aa* and $N - n$ with genotype *AA* or *Aa*, then n/N is an estimate of p^2, where p is the frequency of the *a* allele. This is the maximum-likelihood estimate. The variance of the estimate of p^2 is $p^2(1 - p^2)/N$. Then, by the principle of functional invariance, the estimate of the gene frequency p is $\sqrt{n/N}$. The variance of this estimate is

$$V_p = \frac{p^2(1-p^2)}{N}\left[\frac{dp}{d(p^2)}\right]^2 = \frac{p^2(1-p^2)}{N}\left[\frac{1}{2p}\right]^2 = \frac{1-p^2}{4N}. \tag{A.9.13}$$

If three variables are involved, and Z is a function of X and Y,

$$V_Z = V_X\left(\frac{\partial Z}{\partial X}\right)^2 + V_Y\left(\frac{\partial Z}{\partial Y}\right)^2 + 2\,\mathrm{cov}_{XY}\frac{\partial Z}{\partial X}\frac{\partial Z}{\partial Y}. \tag{A.9.14}$$

In order to get the variance of the estimates when more than one parameter is being estimated, we make use of matrix methods. We first define the *information matrix* as

$$\begin{pmatrix} I_{xx} & I_{xy} \\ I_{xy} & I_{yy} \end{pmatrix} = \mathbf{A}, \qquad \text{A.9.15}$$

where

$$I_{xx} = -E\left(\frac{\partial^2 L}{\partial x^2}\right),$$

$$I_{xy} = -E\left(\frac{\partial^2 L}{\partial x\, \partial y}\right),$$

$$I_{yy} = -E\left(\frac{\partial^2 L}{\partial y^2}\right).$$

In each case the maximum-likelihood estimates are substituted into the formulae. From the information matrix the variances can be found.

The extension to more than two parameters is straightforward, but will not be given here.

Consider again the example of Table A.9.1. Further differentiation of A.9.9 and A.9.9*a*, followed by substitution of the maximum-likelihood estimates from A.9.10 (and recalling that $A + B + C = N$), leads to

$$\begin{aligned} I_{xx} &= \frac{A}{(1-x)^2} + \frac{B}{(x-y)^2} = N^2\left(\frac{A+B}{AB}\right), \\ I_{xy} &= -\frac{B}{(x-y)^2} = -\frac{N^2}{B}, \\ I_{yy} &= \frac{B}{(x-y)^2} + \frac{C}{y^2} = N^2\left(\frac{B+C}{BC}\right). \end{aligned} \qquad \text{A.9.16}$$

The inverse of the information matrix is the *variance-covariance* or *covariance* matrix.

$$\begin{pmatrix} V_X & \mathrm{cov}_{XY} \\ \mathrm{cov}_{XY} & V_Y \end{pmatrix} = \mathbf{C} = \mathbf{A}^{-1}. \qquad \text{A.9.17}$$

In this case

$$\mathbf{A} = N^2 \begin{pmatrix} \dfrac{A+B}{AB} & -\dfrac{1}{B} \\ -\dfrac{1}{B} & \dfrac{B+C}{BC} \end{pmatrix}, \qquad \text{A.9.18}$$

which inverts to

$$\mathbf{C} = \mathbf{A}^{-1} = \frac{1}{N^3}\begin{pmatrix} A(B+C) & AC \\ AC & C(A+B) \end{pmatrix}. \qquad \text{A.9.19}$$

Thus

$$V_x = \frac{A(B+C)}{N^3},$$

$$V_y = \frac{C(A+B)}{N^3}, \qquad \text{A.9.20}$$

$$\text{cov}_{xy} = \frac{AC}{N^3}.$$

To obtain the variance of the gene frequency estimates we again use formulae A.9.12 and A.9.14. The estimates again are

$$p_3 = \sqrt{y} \qquad = \sqrt{\frac{C}{N}},$$

$$p_1 = 1 - \sqrt{x} \quad = 1 - \sqrt{\frac{B+C}{N}},$$

$$p_2 = \sqrt{x} - \sqrt{y} = \sqrt{\frac{B+C}{N}} - \sqrt{\frac{C}{N}}.$$

Using these, along with A.9.6 and A.9.10, we obtain

$$V_{p_3} = V_y\left(\frac{\partial p_3}{\partial y}\right)^2 = \frac{C(A+B)}{N^3}\frac{1}{4y} = \frac{A+B}{4N^2} = \frac{1-p_3^2}{4N},$$

$$V_{p_1} = V_x\left(\frac{\partial p_1}{\partial x}\right)^2 = \frac{A(B+C)}{N^3}\frac{1}{4x} = \frac{A}{4N^2} = \frac{1-(1-p_1)^2}{4N},$$

$$V_{p_2} = V_x\left(\frac{\partial p_2}{\partial x}\right)^2 + V_y\left(\frac{\partial p_2}{\partial y}\right)^2 + 2\,\text{cov}_{xy}\frac{\partial p_2}{\partial x}\frac{\partial p_2}{\partial y} \qquad \text{A.9.21}$$

$$= \frac{A(B+C)}{N^3}\frac{1}{4x} + \frac{C(A+B)}{N^3}\frac{1}{4y} - 2\frac{AC}{N^3}\frac{1}{4\sqrt{xy}}$$

$$= \frac{1}{4N^2}\left(2A + B - 2A\sqrt{\frac{C}{B+C}}\right)$$

$$= \frac{1-p_3^2}{4N} + \frac{1-(1-p_1)^2}{4N} - \frac{p_3[1-(1-p_1)^2]}{2N(1-p_1)}.$$

A.10 Lagrange Multipliers

In maximum or minimum problems it is often required to find a stationary value (maximum or minimum) subject to certain side conditions. The Lagrange method of undetermined multipliers often effects a great simplification in the algebra. We shall not attempt to explain why the procedure works, but will illustrate its use.

Suppose there is a function $f(x, y, z, \ldots)$ in which there are k relations among the variables, $\phi_1(x, y, z, \ldots) = 0$, $\phi_2 = 0, \ldots, \phi_k = 0$. To find the values of $x, y, z, \ldots$ which maximize or minimize f, we equate to 0 the partial derivatives of the function

$$\psi = f + A_1\phi_1 + A_2\phi_2 + \cdots + A_k\phi_k, \qquad \text{A.10.1}$$

where the A_i's are treated as constants. In taking partial derivatives we treat $x, y, z, \ldots,$ as if they were independent.

For example, suppose we wish to find the rectangle of maximum area inscribed in a circle of radius r. The equation of the circle is $x^2 + y^2 = r^2$. The area of the rectangle is $4xy$. So we write

$$\psi(x, y) = 4xy + A(x^2 + y^2 - r^2).$$

Differentiating,

$$\frac{\partial\psi}{\partial x} = 4y + 2Ax = 0,$$

$$\frac{\partial\psi}{\partial y} = 4x + 2Ay = 0.$$

Solving these two equations, $A = -2$ and $x = y$; so the figure is a square.

Of course we could have substituted $\sqrt{r^2 - x^2}$ for y in $4xy$ and differentiated with respect to x only. In this problem the saving of algebra is trivial, but in many problems this procedure effects a great simplification.

To consider a genetic example, what frequency of each of n alleles will maximize the proportion of heterozygotes with random mating? It is easier to ask what minimizes the proportion of homozygotes. So we ask what values of the p_i's minimize $\sum p_i^2$ subject to the side condition that $\sum p_i = 1$. We write

$$\psi = \sum p_i^2 + A(\sum p_i - 1),$$

$$\frac{\partial\psi}{\partial p_i} = 2p_i + A = 0.$$

If we add all k equations, $2\sum p_i + kA = 0$; but since $\sum p_i = 1$, $A = -2/k$. So $p_i = 1/k$ for all k alleles, and the heterozygosity is maximized when the alleles are equally frequent.

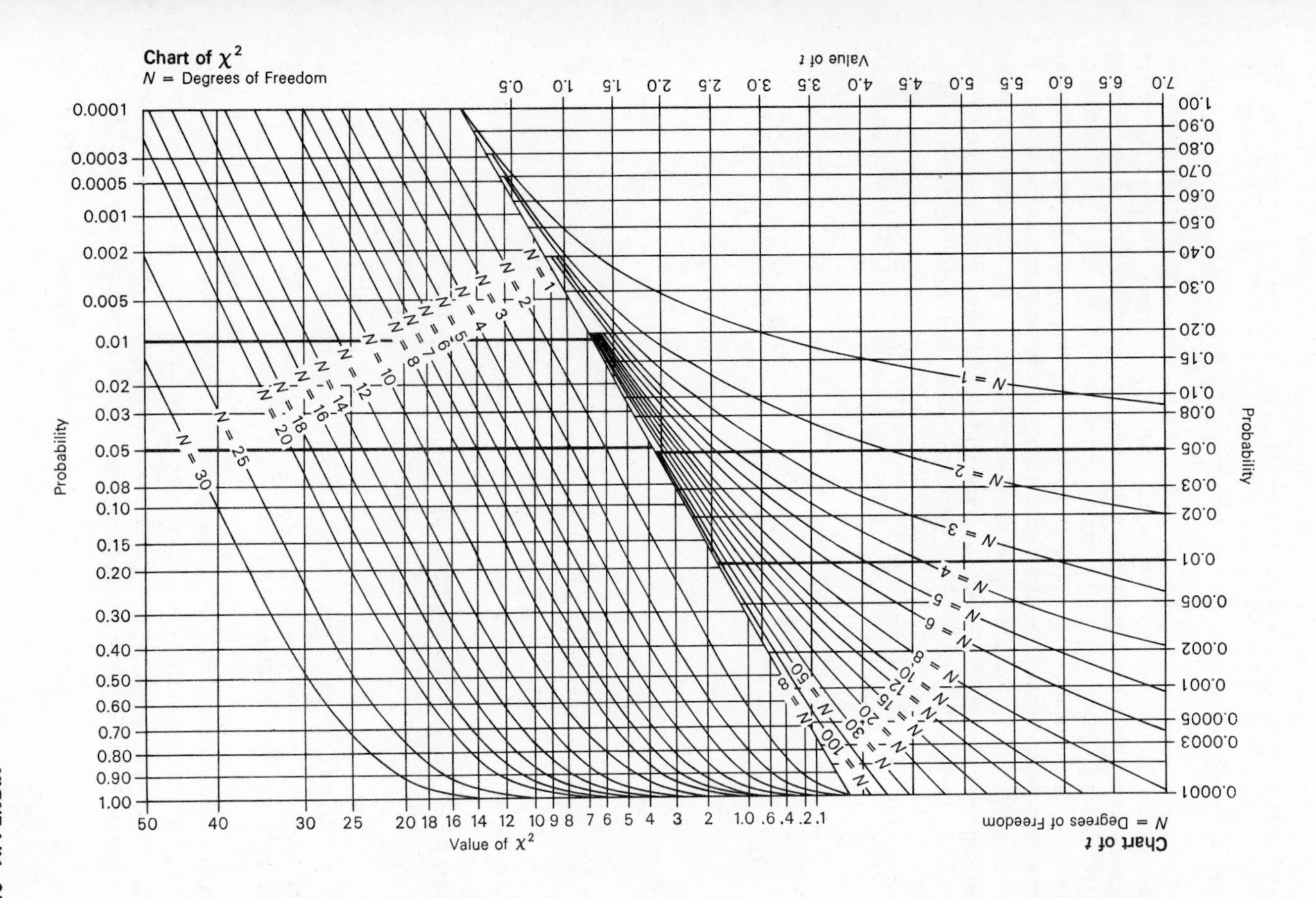

Chart of χ^2
N = Degrees of Freedom
Probability
Value of χ^2
Chart of t
N = Degrees of Freedom
Probability
Value of t

BIBLIOGRAPHY

This list of articles and books includes many that are not referred to in the text; in fact, the majority are in this category. There are also many that we have not studied, but we thought it would be more useful to have a list that is extensive rather than selective. We hope, by so doing, to call attention to the richness and diversity of the literature in this field.

We have not tried to make the list exhaustive, but we have attempted to include most of the papers in the theory of population genetics, particularly those that are in English and which have appeared in readily available journals. Mostly these papers deal with theoretical and mathematical aspects, but we have included several experimental and observational studies, particularly if these include or bear on theoretical points.

Adke, S. R. 1964. A multi-dimensional birth and death process. *Biometrics* 20: 212–216.

Ali, M., and H. H. Hadley. 1955. Theoretical proportion of heterozygosity in populations with various proportions of self- and cross-fertilization. *Agron. J.* 47: 589–590.

Allan, J. S., and A. Robertson. 1964. The effect of initial reverse upon total selection response. *Genet. Res.* 5: 68–79.

Allard, R. W., and J. Adams. 1969. The role of intergenotypic interactions in plant breeding. *Proc. II Intern. Cong. Genet.* 3: 349–370.

Allard, R. W., and P. E. Hansche. 1964. Some parameters of population variability and their implications in plant breeding. *Adv. Agron.* 16: 281–325.

Allard, R. W., and P. E. Hansche. 1965. Population and biometrical genetics in plant breeding. *Proc. XI Int. Cong. Genet.* 3: 665–679.

Allard, R. W., S. K. Jain, and P. L. Workman. 1968. The genetics of inbreeding populations. *Adv. Genet.* 14: 55–131.

Allard, R. W., and C. Wehrhahn. 1963. A theory which predicts stable equilibrium for inversion polymorphism in the grasshopper, *Morabu scurra. Evolution* 18: 129–130.

Allen, G. 1965. Random and nonrandom inbreeding. *Eugen. Quart.* 12: 181–198.

Allen, G. 1966. On the estimation of random inbreeding. *Eugen. Quart.* 13: 67–69.

Allen, R., and A. Fraser. 1968. Simulation of genetic systems. XI. Normalizing selection. *Theor. & Appl. Genet.* 38: 223–225.

Allison, A. C. 1955. Aspects of polymorphism in man. *Cold Spring Harbor Symp. Quant. Biol.* 20: 239–255.

Anderson, F. S. 1960. Competition in populations consisting of one age group. *Biometrics* 16: 19–27.

Anderson, V. L., and O. Kempthorne. 1954. A model for the study of quantitative inheritance. *Genetics* 39: 883–898.

Anderson, W. 1969. Selection in experimental populations. I. Lethal genes. *Genetics* 62: 653–672.

Andrewartha, H. G. 1957. The use of conceptual models in population ecology. *Cold Spring Harbor Symp. Quant. Biol.* 22: 219–232.

Andrewartha, H. G., and L. C. Birch. 1954. *The Distribution and Abundance of Animals.* Univ. Chicago Press, Chicago, Ill.

Andrewartha, H. G., and T. O. Browning. 1961. An analysis of the idea of "resources" in animal ecology. *J. Theor. Biol.* 1: 83–97.

Arellanao, O., T. Ricotti, and A. Diaz. 1964. Über ein Problem der Genetik. *Naturwiss.* 51: 567.

Armitage, P. 1952. The statistical theory of bacterial populations subject to mutation. *J. Roy. Stat. Soc. B* 14: 1–40.

Arnheim, N., and C. E. Taylor. 1969. Non-Darwinian evolution: Consequences for neutral allelic variation. *Nature* 223: 900–903.

Arthur, J. A., and H. Abplanalp. 1964. Studies using computer simulation of reciprocal recurrent selection. *Genetics* 50: 233.

Atkinson, F. V., G. A. Watterson, and P. P. P. Moran. 1960. A matrix inequality. *Quart. J. Math.* 11: 137–140.

Ayala, F. 1969. Evolution of fitness. IV. Genetic evolution of interspecific competitive ability in Drosophila. *Genetics* 61: 737–747.

Bailey, N. T. J. 1961. *Introduction to the Mathematical Theory of Genetic Linkage.* Oxford University Press, Oxford, England.

Baker, G. A., J. Christy, and G. A. Baker. 1964. Analysis of genetic changes in finite populations composed of mixtures of pure lines. *J. Theor. Biol.* 7: 68–85.

Baker, G. A., J. Christy, and G. A. Baker. 1964*a*. Stochastic processes and genotypic frequencies under mixed selfing and random mating. *J. Theor. Biol.* 7: 86–97.

Baker, H. G., and G. L. Stebbins. 1965. *The Genetics of Colonizing Species.* Academic Press, New York.

Barber, H. N. 1965. Selection in natural populations. *Heredity* 20: 551–572.

Barker, J. S. F. 1958. Simulation of genetic systems by automatic digital computers. III. Selection between alleles at an autosomal locus. *Aust. J. Biol. Sci.* 11: 603–612.

Barker, J. S. F. 1958*a*. Simulation of genetic systems by automatic digital computers. IV. Selection between alleles at a sex linked locus. *Aust. J. Biol. Sci.* 11: 613–625.

Barker, J. S. F., and J. C. Butcher. 1966. A simulation study of quasi-fixation of genes due to random fluctuations of selection intensities. *Genetics* 53: 261–268.

Barnett, V. D., 1962. The Monte Carlo solution of a competing species problem. *Biometrics* 18: 76–103.

Barrai, I., M. P. Mi, N. E. Morton, and N. Yasuda. 1965. Estimation of prevalence under incomplete selection. *Amer. J. Hum. Genet.* 17: 221–236.

Barricelli, N. A. 1962, 1963. Numerical testing of evolution theories. *Acta Biotheor.* 16: 70–126.

Bartko, J. J., and G. A. Watterson. 1963. Inference on a genetic model of the Markov chain type. *Biometrika* 50: 251–264.

Bartlett, M. S. 1937. Deviations from expected frequency in the theory of inbreeding *J. Genet.* 35: 83–87.

Bartlett, M. S. 1955. *Stochastic Processes.* Cambridge Univ. Press, Cambridge.

Bartlett, M. S. 1957. On theoretical models for competitive and predatory biological systems. *Biometrika* 44: 27–42.

Bartlett, M. S. 1960. *Stochastic Population Models in Ecology and Epidemiology.* Methuen, London.

Bartlett, M. S., J. C. Gower, and P. H. Leslie. 1960. A comparison of theoretical and empirical results for some stochastic population models. *Biometrika* 47: 1–12.

Bartlett, M. S., and J. B. S. Haldane. 1934. The theory of inbreeding in autotetraploids. *J. Genet.* 29: 175–180.

Bartlett, M. S., and J. B. S. Haldane. 1935. The theory of inbreeding with forced heterozygosis. *J. Genet.* 31: 327–340.

Bateman, A. J. 1950. Is gene-dispersion normal? *Heredity* 4: 353–364.

Bateman, A. J. 1952. Self-incompatibility systems in angiosperms. I. Theory. *Heredity* 6: 285–310.

Bazykin, A. D. 1965. On the possibility of sympatric species formation (in Russian). *Bull. Moscow Soc. Nat., Biol. Ser.* 70: 161–165.

Bellman, R., and R. Kalaba. 1960. Some mathematical aspects of optimal predation in ecology and boviculture. *Proc. Nat. Acad. Sci.* 46: 718–720.

Bennett, J. H. 1953. Junctions in inbreeding. *Genetica* 26: 392–406.

Bennett, J. H. 1953*a*. Linkage in hexasomic inheritance. *Heredity* 7: 265–284.

Bennett, J. H. 1954. Panmixia with tetrasomic and hexasomic inheritance. *Genetics* 39: 150–158.

Bennett, J. H. 1954*a*. On the theory of random mating. *Ann. Eugen.* 18: 311–317.

Bennett, J. H. 1954*b*. The distribution of heterogeneity upon inbreeding. *J. Roy. Stat. Soc. B* 16: 88–99.

Bennett, J. H. 1956. Lethal genes in inbred lines. *Heredity* 10: 263–270.

Bennett, J. H. 1957. Selectively balanced polymorphism at a sex-linked locus. *Nature* 180: 1363–1364.

Bennett, J. H. 1957*a*. The enumeration of genotype-phenotype correspondences. *Heredity* 11: 403–409.

Bennett, J. H. 1958. The existence and stability of selectively balanced polymorphism at a sex-linked locus. *Aust. J. Biol. Sci.* 11: 598–602.

Bennett, J. H. 1963. Random mating and sex linkage. *J. Theor. Biol.* 4: 28–36.

Bennett, J. H. 1968. Mixed self- and cross-fertilization in a tetrasomic species. *Biometrics* 24: 485–500.

Bennett, J. H., and F. E. Binet. 1956. Association between Mendelian factors with mixed selfing and random mating. *Heredity* 10: 51–56.

Bennett, J. H., and C. R. Oertel. 1965. The approach to a random association of genotypes with random mating. *J. Theor. Biol.* 9: 67–76.

Bernstein, F. 1925. Zusammenfassende Betrachtungen über die erblichen Blutstrukturen des Menschen. *Zeit. ind. Abst. Vererb.* 37: 237–269.

Bernstein, F. 1930. Fortgesetzte Untersuchungen aus der Theorie der Blutgruppen. *Zeit. ind. Abst. Vererb.* 56: 233–273.

Beyer, W. H. 1966. *Handbook of Tables for Probability and Statistics.* Chemical Rubber Company, Cleveland.

Binet, F. E., A. M. Clark, and H. T. Clifford. 1959. Correlation due to linkage in certain wild plants. *Genetics* 44: 5–13.

Binet, F. E., and R. T. Leslie. 1960. The coefficient of inbreeding in case of repeated full-sib-matings. *J. Genet.* 57: 127–130.

Binet, F. E., and J. A. Morris. 1962. On total hereditary variance in the case of certain mating systems. *J. Genet.* 58: 108–121.

Birch, L. C. 1948. The intrinsic rate of natural increase of an insect population. *J. Anim. Ecol.* 17: 15–26.

Birch, L. C. 1960. The genetic factor in population ecology. *Amer. Natur.* 94: 5–24.

Bodewig, E. 1936. Mathematische Untersuchungen zum Mendelismus und zur Eugenik. *Genetica* 18: 116–186.

Bodewig, E. 1936*a*. Mathematische Untersuchen zum Mendelismus. *Zeit. ind. Abst. Vererb.* 71: 84–119.

Bodmer, W. F. 1960. Discrete stochastic processes in population genetics. *J. Roy. Stat. Soc. B* 22: 218–244.

Bodmer, W. F. 1963. Natural selection for modifiers of heterozygote fitness. *J. Theor. Biol.* 4: 86–97.

Bodmer, W. F. 1965. Differential fertility in population genetics models. *Genetics* 51: 411–424.

Bodmer, W. F., and L. L. Cavalli-Sforza. 1968. A migration matrix model for the study of random genetic drift. *Genetics* 59: 565–592.

Bodmer, W. F., and A. W. F. Edwards. 1960. Natural selection and the sex ratio. *Ann. Hum. Genet.* 24: 239–244.

Bodmer, W. F., and J. Felsenstein. 1967. Linkage and selection: Theoretical analysis of the deterministic two locus random mating model. *Genetics* 57: 237–265.

Bodmer, W. F., and P. A. Parsons. 1960. The initial progress of new genes with various genetic systems. *Heredity* 15: 283–299.

Bodmer, W. F., and P. A. Parsons. 1962. Linkage and recombination in evolution. *Adv. Genet.* 11: 1–100.

Bogyo, T. P. 1965. Some population parameters as affected by truncation selection. *Genetics* 52: 429.

Bogyo, T. P., and S. W. Ting. 1968. Effect of selection and linkage on inbreeding. *Austr. J. Biol. Sci.* 21: 45–58.

Bohidar, N. R. 1961. Monte Carlo investigations of the effect of linkage on selection. *Biometrics* 17: 506–507.

Bohidar, N. R. 1964. Derivation and estimation of variance and covariance components associated with covariance between relatives under sex-linked transmission. *Biometrics* 20: 505–521.

Bohidar, N. R., and D. G. Patel. 1964. A Monte Carlo investigation of interaction between linkage and selection under stochastic models. *Biometrics* 20: 660.

Bonnier, G. 1947. The genetic effects of breeding in small populations. A demonstration for use in genetic teaching. *Hereditas* 33: 143–151.

Bosso, J. A., O. M. Sorrain, and E. E. A. Favret. 1969. Application of finite absorbent Markov chains to sib mating populations with selection. *Biometrics* 25: 17–26.

Brace, C. L. 1964. The probable mutation effect. *Amer. Natur.* 98: 453–455.

Bradshaw, A. D. 1966. Gene flow and natural selection in closely adjacent populations—a theoretical analysis. *Heredity* 21: 171.

Breese, E. L. 1956. The genetical consequences of assortative mating. *Heredity* 10: 323–343.

Brown, J. L. 1966. Types of group selection. *Nature* 211: 870.

Brown, S. W. 1964. Automatic frequency response in the evolution of male haploidy and other coccid chromosome systems. *Genetics* 49: 797–817.

Bruce, A. B. 1910. The Mendelian theory of heredity and the augmentation of vigor. *Science* 32: 627–628.

Bruck, D. 1957. Male segregation ratio advantage as a factor in maintaining lethal alleles in wild populations of house mice. *Proc. Nat. Acad. Sci.* 43: 152–158.

Brues, Alice M. 1964. The cost of evolution vs. the cost of not evolving. *Evolution* 18: 379–383.

Brues, A. M. 1969. Genetic load and its varieties. *Science* 164: 1130–1136.

Bunak, V. V. 1937. Changes in the mean values of characters in a mixed population. *Ann. Eugen.* 7: 195–206.

Buri, P. 1956. Gene frequency in small populations of mutant Drosophila. *Evolution* 10: 367–402.

Buzzati-Traverso, A. A. 1953. On the role of mutation rate in evolution. *Atti del IX Cong. Intern. di Genet.* 1: 450–462.

Cain, A. J., and P. M. Sheppard. 1954. The theory of adaptive polymorphism. *Amer. Natur.* 88: 321–326.

Campos Rasado, J. M., and A. Robertson. 1966. The genetic control of sex ratio. *J. Theor. Biol.* 13: 324–329.

Cannings, C. 1967. Equilibrium, convergence and stability at a sex-linked locus under natural selection. *Genetics* 56: 613–617.

Cannings, C. 1968. Equilibrium under selection at a multi-allelic sex-linked locus. *Biometrics* 24: 187–189.

Cannings, C. 1969. Unisexual selection at an autosomal locus. *Genetics* 62: 225–229.

Cannings, C., and A. W. F. Edwards. 1969. Expected genotype frequencies in a small sample: Deviation from the Hardy–Weinberg equilibrium. *Amer. J. Hum. Genet.* 21: 245–247.

Carson, H. L. 1955. The genetic characteristics of marginal populations in Drosophila. *Cold Spring Harbor Symp. Quant. Biol.* 20: 276–286.

Carson, H. L. 1959. Genetic conditions which promote or retard the formation of species. *Cold Spring Harbor Symp. Quant. Biol.* 24: 87–105.

Caspari, E., G. S. Watson, and W. Smith. 1966. The influence of cytoplasmic pollen sterility on gene exchange between populations. *Genetics* 53: 741–746.

Castle, W. E. 1903. The law of heredity of Galton and Mendel and some laws governing race improvement by selection. *Proc. Amer. Acad. Sci.* 39: 233–242.

Cavalli-Sforza, L. L. 1950. The analysis of selection curves. *Biometrics* 6: 208–220.

Cavalli-Sforza, L. L. 1952. An analysis of linkage in quantitative inheritance. In *Quantitative Inheritance*. Ed. by E. C. R. Reeve and C. H. Waddington. His Majesty's Stationery Office, London. Pp. 135–144.

Cavalli-Sforza, L. L. 1958. Some data on the genetic structure of human populations. *Proc. X. Inter. Cong. Genet.* 1: 389–407.

Cavalli-Sforza, L. L. 1963. The distribution of migration distances: Models, and applications to genetics. *Entretien de Monaco en Sciences Humanines: Les Deplacements Humains*. Ed. J. Sutter. Pp. 139–158.

Cavalli-Sforza, L. L. 1965. Population structure and human evolution. *Proc. Roy. Soc. B.* 164: 362–379.

Cavalli-Sforza, L. L. 1969. Human diversity. *Proc. II Intern. Cong. Genet.* 3: 405–416.

Cavalli-Sforza, L. L., I. Barrai, and A. W. F. Edwards. 1964. Analysis of human evolution under random genetic drift. *Cold Spring Harbor Symp. Quant. Biol.* 29: 9–20.

Cavalli-Sforza, L. L., and A. W. F. Edwards. 1967. Phylogenetic analysis: Models and estimation procedures. *Evolution* 21: 550–570.

Cavalli-Sforza, L. L., M. Kimura, and I. Barrai. 1966. The probability of consanguineous marriages. *Genetics* 54: 37–60.

Ceppellini, R., M. Siniscalco, and C. A. B. Smith. 1955. The estimation of gene frequencies in a random-mating population. *Ann. Hum. Genet.* 20: 97–115.

Chapman, A. B. 1946. Genetic and nongenetic sources of variation in the weight response of the immature rat ovary to a gonadotropic hormone. *Genetics* 31: 494–507.

Chia, A. B. 1968. Random mating in a population of cyclic size. *J. Appl. Prob.* 5: 21–30.

Chia, A. B., and G. A. Watterson. 1969. Demographic effects on the rate of genetic evolution. I. Constant size populations with two genotypes. *J. Appl. Prob.* 6: 231–248.

Chigusa, S., and T. Mukai. 1964. Linkage disequilibrium and heterosis in experimental populations of *Drosophila melanogaster* with particular reference to the sepia gene. *Jap. J. Genet.* 39: 289–305.

Chung, C. S., and A. B. Chapman. 1958. Comparisons of the predicted with actual gains from selection of parents of inbred progeny of rats. *Genetics* 43: 594–600.

Chung, C. S., O. W. Robison, and N. E. Morton. 1959. A note on deaf mutism. *Ann. Hum. Genet.* 23: 357–366.

Chung, Y. J. 1967. Persistence of a mutant gene in populations of different genetic backgrounds. *Genetics* 57: 957–967.

Clarke, B. C. 1964. Frequency-dependent selection for the dominance of rare polymorphic genes. *Evolution* 18: 364–369.

Clarke, B. C. 1966. The evolution of morph-ratio clines. *Amer. Natur.* 100: 389–402.

Clarke, B. C., and P. O'Donald. 1964. Frequency-dependent selection. *Heredity* 19: 201–206.

Clayton, G. A., G. R. Knight, J. A. Morris, and A. Robertson. 1957. An experimental check on quantitative genetical theory. III. Correlated Responses. *J. Genet.* 55: 171–180.

Clayton, G. A., J. A. Morris, and A. Robertson. 1957. An experimental check on quantitative genetical theory. I. Short-term responses to selection. *J. Genet.* 55: 131–151.

Clayton, G. A., and A. Robertson. 1955. Mutation and quantitative variation. *Amer. Nat.* 89: 151–158.

Clayton, G. A., and A. Robertson. 1957. An experimental check on quantitative genetical theory. II. The long-term effects of selection. *J. Genet.* 55: 152–170.

Coale, A. J., and P. Demeny. 1966. *Regional Model Life Tables and Stable Populations.* Princeton Univ. Press, Princeton, N.J.

Coale, A. J., and C. Y. Tye. 1961. The significance of age patterns of fertility in high-fertility populations. *Milbank Memor. Fund Quart.* 34: 631–646.

Cochran, W. G. 1951. Improvement by means of selection. *Proc. Second Berkeley Symp. Math. Stat. Prob.* Pp. 449–470.

Cockerham, C. C. 1954. An extension of the concept of partitioning hereditary variance for analysis of covariances among relatives when epistasis is present. *Genetics* 39: 859–882.

Cockerham, C. C. 1956. Analysis of quantitative gene action. Genetics in plant breeding. *Brookhaven Symp. Biol.* 9: 53–68.

Cockerham, C. C. 1956*a*. Effects of linkage on the covariances between relatives. *Genetics* 41: 138–141.

Cockerham, C. C. 1959. Partitions of hereditary variance for various genetic models. *Genetics* 44: 1141–1148.

Cockerham, C. C. 1961. Implications of genetic variances in a hybrid breeding program. *Crop Sci.* 1: 47–52.

Cockerham, C. C. 1963. Estimation of genetic variances. *Statistical Genetics and Plant Breeding*. Ed. by W. D. Hanson and H. F. Robinson. *Nat. Acad. Sci.—Nat. Res. Counc. Publ.* 982: 53–93.

Cockerham, C. C. 1967. Group inbreeding and coancestry. *Genetics* 56: 89–104.

Cockerham, C. C. 1969. Variance of gene frequencies. *Evolution* 23: 72–84.

Cockerham, C. C., and B. S. Weir. 1968. Sib mating with two linked loci. *Genetics* 80: 629–640.

Cole, L. C. 1954. The population consequences of life history phenomena. *Quart. Rev. Biol.* 29: 103–137.

Cole, L. C. 1957. Sketches of general and comparative demography. *Cold Spring Harbor Symp. Quant. Biol.* 22: 1–5.

Collins, G. N. 1921. Dominance and the vigor of first generation hybrids. *Amer. Natur.* 55: 116–133.

Collins, R. L. 1967. A general nonparametric theory of genetic analysis. *Genetics* 56: 551.

Comstock, R. E. 1955. Theory of quantitative genetics: Synthesis. *Cold Spring Harbor Symp. Quant. Biol.* 20: 93–109.

Comstock, R. E., and H. F. Robinson. 1952. Estimation of average dominance of genes. In *Heterosis*. Ed. by J. W. Gowen. Iowa State Coll. Press, Ames, Iowa. Pp. 494–516.

Comstock, R. E., H. F. Robinson, and P. H. Harvey. 1949. A breeding procedure designed to make maximum use of both general and specific combining ability. *J. Amer. Soc. Agron.* 41: 360–367.

Connell, J. H., and E. Orias. 1964. The ecological regulation of species diversity. *Amer. Natur.* 98: 399–414.

Constantino, R. F. 1968. The genetical structure of populations and developmental time. *Genetics* 60: 409–418.

Cook, L. M. 1961. The edge effect in population genetics. *Amer Natur.* 95: 295–307.

Cook, L. M. 1965. A note on apostasy. *Heredity* 20: 631–636.

Cormack, R. M. 1964. A boundary problem arising in population genetics. *Biometrics* 20: 785–793.

Cotterman, C. W. 1940. *A Calculus for Statistico-genetics*. Unpublished thesis, Ohio State Univ., Columbus, Ohio.

Cotterman, C. W. 1941. Relatives and human genetic analysis. *Scient. Monthly* 53: 227–234.

Cotterman, C. W. 1953. Regular two-allele and three-allele phenotype systems. *Amer. J. Hum. Genet.* 5: 193–235.

Cotterman, C. W. 1954. Estimation of gene frequencies in nonexperimental populations. In *Statistics and Mathematics in Biology*. Ed. by O. Kempthorne, T. A. Bancroft, J. W. Gowen, and J. L. Lush. Iowa State Coll. Press. Ames, Iowa, Pp. 449–465.

Cotterman, C. W. 1969. Factor-union phenotype systems. *Computer Applications in Genetics*. Ed. by N. E. Morton. Univ. Hawaii Press, Honolulu. Pp. 1–18.

Courant, R., and D. Hilbert. 1962. *Methods of Mathematical Physics*. Vol. I. Interscience Pub., New York.

Cress, C. E. 1966. Heterosis of the hybrid related to gene frequency differences between two populations. *Genetics* 53: 269–274.

Crick, F. H. C. 1967. Origin of the genetic code. *Nature* 213: 119.

Crosby, J. L. 1949. Selection of an unfavorable gene complex. *Evolution* 3: 212–230.

Crosby, J. L. 1960. The use of electronic computation in the study of random fluctuations in rapidly evolving populations. *Proc. Roy. Soc. B* 242: 551–573.

Crosby, J. L. 1961. Teaching genetics with an electronic computer. *Heredity* 16: 255–273.

Crosby, J. L. 1963. The evolution and nature of dominance. *J. Theor. Biol.* 5: 35–51.

Crosby, J. L. 1966. The population genetics of speciation processes. *Heredity* 21: 168.

Crosby, J. L. 1966*a*. Self-incompatibility alleles in the population of *Oenothera organensis*. *Evolution* 20: 567–579.

Crow, J. F. 1945. A chart of the χ^2 and t distributions. *J. Amer. Stat. Assn.* 40: 376.

Crow, J. F. 1948. Alternative hypotheses of hybrid vigor. *Genetics* 33: 477–487.

Crow, J. F. 1952. Dominance and overdominance. In *Heterosis*. Ed. by J. W. Gowen. Iowa State Coll. Press, Ames, Iowa. Pp. 282–297.

Crow, J. F. 1954. Random mating with linkage in polysomics. *Amer. Natur.* 88: 431–434.

Crow, J. F. 1954*a*. Breeding structure of populations. II. Effective population number. *Statistics and Mathematics in Biology*. Iowa State Coll. Press, Ames, Iowa, Pp. 543–556.

Crow, J. F. 1955. General theory of population genetics: Synthesis. *Cold Spring Harbor Symp. Quant. Biol.* 20: 54–59.

Crow, J. F. 1958. Some possibilities for measuring selection intensities in man. *Hum. Biol.* 30: 1–13.

Crow, J. F. 1959. Ionizing radiation and evolution. *Scient. Amer.* 201, Sept. Pp. 138–160.

Crow, J. F. 1961. Population genetics. *Amer. J. Hum. Genet.* 13: 137–150.

Crow, J. F. 1963. The concept of genetic load: A reply. *Amer. J. Hum. Genet.* 15: 310–315.

Crow, J. F. 1966. The quality of people: Human evolutionary changes. *BioScience* 16: 863–867.

Crow, J. F. 1969. Molecular genetics and population genetics. *Proc. XII Int. Cong. Genet.* 3: 105–113.

Crow, J. F. 1970. Genetic loads and the cost of natural selection. *Mathematical Topics in Population Genetics*. Ed. by K. I. Kojima. Springer Verlag, Heidelberg.

Crow, J. F., and Y. J. Chung. 1967. Measurement of effective generation length in Drosophila population cages. *Genetics* 57: 951–955.

Crow, J. F., and J. Felsenstein. 1968. The effect of assortative mating on the genetic composition of a population. *Eugen. Quart.* 15: 85–97.

Crow, J. F., and M. Kimura. 1956. Some genetic problems in natural populations. *Proc. Third Berkeley Symp. Math. Stat. and Prob.* 4: 1–22.

Crow, J. F., and M. Kimura. 1965. The theory of genetic loads. *Proc. XI Int. Cong. Genet.* 3: 495–505.

Crow, J. F., and M. Kimura. 1965*a*. Evolution in sexual and asexual populations. *Amer. Natur.* 99: 439–450.

Crow, J. F., and A. P. Mange. 1965. Measurement of inbreeding from the frequency of marriages between persons of the same surname. *Eugen. Quart.* 12:199–203.

Crow, J. F., and N. E. Morton. 1955. Measurement of gene frequency drift in small populations. *Evolution* 9: 202–214.

Crow, J. F., and N. E. Morton. 1960. The genetic load due to mother-child incompatibility. *Amer. Natur.* 94: 413–419.

Crow, J. F., and W. C. Roberts. 1950. Inbreeding and homozygosis in bees. *Genetics* 35: 612–621.

Crow, J. F., and Rayla G. Temin. 1964. Evidence for the partial dominance of recessive lethal genes in natural populations of Drosophila. *Amer. Natur.* 98: 21–33.

Cruden, Dorothy. 1949. The computation of inbreeding coefficients in closed populations. *J. Hered.* 40: 248–251.

Curnow, R. N. 1964. The effect of continued selection of phenotypic intermediates on gene frequency. *Genet. Res.* 5: 341–353.

Dahlberg, G. 1928. Inbreeding in man. *Genetics* 14: 421–454.

Dahlberg, G. 1938. On rare defects in human populations with particular regard to inbreeding and isolate effects. *Proc. Roy. Soc. Edinb.* 58: 213–232.

Dahlberg, G. 1947. Selection in human populations. *Zool. Bidr. Uppsala* 25: 21–32.

Dahlberg, G. 1947*a*. *Mathematical Methods for Population Genetics.* S. Karger, Basel and New York.

D'Ancona, U. 1954. The struggle for existence. *Bibliotheca Biotheoretica* 6: 1–274.

Daniel, L. 1964. A szelekció biometriai alapja. (The biometrical basis of selection.) *Növenytermelés* 13:369–380.

Dansereau, P. 1952. The varieties of evolutionary opportunity. *Rev. Canad. Biol.* 11: 305–388.

Darwin, C. 1859. *The Origin of Species.* John Murray, London.

Deakin, M. A. B. 1966. Sufficient conditions for genetic polymorphism. *Amer. Natur.* 100: 690–692.

De Finetti, B. 1926. Considerazioni matematiche sul l'ereditarieta mendeliana. *Metron* 6: 1–41.

Dempster, E. R. 1955. Genetic models in relation to animal breeding problems. *Biometrics* 11: 535–536.

Dempster, E. R. 1955*a*. Maintenance of genetic heterogeneity. *Cold Spring Harbor Symp. Quant. Biol.* 20: 25–32.

Dempster, E. R. 1956. Some genetic problems in controlled populations. *Proc. Third Berkeley Symp. Math. Stat. Prob.* 4: 23–40.

Dempster, E. R. 1956*a*. Comments on Professor Lewontin's article. *Amer. Natur.* 90: 385–386.

Dempster, E. R. 1960. The question of stability with positive feedback. *Biometrics* 16: 481–483.

Dempster, E. R., and I. M. Lerner. 1947. The optimum structure of breeding flocks. II. Methods of determination. *Genetics* 32: 567–579.

Dempster, E. R., and I. M. Lerner. 1950. Heritability of threshold characters. *Genetics* 35: 212–236.

Denniston, C. 1967. *Probability and Genetic Relationship*. Unpublished thesis, University of Wisconsin.

Dethier, V. G., and R. H. MacArthur. 1964. A field's capacity to support a butterfly population. *Nature* 201: 728–729.

Dewey, W. J., I. Barrai, N. E. Morton, and M. P. Mi. 1965. Recessive genes in severe mental defect. *Amer. J. Human Genet.* 17: 237–256.

Dickerson, G. E. 1955. Genetic slippage in response to selection for multiple objectives. *Cold Spring Harbor Symp. Quant. Biol.* 20: 213–224.

Dickinson, A. G., and J. L. Jinks. 1956. A generalized analysis of diallel crosses. *Genetics* 41: 65–78.

Dobzhansky, Th. 1951. *Genetics and the Origin of Species*. 3rd Ed. Columbia Univ. Press, N.Y.

Dobzhansky, Th. 1955. A review of some fundamental concepts and problems of population genetics. *Cold Spring Harbor Symp. Quant. Biol.* 20: 1–15.

Dobzhansky, Th. 1956. What is an adaptive trait? *Amer. Natur.* 90: 337–347.

Dobzhansky, Th. 1957. Genetic loads in natural populations. *Science* 126: 191–194.

Dobzhansky, Th. 1957*a*. Mendelian populations as genetics systems. *Cold Spring Harbor Symp. Quant. Biol.* 22: 385–394.

Dobzhansky, Th. 1959. Evolution of genes and genes in evolution. *Cold Spring Harbor Symp. Quant. Biol.* 24: 15–30.

Dobzhansky, Th. 1959*a*. Variation and evolution. *Proc. Amer. Philos. Soc.* 103: 252–263.

Dobzhansky, Th. 1961. Man and natural selection. *Amer. Scientist* 49: 285–299.

Dobzhansky, Th. 1964. How do genetic loads affect the fitness of their carriers in Drosophila populations? *Amer. Natur.* 98: 151–166.

Dobzhansky, Th. 1967. Genetic diversity and diversity of environments. *Proc. Fifth Berkeley Symp. Math. Stat. and Prob.* 4: 295–304.

Dobzhansky, Th., and O. Pavlovsky. 1957. An experimental study of interaction between genetic drift and natural selection. *Evolution* 11: 311–319.

Dobzhansky, Th., and O. Pavlovsky. 1959. How stable is balanced polymorphism? *Proc. Nat. Acad. Sci.* 46: 41–47.

Dobzhansky, Th., and B. Wallace. 1953. The genetics of homeostasis in Drosophila. *Proc. Nat. Acad. Sci.* 39: 162–171.

Dobzhansky, Th., and S. Wright. 1941. Genetics of natural populations. V. Relations between mutation rate and accumulation of lethals in a population of *Drosophila pseudoobscura*. *Genetics* 26: 23–51.

Dodson, E. O. 1962. Note on the cost of natural selection. *Amer. Natur.* 96: 123–126.

Döring, H., and E. Walter. 1959. Über die Berechnung von Inzucht- und Verwandtschaftskoeffizienten. *Biom. Zeit.* 1: 150.

Dowdeswell, W. H. 1955. *The Mechanism of Evolution*. (The Scholarship Series in Biology.) Heinemann, London.

Drobnik, J., and J. Dlouha. 1966. Statistical model of evolution of haploid organisms during simple vegetative reproduction. *J. Theor. Biol.* 11: 418–435.

Dubinin, N. P., and D. D. Romaschoff. 1932. Die genetische Strucktur der Art und ihre Evolution. *Biol. Zhur.* 1: 52–95.

Dunbar, M. J. 1960. The evolution of stability in marine environments. Natural selection at the level of the ecosystem. *Amer. Natur.* 94: 129–136.

Dunn, L. C. 1957. Evidences of evolutionary forces leading to the spread of lethal genes in wild populations of house mice. *Proc. Nat. Acad. Sci.* 43: 158–163.

East, E. M. 1936. Heterosis. *Genetics* 21: 375–397.

Eberhart, S. A. 1964. Theoretical relations among single, three-way, and double cross hybrids. *Biometrics* 20: 522–539.

Eberhart, S. A., W. A. Russell, and L. H. Penny. 1964. Double cross hybrid prediction in maize when epistasis is present. *Crop Sci.* 4:363–366.

Edwards, A. W. F. 1960. On the method of estimating frequencies using the negative binomial distribution. *Ann. Hum. Gen.* 24: 313–318.

Edwards, A. W. F. 1961. The population genetics of "Sex Ratio" in *Drosophila pseudoobscura*. *Heredity* 16: 291–304.

Edwards, A. W. F. 1963. Natural selection and the sex ratio: The approach to equilibrium. *Amer. Natur.* 97: 397–400.

Edwards, A. W. F. 1963*a*. The limitations of population models. *Proc. Second Int. Cong. Hum. Genet.* Pp. 222–223.

Edwards, A. W. F. 1967. Fundamental theorem of natural selection. *Nature* 215: 537–538.

Ellison, B. E. 1965. Limits of infinite populations under random mating. *Proc. Nat. Acad. Sci.* 53: 1266–1272.

Ellison, B. E. 1966. Limit theorems for random mating in infinite populations. *J. Appl. Prob.* 3: 94–114.

Emlen, J. M. 1968. A note on natural selection and the sex ratio. *Amer. Natur.* 102: 94–95.

Emlen, J. M. 1968*a*. Selection for the sex ratio. *Amer. Natur.* 102: 589–591.

Ewens, W. J. 1963. Numerical results and diffusion approximations in a genetic process. Biometrika 50: 241–249.

Ewens, W. J. 1963*a*. Diploid populations with selection depending on gene frequency. *J. Aust. Math. Soc.* 3: 359–374.

Ewens, W. J. 1963*b*. The mean time for absorption in a process of genetic type. *J. Aust. Math. Soc.* 3: 375–383.

Ewens, W. J. 1963*c*. The diffusion equation and a pseudo-distribution in genetics. *J. Roy. Stat. Soc. B* 25: 405–412.

Ewens, W. J. 1964. The maintenance of alleles by mutations. *Genetics* 50: 891–898.

Ewens, W. J. 1964*a*. On the problem of self-sterility alleles. *Genetics* 50: 1433–1438.

Ewens, W. J. 1964*b*. The pseudo-transient distribution and its uses in genetics. *J. Appl. Prob.* 1: 141–156.

Ewens, W. J. 1965. A note on Fisher's theory of the evolution of dominance. *Ann. Hum. Genet.* 29: 85–88.

Ewens, W. J. 1965*a*. The adequacy of the diffusion approximation to certain distributions in genetics. *Biometrics* 21: 386–394.

Ewens, W. J. 1966. Further notes on the evolution of dominance. *Heredity* 20: 443–450.

Ewens, W. J. 1966*a*. Linkage and the evolution of dominance. *Heredity* 21: 363–370.

Ewens, W. J. 1967. A note on the mathematical theory of the evolution of dominance. *Amer. Natur.* 101: 35–40.

Ewens, W. J. 1967*a*. The probability of survival of a mutant. *Heredity* 22: 307–310.

Ewens, W. J. 1967*b*. The probability of survival of a new mutant in a fluctuating environment. *Heredity* 22: 438–443.

Ewens, W. J. 1967*c*. The probability of fixation of a mutant: The two-locus case. *Evolution* 21: 532–540.

Ewens, W. J. 1967*d*. Random sampling and the rate of gene replacement. *Evolution* 21: 657–663.

Ewens, W. J. 1968. A genetic model having complex linkage behavior. *Theor. & Appl. Genet.* 38: 140–143.

Ewens, W. J. 1969. *Population Genetics*. Methuen, London.

Ewens, W. J., and P. M. Ewens. 1966. The maintenance of alleles by mutation—Monte Carlo results for normal and self-fertility populations. *Heredity* 21: 371–378.

Fadeev, D. K., and V. N. Fadeeva. 1963. *Computation Methods in Linear Algebra*. Freeman Pub. C., San Francisco.

Falconer, D. S. 1960. *Introduction to Quantitative Genetics*. The Ronald Press Co., New York.

Falconer, D. S. 1967. The inheritance of liability to diseases with variable age of onset, with particular reference to diabetes mellitus. *Ann. Hum. Genet.* 31: 1–20.

Falk, C., and C. C. Li. 1969. Negative assortative mating: Exact solution to a simple model. *Genetics* 62: 215–223.

Falk, H., and C. T. Falk. 1969. Stability of solutions to certain nonlinear difference equations of population genetics. *Biometrics* 25: 27–37.

Feldman, M. W. 1966. On the offspring number distribution in a genetic population. *J. Appl. Prob.* 3: 129–141.

Feldman, M. W., M. Nabholz, and W. F. Bodmer. 1969. Evolution of the Rh polymorphism: A model for the interaction of incompatibility, reproductive compensation, and heterozygote advantage. *Amer. J. Hum. Genet.* 21: 171–193.

Feller, W. 1950. *An Introduction to Probability Theory and its Applications*. Vol. I. 3rd Ed. 1968. Wiley, New York.

Feller, W. 1951. Diffusion processes in genetics. *Proc. Second Berkeley Symp. Math. Stat. Prob.* Pp. 227–246.

Feller, W. 1952. The parabolic differential equations and the associated semi-group of transformations. *Ann. Math.* 55: 468–519.

Feller, W. 1954. Diffusion processes in one dimension. *Trans. Amer. Math. Soc.* 77: 1–31.

Feller, W. J. 1966. On the influence of natural selection on population size. *Proc. Nat. Acad. Sci.* 55: 733–737.

Feller, W. 1966*a*. *An Introduction to Probability Theory and Its Applications*. Vol. II. Wiley, New York.

Feller, W. 1967. On fitness and the cost of natural selection. *Genet. Res.* 9: 1–15.

Felsenstein, J. 1965. The effect of linkage on directional selection. *Genetics* 52: 349–363.

Finney, D. J. 1952. The equilibrium of a self-incompatible polymorphic species. *Genetica* 26: 33–64.

Finney, D. J. 1962. Genetic gains under three methods of selection. *Genet. Res.* 3: 417–423.

Fish, H. D. 1914. On the progressive increase of homozygosis in brother-sister matings. *Amer. Natur.* 48: 759–761.

Fisher, R. A. 1918. The correlation between relatives on the supposition of Mendelian inheritance. *Trans. Roy. Soc. Edinb.* 52: 399–433.

Fisher, R. A. 1922. On the dominance ratio. *Proc. Roy. Soc. Edinb.* 52: 321–341.

Fisher, R. A. 1925. *Statistical Methods for Research Workers*. 13th Ed. 1958. Oliver and Boyd, London.

Fisher, R. A. 1928. The possible modification of the response of the wild type to recurrent mutations. *Amer. Natur.* 62: 115–126.

Fisher, R. A. 1928*a*. Two further notes on the origin of dominance. *Amer. Natur.* 62: 571–574.

Fisher, R. A. 1929. The evolution of dominance; reply to Professor Sewall Wright. *Amer. Natur.* 63: 553–556.

Fisher, R. A. 1930. *The Genetical Theory of Natural Selection*. Clarendon Press, Oxford.

Fisher, R. A. 1930*a*. The evolution of dominance in certain polymorphic species. *Amer. Natur.* 64: 385–406.

Fisher, R. A. 1930*b*. Mortality among plants and its bearing on natural selection. *Nature* 125: 972–973.

Fisher, R. A. 1930*c*. Biometry and evolution. *Nature* 126: 246–247.

Fisher, R. A. 1930*d*. Genetics, mathematics, and natural selection. *Nature* 126: 805–806.

Fisher, R. A. 1930*e*. The distribution of gene ratios for rare mutations. *Proc. Roy. Soc. Edinb.* 50: 205–220.

Fisher, R. A. 1931. The evolution of dominance. *Biol. Rev.* 6: 345–368.

Fisher, R. A. 1932. Inheritance of acquired characters. *Nature* 130: 579.

Fisher, R. A. 1932*a*. The evolutionary modification of genetic phenomena. *Proc. Sixth Int. Cong. Genet.* 1: 165–172.

Fisher, R. A. 1932*b*. The bearing of genetics on theories of evolution. *Sci. Prog. Twent. Cent.* 26: 273–287.

Fisher, R. A. 1933. Selection in the production of ever-sporting stocks. *Ann. Bot.* 188: 727–733.

Fisher, R. A. 1933*a*. Number of Mendelian factors in quantitative inheritance. *Nature* 131: 400–401.

Fisher, R. A. 1933*b*. Protective adaptations of animals, especially insects. *Proc. Entom. Soc. Lond.* 7: 87–89.

Fisher, R. A. 1934. Professor Wright on the theory of dominance. *Amer. Natur.* 68: 370–374.

Fisher, R. A. 1934*a*. Indeterminism and natural selection. *Phil. Sci.* 1: 99–117.

Fisher, R. A. 1934*b*. Adaptation and mutations. *School Sci. Rev.* 59: 294–301.

Fisher, R. A. 1935. The sheltering of lethals. *Amer. Natur.* 69: 446–455.

Fisher, R. A. 1935*a*. On the selective consequences of East's (1927) theory of heterostylism in *Lythrum. J. Genet.* 30: 369–382.

Fisher, R. A. 1936. The measurement of selective intensity. *Proc. Roy. Soc. Lond. B* 121: 58–62.

Fisher, R. A. 1937. The wave of advance of advantageous genes. *Ann. Eugen.* 7: 355–369.

Fisher, R. A. 1939. Selective forces in wild populations of *Paratettix texanus. Ann. Eugen.* 9: 109–122.

Fisher, R. A. 1939*a*. Stage of enumeration as a factor influencing the variance in the number of progeny, frequency of mutants and related quantities. *Ann. Eugen.* 9: 406–408.

Fisher, R. A. 1940. Non-lethality of the mid factor in *Lythrum salicaria. Nature* 146: 521.

Fisher, R. A. 1941. The theoretical consequence of polyploid inheritance for the mid style form of *Lythrum salicaria. Ann. Eugen.* 11: 31–38.

Fisher, R. A. 1941*a*. Average excess and average effect of a gene substitution. *Ann. Eugen.* 11: 53–63.

Fisher, R. A. 1942. The polygene concept. *Nature* 150: 154.

Fisher, R. A. 1943. Allowance for double reduction in the calculation of genotypic frequencies with polysomic inheritance *Ann. Eugen.* 12: 169–171.

Fisher R. A. 1947. Number of self-sterility alleles. *Nature* 160: 797–798.

Fisher, R. A. 1947*a*. The theory of linkage in polysomic inheritance. *Phil. Trans. Roy. Soc. B* 233: 55–87.

Fisher, R. A. 1949. *The Theory of Inbreeding*. 2nd Ed. 1965. Oliver and Boyd, London.

Fisher, R. A. 1949*a*. A theoretical system of selection for homostyle *Primula. Sankhya* 9: 325–342.

Fisher, R. A. 1950. A class of enumerations of importance in genetics. *Proc. Roy. Soc. B* 136: 509–520.

Fisher, R. A. 1950*a*. Gene frequencies in a cline determined by selection and diffusion. *Biometrics* 6: 353–361.

Fisher, R. A. 1952. Statistical methods in genetics. The Bateson Lecture, 1951. *Heredity* 6: 1–12.

Fisher, R. A. 1953. Population genetics. *Proc. Roy. Soc. B* 141: 510–523.

Fisher, R. A. 1954. A fuller theory of "junctions" in inbreeding. *Heredity* 8: 187–199.

Fisher, R. A. 1956. *Statistical Methods and Scientific Inference*. Oliver and Boyd, London.

Fisher, R. A. 1958. *The Genetical Theory of Natural Selection*. 2nd ed. Dover Press, New York.

Fisher, R. A. 1958*a*. Polymorphism and natural selection. *J. Ecol.* 46: 289–293.

Fisher, R. A. 1959. Natural selection from the genetical standpoint. *Austr. J. Sci.* 22: 16–17.

Fisher, R. A. 1959*a*. An algebraically exact examination of junction formation and transmission in parent-offspring inbreeding. *Heredity* 13: 179–186.

Fisher, R. A. 1961. A model for the generation of self-sterility alleles. *J. Theor. Biol.* 1: 411–414.

Fisher, R. A. 1962. Enumeration and classification in polysomic inheritance. *J. Theor. Biol.* 2: 309–311.

Fisher, R. A., and E. B. Ford. 1928. The variability of species in the *Lepidoptera*, with reference to abundance and sex. *Trans. Entom. Soc. London* 2: 367–384.

Fisher, R. A., and E. B. Ford. 1947. The spread of a gene in natural conditions in a colony of the moth *Panaxia dominula L. Heredity* 1: 143–174.

Fisher, R. A., and E. B. Ford. 1950. The "Sewall Wright" effect. *Heredity* 4: 117–119.

Fisher, R. A., F. R. Immer, and O. Tedin. 1932. The genetical interpretation of statistics of the third degree in the study of quantitative inheritance. *Genetics* 17: 107–124.

Fisher, R. A., and F. Yates. 1963. *Statistical Tables for Biological Agricultural, and Medical Research.* 6th Ed. Hafner Pub. Co., New York.

Fitch, W. M. 1966. An improved method for testing for evolutionary homology. *J. Mol. Biol.* 16: 9–16.

Fitch, W. M., and E. Margoliash. 1967. Construction of phylogenetic trees. *Science* 155: 279–284.

Ford, E. B. 1964. *Ecological Genetics.* Methuen, London; John Wiley, New York.

Ford, E. B. 1965. *Genetic Polymorphism.* Faber & Faber, London.

Ford, E. B., and P. M. Sheppard. 1965. Natural selection and the evolution of dominance. *Heredity* 21: 139–146.

Frank, P. W. 1960. Prediction of population growth form in *Daphnia pulex* cultures. *Amer. Natur.* 94: 357–372.

Fraser, A. S. 1957. Simulation of genetic systems by automatic digital computers. I. Introduction. *Aust. J. Biol. Sci.* 10: 484–491.

Fraser, A. S. 1957*a*. Simulation of genetic systems by automatic digital computers. II. Effects of linkage on rates of advance under selection. *Aust. J. Biol. Sci.* 10: 492–499.

Fraser, A. S. 1958. Monte Carlo analyses of genetic models. *Nature* 181: 208–209.

Fraser, A. S. 1960. Simulation of genetic systems by automatic digital computers. V. Linkage, dominance, and epistasis. *Biometrical Genetics.* Ed. by O. Kempthorne, Pergamon Press, New York. Pp. 70–83.

Fraser, A. S. 1960*a*. Simulation of genetic systems by automatic digital computers. VI. Epistasis. *Aust. J. Biol. Sci.* 13: 150–162.

Fraser, A. S. 1960*b*. Simulation of genetic systems by automatic digital computers. VII. Effects of reproductive rate and intensity of selection on genetic structure. *Aust. J. Biol. Sci.* 13: 344–350.

Fraser, A. S. 1962. Simulation of genetic systems. *J. Theor. Biol.* 2: 329–346.

Fraser, A. S. 1967. Gametic disequilibrium in multigenic systems under normalizing selection. *Genetics* 55: 507–512.

Fraser, A. S., and D. Burnell. 1967. Simulation of genetic systems. XI. Inversion polymorphism. *Amer. J. Hum. Genet.* 19: 270–287.

Fraser, A. S., D. Burnell, and D. Miller. 1966. Simulation of genetic systems. X. Inversion polymorphism. *J. Theor. Biol.* 13: 1–14.

Fraser, A. S., and P. E. Hansche. 1964. Simulation of genetic systems. Major and minor loci. *Genetics Today*. Pergamon Press, New York. Pp. 507–516.

Fraser, A. S., D. Miller, and D. Burnell, 1965. Polygenic balance. *Nature* 206: 114.

Freese, E. 1962. On the evolution of base composition of DNA. *J. Theor. Biol.* 3: 82–101.

Freire–Maia, N. 1964. On the methods available for estimating the load of mutations disclosed by inbreeding. *Cold Spring Harbor Symp. Quant. Biol.* 29: 31–39.

Frota-Pessoa, O. 1957. The estimation of the size of isolates based on census data. *Amer. J. Hum. Genet.* 2: 9–16.

Frydenberg, O. 1963. Population studies of a lethal mutant in *Drosophila melanogaster*. I. Behaviour in populations with discrete generations. *Hereditas* 50: 89–116.

Gabriel, M. L. 1965. Primitive genetic mechanisms and the origin of chromosomes. *Amer. Natur.* 94: 257–269.

Gale, J. S. 1964. Some applications of the theory of junctions. *Biometrics* 20: 85–117.

Galton, F. 1889. *Natural Inheritance*. Macmillan & Co., London.

Garber, M. J. 1951. Approach to genotypic equilibrium with varying percentage of self-fertilization, *J. Hered.* 42: 299–300.

Garfinkel, David. 1962. Digital computer simulation of ecological systems. *Nature* 194: 856–857.

Gates, C. E., R. E. Comstock, and H. F. Robinson. 1957. Generalized genetic variance and covariance formulae for self-fertilized crops assuming linkage. *Genetics* 42: 749–763.

Gause, G. F. 1934. *The Struggle for Existence*. Williams and Wilkins, Baltimore.

Geiringer, H. 1944. On the probability theory of linkage in Mendelian heredity. *Ann. Math. Stat.* 15: 25–57.

Geiringer, H. 1945. Further remarks on linkage theory in Mendelian heredity. *Ann. Math. Stat.* 16: 390–393.

Geiringer, H. 1947. Contribution to the heredity theory of multivalents. *J. Math. Phys.* 26: 246–278.

Geiringer, H. 1948. On the mathematics of random mating in case of different recombination values for males and females. *Genetics* 33: 548–564.

Geiringer, H. 1949. On some mathematical problems arising in the development of Mendelian genetics. *J. Amer. Stat. Assoc.* 44: 526–547.

Geiringer, H. 1949*a*. Chromatid segregation in tetraploids and hexaploids. *Genetics* 34: 665–684.

Geiringer, H. 1949*b*. Contribution to the linkage theory of autopolyploids. *Bull. Math. Biophys.* 11: 59–82, 197–219.

Ghat, G. L. 1964. The genotypic composition and variability in plant populations under mixed self-fertilization and random mating. *J. Indian Soc. Agri. Stat.* 16: 94–125.

Ghat, G. L. 1967. Loss of heterozygosity in populations under mixed random mating and selfing. *J. Indian Soc. Agri. Stat.* 18: 73–81.

Gilbert, N. E. G. 1960. Predicting performance in F_1 and F_2 generations. *Heredity* 13: 146–149.

Gilbert, N. E. G. 1960*a*. Polygene analysis. *Genet. Res.* 2: 96–105.

Gilbert, N. E. G. 1961. Polygene analysis. II. Selection. *Genet. Res.* 2: 456–460.

Gill, J. L. 1965. Effects of finite size on selection advances in simulated genetic populations. *Aust. J. Biol. Sci.* 18: 599–617.

Gill, J. L. 1965*a*. A Monte Carlo evaluation of predicted selection response. *Aust. J. Biol. Sci.* 18: 999–1007.

Gill, J. L. 1965*b*. Selection and linkage in simulated genetic populations. *Aust. J. Biol. Sci.* 18: 1171–1187.

Gill, J. L., and B. A. Clemmer. 1966. Effects of selection and linkage on degree of inbreeding. *Aust. J. Biol. Sci.* 19: 307–317.

Gillois, M. 1964. *La Relation d'Identité en Génétique.* Unpublished thesis, Faculty of Science, Univ. of Paris.

Gillois, M. 1966. Le concept d'indentité et son importance en génétique. *Annales de Génétique* 9: 58–65.

Goldberg, S. 1950. *On a Singular Diffusion Equation.* Ph.D. thesis. Cornell University, Ithaca, N.Y.

Goodhart, C. B. 1963. The Sewall Wright effect. *Amer. Natur.* 97: 407–409.

Goodman, L. A. 1967. The probabilities of extinction for birth-and-death processes that are age-dependent or phase-dependent. *Biometrika* 54: 579–596.

Goodman, L. A. 1968. Stochastic models for the population growth of sexes. *Biometrika* 55: 469–487.

Gowe, R. S. A., A. Robertson, and B. D. H. Latter. 1959. Environment and poultry breeding problems. 5. The design of poultry control strains. *Poult. Sci.* 38: 462–471.

Gowen, J. W., Ed. 1952. *Heterosis.* Iowa State Coll. Press, Ames, Iowa.

Grant, V. 1963. *The Origin of Adaptations.* Columbia University Press, New York.

Griffing, B. 1950. Analysis of quantitative gene action by constant parent regression and related techniques. *Genetics* 35: 303–321.

Griffing, B. 1956. Concept of general and specific combining ability in relation to diallel crossing systems. *Aust. J. Biol. Sci.* 9: 463–493.

Griffing, B. 1956*a*. A generalized treatment of the use of diallel crosses in quantitative inheritance. *Heredity* 10: 31–50.

Griffing, B. 1957. Linkage in trisomic inheritance. *Heredity* 11: 67–92.

Griffing, B. 1960. Theoretical consequences of truncation selection based on the individual phenotype. *Aust. J. Biol. Sci.* 13: 309–343.

Griffing, B. 1960*a*. Accommodation of linkage in mass selection theory. *Aust. J. Biol. Sci.* 13: 501–526.

Griffing, B. 1961. Accommodation of gene-chromosome configuration effects in quantitative inheritance and selection theory. *Aust. J. Biol. Sci.* 14: 402–414.

Griffing, B. 1962. Consequences of truncation selection based on combinations of individual performance and general combining ability. *Aust. J. Biol. Sci.* 15: 333–351.

Griffing, B. 1962*a*. Prediction formulae for general combining ability selection methods utilizing one or two random-mating populations. *Aust. J. Bio. Sci.* 15: 650–665.

Griffing, B. 1963. Comparisons of potentials for general combining ability selection methods utilizing one or two random-mating populations. *Aust. J. Biol. Sci.* 16: 838–862.

Griffing, B. 1965. Influence of sex on selection. I. Contribution of sex-linked genes. *Aust. J. Biol. Sci.* 18: 1157–1170.

Griffing, B. 1966. Influence of sex on selection. II. Contribution of autosomal genotypes having different values in the two sexes. *Aust. J. Biol. Sci.* 19: 593–606.

Griffing, B. 166*a*. Influence of sex on selection. III. Joint contributions of sex-linked and autosomal genes. *Aust. J. Biol. Sci.* 19: 775–794.

Griffing, B. 1967. Selection in reference to biological groups. *I.* Individual and group selection applied to populations of unordered groups. *Aust. J. Biol. Sci.* 20: 127–139.

Griffing, B. 1968. Selection in reference to biological groups. II. Consequences of selection in groups of one size when evaluated in groups of a different size. *Aust. J. Biol. Sci.* 21: 1163–1170.

Hagedoorn, A. L., and A. C. Hagedoorn. 1921. *The Relative Value of the Processes Causing Evolution.* Martinus Nijhoff, The Hague.

Haigh, J. 1969. An enumeration problem in self-sterility. *Biometrics* 25: 39–47.

Hairston, N. G. 1959. Species abundance and community organization. *Ecology* 40: 404–416.

Hairston, N. G., F. E. Smith, and L. B. Slobodkin. 1960. Community structure, population control, and competition. *Amer. Natur.* 94: 421–425.

Hajnal, J. 1963. Concepts of random mating and the frequency of consanguineous marriages. *Proc. Roy. Soc. B* 159: 125–177.

Haldane, J. B. S. 1919. The combination of linkage values, and the calculation of distance between loci of linked factors. *J. Genet.* 8: 299–309.

Haldane, J. B. S. 1924. A mathematical theory of natural and artificial selection. Part I. *Trans. Camb. Phil. Soc.* 23: 19–41.

Haldane, J. B. S. 1924*a*. A mathematical theory of natural and artificial selection. Part II. *Biol. Proc. Camb. Phil. Soc., Biol. Sci.* 1: 158–163.

Haldane, J. B. S. 1924*b*. A mathematical theory of natural and artificial selection. Part III. *Proc. Camb. Phil. Soc.* 23: 363–372.

Haldane, J. B. S. 1924*c*. A mathematical theory of natural and artificial selection. Part IV. *Proc. Camb. Phil. Soc.* 23: 235–243.

Haldane, J. B. S. 1927. A mathematical theory of natural and artificial selection. Part V. Selection and mutation. *Proc. Camb. Phil. Soc.* 28: 838–844.

Haldane, J. B. S. 1930. A mathematical theory of natural and artificial selection. VI. Isolation. *Proc. Camb. Phil. Soc.* 26: 220–230.

Haldane, J. B. S. 1930*a*. A mathematical theory of natural and artificial selection. VII. Selection intensity as a function of mortality rate. *Proc. Camb. Phil. Soc.* 27: 131–136.

Haldane, J. B. S. 1930*b*. A mathematical theory of natural and artificial selection. VIII. Metastable populations. *Proc. Camb. Phil. Soc.* 27: 137–142.

Haldane, J. B. S. 1930*c*. A note on Fisher's theory of the origin of dominance and on a correlation between dominance and linkage. *Amer. Natur.* 64: 87–90.

Haldane, J. B. S. 1930*d*. The theoretical genetics of autopolyploids. *J. Genet.* 22: 359–372.

Haldane, J. B. S. 1932. A mathematical theory of natural and artificial selection. IX. Rapid selection. *Proc. Camb. Phil. Soc.* 28: 244–248.

Haldane, J. B. S. 1932*a*. *The Causes of Evolution*. Harper & Row, New York.

Haldane, J. B. S. 1936. The amount of heterozygosis to be expected in an approximately pure line. *J. Genet.* 32: 375–391.

Haldane, J. B. S. 1937. The effect of variation on fitness. *Amer. Natur.* 71: 337–349.

Haldane, J. B. S. 1937*a*. Some theoretical results of continued brother-sister mating. *J. Genet.* 34: 265–274.

Haldane, J. B. S. 1938. Indirect evidence for the mating system in natural populations. *J. Genet.* 36: 213–220.

Haldane, J. B. S. 1939. The spread of harmful autosomal recessive genes in human populations. *Ann. Eugen.* 9: 232–237.

Haldane, J. B. S. 1939*a*. The equilibrium between mutation and random extinction. *Ann. Eugen.* 9: 400–405.

Haldane, J. B. S. 1939*b*. The theory of the evolution of dominance. *J. Genet.* 37: 365–374.

Haldane, J. B. S. 1940. The conflict between selection and mutation of harmful recessive genes. *Ann. Eugen.* 10: 417–421.

Haldane, J. B. S. 1941. Selection against heterozygosis in man. *Ann. Eugen.* 11: 333–340.

Haldane, J. B. S. 1946. The interaction of nature and nurture. *Ann. Eugen.* 13: 197–205.

Haldane, J. B. S. 1947. The dysgenic effect of induced recessive mutations. *Ann. Eugen.* 14: 35–43.

Haldane, J. B. S. 1948. The theory of a cline. *J. Genet.* 48: 277–284.

Haldane, J. B. S. 1948*a*. The number of genotypes which can be formed with a given number of genes. *J. Genet.* 49: 117–119.

Haldane, J. B. S. 1949. Human evolution: past and future. *Genetics, Paleontology, and Evolution*. Ed. by G. L. Jepsen, E. Mayr, and G. G. Simpson. Princeton Univ. Press, Princeton, N.J. Pp. 405–418.

Haldane, J. B. S. 1949*a*. Disease and evolution. *La Ricerca Scient.* 19: 1–11.

Haldane, J. B. S. 1949*b*. Parental and fraternal correlations in fitness. *Ann. Eugen.* 14: 288–292.

Haldane, J. B. S. 1949*c*. The association of characters as a result of inbreeding and linkage. *Ann. Eugen.* 15: 15–23.

Haldane, J. B. S. 1949*d*. Suggestions as to quantitative measurement of rates of evolution. *Evolution* 3: 51–56.

Haldane, J. B. S. 1949*e*. Some statistical problems arising in genetics. *J. Roy. Stat. Soc. B* 11: 1–14.

Haldane, J. B. S. 1949*f*. The rate of mutation of human genes. *Proc. Eighth Int. Cong. Genet., Stockholm*. Pp. 267–273.

Haldane, J. B. S. 1951. The mathematics of biology. *Sci. J. Roy. Coll. Sci.* 22: 1–11.

Haldane, J. B. S. 1953. Some animal life tables. *J. Inst. Actuaries* 79: 83–89.

Haldane, J. B. S. 1953*a*. Animal populations and their regulation. *New Biology* 15: 9–24.

Haldane, J. B. S. 1954. The measurement of natural selection. *Caryologia* 6: 480–487 (Suppl.).

Haldane, J. B. S. 1954*a*. The statics of evolution. *Evolution as a Process*. Ed. by J. Huxley, A. C. Hardy, and E. B. Ford. Allen and Unwin, London. Pp. 109–121.

Haldane, J. B. S. 1954*b*. An exact test for randomness of mating. *J. Genet.* 52: 631–635.

Haldane, J. B. S. 1955. Population genetics. *New Biology* 18: 34–51.

Haldane, J. B. S. 1955*a*. The complete matrices for brother-sister and alternate parent-offspring mating involving one locus. *J. Genet.* 53: 315–324.

Haldane, J. B. S. 1955*b*. On the biochemistry of heterosis, and the stabilization of polymorphism. *Proc. Roy. Soc. B* 144: 217–220.

Haldane, J. B. S. 1956. The estimation of viabilities. *J. Genet.* 54: 294–296.

Haldane, J. B. S. 1956*a*. The relation between density regulation and natural selection. *Proc. Roy. Soc. London B* 145: 306–308.

Haldane, J. B. S. 1956*b*. The conflict between inbreeding and selection. I. Self-fertilization. *J. Genet.* 54: 56–63.

Haldane, J. B. S. 1956*c*. The theory of selection for melanism in Lepidoptera. *Proc. Roy. Soc. B* 145: 303–308.

Haldane, J. B. S. 1957. The cost of natural selection. *J. Genet.* 55: 511–524.

Haldane, J. B. S. 1957*a*. The conditions for co-adaptation in polymorphism for inversions. *J. Genet.* 55: 218–225.

Haldane, J. B. S. 1958. The theory of evolution before and after Bateson. *J. Genet.* 56: 11–27.

Haldane, J. B. S. 1960. More precise expressions for the cost of natural selection. *J. Genet.* 57: 351–360.

Haldane, J. B. S. 1961. Some simple systems of artificial selection. *J. Genet.* 56: 345–350.

Haldane, J. B. S. 1961*a*. Natural selection in man. *Prog. Med. Gen.* 1: 27–37.

Haldane, J. B. S. 1962. Conditions for stable polymorphism at an autosomal locus. *Nature* 193: 1108.

Haldane, J. B. S. 1962*a*. Natural selection in a population with annual breeding but overlapping generations. *J. Genet.* 58: 122–124.

Haldane, J. B. S. 1962*b*. The selection of double heterozygotes. *J. Genet.* 58: 125–128.

Haldane, J. B. S. 1963. Tests for sex-linked inheritance on population samples. *Ann. Hum. Genet.* 27: 107–111.

Haldane, J. B. S. 1964. A defense of beanbag genetics. *Persp. in Biol. and Med.* 7: 343–359.

Haldane, J. B. S., and S. D. Jayakar. 1963. The solution of some equations occurring in population genetics. *J. Genet.* 58: 291–317.

Haldane, J. B. S., and S. D. Jayakar. 1963*a*. Polymorphism due to selection of varying direction. *J. Genet.* 58: 237–242.

Haldane, J. B. S., and S. D. Jayakar. 1963*b*. Polymorphism due to selection depending on the composition of a population. *J. Genet.* 58: 318–323.

Haldane, J. B. S., and S. D. Jayakar. 1963*c*. The elimination of double dominants in large random mating populations. *J. Genet.* 58: 243–251.

Haldane, J. B. S., and S. D. Jayakar. 1964. Equilibria under natural selection at a sex-linked locus. *J. Genet.* 59: 29–36.

Haldane, J. B. S., and S. D. Jayakar. 1965. The nature of human genetic loads. *J. Genet.* 59: 143–149.

Haldane, J. B. S., and S. D. Jayakar. 1965*a*. Selection for a single pair of allelomorphs with complete replacement. *J. Genet.* 59: 171–177.

Haldane, J. B. S., and P. Moshinsky. 1939. Inbreeding in Mendelian populations with special reference to human cousin marriage. *Ann. Eugen.* 9: 321–340.

Haldane, J. B. S., and C. H. Waddington. 1931. Inbreeding and linkage. *Genetics* 16: 357–374.

Hamilton, W. D. 1963. The evolution of altruistic behavior. *Amer. Natur.* 97: 354–356.

Hamilton, W. D. 1964. The genetical evolution of social behavior. *I. J. Theor. Biol.* 7: 1–16. II. *J. Theor. Biol.* 7: 17–52.

Hamilton, W. D. 1967. Extraordinary sex ratios. *Science* 155: 477–488.

Hansche, P. E., S. K. Jain, and R. W. Allard, 1966. The effect of epistasis and gametic unbalance on genetic load. *Genetics* 54: 1027–1040.

Hanson, W. D. 1958. The theoretical distribution of lengths of undisturbed chromosome segments in F_1 gametes. *Biometrics* 14: 135–136.

Hanson, W. D. 1959. The theoretical distribution of lengths of parental gene blocks in the gametes of an F_1 individual. *Genetics* 44: 197–209.

Hanson, W. D. 1959*a*. Early generation analysis of lengths of heterozygous chromosome segments around a locus held heterozygous with backcrossing or selfing. *Genetics* 44: 833–838.

Hanson, W. D. 1959*b*. Theoretical distribution of the initial linkage block lengths intact in the gametes of a population intermated for *n* generations. *Genetics* 44: 839–846.

Hanson, W. D. 1959*c*. The breakup of initial linkage blocks under selected mating systems. *Genetics* 44: 857–868.

Hanson, W. D., and C. R. Weber. 1961. Resolution of genetic variability in self-pollinated species with an application to the soybean. *Genetics* 46: 1425–1434.

Hanson, W. D. 1962. Average recombination per chromosome. *Genetics* 47: 407–415.

Hanson, W. D. 1966. Effects of partial isolation (distance), migration, and different fitness requirements among environmental pockets upon steady state gene frequencies. *Biometrics* 22: 453–468.

Hanson, W. D., and B. I. Hayman. 1963. Linkage effects on additive genetic variance among homozygous lines arising from a cross between two homozygous parents. *Genetics* 48: 755–766.

Hardin, G. 1960. The competitive exclusion principle. *Science* 131: 1292–1298.

Hardy, G. H. 1908. Mendelian proportions in a mixed population. *Science* 28: 49–50.

Harris, D. L. 1964. Expected and predicted progress from index selection involving estimates of population parameters. *Biometrics* 20: 46–72.

Harris, D. L. 1964*a*. Biometrical parameters of self-fertilizing diploid populations. *Genetics* 50: 931–956.

Harris, D. L. 1964*b*. Genotypic covariances between inbred relatives. *Genetics* 50: 1319–1348.

Harris, D. L. 1965. Biometrical genetics in man. *Methods and Goals in Human Behavior Genetics*. Academic Press, New York. Pp. 81–94.

Harris, T. E. 1963. *The Theory of Branching Processes*. Prentice-Hall, Englewood, N.J.

Hartl, D. L., Y. Hiraizumi, and J. F. Crow. 1967. Evidence for sperm dysfunction as the mechanism of segregation distortion in *Drosophila melanogaster*. *Proc. Nat. Acad. Sci.* 58: 2240–2245.

Hartl, D. L., and T. Maruyama. 1968. Phenogram enumeration: The number of regular genotype-phenotype correspondences in genetic systems. *J. Theor. Biol.* 20: 129–163.

Hashiguchi, S., and H. Morishima. 1969. Estimation of genetic contribution of principal components to individual variates concerned. *Biometrics* 25: 9–16.

Hasofer, A. M. 1966. A continuous-time model in population genetics. *J. Theor. Biol.* 11: 150–163.

Hayman, B. I. 1953. Mixed selfing and random mating when homozygotes are at a disadvantage. *Heredity* 7: 185–192.

Hayman, B. I. 1954. The analysis of variance of diallel crosses. *Biometrics* 10: 235–244.

Hayman, B. I. 1954*a*. The theory and analysis of diallel crosses. *Genetics* 39: 789–809.

Hayman, B. I. 1955. Description and analysis of gene action and interaction. *Cold Spring Harbor Symp. Quant. Biol.* 20: 79–86.

Hayman, B. I. 1958. The theory and analysis of diallel crosses. *Genetics* 43: 63–85.

Hayman, B. I. 1958*a*. The separation of epistatic from additive and dominance variation in generation means. *Heredity* 12: 371–390.

Hayman, B. I. 1960. Maximum likelihood estimation of genetic components of variation. *Biometrics* 16: 369–381.

Hayman, B. I. 1960*a*. The separation of epistatic from additive and dominance variation in generation means. II. *Genetica* 31: 133–146.

Hayman, B. I. 1960*b*. The theory and analysis of diallel crosses. III. *Genetics* 45: 155–172.

Hayman, B. I. 1960*c*. Heterosis and quantitative inheritance. *Heredity* 15: 324–327.

Hayman, B. I. 1962. The gametic distribution in Mendelian heredity. *Aust. J. Biol. Sci.* 15: 166–182.

Hayman, B. I., and K. Mather. 1953. The progress of inbreeding when homozygotes are at a disadvantage. *Heredity* 7: 165–183.

Hayman, B. I., and K. Mather. 1955. The description of genic interactions in continuous variation. *Biometrics* 11: 69–82.

Hayman, B. I., and K. Mather. 1956. Inbreeding when homozygotes are at a disadvantage: A reply. *Heredity* 10: 271–274.

Hazel, L. N. 1943. The genetic basis for constructing selection indexes. *Genetics* 28: 476–490.

Hazel, L. N., and J. L. Lush. 1942. The efficiency of three methods of selection. *J. Hered.* 33: 393–399.

Heidhues, T. 1961. Anwendung statistischer Methoden in der modernen Tierzüchtung. *Züchtungskunde* 33: 1–12.

Hellwig, G. 1964. Über ein enfaches Prinzip, welches die Entropierzeugung von Lebeswesens betrifft. *J. Theor. Biol.* 6: 258–274.

Herbst, W. 1927. Variation, Mendelismus und Selektion in mathematischer Behandlung. *Zeit. ind. Abst. Vererb.* 44: 110–125.

Highton, R. 1966. The effect of mating frequency on phenotypic ratios in sibships when only one parent is known. *Genetics* 54: 1019–1025.

Hill, W. G. 1968. Population dynamics of linked genes in finite populations. *Proc. XII Intern. Congr. Genetics* 2: 146–147.

Hill, W. G. 1969. On the theory of artificial selection in finite populations. *Genet. Res.* 13: 143–163.

Hill, W. G. 1969*a*. The rate of selection advance for non-additive loci. *Genet. Res.* 13: 165–173.

Hill, W. G., and A. Robertson. 1966. The effect of linkage on limits to artificial selection. *Genet. Res.* 8: 269–294.

Hill, W. G., and A. Robertson. 1968. Linkage disequilibrium in finite populations. *Theor. and Appl. Genet.* 38: 226–231.

Hill, W. G., and A. Robertson. 1968*a*. The effects of inbreeding at loci with heterozygote advantage. *Genetics* 60: 615–628.

Hiraizumi, Y., L. Sandler, and J. F. Crow. 1960. Meiotic drive in natural populations of *Drosophila melanogaster*. III. Populational implications of the segregation-distorter locus. *Evolution* 14: 433–444.

Hoen, K., and A. H. E. Grandage. 1960. Calculation of inbreeding in family selection studies on the IBM 650 data processing machine. *Biometrics* 16: 292–296.

Hogben, L. 1932. Filial and fraternal correlations in sex-linked inheritance. *Proc. Roy. Soc. Edinb.* 52: 331–336.

Hogben, L. 1933. A matrix notation for Mendelian populations. *Proc. Roy. Soc. Edinb.* 53: 7–25.

Hogben, L. 1933*a*. *Nature and Nurture*. Norton, New York.

Hogben, L. 1946. *An Introduction to Mathematical Genetics*. Norton, New York.

Holgate, P. 1964. Genotype frequencies in a section of a cline. *Heredity* 19: 501–509.

Holgate, P. 1966. A mathematical study of the founder principle of evolutionary genetics. *J. Appl. Prob.* 3: 115–128.

Holgate, P. 1966*a*. Two limit distributions in evolutionary genetics. *J. Theor. Biol.* 11: 362–369.

Holgate, P. 1967. Divergent population processes and mammal outbreaks. *J. Appl. Prob.* 4: 1–8.

Holgate, P. 1968. Interaction between migration and breeding studied by means of genetic algebras. *J. Appl. Prob.* 5: 1–8.

Holgate, P. 1968*a*. The genetic algebra of k linked loci. *Proc. London Math. Soc.* (3), 18: 315–327.
Horner, T. W. 1956. Parent-offspring and full-sib correlations under a parent-offspring mating system. *Genetics* 41: 460–468.
Horner, T. W., and C. R. Weber. 1956. Theoretical and experimental study of self fertilized populations. *Biometrics* 12: 404–414.
House, V. L. 1953. The use of the binomial expansion for a classroom demonstration of drift in small populations. *Evolution* 7: 84–88.
Hubby, J. L., and R. C. Lewontin. 1966. A molecular approach to the study of genic heterozygosity in natural populations. I. The number of alleles at different loci in *Drosophila pseudoobscura*. *Genetics* 54: 577–594.
Hull, P. 1964. Equilibrium of gene frequency produced by partial incompatibility of offspring with dam. *Proc. Nat. Acad. Sci.* 51: 461–464.
Hulse, F. S. 1957. Exogamie et heterosis. *Arch. Suisses Anthro. Gener.* 22: 103–125.
Hutchinson, G. E. 1948. Circular causal systems in ecology. *Ann. N. Y. Acad. Sci.* 50: 221–246.
Hutchinson, G. E. 1954. Notes on oscillatory populations. *J. Wildl. Manag.* 18: 107–109.
Hutchinson, G. E. 1959. Homage to Santa Rosalia or why are there so many kinds of animals? *Amer. Natur.* 93: 145–159.
Hutchinson, G. E., and R. H. MacArthur. 1959. A theoretical ecological model of size distribution among species of animals. *Amer. Natur.* 93: 117–125.
Huxley, J. (ed.) 1940. *The New Systematics*. The Clarendon Press, Oxford, England.

Ivlev, V. S. 1965. On the quantitative relationship between survival rate of larvae and their food supply. *Bull. Math. Biophys.* 27: 215–222.

Jacquard, A. 1969. Evolution of genetic structure of small populations. *Social Biol.* 16: 143–157.
Jagers, P. 1969. The proportions of individuals of different kinds in two-type populations. A branching problem arising in biology. *J. Appl. Prob.* 6: 249–260.
Jain, H. K., and S. K. Jain. 1961. Differential non-genetic variability in the expression of major genes and polygenes. *Amer. Natur.* 95: 385–387.
Jain, S. K. 1961*a*. On the possible adaptive significance of male sterility in predominantly inbreeding populations. *Genetics* 46: 1237–1240.
Jain, S. K. 1963. Sex ratios under natural selection. *Nature* 200: 1340–41.
Jain, S. K. 1968. Simulation of models involving mixed selfing and random mating. II. Effects of selection and linkage in finite populations. *Theor. & Appl. Genet.* 38: 232–242.
Jain, S. K., and R. W. Allard. 1965. The nature and stability of equilibria under optimizing selection. *Proc. Nat. Acad. Sci.* 54: 1436–1443.
Jain, S. K., and R. W. Allard. 1966. The effects of linkage, epistasis, and inbreeding on population changes under selection. *Genetics* 53: 633–659.
Jain, S. K., and P. L. Workman. 1967. Generalized F-statistics and the theory of inbreeding and selection. *Nature* 214: 674–678.
James, J. W. 1961. Selection in two environments. *Heredity* 16: 145–152.

James, J. W. 1962. The spread of genes in random mating control populations. *Genet. Res.* 3: 1–10.

James, J. W. 1962*a*. The spread of genes in populations under selection. *Proc. World Poult. Cong.* 12: 14–16.

James, J. W. 1962*b*. On Schwartz and Wearden's method of estimating heritability. *Biometrics* 18: 123–125.

James, J. W. 1962*c*. Conflict between directional and centripetal selection. *Heredity* 17: 487–499.

James, J. W. 1965. Simultaneous selection for dominant and recessive mutants. *Heredity* 20: 142–144.

James, J. W. 1965*b*. Response curves in selection experiments. *Heredity* 20: 57–63.

James, J. W., and G. McBride. 1958. The spread of genes by natural and artificial selection in a closed poultry flock. *J. Genet.* 56: 55–62.

Jennings, H. S. 1912. The production of pure homozygotic organisms from heterozygotes by self-fertilization. *Amer. Natur.* 45: 487–491.

Jennings, H. S. 1914. Formula for the results of inbreeding. *Amer. Natur.* 48: 693–696.

Jennings, H. S. 1916. The numerical results of diverse systems of breeding. *Genetics* 1: 53–89.

Jennings, H. S. 1917. The numerical results of diverse systems of breeding, with respect to two pairs of characters, linked or independent, with special relation to the effects of linkage. *Genetics* 2: 97–154.

Jensen, L., and E. Pollak. 1969. Random selective advantages of a gene in a finite population. *J. Appl. Prob.* 6: 19–37.

Jepson, G. L., Ernst Mayr, and G. G. Simpson. 1949. *Genetics, Paleontology, and Evolution.* Princeton Univ. Press, Princeton, N.J.

Jones, D. F. 1917. Dominance of linked factors as a means of accounting for heterosis. *Genetics* 2: 466–479.

Jones, L. P. 1969. Effects of artificial selection on rates of inbreeding in populations of *Drosophila melanogaster*. *Aust. J. Biol. Sci.* 22: 143–169.

Jones, R. M., and K. Mather. 1958. Interaction of genotype and environment in continuous variation. II. Analysis. *Biometrics* 14: 489–498.

Jones, R. M. 1960. Linkage distributions and epistacy in quantitative inheritance. *Heredity* 15: 153–159.

Kalmus, H., and S. Maynard Smith. 1967. Some evolutionary consequences of pegmatypic mating systems. *Amer. Natur.* 100: 619–633.

Kalmus, H., and C. A. B. Smith. 1948. Production of pure lines in bees. *J. Genetics* 49: 153–158.

Karlin, S. 1966. *A First Course in Stochastic Processes.* Academic Press, N.Y.

Karlin, S. 1968. Equilibrium behavior of population genetic models with non-random mating. Part I. Preliminaries and special mating systems. *J. Appl. Prob.* 5: 231–313.

Karlin, S. 1968*a*. Equilibrium behavior of population genetic models with non-random mating. Part II. Pedigrees, homozygosity, and stochastic models. *J. Appl. Prob.* 5: 487–566.

Karlin, S. 1968*b*. Rates of approach to homozygosity for finite stochastic models with variable population size. *Amer. Natur.* 102: 443–455.

Karlin, S., and M. W. Feldman. 1968. Analysis of models with homozygote X heterozygote matings. *Genetics* 59: 105–116.

Karlin, S., and M. W. Feldman. 1968*a*. Further analysis of negative assortative mating. *Genetics* 59: 117–136.

Karlin, S., and J. L. McGregor. 1961. The Hahn polynomials, formulas and an application. *Scripta Mathematica* 26: 33–46.

Karlin, S., and J. McGregor. 1962. On a genetics model of Moran. *Proc. Camb. Phil. Soc.* 58: 299–311.

Karlin, S., and J. McGregor. 1964. On some stochastic models in genetics. *Stochastic Models in Medicine and Biology*. Ed. by J. Gurland. Univ. of Wisconsin Press, Madison, Wisc. Pp. 245–279.

Karlin, S., and J. McGregor. 1964*a*. Direct product branching processes and related Markov chains. *Proc. Nat. Acad. Sci.* 51: 598–602.

Karlin, S., and J. L. McGregor. 1965. Direct product branching processes and related induced Markoff chains. I. Calculations of rates of approach to homozygosity. *Bernoulli, Bayes, Laplace Anniversary Volume*. Springer Verlag, New York, Pp. 111–145.

Karlin, S., and J. McGregor. 1967. The number of mutant forms maintained in a population. *Proc. Fifth Berkeley Symp. Math. Stat. Prob.* 4: 415–438.

Karlin, S., and J. McGregor. 1968. The role of the Poisson progeny distribution in population genetics models. *Math. Biosciences* 2: 11–17.

Karlin, S., and J. McGregor. 1968*a*. Rates and probabilities of fixation for two locus random mating finite populations without selection. *Genetics* 58: 141–159.

Karlin, S., J. McGregor, and W. F. Bodmer. 1967. The rate of production of recombinants between linked genes in finite populations. *Proc. Fifth Berkeley Symp. Math Stat. Prob.* 4: 403–414.

Karlin, S., and F. M. Scudo. 1969. Assortative mating based on phenotype. II. Two autosomal alleles without dominance. *Genetics* 63: 499–510

Kemp, W. B. 1929. Genetic equilibrium and selection. *Genetics* 14: 85–127.

Kempthorne, O. 1954. The correlation between relatives in a random mating population. *Proc. Roy. Soc. B* 143: 103–113.

Kempthorne, O. 1955. The theoretical values of correlations between relatives in random mating populations. *Genetics* 40: 153–167.

Kempthorne, O. 1955*a*. The correlations between relatives in random mating populations. *Cold Spring Harbor Symp. Quant. Biol.* 20: 60–75.

Kempthorne, O. 1955*b*. The correlation between relatives in a simple autotetraploid population. *Genetics* 40: 168–174.

Kempthorne, O. 1955*c*. The correlations between relatives in inbred populations. *Genetics* 40: 681–691.

Kempthorne, O. 1956. The theory of the diallel cross. *Genetics* 41: 451–459.

Kempthorne, O. 1957. *An Introduction to Genetic Statistics*. John Wiley and Sons, New York.

Kempthorne, O. 1960. *Biometrical Genetics*. Pergamon Press, New York.

Kempthorne, O. 1967. The concept of identity of genes by descent. *Proc. Fifth Berkeley Symp. Math. Stat. Prob.* 4: 333–348.

Kempthorne, O., and O. B. Tandon. 1953. The estimation of heritability by regression of offspring on parent. *Biometrics* 9: 90–100.

Kerner, E. H. 1957. A statistical mechanics of interacting biological species. *Bull. Math. Biophys.* 19: 121–141.

Kerner, E. H. 1959. Further considerations on the statistical mechanics of biological associations. *Bull. Math. Biophys.* 21: 217–255.

Kerner, E. H. 1961. On the Volterra–Lotka principle. *Bull. Math. Biophys.* 23: 141–157.

Kerr, W. E. 1967. Multiple alleles and genetic load in bees. *J. Apicult. Res.* 6: 61–64.

Kerr, W. E., and S. Wright. 1954. Experimental studies of the distribution of gene frequencies in very small populations of *Drosophila melanogaster*. I. Forked. *Evolution* 8: 172–177.

Kerr, W. E., and S. Wright. 1954*a*. Experimental studies of the distribution of gene frequencies in very small populations of *Drosophila melanogaster*. III. Aristapedia and spineless. *Evolution* 8: 293–301.

Kerster, Harold W. 1964. Neighborhood size in the rusty lizard, *Sceloporus olivaceus*. *Evolution* 18: 445–457.

Keyfitz, N. 1964. The population projection as a matrix operator. *Demography* 1: 56–73.

Keyfitz, N. 1967. Estimating the trajectory of a population. *Proc. Fifth Berkeley Symp. Math. Stat. Prob.* 4: 81–114.

Keyfitz, N., and E. M. Murphy. 1967. Matrix and multiple decrement in population analysis. *Biometrics* 23: 485–503.

Khanzanie, R. G. 1968. An indication of the asymptotic nature of the Mendelian Markov process. *J. Appl. Prob.* 5: 350–356.

Khazanie, R. G., and H. E. McKean. 1966. A Mendelian Markov process with binomial transition probabilities. *Biometrika* 53: 37–48.

Kimura, M. 1954. Process leading to quasi-fixation of genes in natural populations due to random fluctuations of selection intensities. *Genetics* 39: 280–295.

Kimura, M. 1955. Stochastic processes and distribution of gene frequencies under natural selection. *Cold Spring Harbor Symp. Quant. Biol.* 20: 33–53.

Kimura, M. 1955*a*. Solution of a process of random genetic drift with a continuous model. *Proc. Nat. Acad. Sci.* 41: 144–150.

Kimura, M. 1955*b*. Random genetic drift in a multi-allelic locus. *Evolution* 9: 419–435.

Kimura, M. 1956. Random genetic drift in a tri-allelic locus; exact solution with a continuous model. *Biometrics* 12: 57–66.

Kimura, M. 1956*a*. Rules for testing stability of a selective polymorphism. *Proc. Nat. Acad. Sci.* 42: 336–340.

Kimura, M. 1956*b*. A model of a genetic system which leads to closer linkage by natural selection. *Evolution* 10: 278–287.

Kimura, M. 1957. Some problems of stochastic processes in genetics. *Ann. Math. Stat.* 28: 882–901.

Kimura, M. 1958. On the change of population fitness by natural selection. *Heredity* 12: 145–167.
Kimura, M. 1958*a*. Theoretical basis for the study of inbreeding in man (in Japanese with English summary). *Jap. J. Hum. Genet.* 3: 51–70.
Kimura, M. 1958*b*. Zygotic frequencies in a partially self-fertilizing population. *Ann. Rep. Nat. Inst. Genet., Japan.* 8: 104–105.
Kimura, M. 1959. Conflict between self-fertilization and outbreeding in plants. *Ann. Rep. Nat. Inst. Genet.* Japan 9: 87–88.
Kimura, M. 1960. Evolution of epistasis between closely linked loci (in Japanese). *Jap. J. Genet.* 35: 274.
Kimura, M. 1960*a*. *Outline of Population Genetics*. Baifukan, Tokyo, (in Japanese).
Kimura, M. 1960*b*. Optimum mutation rate and degree of dominance as determined by the principle of minimum genetic load. *J. Genet.* 57: 21–34.
Kimura, M. 1960*c*. Relative applicability of the classical and the balance hypothesis to man, especially with respect to quantitative characters. *J. Rad. Res.* 1–2: 155–164.
Kimura, M. 1960*d*. Genetic load of a population and its significance in evolution (in Japanese). *Jap. J. Genet.* 35: 7–33.
Kimura, M. 1961. Some calculations on the mutational load. *Jap. J. Genet.* 36 suppl: 179–190.
Kimura, M. 1961*a*. Natural selection as the process of accumulating genetic information in adaptive evolution. *Genet. Res.* 2: 127–140.
Kimura, M. 1962. On the probability of fixation of mutant genes in a population. *Genetics* 47: 713–719.
Kimura, M. 1963. A probability method for treating inbreeding systems, especially with linked genes. *Biometrics* 19: 1–17.
Kimura, M. 1964. Diffusion models in population genetics. *J. App. Prob.* 1: 177–232.
Kimura, M. 1965. Some recent advances in the theory of population genetics. *Jap. J. Hum. Genet.* 10(2): 43–48.
Kimura, M. 1965*a*. A stochastic model concerning the maintenance of genetic variability in quantitative characters. *Proc. Nat. Acad. Sci.* 54: 731–736.
Kimura, M. 1965*b*. Attainment of quasi linkage equilibrium when gene frequencies are changing by natural selection. *Genetics* 52: 875–890.
Kimura, M. 1967. On the evolutionary adjustment of spontaneous mutation rates. *Genet. Res.* 9: 23–34.
Kimura, M. 1968. Genetic variability maintained in a finite population due to mutational production of neutral and nearly neutral isoalleles. *Genet. Res.* 11: 247–269.
Kimura, M. 1968*a*. Evolutionary rate at the molecular level. *Nature* 217: 624–626.
Kimura, M. 1969. The number of heterozygous nucleotide sites maintained in a finite population due to steady flux of mutations. *Genetics* 61: 893–903.
Kimura, M. 1969*a*. The length of time required for a selectively neutral mutant to reach fixation through random frequency drift in a finite population. *Genet. Res.* (in press).
Kimura, M. 1969*b*. The rate of molecular evolution considered from the standpoint of population genetics. *Proc. Nat. Acad. Sci.* 63: 1181–1188.

Kimura, M., and J. F. Crow. 1963. On the maximum avoidance of inbreeding. *Genet. Res.* 4: 399–415.

Kimura, M., and J. F. Crow. 1963*a*. The measurement of effective population number. *Evolution* 17: 279–288.

Kimura, M., and J. F. Crow. 1964. The number of alleles that can be maintained in a finite population. *Genetics* 49: 725–738.

Kimura, M., and J. F. Crow. 1969. Natural selection and gene substitution. *Genet. Res.* 13: 127–142.

Kimura, M., and H. Kayano. 1961. The maintenance of supernumerary chromosomes in wild populations of *Lilium callosum* by preferential segregation. *Genetics* 46: 1699–1712.

Kimura, M., and T. Maruyama. 1966. The mutational load with epistatic gene interactions in fitness. *Genetics* 54: 1337–1351.

Kimura, M., and T. Maruyama. 1969. The substitutional load in a finite population. *Heredity* 24: 101–114.

Kimura, M., T. Maruyama, and J. F. Crow. 1963. The mutation load in small populations. *Genetics* 48: 1303–1312.

Kimura, M., and T. Ohta. 1969. The average number of generations until fixation of a mutant gene in a finite population. *Genetics* 61: 763–771.

Kimura, M., and T. Ohta. 1970. Genetic loads at a polymorphic locus which is maintained by frequency-dependent selection. *Genet. Res.* (in press).

Kimura, M., and G. H. Weiss. 1964. The stepping stone model of population structure and the decrease of genetic correlation with distance. *Genetics* 49: 561–576.

King, J. C. 1961. Inbreeding, heterosis, and information theory. *Amer. Natur.* 95: 345–364.

King, J. L. 1965. The effect of litter culling—of family planning—on the rate of natural selection. *Genetics* 51: 425–429.

King, J. L. 1966. The gene interaction component of the genetic load. *Genetics* 53: 403–413.

King, J. L. 1967. Continuously distributed factors affecting fitness. *Genetics* 55: 483–492.

King, J. L., and T. H. Jukes. 1969. Non-Darwinian evolution. *Science* 164: 788–798.

Kingman, J. F. C. 1961. A matrix inequality. *Quart. J. Math.* 12: 78–80.

Kingman, J. F. C. 1961*a*. A mathematical problem in population genetics. *Proc. Camb. Phil. Soc.* 57: 574–582.

Kirkman, H. N. 1966. Properties of X-linked alleles during selection. *Amer. J. Hum. Genet.* 18: 424–432.

Klopfer, P. H., and R. H. MacArthur. 1960. Niche size and faunal diversity. *Amer. Natur.* 94: 293–300.

Klopfer, P. H., and R. H. MacArthur. 1961. On the causes of tropical species diversity: Niche overlap. *Amer. Natur.* 95: 223–226.

Knight, G. R., and A. Robertson. 1957. Fitness as a measurable character in Drosophila. *Genetics* 42: 524–530.

Kojima, K. 1959. Role of epistasis and overdominance in stability of equilibria with selection. *Proc. Nat. Acad. Sci.* 45: 984–989.

Kojima, K. 1959*a*. Stable equilibria for the optimum model. *Proc. Nat. Acad. Sci.* 45: 989–993.

Kojima, K. I. 1961. Effects of dominance and size of population on response to mass selection. *Genet. Res.* 2: 177–188.

Kojima, K. I. 1965. The evolutionary dynamics of two-gene systems. *Computers in Biomedical Research. Vol. I.* Acad. Press, New York. Pp. 197–220.

Kojima, K. I., and T. M. Kelleher. 1961. Changes of mean fitness in random mating populations when epistasis and linkage are present. *Genetics* 46: 527–540.

Kojima, K. I., and T. M. Kelleher. 1962. Survival of mutant genes. *Amer. Natur.* 96: 329–346.

Kojima, K. I., and H. E. Schaffer. 1964. Accumulation of epistatic gene complexes. *Evolution* 18: 127–129.

Kojima, K. I., and H. E. Schaffer. 1967. Survival process of linked mutant genes. *Evolution* 21: 518–531.

Kolman, W. 1961. The mechanism of natural selection for the sex ratio. *Amer. Natur.* 95: 373–377.

Kolmogorov, A. 1931. Über die analytischen Methoden in der Wahrscheinlichkeitsrechnung. *Math. Ann.* 104: 415–458.

Kolmogorov, A. 1935. Deviations from Hardy's formula in partial isolation. *C. R. (Doklady) Acad. Sci. URSS.* 3(63): 129–132.

Kolmogorov, A. N. 1959. Transition of branching processes into diffusion processes and some problems in genetics (in Russian). *Teor. Veroy. i. Primeonon* 4: 233–236.

Komai, T., M. Chico, and Y. Hosino. 1950. Contributions to the evolutionary genetics of the lady-beetle, *Harmonia*, I. Geographical and temporal variations in the relative frequency of the elytral pattern types and in the frequency of elytral ridge. *Genetics* 35: 589–601.

Korde, V. T. 1960. The correlations between relatives for a sex-linked character under inbreeding. *Heredity* 14: 401–409.

Korn, G. A., and T. M. Korn. 1968. *Mathematical Handbook for Scientists and Engineers*. 2nd Ed. McGraw-Hill, New York.

Kosambi, D. D. 1943. The estimation of map distances from recombination values. *Ann. Eugen.* 12: 172–175.

Kosswig, C. 1960. Genetische Analyse stammesgeschichtlicher Einheiten. *Verb. deutsch. Zool. Ges.* 1959: 42–73. *Zool. Anz., Suppl.* 23: 42–73.

Kostitzin, V. A. 1939. *Mathematical Biology*. Harrap, London.

Krieger, H., and N. Friere-Maia. 1961. Estimate of the load of mutations in homogeneous populations from data on mixed samples. *Genetics* 46: 1565–1566.

Kudo, A. 1962. A method for calculating the inbreeding coefficient. *Am. J. Hum. Genet.* 14: 426–432.

Kudo, A., and K. Sakaguchi. 1963. A method for calculating the inbreeding coefficient. II. Sex-linked genes. *Amer. J. Hum. Genet.* 15: 476–480.

Kyle, W. H., and A. B. Chapman. 1953. Experimental check of the effectiveness of selection for a quantitative character. *Genetics* 38: 421–433.

Lack, D. 1954. The evolution of reproductive rates. In *Evolution as a Process*. Ed. by J. Huxley, A. C. Hardy, and E. B. Ford. Allen and Unwin, London. Pp. 172–187.

Lagervall, P. M. 1960. Quantitative inheritance and dominance I. The coefficient of relationship caused by dominance. *Hereditas* 46: 481–496.

Lagervall, P. M. 1961. Quantitative inheritance and dominance. II. The genic and the dominance variance under inbreeding. *Hereditas* 47: 111–130.

Lagervall, P. M. 1961*a*. Quantitative inheritance and dominance. III. The genetic covariance of relatives in inbred populations. *Hereditas* 47: 131–159.

Lagervall, P. M. 1961*b*. Quantitative inheritance and dominance. IV. The average degree of dominance. *Hereditas* 47: 197–202.

Laidlaw, H. H., F. P. Gomes, and W. E. Kerr. 1956. Estimation of the number of lethal alleles in a panmictic population of *Apis mellifera* L. *Genetics* 41: 179–188.

Latter, B. D. H. 1959. Genetic sampling in a random mating population of constant size and sex ratio. *Aust. J. Biol. Sci.* 12: 500–505.

Latter, B. D. H. 1960. Natural selection for an intermediate optimum. *Aust. J. Biol. Sci.* 13: 30–35.

Latter, B. D. H. 1964. The evolution of non-additive genetic variance under artificial selection. I. Modification of dominance at a single autosomal locus. *Aust. J. Biol. Sci.* 17: 427–435.

Latter, B. D. H. 1965. The response to artificial selection due to autosomal genes of large effect. I. Changes in gene frequency at an additive locus. *Aust. J. Biol. Sci.* 18: 585–598.

Latter, B. D. H. 1965*a*. The response to artificial selection due to autosomal genes of large effect. II. The effects of linkage on limits to selection in finite populations. *Aust. J. Biol. Sci.* 18: 1009–1024.

Latter, B. D. H. 1965*b*. Quantitative genetic analysis in *Phalaris tuberosa*. I. The statistical theory of open-pollinated progeny. *Genet. Res.* 6: 360–370.

Latter, B. D. H. 1966. The response to artificial selection due to autosomal genes of large effect. III. The effects of linkage on the rate of advance and approach to fixation in finite populations. *Aust. J. Biol. Sci.* 19: 131–146.

Latter, B. D. H. 1966*a*. The interaction between effective population size and linkage intensity under artificial selection. *Genet. Res.* 7: 313–323.

Latter, B. D. H., and C. E. Novitski. 1969. Selection in finite populations with multiple alleles. I. Limits to directional selection. *Genetics* 62: 859–876.

Latter, B. D. H., and A. Robertson. 1962. The effects of inbreeding and artificial selection on reproductive fitness. *Genet. Res.* 3: 110–138.

Lea, D. E., and C. A. Coulson. 1949. The distribution of the numbers of mutants in bacterial populations. *J. Genet.* 49: 264–285.

Lee, B. T. O., and P. A. Parsons. 1968. Selection, prediction and response. *Biol. Rev.* 43: 139–174.

Lefkovitch, L. P. 1965. The study of population growth in organisms grouped by stages. *Biometrics* 21: 1–18.

Lefkovitch, L. P. 1966. A population model incorporating delayed responses. *Bull. Math. Biophys.* 28: 219–233.

Leigh, E. G. 1965. On the relation between the productivity, biomass, diversity, and stability of a community. *Proc. Nat. Acad. Sci.* 53: 777–783.

Leigh, E. 1966. *Ecological Aspects of Population Genetics.* Unpublished thesis, Yale University.

Lerner, I. M. 1950. *Population Genetics and Animal Improvement.* The University Press, Cambridge.

Lerner, I. M. 1954. *Genetic Homeostasis.* Oliver and Boyd, London.

Lerner, I. M. 1958. *The Genetic Basis of Selection.* Wiley, New York.

Lerner, I. M., and E. R. Dempster. 1962. Indeterminism and interspecific competition. *Proc. Nat. Acad. Sci.* 48: 821–826.

Le Roy, H. L. 1960. *Statistische Methoden der Populationsgenetik.* Birkhäusen Verlag, Basel.

Leslie, P. H. 1945. On the use of matrices in certain population mathematics. *Biometrika* 33: 183–212.

Leslie, P. H. 1948. Some further remarks on the use of matrices in population mathematics. *Biometrika* 35: 213–245.

Leslie, P. H. 1958. A stochastic model for studying the properties of certain biological systems by numerical methods. *Biometrika* 45: 16–31.

Leslie, P. H., and J. C. Gower. 1958. The properties of a stochastic model for two competing species. *Biometrika* 45: 316–330.

Leslie, P. H., and J. C. Gower. 1960. The properties of a stochastic model for the predator–prey type of interaction between two species. *Biometrika* 47: 219–234.

Levene, H. 1949. On a matching problem arising in genetics. *Ann. Math. Stat.* 20: 91–94.

Levene, H. 1949*a*. A new measure of sexual isolation. *Evolution* 3: 315–321.

Levene, H. 1953. Genetic equilibrium when more than one ecological niche is available. *Amer Natur.* 87: 311–313.

Levene, H. 1963. Inbred genetic loads and the determination of population structure. *Proc. Nat. Acad. Sci.* 50: 587–592.

Levene, H. 1967. Genetic diversity and diversity of environment: Mathematical aspects. *Proc. Fifth Berkeley Symp. Math. Stat. Prob.* 4: 305–316.

Levings, C. S. 1964. Genetic relationships among autotetraploid relatives. *J. Hered.* 55: 262–266.

Levins, R. 1962. Theory of fitness in a heterogeneous environment. I. The fitness set and adaptive function. *Amer. Natur.* 96: 361–373.

Levins, R. 1963. Theory of fitness in a heterogeneous environment. II. Developmental Flexibility and niche selection. *Amer. Natur.* 97: 75–90.

Levins, R. 1964. The theory of fitness in a heterogeneous environment. IV. The adaptive significance of gene flow. *Evolution* 18: 635–638.

Levins, R. 1965. Theory of fitness in a heterogeneous environment. III. The response to selection. *J. Theor. Biol.* 7: 224–240.

Levins, R. 1965*a*. Theory of fitness in a heterogeneous environment. V. Optimal genetic systems. *Genetics* 52: 891–904.

Levins, R. 1965*b*. Genetic consequences of natural selection. *Theoretical and Mathematical Biology.* Ed. by T. H. Waterman and H. J. Morowitz. Blaisdell, Waltham Mass. Pp. 388–397.

Levins, R. 1966. The strategy of model building in population biology. *Amer. Sci.* 54: 421–431.

Levins, R. 1967. Theory of fitness in a heterogeneous environment. VI. The adaptive significance of mutation. *Genetics* 56: 163–178.

Levins, R. 1968. *Evolution in Changing Environments.* Princeton University Press, Princeton, N.J.

Lewis, E. G. 1942. On the generation and growth of a population. *Sankhya* 6: 93–96.

Lewontin, R. C. 1953. The effect of compensation on populations subject to natural selection. *Amer. Natur.* 87: 375–381.

Lewontin, R. C. 1957. The adaptation of populations to varying environments. *Cold Spring Harbor Symp. Quant. Biol.* 22: 395–408.

Lewontin, R. C. 1958. A general method for investigating the equilibrium of gene frequency in a population. *Genetics* 43: 421–433.

Lewontin, R. C. 1961. Evolution and the theory of games. *J. Theor. Biol.* 1: 382–403.

Lewontin, R. C. 1962. Interdeme selection controlling a polymorphism in the house mouse. *Amer. Natur.* 96: 65–78.

Lewontin, R. C. 1964. The role of linkage in natural selection. *Genetics Today.* Pergamon Press, New York. Pp. 517–525.

Lewontin, R. C. 1964*a*. The interaction of selection and linkage. I. General considerations of heterotic models. *Genetics* 49: 49–67.

Lewontin, R. C. 1964*b*. The interaction of selection and linkage. II. Optimum models. *Genetics* 50: 757–782.

Lewontin, R. C. 1965. Selection for colonizing ability. *The Genetics of Colonizing Species.* Ed. by H. G. Baker and G. L. Stebbins. Academic Press, New York. Pp. 77–94.

Lewontin, R. C. 1965*a*. Selection in and of populations. Ideas in Modern Biology. *Proc. XVI Intern. Cong. Zool.*

Lewontin, R. C. 1967. The genetics of complex systems. *Proc. Fifth Berkeley Symp. Math. Stat. Prob.* 4: 439–456.

Lewontin, R. C. 1967*a*. Population genetics. *Ann. Rev. Genet.* 1: 37–70.

Lewontin, R. C. 1968. The effect of differential viability on the population dynamics of *t* alleles in the house mouse. *Evolution* 22: 262–273.

Lewontin, R. C., and C. C. Cockerham. 1959. The goodness-of-fit test for detecting selection in random mating populations. *Evolution* 13: 561–564.

Lewontin, R. C., and L. C. Dunn. 1960. The evolutionary dynamics of a polymorphism in the house mouse. *Genetics* 45: 705–722.

Lewontin, R. C., and J. L. Hubby. 1966. A molecular approach to the study of genic heterozygosity in natural populations. II. Amount of variation and degree of heterozygosity in natural populations of *Drosophila pseudoobscura.* *Genetics* 54: 595–609.

Lewontin, R. C., and P. Hull. 1967. The interaction of selection and linkage. III. Synergistic effect of blocks of genes. *Der Züchter* 37: 93–98.

Lewontin, R. C., and K. Kojima. 1960. The evolutionary dynamics of complex polymorphisms. *Evolution* 14: 458–472.

Lewontin, R. C., and M. J. D. White. 1960. Interaction between inversion polymorphisms of two chromosome pairs in the grasshopper, *Moraba scurra*. *Evolution* 14: 116–129.

Li, C. C. 1953. Some general properties of recessive inheritance. *Amer. J. Hum. Genet.* 5: 269–279.

Li, C. C. 1953*a*. Is Rh facing a crossroad? A critique of the compensation effect. *Amer. Natur.* 87: 257–261.

Li, C. C. 1953*b*. A direct proof of the relation between genotypic mating correlation and the gametic uniting correlation in equilibrium populations. *J. Hered.* 44: 39–40.

Li, C. C. 1955. The stability of an equilibrium and the average fitness of a population. *Amer. Natur.* 89: 281–296.

Li, C. C. 1955*a*. *Population Genetics*. Univ. Chicago Press, Chicago, Ill.

Li, C. C. 1957. Repeated linear regression and variance components of a population with binomial frequencies. *Biometrics* 13: 225–233.

Li, C. C. 1957*a*. The genetic variance of autotetraploids with two alleles. *Genetics* 42: 583–592.

Li, C. C. 1959. Notes on relative fitness of genotypes that form a geometric progression. *Evolution* 13: 564–567.

Li, C. C. 1961. *Human Genetics: Principles and Methods*. McGraw–Hill, New York.

Li, C. C. 1962. On "reflexive selection." *Science* 136: 1055–1056.

Li, C. C. 1963. Decrease of population fitness upon inbreeding. *Proc. Nat. Acad. Sci.* 49: 439–445.

Li, C. C. 1963*a*. Genetic aspects of consanguinity. *Amer. J. Med.* 34: 702–714.

Li, C. C. 1963*b*. The way the load ratio works. *Amer. J. Human Gen.* 15: 316–321.

Li, C. C. 1963*c*. Equilibrium under differential selection in the sexes. *Evolution* 17: 493–496.

Li, C. C. 1967. The maximization of average fitness by natural selection for a sex-linked locus. *Proc. Nat. Acad. Sci.* 57: 1260–1261.

Li, C. C. 1967*a*. Fundamental theorem of natural selection. *Nature* 214: 505–506.

Li, C. C. 1967*b*. Genetic equilibrium under selection. *Biometrics* 23: 397–484.

Li, C. C., and D. G. Horvitz. 1953. Some methods of estimating the inbreeding coefficient. *Amer. J. Human Genet.* 5: 107–117.

Li, C. C., and L. Sacks. 1954. The derivation of joint distribution and correlation between relatives by the use of stochastic matrices. *Biometrics* 10: 347–360.

Lillestøl, J. 1968. Another approach to some Markov chain models in population genetics. *J. Appl. Prob.* 5: 9–20.

Lloyd, M. 1964. Weighting individuals by reproductive value in calculating species diversity. *Amer. Natur.* 98: 190–192.

Lloyd, M., and R. J. Chelardi. 1964. A table for calculating the "equitability" component of species diversity. *J. Anim. Ecol.* 33: 217–225.

Lotka, A. J. 1922. The stability of the normal age distribution. *Proc. Nat. Acad. Sci.* 8: 339–345.

Lotka, A. J. 1925. *Elements of Physical Biology*. Williams and Wilkins, Baltimore.

Lotka, A. J. 1931. The extinction of families. *J. Wash. Acad. Sci.* 21: 377.

Lotka, A. J. 1945. Population analysis as a chapter in the mathematical theory of evolution. *Essays in Growth and Form*. Ed. by W. E. Le Gros Clark and P. B. Medawar. Oxford, England. Pp. 355–385.

Lotka, A. J. 1956. *Elements of Mathematical Biology*. Dover Publications, New York. (Revised edition of *Elements of Physical Biology*, 1925.)

Ludwig, W., and H. V. Schelling. 1948. Der Inzuchtgrad in endlichen panmiktischen Populationen. *Zool. Zent.* 67: 268–275.

Lush, J. L. 1940. Intrasire correlations or regressions of offspring on dam as a method of estimating heritability of characteristics. *Proc. Amer. Soc. Animal Prod.* 1940: 293–301.

Lush, J. L. 1945. *Animal Breeding Plans*. 3rd Ed. Iowa State College Press, Ames, Iowa.

Lush, J. L. 1946. Chance as a cause of changes in gene frequency within pure breeds of livestock. *Amer. Natur.* 80: 318–342.

Lush, J. L. 1947. Family merit and individual merit as bases for selection. *Amer. Natur.* 81: 241–261, 362–379.

MacArthur, R. H. 1955. Fluctuations of animal populations, and a measure of community stability. *Ecology* 36: 553–536.

MacArthur, R. H. 1957. On the relative abundance of bird species. *Proc. Nat. Acad. Sci.* 43: 293–295.

MacArthur, R. H. 1958. A note on stationary age distributions in single species populations and stationary species populations in a community. *Ecology* 39: 146–147.

MacArthur, R. H. 1960. On the relative abundance of species. *Amer. Natur.* 94: 25–36.

MacArthur, R. H. 1960*a*. On the relation between reproductive value and optimal predation. *Proc. Nat. Acad. Sci.* 46: 143–145.

MacArthur, R. H. 1961. Population effects of natural selection. *Amer. Natur.* 95: 195–199.

MacArthur, R. H. 1962. Some generalized theorems of natural selection. *Proc. Nat. Acad. Sci.* 38: 1893–1897.

MacArthur, R. H. 1964. Environmental factors affecting bird species diversity. *Amer. Natur.* 98: 387–397.

MacArthur, R. H. 1965. Patterns of species diversity. *Biol. Rev.* 40: 510–533.

MacArthur, R. H. 1965*a*. Ecological consequences of natural selection. *Theoretical and Mathematical Biology*. Ed. by T. H. Waterman and H. J. Morowitz. Blaisdell, Waltham, Mass. Pp. 388–397.

MacArthur, R., and R. Levins. 1964. Competition, habitat selection, and character displacement in a patchy environment. *Proc. Nat. Acad. Sci.* 51: 1207–1210.

MacArthur, R. H., and E. O. Wilson. 1963. An equilibrium theory of insular zoogeography. *Evolution* 17: 373–387.

MacArthur, R. H., and E. O. Wilson. 1967. *The Theory of Island Biogeography*. Princeton University Press, Princeton, N.J.

McBride, G., and A. Robertson. 1963. Selection using assortative mating in *D. melanogaster*. *Genet. Res.* 4: 356–369.

McPhee, H. C., and S. Wright. 1925. Mendelian analysis of the pure breeds of livestock. III. The shorthorns. *J. Hered.* 16: 205–215.

McPhee, H. C., and S. Wright. 1926. Mendelian analysis of the pure breeds of livestock. IV. The British dairy shorthorns. *J. Hered.* 17: 397–401.

Malécot, G. M. 1944. Sur un problème de probabilités en chaîne que pose la génetique. *Compt. Rend. de l'Acad. des Sci.* 219: 379–381.

Malécot, G. 1948. *Les mathématiques de l'hérédité.* Masson et Cie, Paris.

Malécot, G. 1955. Decrease of relationship with distance. *Cold Spring Harbor Symp. Quant. Biol.* 20: 52–53.

Malécot, G. 1959. Les modèles stochastiques en génétique de population. *Pub. Inst. Stat. Univ. de Paris* 8(3): 173–210.

Malécot, G. 1966. *Probabilités et Hérédité.* Presses Universitaires de France.

Malécot, G. 1967. Identical loci and relationship. *Proc. Fifth Berkeley Symp. Math. Stat. Prob.* 4: 317–332.

Mandel, S. P. H. 1959. The stability of a multiple allelic system. *Heredity* 13: 289–302.

Mandel, S. P. H. 1959*a*. Stable equilibrium at a sex-linked locus. *Nature* 183: 1347–1348.

Mandel, S. P. H., and I. M. Hughes. 1958. Change in mean viability at a multi-allelic locus in a population under random mating. *Nature* 182: 63–64.

Margalef, R. 1957. Information theory in ecology. *General Systems* 3: 36–71.

Margalef, R. 1963. On certain unifying principles in ecology. *Amer. Natur.* 97: 357–374.

Martin, F. G., and C. C. Cockerham. 1960. High speed selection studies. *Biometrical Genetics*. Ed. by O. Kempthorne. Pergamon Press, London. Pp. 35–45.

Maruyama, T. 1969. Genetic correlation in the stepping stone model with non-symmetrical migration rates. *Jour. Appl. Prob.* 6: 463–477.

Maruyama, T. 1970. On the fixation probability of mutant genes in a subdivided population. *Genet. Res.* (in press).

Maruyama, T. 1970*a*. Rate of decrease of genetic variability in a subdivided population. *Biometrika* (in press).

Mather, K. 1941. Variation and selection of polygenic characters. *J. Genet.* 41: 159–193.

Mather, K. 1942. The balance of polygenic combinations. *J. Genet.* 43: 309–336.

Mather, K. 1943. Polygenic inheritance and natural selection. *Biol. Rev.* 18: 32–64.

Mather, K. 1946. Dominance and heterosis. *Amer. Natur.* 80: 91–96.

Mather, K. 1949. *Biometrical Genetics.* Dover Pub., New York.

Mather, K. 1953. The genetical structure of populations. *Symp. Soc. Exp. Biol.* 7: 66–95.

Mather, K. 1955. Response to selection. *Cold Spring Harbor Symp. Quant. Biol.* 20: 158–165.

Mather, K. 1955*a*. Polymorphism as an outcome of disruptive selection. *Evolution* 9: 52–61.

Mather, K. 1963. Genetical demography. *Proc. Roy. Soc. B* 159: 106–125.

Mather, K. 1966. Variability and selection. *Proc. Roy. Soc. B.* 164: 328–340.

Mather, K. 1967. Complementary and duplicate gene interactions in biometrical genetics. *Heredity* 22: 97–103.

Matsunaga, E. 1966. Possible genetic consequences of family planning. *J. Amer. Med. Assn.* 198: 533–540.

Matzinger, D. F., and O. Kempthorne. 1956. The modified diallel table with partial inbreeding and interactions with environment. *Genetics* 41:822–833.

Maynard Smith, J. 1958. *The Theory of Evolution*. Penguin Books, London.

Maynard Smith, J. 1962. Disruptive selection, polymorphism, and sympatric speciation. *Nature* 195: 60–62.

Maynard Smith, J. 1964. Kin selection and group selection. *Nature* 201: 1145–1147.

Maynard Smith, J. 1965. The evolution of alarm calls. *Amer. Natur.* 99: 59–63.

Maynard Smith, J. 1966. Sympatric speciation. *Amer. Natur.* 100: 637–650.

Maynard Smith, J. 1968. Evolution in sexual and asexual populations. *Amer. Natur.* 102: 469–473.

Maynard Smith, J. 1968*a*. "Haldane's dilemma" and the rate of evolution. *Nature* 29: 1114–1116.

Maynard Smith, J. 1968*b*. *Mathematical Ideas in Biology*. Cambridge Univ. Press, Cambridge, England.

Mayo, O. 1966. On the problem of self-incompatibility alleles. *Biometrics* 22: 111–120.

Mayr, E. 1947. Ecological factors in speciation. *Evolution* 1: 263–287.

Mayr, E. 1954. Change of genetic environment and evolution. *Evolution as a Process*. Ed. by J. Huxley, A. C. Hardy, and E. B. Ford. Allen and Unwin, London, Pp. 157–180.

Mayr, E. 1956. Geographical gradients and climatic adaptation. *Evolution* 10: 105–108.

Mayr, E. 1963. *Animal Species and Evolution*. Harvard Univ. Press, Cambridge, Mass.

Mérat, P. 1967. Les gènes influant sur la variance d'un charactère quantitif et leurs répercussions possibles sur la sélection. *Ann. Genetique* 10: 212–230.

Mettler, L. E., and T. G. Gregg. 1969. *Population Genetics and Evolution*. Prentice-Hall, Englewood Cliffs, N.J.

Milkman, R. D. 1967. Heterosis as a major cause of heterozygosity in nature. *Genetics* 55: 493–495.

Miller, G. F. 1962. The evaluation of eigenvalues of a differential equation arising in a problem in genetics. *Proc. Camb. Phil. Soc.* 58: 588–593.

Mode, C. J. 1958. A mathematical model for the co-evolution of obligate parasites and their hosts. *Evolution* 12: 158–165.

Mode, C. J. 1960. A model of a host-pathogen system with particular reference to the rusts of cereals. *Biometrical Genetics*. Ed. by O. Kempthorne, Pergamon Press, New York. Pp. 84–96.

Mode, C. J. 1961. A generalized model of a host-pathogen system. *Biometrics* 17: 386–404.

Mode, C. J. 1962. Some multi-dimensional birth and death processes and their applications in population genetics. *Biometrics* 18: 543–567; 19: 667.

Mode, C. J. 1966. Some multi-dimensional branching processes as motivated by a class of problems in mathematical genetics. *Bull. Math. Biophys.* 28: 25–50; 28: 181–190.

Mode, C. J. 1966*a*. A stochastic calculus and its application to some fundamental theorems of natural selection. *J. Appl. Prob.* 3: 327–352.

Mode, C. J. 1967. On the probability a line becomes extinct before a favorable mutation appears. *Bull. Math. Biophys.* 29: 343–348.

Mode, C. J. 1968. A multidimensional age-dependent branching process with applications to natural selection I. *Math. Bioscience* 3: 1–18.

Mode, C. J. 1968*a*. A multidimensional age-dependent branching process with applications to natural selection II. *Math. Bioscience* 3: 231–247.

Mode, C. J., and H. F. Robinson. 1959. Pleiotropism and the genetic variance and covariance. *Biometrics* 15: 518–537.

Moment, G. B. 1962. Reflexive selection: A possible answer to an old puzzle. *Science* 136: 262–263.

Moment, G. B. 1962*a*. On "reflexive selection." *Science* 136: 1056.

Moody, P. A. 1947. A simple model of "drift" in small populations. *Evolution* 3: 217–218.

Moran, P. A. P. 1958. Random processes in genetics. *Proc. Camb. Phil. Soc.* 54: 60–71.

Moran, P. A. P. 1958*a*. The effect of selection in a haploid genetic population. *Proc. Camb. Phil. Soc.* 54: 463–467.

Moran, P. A. P. 1958*b*. The distribution of gene frequency in a bisexual diploid population. *Proc. Camb. Phil. Soc.* 54: 468–474.

Moran, P. A. P. 1958*c*. A general theory of the distribution of gene frequencies. I. Overlapping generations. II. Non-overlapping generations. *Proc. Roy. Soc. B* 149: 102–112, 113–116.

Moran, P. A. P. 1958*d*. The rate of approach to homozygosity. *Ann. Hum. Genet.* 23: 1–5.

Moran, P. A. P. 1959. The theory of some genetical effects of population subdivision. *Aust. J. Biol. Sci.* 12: 109–116.

Moran, P. A. P. 1959*a*. The survival of a mutant gene under selection. *J. Aust. Math. Soc.* 1: 121–126.

Moran, P. A. P. 1960. The survival of a mutant gene under selection. II. *J. Aust. Math. Soc.* 1: 485–491.

Moran, P. A. P. 1961. The survival of a mutant under general conditions. *Proc. Camb. Phil. Soc.* 57: 304–314.

Moran, P. A. P. 1962. *The Statistical Processes of Evolutionary Theory.* The Clarendon Press, Oxford.

Moran, P. A. P. 1963. On the measurement of natural selection dependent on several loci. *Evolution* 17: 182–186.

Moran, P. A. P. 1963*a*, Some general results on random walks, with genetic applications. *J. Aust. Math. Soc.* 3: 468–479.

Moran, P. A. P. 1963*b*. Balanced polymorphisms with unlinked loci. *Aust. J. Biol. Sci.* 16: 1–5.

Moran, P. A. P. 1964. On the nonexistence of adaptive topographies. *Ann. Human Genet.* 27: 383–393.

Moran, P. A. P. 1967. Unsolved problems in evolutionary biology. *Proc. Fifth Berkeley Symp. Math. Stat. Prob.* 4: 457–480.

Moran, P. A. P., and C. A. B. Smith. 1966. Commentary on R. Fisher's Paper on the Correlation Between Relatives on the Supposition of Mendelian Inheritance. Cambridge Univ. Press, Cambridge, England.

Moran, P. A. P., and G. A. Watterson. 1958. The genetic effects of family structure in natural populations. *Aust. J. Biol. Sci.* 12: 1–15.

Morishima, H. 1969. Phenetic similarity and phylogenetic relationships among strains of *Oryza perennis*, estimated by method of numerical taxonomy. *Evolution* 17: 170–181.

Morris, R. F. 1959. Single-factor analysis in population dynamics. *Ecology* 40: 580–588.

Morse, P. M., and H. Feshbach. 1953. *Methods of Theoretical Physics.* Part I and II. McGraw-Hill, New York.

Morton, N. E. 1955. Non-randomness in consanguineous marriage. *Ann. Hum. Genet.* 20: 116–124.

Morton, N. E. 1960. The mutational load due to detrimental genes in man. *Amer. J. Hum. Genet.* 12: 348–364.

Morton, N. E. 1965. Models and evidence in human population genetics. *Proc. XI Int. Cong. Genet.* Pp. 935–950.

Morton, N. E. 1969. Human population structure. *Ann. Rev. Genet.* 3: 53–74.

Morton, N. E., Ed. 1969*a*. *Computer Applications in Genetics.* Univ. Hawaii Press, Honolulu.

Morton, N. E., C. S. Chung, and M. P. Li. 1967. *Genetics of Interracial Crosses in Hawaii.* S. Karger, New York.

Morton, N. E., J. F. Crow, and H. J. Muller. 1956. An estimate of the mutational damage in man from data on consanguineous marriages. *Proc. Nat. Acad. Sci.* 42: 855–863.

Morton, N. E., and S. Wright. 1968. Genetic studies of cystic fibrosis in Hawaii. *Amer. J. Hum. Genet.* 20: 157–169.

Morton, N. E., and N. Yasuda. 1962. The genetical structure of human populations. *Entretien de Monaco en Sciences Humaines: Les Deplacements Humains.* Ed. by J. Sutter. Pp. 185–202.

Morton, N. E., N. Yasuda, C. Miki, and S. Yee. 1968. Bioassay of population structure under isolation by distance. *Amer. J. Human. Genet.* 20: 411–419.

Moser, H. 1958. The dynamics of bacterial populations maintained in the chemostat. *Carnegie Inst. Pub.* 614.

Mukai, T. 1964. The genetic structure of natural populations of *Drosophila melanogaster*. I. Spontaneous mutation rate of polygenes controlling viability. *Genetics* 50: 1–19.

Mukai, T. 1969. The genetic structure of natural populations of *Drosophila melanogaster*. VII. Synergistic interaction of spontaneous mutant polygenes controlling viability. *Genetics* 61: 749–761.

Mukai, T., and A. B. Burdick. 1959. Single gene heterosis associated with a second chromosome recessive lethal in *Drosophila melanogaster*. *Genetics* 44: 211–232.

Mulholland, H. P., and C. A. B. Smith. 1959. An inequality arising in genetical theory. *Amer. Math. Mon.* 66: 673–683.

Muller, H. J. 1914. The bearing of the selection experiments of Castle and Phillips on the variability of genes. *Amer. Nat.* 48: 567–576.

Muller, H. J. 1925. Why polyploidy is rarer in animals than in plants. *Amer. Nat.* 59: 346–353.

Muller, H. J. 1929. The gene as the basis of life. *Proc. Int. Congr. Plant. Sci.* 1: 897–921.

Muller, H. J. 1932. Some genetic aspects of sex. *Amer. Natur.* 68: 118–138.

Muller, H. J. 1936. On the variability of mixed races. *Amer. Nat.* 70: 409–442.

Muller, H. J. 1939. Reversibility in evolution considered from the standpoint of genetics. *Biol. Rev.* 14: 261–280.

Muller, H. J. 1942. Isolating mechanisms, evolution, and temperature. *Biol. Symp.* 6: 71–125.

Muller, H. J. 1949. The Darwinian and modern conceptions of natural selection. *Proc. Amer. Philos. Soc.* 93: 459–470.

Muller, H. J. 1950. Evidence of the precision of genetic adaptation. *The Harvey Lectures* 18: 165–229.

Muller, H. J. 1950*a*. Our load of mutations. *Amer. J. Human Genet.* 2: 111–176.

Muller, H. J. 1958. Evolution by mutation. *Bull. Amer. Math. Soc.* 64: 137–160.

Muller, H. J. 1964. The relation of recombination to mutational advance. *Mutation Res.* 1: 2–9.

Muller, H. J. 1967. What genetic course will man steer? *Proc. Third Int. Cong. Hum. Genet.* Pp. 521–543.

Murray, M. 1964. Multiple mating and effective population size in *Cepaea nemoralis*. *Evolution* 18: 283–291.

Nair, K. R. 1954. The fitting of growth curves. *Statistics and Mathematics in Biology*. Ed. by O. Kempthorne, T. A. Bancroft, J. W. Gowen, and J. L. Lush. Iowa State College, Ames, Iowa. Pp. 119–132.

Narain, P. 1963. On mathematical representation of gene action and interaction. *J. Indian Soc. Agri. Stat.* 15: 270.

Narain, P. 1965. The description of gene action and interaction with multiple alleles in continuous variation. *Genetics* 52: 43–53.

Narain, P. 1965*a*. Homozygosity in a selfed population with an arbitrary number of linked loci. *J. Genet.* 59: 1–13.

Narain, P. 1966. Effect of linkage on homozygosity of a population under mixed selfing and random mating. *Genetics* 54: 303–314.

Narain, P. 1969. A note on the diffusion approximation for the variance of the number of generations until fixation of a neutral mutant gene. Submitted to *Genet. Res.*

Nassar, R. F. 1965. Effect of correlated gene distribution due to sampling on diallel analysis. *Genetics* 52: 9–20.

Nassar, R. F. 1969. Distribution of gene frequencies under the case of random genetic drift with and without selection. *Theoret. App. Genetics* 39: 145–149.

Naylor, A. F. 1962. Mating systems which could increase heterozygosity for a pair of alleles. *Amer. Natur.* 96: 51–60.

Naylor, A. F. 1963. A theorem on possible kinds of mating systems which tend to increase heterozygosity. *Evolution* 17: 369–370.

Naylor, A. F. 1964. Natural selection through maternal influence. *Heredity* 19: 509–511.

Neal, N. P. 1935. The decrease in yielding capacity in advanced generations of hybrid corn. *J. Amer. Soc. Agron.* 27: 666–670.

Neel, J. V., and W. J. Schull. 1954. *Human Heredity*. The Univ. of Chicago Press, Chicago.

Nei, M. 1964. Effects of linkage and epistasis on the equilibrium frequencies of lethal genes. I. Linkage equilibrium. *Jap. J. Genet.* 39: 1–6.

Nei, M. 1964*a*. Effects of linkage and epistasis on the equilibrium frequencies of lethal genes. II. Numerical solutions. *Jap. J. Genet.* 39: 7–25.

Nei, M. 1965. Effect of linkage on the genetic load manifested under inbreeding. *Genetics* 51: 679–688.

Nei, M. 1965*a*. Variation and covariation of gene frequencies in subdivided populations. *Evolution* 19: 256–258.

Nei, M. 1967. Modification of linkage intensity by natural selection. *Genetics* 57: 625–641.

Nei, M. 1968. Evolutionary change of linkage intensity. *Nature* 218: 1160–1161.

Nei, M. 1968*a*. The frequency distribution of lethal chromosomes in finite populations. *Proc. Nat. Acad. Sci.* 60: 517–524.

Nei, M., and Y. Imaizumi. 1966. Genetic structure of human populations I. Local differentiation of blood group gene frequencies in Japan. *Heredity* 21: 9–35.

Nei, M., and Y. Imaizumi. 1966*a*. Genetic structure of human populations. II. Differentiation of blood group gene frequencies among isolated populations. *Heredity* 21: 183–190, 344.

Nei, M., and Y. Imaizumi. 1966*b*. Effects of restricted population size and increase in mutation rate on the genetic variation of quantitative characters. *Genetics* 54: 763–782.

Nei, M., K. I. Kojima, and H. E. Schaffer. 1967. Frequency changes of new inversions in populations under mutation-selection equilibria. *Genetics* 57: 741–750.

Nei, M., and M. Murata. 1966. Effective population size when fertility is inherited. *Genet. Res.* 8: 257–260.

Nelder, J. A. 1952. Some genotypic frequencies and variance components occurring in biometrical genetics. *Heredity* 6: 387–394.

Nelder, J. A. 1953. Statistical methods in biometrical genetics. *Heredity* 7: 111–119.

Nicholson, A. J. 1933. The balance of animal populations. *J. Anim. Ecol.* 2(suppl.): 132–178.

Nicholson, A. J. 1950. Population oscillation caused by competition for food. *Nature* 165: 476–477.

Nicholson, A. J. 1954. An outline of the dynamics of animal population. *Aust. J. Zool.* 2: 9–65.

Nicholson, A. J. 1957. The self-adjustment of populations to change. *Cold Spring Harbor Symp. Quant. Biol.* 22: 153–173.

Nicholson, A. J., and V. A. Bailey. 1935. The balance of animal populations. Part I. *Proc. Zool. Soc. London* 551–598.

Nikoro, Z. S. 1964. Alteration of population structure as a result of selection in the case of overdominance (in Russian). *Bull. Moscow Soc. Nat. Bio. Ser.* 49: 5–21.

Norton. H. T. J. 1928. Natural selection and Mendelian variation. *Proc. Lond. Math. Soc.* 28: 1–45.

Novick, A., and L. Szilard. 1950. Experiments with the chemostat on spontaneous mutations in bacteria. *Proc. Nat. Acad. Sci.* 36: 708–719.

O'Donald, P. 1960. Inbreeding as a result of imprinting. *Heredity* 15: 79–85.

O'Donald, P. 1960*a*. Assortive mating in a population in which two alleles are segregating. *Heredity* 15: 389–396.

O'Donald, P. 1962. The theory of sexual selection. *Heredity* 17: 541–552.

O'Donald, P. 1963. Sexual selection and territorial behavior. *Heredity* 18: 361–364.

O'Donald, P. 1963*a*. Sexual selection for dominant and recessive genes. *Heredity* 18: 451–457.

O'Donald, P. 1967. A general model of sexual and natural selection. *Heredity* 22: 499–518.

O'Donald, P. 1968. Measuring the intensity of natural selection. *Nature* 220: 197–198.

O'Donald, P. 1968*a*. The evolution of dominance by selection for an optimum. *Genetics* 58: 451–460.

O'Donald, P. 1968*b*. Models of the evolution of dominance. *Proc. Roy. Soc. B.* 171: 127–143.

O'Donald, P. 1969. "Haldane's dilemma" and the rate of natural selection. *Nature* 221: 815–816.

O'Donald, P. 1969*a*. The selective coefficients that keep modifying genes in a population. *Genetics* 62: 435–444.

Ohta, T. 1967. Probability of fixation of mutant genes and the theory of limits in artificial selection. *Jap. J. Genet.* 42: 353–360.

Ohta, T. 1968. Effect of initial linkage disequilibrium and epistasis on fixation probability in a small population, with two segregating loci. *Theor. Appl. Genet.* 38: 243–248.

Ohta, T., and M. Kimura. 1969. Linkage disequilibrium due to random genetic drift. *Genet. Res.* 13: 47–55.

Ohta, T., and M. Kimura. 1969*a*. Linkage disequilibrium at steady state determined by random genetic drift and recurrent mutation. *Genetics* 63: 229–238.

Ohta, T., and K. Kojima. 1968. Survival probabilities of new inversions in large populations. *Biometrics* 24: 501–516.

Opsahl, B. 1956. The discrimination of interactions and linkage in continuous variation. *Biometrics* 12: 415–432.

Orians, G. H. 1962. Natural selection and ecological theory. *Amer. Natur.* 96: 257–263.

Orias, E., and F. J. Rohlf. 1964. Population genetics of the mating type locus in *Tetrahymena pyriformis*, variety 8. *Evolution* 18: 620–629.

Osborn, R., and W. S. B. Paterson. 1952. On the sampling variance of heritability estimates derived from variance analysis. *Proc. Roy. Soc. Edinb. B.* 64: 456–461.

Osborne, R. 1957. Correction for regression on a secondary trait as a method of increasing the efficiency of selective breeding. *Aust. J. Biol. Sci.* 10: 365–366.

Owen, A. R. G. 1952. A genetical system admitting of two stable equilibria. *Nature* 170: 1127.

Owen, A. R. G. 1953. A genetical system admitting of two distinct stable equilibria under natural selection. *Heredity* 7: 97–102.

Owen, A. R. G. 1954. Balanced polymorphism of a multiple allelic series. *Caryologia* 6 (suppl.) 1240–1241.

Owen, A. R. G. 1959. Mathematical models for selection. *Symp. Soc. Study Hum. Biol.* 2: 11–16.

Page, A. R., and B. I. Hayman. 1960. Mixed sib and random mating when homozygotes are at a disadvantage. *Heredity* 14: 187–196.

Parsons, P. A. 1957. Selfing under conditions favouring heterozygosity. *Heredity* 11: 411–421.

Parsons, P. A. 1959. Equilibria in auto-tetraploids under natural selection for a simplified model of viabilities. *Biometrics* 15: 20–29.

Parsons, P. A. 1961. The initial progress of new genes with viability differences between sexes and with sex linkage. *Heredity* 16: 103–107.

Parsons, P. A. 1962. The initial increase of a new gene under positive assortative mating. *Heredity* 17: 267–276.

Parsons, P. A. 1963. Complex polymorphisms where the coupling and repulsion double heterozygote viabilities differ. *Heredity* 18: 369–374.

Parsons, P. A. 1963*a*. Polymorphism and the balanced polygenic combination. *Evolution* 17: 564–574.

Parsons, P. A. 1963*b*. Migration as a factor in natural selection. *Genetica* 33: 184–206.

Parsons, P. A. 1964. Polymorphism and the balanced polygenic complex—a comment. *Evolution* 18: 512.

Parsons, P. A. 1964*a*. Complex polymorphisms with recombination differing between sexes. *Aust. J. Biol. Sci.* 17: 317–322.

Parsons, P. A. 1964*b*. Interactions within and between chromosomes. *J. Theor. Biol.* 6: 208–216.

Parsons, P. A., and W. F. Bodmer. 1961. The evolution of overdominance: Natural selection and heterozygote advantage. *Nature* 190: 7–12.

Patau, K. 1938. Die mathematische Analyse der Evolutionsvorgange. *Zeits. f. Abst. u. Vererb.* 76: 220–228.

Patlak, C. S. 1953. The effect of the previous generations on the distribution of gene frequencies in populations. *Proc. Nat. Acad. Sci.* 39: 1063–1068.

Patten, B. C. 1959. An introduction to the cybernetics of the eosystem: The trophic-dynamic aspect. *Ecology* 40: 221–231.

Pearl, R. 1913. A contribution towards an analysis of the problem of inbreeding. *Amer. Natur.* 47: 577–614.

Pearl, R. 1914. On the results of inbreeding a Mendelian population; a correction and extension of previous conclusions. *Amer. Natur.* 48: 57–62.

Pearl, R. 1914*a*. On a general formula for the constitution of the *n*th generation of a Mendelian population in which all matings are of brother X sister. *Amer. Natur.* 48: 491–494.

Pearl, R. 1940. *Medical Biometry and Statistics.* W. B. Saunders, Philadelphia.

Pearson, E. S., and H. O. Hartley. 1958. *Biometrika Tables for Statisticians. Vol. I.* The University Press, Cambridge.

Pearson, K. 1904. On a generalized theory of alternative inheritance, with special references to Mendel's laws. *Phil. Trans. Roy. Soc. A* 203: 53–86.

Pearson, K. 1909. The theory of ancestral contributions in heredity. *Proc. Roy. Soc. B.* 81: 219–224.

Pearson, K. 1909*a*. On the ancestral gemetic correlations of a Mendelian population mating at random. *Proc. Roy. Soc. B.* 81: 225–229.

Pearson, K., and A. Lee. 1903. On the laws of inheritance in man. I. Inheritance of physical characters. *Biometrika* 2: 357–462.

Pederson, D. G. 1966. The expected degree of heterozygosity in a double-cross hybrid population. *Genetics* 53: 669–674.

Pederson, D. G. 1969. The prediction of selection response in a self-fertilizing species. *Aust. J. Biol. Sci.* 22: 117–129.

Penrose, L. S. 1949. The meaning of "fitness" in human populations. *Ann. Eugen.* 14: 301–304.

Penrose, L. S. 1964. Some formal consequences of genes in stable equilibrium. *Ann. Hum. Genet.* 28: 159–166.

Penrose, L. S., S. M. Smith, and D. A. Sprott. 1957. On the stability of allelic systems, with special reference to haemoglobins A, S, and C. *Ann. Hum. Genet.* 21: 90–93.

Pimentel, D. 1961. Animal population regulation by the genetic feed-back mechanism. *Amer. Natur.* 95: 65–79.

Pimentel, D., E. H. Feinberg, P. W. Wood, and J. T. Hayes. 1965. Selection, spatial distribution, and the coexistence of competing fly species. *Amer. Natur.* 99: 97–109.

Pirchner, F. 1969. *Population Genetics and Animal Breeding.* W. H. Freeman, San Francisco.

Planck, M. 1917. Über einen Satz der statistischen Dynamik und seine Erweiterung in der Quantentheorie. *Sitz. der preuss. Akad.* Pp. 324–341.

Plum, M. 1954. Computation of inbreeding and relationship coefficients. *J. Hered.* 45: 92–94.

Pollak, E. 1966. Some consequences of selection by culling when there is superiority of heterozygotes. *Genetics* 53: 977–988.

Pollak, E. 1966*a*. On the survival of a gene in a subdivided population. *J. Appl. Prob.* 3: 142–155.

Pollak, E. 1966*b*. Some effects of fluctuating offspring distributions on the survival of a gene. *Biometrika* 53: 391–396.

Pollak, E. 1968. On random genetic drift in a subdivided population. *J. Appl. Prob.* 5: 314–333.

Pollard, J. H. 1966. On the use of the direct matrix product in analysing certain stochastic population models. *Biometrika* 53: 397–415.

Pollard, J. H. 1968. The multi-type Galton–Watson process in a genetical context. *Biometrics* 24: 147–158.

Powers, L. 1944. An expansion of Jones's theory for the explanation of heterosis. *Amer. Natur.* 78: 275–280.

Preston, F. W. 1962. The canonical distribution of commoness and rarity: Part I. *Ecology* 43: 185–215. Part II. *Ibid.* 43: 410–432.

Prout, T. 1953. Some effects of the variations in segregation ratio and of selection on the frequency of alleles under random mating. *Acta Genet. Stat. Med.* 4: 148–151.

Prout, T. 1962. The effects of stabilizing selection on the time of development in *Drosophila melanogaster*. *Genet. Res.* 3: 364–382.

Prout, T. 1965. The estimation of fitness from genotypic frequencies. *Evolution* 19: 546–551.

Prout, T. 1968. Sufficient conditions for multiple niche polymorphism. *Amer. Natur.* 102: 493–496.

Purser, A. F. 1966. Increase in heterozygote frequency with differential fertility *Heredity* 21: 322–327.

Quastler, H. 1959. Information theory of biological integration. *Amer. Natur.* 93: 245–254.

Qureshi, A. W., and O. Kempthorne. 1968. On the fixation of genes of large effects due to continued truncation selection in small populations of polygenic systems with linkage. *Theor. & Appl. Genet.* 38: 249–255.

Qureshi, A. W., O. Kempthorne, and L. N. Hazel. 1968. The role of finite size and linkage in response to continued truncation selection. 1. Additive gene action. 2. Dominance and overdominance. *Theor. & Appl. Genet.* 38: 256–276.

Race, R. R., and R. Sanger. 1962. *Blood Groups in Man.* Fourth Ed. F. A. Davis Co., Philadelphia.

Rajagoplan, M. 1958. Effect of linkage on the homozygosity of a selfed population. *J. Indian Soc. Agri. Stat.* 10: 64–66.

Rasmussen, D. I. 1964. Blood group polymorphism and inbreeding in natural populations of the deer mouse *Peromyscus maniculatus*. *Evolution* 18: 219–229.

Rasmuson, M., 1961. *Genetics on the Population Level.* Svenska Bokförlaget, Stockholm.

Rawlings, J. O., and C. C. Cockerham. 1962. Analysis of double cross hybrid populations. *Biometrics* 18: 229–244.

Reed, J., R. Toombs, and N. A. Barricelli. 1967. Simulation of biological evolution and machine learning. *J. Theor. Biol.* 17: 319–342.

Reed, T. E. 1959. The definition of relative fitness of individuals with specific genetic traits. *Amer. J. Hum. Genet.* 11: 137–155.

Reeve, E. C. R. 1955. Inbreeding with the homozygotes at a disadvantage. *Ann. Hum. Genet.* 19: 332–346.

Reeve, E. C. R. 1955*a*. The variance of the genetic correlation coefficient. *Biometrics* 11: 357–374.

Reeve, E. C. R. 1957. Inbreeding with selection and linkage. I. Selfing. *Ann. Hum. Genet.* 21: 277–288.

Reeve, E. C. R. 1961. A note on non-random mating in progeny tests. *Genet. Res.* 2: 195–203.

Reeve, E. C. R., and J. C. Gower. 1958. Inbreeding with selection and linkage. II. Sib-mating. *Ann. Hum. Genet.* 23: 36–49.

Rehfeld, C. E., J. W. Bacus, J. A. Pagels, and M. H. Dipert. 1967. Computer calculation of Wright's inbreeding coefficient. *J. Hered.* 58: 81–84.

Reiersöl, O. 1962. Genetic algebras studied recurisvely and by means of differential operators. *Math. Scand.* 10: 25–44.

Rendel, J. M. 1953. Heterosis. *Amer. Natur.* 87: 129–138.

Rendel, J. M. 1958. Optimum group size in half-sib family selection. *Biometrics* 15: 376–381.

Rendel, J. M. 1959. Evolution of dominance. The evolution of living organisms. *Royal Soc. Victoria, Melbourne.* Pp. 102–110.

Rhodes, E. C. 1940. Population mathematics. *J. Roy. Stat. Soc.* 103: 68–89, 218–245, 362–387.

Richardson, R. H., and K. I. Kojima. 1965. The kinds of genetic variability in relation to selection responses in Drosophila fecundity. *Genetics* 52: 583–598.

Richardson, W. H. 1964. Frequencies of genotypes of relatives as determined by stochastic matrices. *Genetics* 35: 323–354.

Robbins, R. B. 1917. Some applications of mathematics to breeding problems. *Genetics* 2: 489–504.

Robbins, R. B. 1918. Some applications of mathematics to breeding problems. II. *Genetics* 3: 73–92.

Robbins, R. B. 1918*a*. Some applications of mathematics to breeding problems. III. *Genetics* 3: 375–389.

Robbins, R. B. 1918*b*. Random mating with the exception of sister by brother mating. *Genetics* 3: 390–396.

Robertson, A. 1952. The effects of inbreeding on the variation due to recessive genes. *Genetics* 37: 189–207.

Robertson, A. 1953. A numerical description of breed structure. *J. Agric. Sci.* 43: 334–336.

Robertson, A. 1955. Prediction equations in quantitative genetics. *Biometrics* 11: 95–98.

Robertson, A. 1956. The effect of selection against extreme deviants based on deviation or on homozygosis. *J. Genet.* 54: 236–248.

Robertson, A. 1957. Optimum group size in progeny testing and family selection. *Biometrics* 13: 442–450.

Robertson, A. 1960. On optimum family size in selection programmes. *Biometrics* 16: 296–298.

Robertson, A. 1960*a*. A theory of limits in artificial selection. *Proc. Roy. Soc. B* 153: 234–249.

Robertson, A. 1961. Inbreeding in artificial selection programmes. *Genet. Res.* 2: 189–194.

Robertson, A. 1962. Selection for heterozygotes in small populations. *Genetics* 47: 1291–1300.

Robertson, A. 1964. The effect of non-random mating within inbred lines on the rate of inbreeding. *Genet. Res.* 5: 164–167.

Robertson, A. 1965. The interpretation of genotypic ratios in domestic animal populations. *Animal Prod.* 7: 319–324.

Robertson, A. 1967. Animal breeding. *Ann. Review of Genet.* 1: 295–312.

Robertson, A. 1967*a*. The nature of quantitative genetic variation. *Heritage From Mendel*, Ed. R. A. Brink. University of Wisconsin Press, Madison. Pp. 265–280.

Robertson, A. 1969. The theory of animal breeding. *Proc. XII Intern. Cong. Genet.* 3: 371–377.

Robertson, A., and I. M. Lerner. 1949. The heritability of all-or-none traits: viability of poultry. *Genetics* 34: 395–411.

Robinson, P., and D. F. Bray. 1965. Expected effects on the inbreeding coefficient and rate of gene loss of four methods of reproducing finite diploid populations. *Biometrics* 21: 447–458.

Rosado, J. M. C., and A. Robertson. 1966. The genetic control of sex ratio. *J. Theor. Biol.* 13: 324–329.

Rosenzweig, M. L., and R. H. MacArthur. 1963. Graphical representation and stability conditions of predator-prey interactions. *Amer. Natur.* 97: 209–223.

Ryan, F. J. 1953. Natural selection in bacterial populations. *Atti del VI Cong. Int. Microbiol.* 1: 1–9.

Sacks, J. M. 1967. A stable equilibrium with minimum average fitness. *Genetics* 56: 705–708.

Sager, R. 1966. Mendelian and non-Mendelian heredity; a reappraisal. *Proc. Roy. Soc. B* 164: 290–297.

Sakai, K. 1955. Competition in plants and its relation to selection. *Cold Spring Harbor Symp. Quant. Biol.* 20: 137–157.

Sandler, L., and E. Novitski. 1957. Meiotic drive as an evolutionary force. *Amer. Natur.* 91: 105–110.

Sanghvi, L. D. 1963. The concept of genetic load: a critique. *Amer. J. Hum. Genet.* 15: 298–309.

Schäfer, W. 1937. Über die Zunahme der Isozygotie (Gleicherbarkeit) bei fortgesetzer Bruder-Schwester-Inzucht. *Zeit. ind. Abst. Vererb.* 72: 50–78.

Scheuer, P. A. G., and S. P. H. Mandel. 1959. An inequality in population genetics. *Heredity* 13: 519–524.

Schmalhausen, I. I. 1949. *Factors of Evolution.* Blakiston, Philadelphia.

Schmalhausen, I. I. 1960. Evolution and cybernetics. *Evolution* 14: 509–524.

Schnell, F. W. 1961. Some general formulations of linkage effects in inbreeding. *Genetics* 46: 947–957.

Schnell, F. W. 1963. The covariances between relatives in the presence of linkage. *Statistical Genetics and Plant Breeding*. Ed. W. D. Hanson and H. F. Robinson. Humphrey, New York. Pp. 468–483.

Schnell, F. W. 1965. Die Covarianz zwischen Verwandten in einer gen-orthogonal Population. I. Allgemeiner Theorie. *Biom. Zeit.* 7: 1–54.

Schull, W. J., and J. V. Neel. 1965. *The Effects of Inbreeding on Japanese Children.* Harper & Row, New York.

Scudo, F. M. 1964. Sex population genetics. *La Ricerca Scient.* 34: 93–146.

Scudo, F. M. 1967. L'accoppiamento assortativo basato sul tenotipo di parenti; elcune consequenze in popoluzioni. *Atti. Ast. Lomb. B* 101: 435–455.

Scudo, F. M. 1967*a*. The adaptive value of sexual dimorphism. I. Anisogamy. *Evolution* 21: 285–291.

Scudo, F. M. 1967*b*. Selection on both haplo and diplophase. *Genetics* 56: 693–704.

Scudo, F. M. 1967*c*. Criteria for the analysis of multifactorial sex-determination. *Ital. J. Zool.* 1: 1–21.

Scudo, F. M. 1968. On mixtures of inbreeding systems. *Heredity* 23: 142–143.

Scudo, F. M. 1969. On the adaptive value of sexual dimorphism. II. Unisexuality. *Evolution* 23: 36–49.

Scudo, F. M., and S. Karlin. 1969. Assortative mating based on phenotype. I. Two alleles with dominance. *Genetics* 63: 479–498.

Searle, S. R. 1966. *Matrix Algebra for the Biological Sciences.* Wiley, New York.

Searle, S. R. 1961. Phenotypic, genetic, and environmental correlations. *Biometrics* 17: 474–480.

Searle, S. R. 1965. The value of indirect selection: I. Mass selection. *Biometrics* 21: 682–707.

Seiger, M. B. 1967. A computer simulation of the influence of imprinting on population structure. *Amer. Natur.* 101: 47–57.

Sen, S. N. 1960. Complete selection with partial self-fertilization. *J. Genet.* 57: 339–344.

Sen, S. N. 1964. Selection after crossing two homozygous stocks in a partially self-fertilized population. *J. Genet.* 59: 69–76.

Sen, S. N. 1966. Selection in a mixed population subjected to apartheid. *J. Genet.* 59: 250–253.

Seyffert, W. 1960. Theoretische Untersuchungen über die Zusammensetzung tetrasomer Population. I. Panmixie. *Biom. Zeit.* 2: 1–44.

Seyffert, W. 1960*a*. Theoretische Untersuchungen über die Zusammensetzung tetrasomer Populationen. II. Selbstbefruchtung. *Zeit. Vererb.* 90: 356–374.

Seyffert, W. 1966. Die Simulation quantitativer Merkmale durch Gene mit biochemisch definierbarer Wirkung. I. Ein einfaches Modell. *Der Zuchter* 36: 195–163.

Shaw, R. F. 1958. The theoretical genetics of the sex ratio. *Genetics* 43: 149–163.

Shaw, R. F., and J. D. Mohler. 1953. The selective significance of the sex ratio. *Amer. Natur.* 87: 337–342.

Sheppard, P. M. 1953. Polymorphism, linkage, and the blood groups. *Amer. Natur.* 87: 283–294.

Sheppard, P. M. 1956. Ecology and its bearing on population genetics. *Proc. Roy. Soc. B.* 145: 308–315.

Sheppard, P. M. 1958. *Natural Selection and Heredity.* Hutchinson and Co., London.

Sheppard, P. M., and E. B. Ford. 1966. Natural selection and the evolution of dominance. *Heredity* 21: 139–147.

Shikata, M. 1963. Representation and calculation of selfed population by group ring. *J. Theor. Biol.* 5: 142–160.

Shikata, M. 1964. Interference effect of crossovers in selfed populations. *J. Theor. Biol.* 7: 181–223.

Shikata, M. 1965. A generalization of the inbreeding coefficient. *Biometrics* 21: 665–681.

Shikata, M. 1966. Transformation of generalized inbreeding coefficient in components of pedigree. *J. Theor. Biol.* 10: 11–14.

Shikata, M. 1966*a*. Calculations of inbreeding coefficient and gene-set probability in diploid populations. *J. Theor. Biol.* 10: 15–27.

Shikata, M. 1966*b*. Cross-over effect in homozygosity by descent. *J. Theor. Biol.* 10: 196–208.

Shikata, M. 1968. Recombination effect on consanguinity: Generalized inbreeding coefficient in first cousin mating with two linked loci. *Jap. J. Human Genet.* 13: 1–9.

Shimbel, A. 1965. Information theory and genetics. *Bull. Math. Biophys.* 27: 177–181.

Singh, M., and R. C. Lewontin. 1966. Stable equilibria under optimizing selection. *Proc. Nat. Acad. Sci.* 56: 1345–1348.

Skellam, J. G. 1948. The probability distribution of gene-differences in relation to selection, and random extinction. *Proc. Camb. Phil. Soc.* 45: 364–367.

Skellam, J. G. 1949. The probability distribution of gene differences in relation to selection, mutation, and random extinction. *Proc. Camb. Phil. Soc.* 45: 364–367.

Skellam, J. G. 1951. Random dispersal in theoretical populations. *Biometrika* 38: 196–218.

Skellam, J. G. 1951*a*. Gene dispersion in heterogeneous populations. *Heredity* 5: 433–435.

Slatis, H. M. 1960. An analysis of inbreeding in the European bison. *Genetics* 45: 275–287.

Slobodkin, L. B. 1953. An algebra of population growth. *Ecology* 34: 513–519.

Slobodkin, L. B. 1958. Meta-models in theoretical ecology. *Ecology* 39: 550–551.

Slobodkin, L. B. 1960. Ecological energy relationships at the population level. *Amer. Natur.* 94: 213–236.

Slobodkin, L. B. 1961. Preliminary ideas for a predictive theory of ecology. *Amer. Natur.* 95: 147–153.

Slobodkin, L. B. 1962. *Growth and Regulation of Animal Populations.* Holt, Rinehart and Winston, New York.

Slobodkin, L. B. 1964. The strategy of evolution. *Amer. Sci.* 52: 342–357.

Smith, C. A. B. 1966. *Biomathematics.* Hafner, New York.

Smith, C. A. B. 1967. Notes on the gene frequency estimation with multiple alleles. *Ann. Hum. Genet.* 31: 99–107.

Smith, C. A. B. 1969. Local fluctuations in gene frequencies. *Ann. Hum. Genetics* 32: 251–260.

Snyder, L. H. 1947. The principles of gene distribution in human populations. *Yale J. Biol. Med.* 19: 817–833.

Snyder, L. H., and C. W. Cotterman. 1936. Studies in human inheritance XIII. A table to determine the expected proportion of females showing a sex-influenced character corresponding to any given proportion of males showing the character. *Genetics* 21: 79–83.

Sperlich, D. 1966. Equilibria for inversions induced by X-rays in isogenic strains of *Drosophila pseudoobscura*. *Genetics* 53: 835–842.

Spetner, L. M. 1964. Natural selection: an information-transmission mechanism for evolution. *J. Theor. Biol.* 7: 412–429.

Spiegelman, S. *et al.* 1969. Chemical and mutational studies of a replicating RNA molecule. *Proc. XII Intern. Congr. Genetics* 3: 127–154.

Spiess, E. 1968. Experimental population genetics. *Ann. Rev. Genet.* 2: 165–208.

Spofford, J. B. 1969. Heterosis and the evolution of duplications. *Amer. Natur.* 103: 407–432.

Sprott, D. A. 1957. The stability of a sex-linked allelic system. *Ann. Hum. Genet.* 22: 1-6.

Stahl, F. W., and N. E. Murray. 1966. The evolution of gene clusters and genetic circularity. *Genetics* 53: 569–576.

Stanton, R. G. 1946. Filial and fraternal correlations in successive generations. *Ann. Eugen.* 13: 18–24.

Stanton, R. G. 1960. Genetic correlations with mutiple alleles. *Biometrics* 16: 235–244.

Stern, C. 1943. The Hardy–Weinberg Law. *Science* 97: 137–138.

Stratton, J. A., P. M. Morse, L. J. Chu, and P. A. Hutner. 1941. *Elliptic, Cylinder and Spheroidal Wave Functions.* Wiley, New York.

Stratton, J. A., P. M. Morse, L. J. Chu, J. D. C. Little, and F. J. Corbató. 1956. *Spheroidal Wave Functions.* Technology Press of M.I.T. & Wiley, New York.

Streams, F. A., and D. Pimentel. 1961. Effects of immigration on the evolution of populations. *Amer. Natur.* 95: 201–210.

Stuber, C. W., and C. C. Cockerham. 1966. Gene effects and variants in hybrid populations. *Genetics* 54: 1279–1286.

"Student." 1929. Evolution by selection. *J. Agric. Res.* 39: 451–476.

Sturtevant, A. H. 1918. An analysis of the effects of selection. *Carn. Inst. Wash. Publ.* No. 264, pp. 1–68.

Sturtevant, A. H. 1948. The evolution and function of genes. *Amer. Scient.* 36: 225–236.

Sturtevant, A. H. 1937, 1938. Essays on evolution. I. On the effects of selection on mutation rate. II. On the effects of selection on social insects. III. On the origin of interspecific sterility. *Quart. Rev. Biol.* 12: 464–467; 13: 74–76, 333–335.

Sturtevant, A. H., and K. Mather. 1938. The interrelations of inversions, heterosis, and recombination. *Amer. Natur.* 72: 447–452.

Sueoka, N. 1962. On the genetic basis of variation and heterogeneity of DNA base composition. *Proc. Nat. Acad. Sci.* 48: 582–592.

Sutter, J., and L. Tabah. 1951. La mesure de l'endogamie et ses applications demographiques. *J. Soc. Stat. Paris.* 92: 243–267.

Sved, J. A. 1964. The average recombination frequency per chromosome. *Genetics* 49: 367–371.

Sved, J. A. 1968. The stability of linked systems of loci with a small population size. *Genetics* 59: 543–563.

Sved, J. A. 1968*a*. Possible rates of gene substitution in evolution. *Amer. Natur.* 102: 283–292.

Sved, J. A., T. E. Reed, and W. F. Bodmer. 1967. The number of balanced polymorhpisms that can be maintained in a natural population. *Genetics* 55: 469–481.

Tallis, G. M. 1962. A selection index for optimum genotype. *Biometrics* 18: 120–122.

Tallis, G. M. 1966. Equilibria under selection for k alleles. *Biometrics* 22: 121–127.

Teissier, G. 1944. Équilibre des gènes léthaux dans les populations stationnaires panmictique. *Rev. Scient.* 82: 145–159.

Teissier, G. 1954. Conditions d'équilibre d'un couple d'alleles et supériorité des hétérozygotes. *Comp. Rend. Acad. Sci.* 238: 621–623.

Thoday, J. M. 1953. Components of fitness. *Symp. Soc. Exp. Biol.* 7: 96–113.

Thompson, J. B., and H. Rees. 1956. Selection for heterozygotes during inbreeding. *Nature* 177: 385–386.

Thorpe, W. H. 1945. The evolutionary significance of habitat selection. *J. Anim. Ecol.* 14: 67–70.

Tietze, H. 1923. Über das Schicksal gemischter Populationen nach den mendelischen Vererbungsgesetzen. *Zeit. angew. Math. Mech.* 3: 362–393.

Tomlinson, J. 1966. The advantages of hermaphroditism and parthenogenesis. *Theor. Biol.* 11: 54–58.

Trustrum, G. B. 1961. The correlation between relatives in a random mating diploid population. *Proc. Camb. Phil. Soc.* 57: 315–320.

Tukey, J. W. 1954. Causation, regression and path analysis. *Statistics and Mathematics in Biology*. Ed. by O. Kempthorne, T. A. Bancroft, J. W. Gowen, and J. L. Lush. Iowa State College Press, Ames, Iowa. Pp. 35–66.

Turner, J. R. G. 1967. Fundamental theorem of natural selection. *Nature* 215: 1080.

Turner, J. R. G. 1967*a*. Why does the genotype not congeal? *Evolution* 21: 645–656.

Turner, J. R. G. 1967*b*. Mean fitness and the equilibria in multilocus polymorphisms. *Proc. Roy. Soc. London B.* 169: 31–58.

Turner, J. R. G. 1967*c*. The effect of mutation on fitness in a system of two coadapted loci. *Ann. Hum. Genet.* 30: 329–334.

Turner, J. R. G. 1967*d*. On supergenes. I. The evolution of supergenes. *Amer. Natur.* 101: 195–221.

Turner, J. R. G. 1968. Natural selection for and against a polymorphism which interacts with sex. *Evolution* 22: 481–495.

Turner, J. R. G. 1968*a*. On supergenes. II. The estimation of gametic excess in natural populations. *Genetics* 39: 82–93.

Turner, J. R. G. 1968*b*. How does treating congenital diseases affect the genetic load? *Eugen. Quart.* 15: 191–197.

Turner, J. R. G. 1969. The basic theorems of natural selection: A naive approach. *Heredity* 24: 75–84.

Turner, J. R. G., and M. H. Williamson. 1968. Population size, natural selection, and the genetic load. *Nature* 218: 700.

Utida, S. 1957. Population fluctuation, an experimental and theoretical approach. *Cold Spring Harbor Symp. Quant. Biol.* 22: 139–151.

Vandermeer, J. H. 1968. Reproductive value in a population of arbitrary age distribution. *Amer. Natur.* 102: 586–589.

Vandermeer, J. H., and M. H. MacArthur. 1966. A reformulation of alternative (b) of the broken stick model of species abundance. *Ecology* 47: 139–140.

Van der Veen, J. H. 1960. Heterozygote superiority, selection intensity, and plateauing. *Heredity* 15: 321–323.

Van der Veen, J. H. 1960*a*. Ein geinduceerd suboptimaal evenwicht bij massaselektie. *Genen en Phaenen* 5: 49–52.

Van Valen, L. 1960. Nonadaptive aspects of evolution. *Amer. Natur.* 94: 305–308.

Van Valen, L. 1963. Haldane's dliemma, evolutionary rates and heterosis. *Amer. Natur.* 97: 185–190.

Van Valen, L. 1965. Selection in natural populations. III. Measurement and estimation. *Evolution* 19: 514–528.

Van Valen, L., and R. Levins. 1968. The origins of inversion polymorphisms. *Amer. Natur.* 102: 5–24.

Varley, G. C. 1947. The natural control of population balance in the knapweed gall-fly (*Urophora jaceana*). *J. Anim. Ecol.* 16: 139–187.

Verner, J. 1965. Selection for sex ratio. *Amer. Natur.* 99: 419–421.

Visconti, N., and M. Delbruck. 1953. The mechanism of genetic recombination in phage. *Genetics* 38: 5–33.

Volterra, V. 1926. *La Lutte Pour La Vie.* Gauthier, Paris.

Von Hofstein, N. 1951. The genetic effect of negative selection in man. *Hereditas* 37: 157–265.

Waaler, G. H. M. 1927. Über die Erbishkeitsverhältnisse der verschiedenen Arten von angeborener Rotgrünblindheit. *Zeit. ind. Abst. Vererb.* 45: 279–333.

Waddington, C. H. 1957. *The Strategy of the Genes.* Allen and Unwin, London.

Wahlund, S. 1928. Zuzammensetzung von Populationen und Korrelationserscheinungen vom Standpunkt der Vererbungslehre aus betrachtet. *Hereditas* 11: 65–106.

Wallace, B. 1953. On coadaptation in Drosophila. *Amer. Natur.* 87: 343–358.

Wallace, B. 1958. The role of heterozygosity in Drosophila populations. *Proc. Xth Intern. Cong. Genet.* 1: 408–419.

Wallace, B. 1958*a*. The comparison of observed and calculated zygotic distributions. *Evolution* 12: 113–115.

Wallace, B. 1963. Genetic diversity, genetic uniformity, and heterosis. *Canad. J. Genet. Cytol.* 5: 239–253.

Wallace, B. 1968. *Topics in Population Genetics.* W. W. Norton, New York.

Wangersky, P. J., and W. J. Cunningham. 1956. On time lags in equations of growth. *Proc. Nat. Acad. Sci.* 42: 699–702.

Wangersky, P. J., and W. J. Cunningham. 1957. Time lag in population models. *Cold Spring Harbor Symp. Quant. Biol.* 22: 329–338.

Warburton, F. E. 1967. Increase in the variance of fitness due to selection. *Evolution* 21: 197–198.

Warren, H. D. 1917. Numerical effects of natural selection acting on Mendelian characters. *Genetics* 2: 305–312.

Watson, G. S., and E. Caspari. 1960. The behavior of cytoplasmic pollen sterility in populations. *Evolution* 14: 56–63.

Watt, K. E. F. 1959. A mathematical model for the effect of densities of attacked and attacking species on the number attacked. *Canad. Entom.* 91: 129–144.

Watt, K. E. F. 1962. Use of mathematics in population ecology. *Ann. Rev. Entom.* 7: 243–260.

Watterson, G. A. 1959. Non-random mating, and its effect on the rate of approach to homozygosity. *Ann. Hum. Genet.* 23: 204–220.

Watterson, G. A. 1959*a*. A new genetic population model, and its approach to homozygosity. *Ann. Hum. Genet.* 23: 221–232.

Watterson, G. A. 1961. Markov chains with absorbing states: a genetic example. *Ann. Math. Stat.* 32: 716–729.

Watterson, G. A. 1962. Some theoretical aspects of diffusion theory in population genetics. *Ann. Math. Stat.* 33: 939–957.

Watterson, G. A. 1964. The application of diffusion theory to two population genetic models of Moran. *J. Appl. Prob.* 1: 233–246.

Weber, E. 1959. The genetical analysis of characters with continuous variability on a mendelian basis. *Genetics* 44: 1131–1139.

Weber, E. 1960. The genetical analysis of characters with continuous variability on a mendelian basis. II. Monohybrid segregation and linkage analysis. *Genetics* 45: 459–466.

Weber, E. 1960*a*. The genetical analysis of characters with continuous variability on a mendelian basis. III. Dihybrid segregation. *Genetics* 45: 567–572.

Weinberg, W. 1908. Über den Nachweis der Vererbung beim Menschen. *Jahresh. Verein f. vaterl. Naturk. Württem.* 64: 368–382.

Weinberg, W. 1909. Über Vererbungsgesetze beim Menschen. *Zeit. ind. Abst. Vererb.* 1: 277–330. 1 :377–392, 440–460; 2: 276–330.

Weinberg, W. 1910. Weiteres Beiträge zur Theorie der Vererbung. *Arch. Rass. Ges. Biol.* 7: 35–49, 169–173.

Weiss, G. H. 1963. Comparison of a deterministic and a stochastic model for interaction between antagonistic species. *Biometrics* 19: 595–602.

Weiss, G. H., and M. Kimura. 1965. A mathematical analysis of the stepping stone model of genetic correlation. *J. Appl. Prob.* 2: 129–149.

Weldon, W. F. R. 1901. A first study of natural selection in *Clausilia laminata* (Montagu). *Biometrika* 1: 109–124.

Wentworth, E. N., and B. L. Remick. 1916. Some breeding properties of the generalized Mendelian population. *Genetics* 1: 608–616.

Wigan, L. G. 1944. Balance and potence in natural populations. *J. Genet.* 46: 150–160.

Willham, R. L. 1963. The covariance between relatives for characters composed of components contributed by related individuals. *Biometrics* 19: 18–27.

Williams, E. J. 1961. The growth and age-distribution of a population of insects under uniform conditions. *Biometrics* 17: 349–358.

Williams, G. C. 1957. Pleitropy, natural selection, and the evolution of senescence. *Evolution* 11: 398–412.

Williams, G. C. 1966. *Adaptation and Natural Selection*. Princeton University Press, Princeton, N.J.

Williams, G. C. 1966*a*. Natural selection, the costs of reproduction, and a refinement of Lack's principle. *Amer. Natur.* 100: 687–690.

Williams, G. C., and D. C. Williams. 1957. Natural selection of individually harmful social adaptations among sibs with special reference to social insects. *Evolution* 11: 32–39.

Williamson, M. H. 1958. Selection, controlling factors, and polymorphism. *Amer. Natur.* 92: 329–335.

Wills, C., J. Crenshaw, and J. Vitale. 1969. A computer model allowing maintenance of large amounts of genetic variability in mendelian populations. 1. Assumptions and results for large populations. *Genetics* (in press).

Willson, M. F., and E. R. Pianka. 1963. Sexual selection, sex ratio and mating system. *Amer. Natur.* 97: 405–407.

Wilson, S. P., W. H. Kyle, and A. E. Bell. 1966. The influence of mating systems on parent-offspring regression. *J. Hered.* 57: 124–125.

Woodger, J. H. 1965. Theorems on random evolution. *Bull. Math. Biophys.* 27: 145–150.

Workman, P. L. 1964. The maintenance of heterozygosity by partial negative assortative mating. *Genetics* 50: 1369–1382.

Workman, P. L., and R. W. Allard. 1962. Population studies in predominantly self-pollinated species. III. A matrix model for mixed selfing and random outcrossing. *Proc. Nat. Acad. Sci.* 48: 1318–1325.

Workman, P. L., and S. K. Jain. 1966. Zygotic selection under mixed random mating and self-fertilization: Theory and problems of estimation. *Genetics* 54: 159–171.

Wright, S. 1917. The average correlation within subgroups of a population. *J. Wash. Acad. Sci.* 7:532–535.

Wright, S. 1921. Systems of mating. I. The biometric relations between parent and offspring. *Genetics* 6: 111–123.

Wright, S. 1921*a*. Systems of mating. II. The effects of inbreeding on the genetic composition of a population. *Genetics* 6: 124–143.

Wright, S. 1921*b*. System of mating. III. Assortative mating based on somatic resemblance. *Genetics* 6: 144–161.

Wright, S. 1921*c*. Systems of mating. IV. The effects of selection. *Genetics* 6: 162–166.

Wright, S. 1921*d*. Systems of mating. V. General considerations. *Genetics* 6: 167–178.

Wright, S. 1921*e*. Correlation and causation. *J. Agric. Res.* 20: 557–585.

Wright, S. 1922. Coefficients of inbreeding and relationship. *Amer. Natur.* 56: 330–338.

Wright, S. 1922*a*. The effects of inbreeding and crossbreeding on guinea pigs. *Bull. U.S. Dept. Agric.* 1121: 1–59.

Wright, S. 1923. The theory of path coefficients. *Genetics* 8: 239–255.

Wright, S. 1923*a*. Mendelian analysis of the pure breeds of livestock. I. The measurement of inbreeding and relationship. *J. Hered.* 14: 339–348.

Wright, S. 1923*b*. Mendelian analysis of the pure breeds of livestock. II. The Duchess family of shorthorns as bred by Thomas Bates. *J. Hered.* 14: 405–422.

Wright, S. 1929. The evolution of dominance. Comment on Dr. Fisher's reply. *Amer. Natur.* 63: 1–5.

Wright, S. 1929*a*. Fisher's theory of dominance. *Amer. Natur.* 63: 274–279.

Wright, S. 1929*b*. The evolution of dominance. *Amer. Natur.* 63: 556–561.

Wright, S. 1931. Evolution in Mendelian populations. *Genetics* 16: 97–159.

Wright, S. 1932. The roles of mutation, inbreeding, cross-breeding and selection in evolution. *Proc. VI Int. Cong. Genet.* 1: 356–366.

Wright, S. 1933. Inbreeding and recombination. *Proc. Nat. Acad. Sci.* 19: 420–433.

Wright, S. 1933*a*. Inbreeding and homozygosis. *Proc. Nat. Acad. Sci.* 19: 411–420.

Wright, S. 1934. Physiological and evolutionary theories of dominance. *Amer. Natur.* 68: 25–53.

Wright, S. 1934*a*. The method of path coefficients. *Ann. Math. Stat.* 5: 161–215.

Wright, S. 1934*b*. Professor Fisher on the theory of dominance. *Amer. Natur.* 68: 562–565.

Wright, S. 1935. The analysis of variance and the correlations between relatives with respect to deviations from an optimum. *J. Genet.* 30: 243–256.

Wright, S. 1935*a*. Evolution in populations in approximate equilibrium. *J. Genet.* 30: 257–266.

Wright, S. 1937. The distribution of gene frequencies in populations. *Proc. Nat. Acad. Sci.* 23: 307–320.

Wright, S. 1938. The distribution of gene frequencies under irreversible mutation. *Proc. Nat. Acad. Sci.* 24: 253–259.

Wright, S. 1938*a*. The distribution of gene frequencies in populations of polyploids. *Proc. Nat. Acad. Sci.* 24: 372–377.

Wright, S. 1938*b*. Size of population and breeding structure in relation to evolution. *Science* 87: 430–431.

Wright, S. 1939. The distribution of self-sterility alleles in populations. *Genetics* 24: 538–552.

Wright, S. 1939*a*. Statistical genetics in relation to evolution. *Act. scient. et indus.*, 802: 1–63.

Wright, S. 1940. Breeding structure of populations in relation to speciation. *Amer. Natur.* 74: 232–248.

Wright, S. 1941. The probability of fixation of reciprocal translocations. *Amer. Natur.* 75: 513–522.

Wright, S. 1942. Statistical genetics and evolution. *Bull. Amer. Math. Soc.* 48: 223–246.

Wright, S. 1943. Isolation by distance. *Genetics* 28: 114–138.

Wright, S. 1945. Tempo and mode in evolution: A critical review. *Ecology* 26: 415–419.

Wright, S. 1945*a*. The differential equation of the distribution of gene frequencies. *Proc. Nat. Acad. Sci.* 31: 382–389.

Wright, S. 1946. Isolation by distance under diverse systems of mating. *Genetics* 31: 39–59.

Wright, S. 1948. On the roles of directed and random changes in gene frequency in the genetics of populations. *Evolution* 2: 279–294.

Wright, S. 1948*a*. Evolution, organic. *Encyclopedia Britannica* 8: 915–929.

Wright, S. 1948*b*. Genetics of populations. *Encyclopedia Britannica* 10: 111–112.

Wright, S. 1949. Adaptation and selection. *Genetics, Paleontology, and Evolution.* Ed. by G. L. Jepson, G. G. Simpson, and E. Mayr. Princeton Univ. Press, Princeton, N.J. Pp. 365–389.

Wright, S. 1949*a*. Population structure and evolution. *Proc. Amer. Phil. Soc.* 93: 471–478.

Wright, S. 1951. The genetical structure of populations. *Ann. Eugen.* 15: 323–354.

Wright, S. 1951*a*. Fisher and Ford on "The Sewall Wright Effect." *Amer. Scient.* 39: 452–479.

Wright, S. 1952. The genetics of quantitative variability. *Quantitative Inheritance.* Her Majesty's Stationery Office, London, Pp. 5–41.

Wright, S. 1952*a*. The theoretical variance within and among subdivisions of a population that is in a steady state. *Genetics* 37: 313–321.

Wright, S. 1953. Gene and organism. *Amer. Natur.* 87: 5–18.

Wright, S. 1954. The interpretation of multivariate systems. *Statistics and Mathematics in Biology.* Ed. by O. Kempthorne, W. A. Bancroft, J. W. Gowen. and J. L. Lush. Iowa State College Press, Ames, Iowa. Pp. 11–33.

Wright, S. 1955. Classification of the factors of evolution. *Cold Spring Harbor Symp. Quant. Biol.* 20: 16–24.

Wright, S. 1956. Modes of selection. *Amer. Natur.* 40: 5–24.

Wright, S. 1960. Physiological genetics, ecology of populations, and natural selection. *The Evolution of Life.* Ed. by Sol Tax. Univ. of Chicago Press, Chicago, Ill. 1: 429–475.

Wright, S. 1960*a*. On the number of self-incompatibility alleles maintained in equilibrium by a given mutation rate in a population of given size: a reexamination. *Biometrics* 16: 61–85.

Wright, S. 1960*b*. Path coefficients and path regressions: alternative or complementary concepts? *Biometrics* 16: 189–202.

Wright, S. 1960*c*. The treatment of reciprocal interaction, with or without lag, in path analysis. *Biometrics* 16: 423–445.

Wright, S. 1963. Discussion of systems of mating used in mammalian genetics. *Methodology in Mammalian Genetics.* Ed. by W. J. Burdette. Holden-Day, San Francisco. Pp. 41–53.

Wright, S. 1964. Pleiotropy in the evolution of structural reduction and dominance. *Amer. Natur.* 98: 65–69.

Wright, S. 1964*a*. Stochastic processes in evolution. *Stochastic Models in Medicine and Biology.* Ed. by John Gurland. Univ. of Wisconsin Press, Madison, Wisc. Pp. 199–244.

Wright, S. 1964*b*. The distribution of self-incompatibility alleles in populations. *Evolution* 18: 609–619.

Wright, S. 1965. Factor interaction and linkage in evolution. *Proc. Roy. Soc. B* 162: 80–104.

Wright, S. 1965*a*. The interpretation of population structure by F-statistics with special regard to systems of mating. *Evolution* 19: 395–420.

Wright, S. 1966. Polyallelic random drift in relation to evolution. *Proc. Nat. Acad. Sci.* 55: 1074–1080.

Wright, S. 1967. "Surfaces" of selective value. *Proc. Nat. Acad. Sci.* 58: 165–172.

Wright, S. 1967*a*. The foundations of population genetics. *Heritage From Mendel*, Ed. R. A. Brink. Univ. of Wisconsin Press, Madison. Pp. 245–264.

Wright, S. 1968. *Evolution and the Genetics of Populations. Vol. I. Genetic and Biometric Foundations*. Univ. of Chicago Press, Chicago.

Wright, S. 1969. *Evolution and the Genetics of Populations. Vol. 2. The Theory of Gene Frequencies*. University of Chicago Press, Chicago.

Wright, S., and Th. Dobzhansky. 1946. Genetics of natural populations. XII. Experimental reproduction of some of the changes caused by natural selection in certain populations of *Drosophila pseudoobscura*. *Genetics* 31: 125–156.

Wright, S., Th. Dobzhansky, and W. Hovanitz. 1942. Genetics of natural populations. VII. The allelism of lethals in the third chromosome of *Drosophila pseudoobscura*. *Genetics* 27: 363–394.

Wright, S., and W. E. Kerr. 1954. Experimental studies of the distribution of gene frequencies in very small populations of *Drosophila melangaster*. II. Bar. *Evolution* 8: 225–240.

Wright, S., and H. C. McPhee. 1925. An approximate method of calculating coefficients of inbreeding and relationship. *J. Agric. Res.* 31: 377–383.

Yanase, T. 1964. A note on the patterns of migration in isolated populations. *Jap. J. Hum. Genet.* 9: 136–152.

Yasuda, N. 1968. Estimation of inbreeding coefficient from phenotype frequencies by a method of maximum likelihood scoring. *Biometrics* 24: 915–936.

Yasuda, N. 1968*a*. An extension of Wahlund's principle to evaluate mating type frequency. *Amer. J. Hum. Genet.* 20: 1–23.

Yasuda, N. 1969. The estimation of the variance effective population number based on gene frequency. *Jap. J. Hum. Genet.* 14: 10–16.

Yasuda, N., and M. Kimura. 1968. A gene-counting method of maximum likelihood for estimating gene frequencies in ABO and ABO-like systems. *Ann. Hum. Genet.* 31: 409–420.

Yntema, L. 1952. *Mathematical Models of Demographic Analysis*. J. J. Groen and Zoon, Leiden.

Young, S. S. Y. 1961. A further examination of the relative efficiency of three methods of selection for genetic gains under less restricted conditions. *Genet. Res.* 2: 106–121.

Young, S. S. Y. 1966. Computer simulation of directional selection in large populations. I. The programme, the additive and dominance models. *Genetics* 53: 189–205.

Young, S. S. Y., and H. Weiler. 1960. Selection for two correlated traits by independent culling levels. *J. Genet.* 57: 329–358.

Yule, G. U. 1902. Mendel's laws and their probable relation to intra-racial heredity. *New Phytol.* 1: 192–207, 222–238.

Zirkle, C. 1926. Some numerical results of selection upon polyhybrids. *Genetics* 11: 531–583.

GLOSSARY AND INDEX OF SYMBOLS

This list gives the symbol, a brief identification or definition, and the page where it is defined or introduced.

1. Latin Symbols

Symbol	Definition	Page
A	ancestor	70
A	correction between the genic values of mates	156
$\mathbf{A}$	matrix	500
$\|\mathbf{A}\|$	determinant	502
$\mathbf{A}^{-1}$	inverse of matrix $\mathbf{A}$	
A_1, A_2, A_i, A_j	allelic genes	33
a_{ij}	average excess, or deviation from the mean, of genotype A_iA_j	206
a_i	average excess of allele A_i	211
B	Beta function	437

Symbol	Meaning	Page
B_1, B_2, B_k, B_l	allelic genes, non alleles of genes at the A locus	47
b	birth rate	7
b_x	birth rate at age x	11
b_{yx}	regression of y on x	488, 140
c	recombination or "crossover" fraction	47
C	covariance	136
C	Haldane's cost of natural selection	246
Cov	covariance	484, 108
D	deviation of heterozygote from mean of homozygotes	99
D	$p_1p_4 - p_2p_3$, a measure of linkage disequilibrium	197
E	epistatic parameter	198
E	expected value operator	107
f, f_I	inbreeding coefficient of individual I, probability of autozygosity	64
f_{IJ}	coefficient of consanguinity or kinship of I and J	
f_{IS}, f_{ST}, f_{IT}	inbreeding coefficients in a subdivided population	105
$f(x)$	probability generating function	419
f_{ij}	fertility of genotype A_iA_j	185
F	two-locus inbreeding coefficient, probability of double autozygosity	96
F	hypergeometric function	383
G	geometric mean	480
G_{ij}	genic or additive value of genotype A_iA_j	210
h	a measure of dominance	260
h^2	heritability, V_g/V_t	124
H^2	heritability, V_h/V_t	124
H	heritability, same as h^2	156
H	harmonic mean	481
H	proportion of heterozygotes	63
H_0	initial proportion of heterozygotes	64
H_t	proportion of heterozygotes at time t	64
I	selection differential	226
$\mathbf{I}$	identity matrix	501
K	carrying capacity	23
K	number of possible allelic states	453
K	measure of segregation distortion	311
K	proportion of homozygous recessives from consanguineous matings	74

k	the number of progeny from a parent	346
k_0, k_1, k_2	Cotterman k-coefficients	132
L	genetic load	299
L_u	upper confidence limit	497
L_l	lower confidence limit	497
$l(x)$	probability of survival to age x	17
M	migration coefficient (same as m on p. 390)	267
M	arithmetic mean	480
$M_{\delta x}$	mean gene frequency change in one generation	372
m	migration coefficient	390
m	proportion of males	42
m	fitness in Malthusian parameters	7
m_i	fitness of allele A_i	191
m_{ij}	fitness of genotype A_iA_j	191
$\dot{N}$	population number	5
N_e	effective population number	109, 345
$N_{e(f)}$	inbreeding effective population number	347
$N_{e(v)}$	variance effective population number	357
N_{ij}	number of genotype A_iA_j in the population	
n	number of alleles	168
n_e	effective number of alleles	324
n_e	effective number of gene loci	154
n_{xt}	number of individuals of age x at time t	11
P	panmictic index	64
P	Legendre polynomial	386
P_{ij}	proportion of genotype A_iA_j	
p	allele frequency	38
p	initial allele frequency	328
p_i	proportion of allele A_i	33
p_1, p_2, p_3, p_4	frequencies of four chromosome types	196
p_x	probability of surviving from age x to age $x+1$	
Q	a quantity to be minimized	487
q	an allele frequency, or a probability $(=1-p)$	
r	intrinsic rate of increase	23
r	correlation coefficient	487, 137
r	Wright's coefficient of relationship	69, 138
s	selective advantage	8, 182, 192
T	Gegenbauer polynomial	383

t	relative deviate	498
t	time	5
u	mutation rate	259
u_{ij}	mutation rate from A_i to A_j	263
u	probability of ultimate fixation of a mutant gene	421
$u(p,t)$	probability of fixation at generation t, given frequency p at generation 0	423
V	variance	10, 54, 482
V_d	dominance variance	120
V_e	environmental variance	120
V_g	genic variance	120
V_h	genotypic variance	120
V_i	interaction or epistatic variance	120
V_t	total variance	120
V_{gam}	gametic variance	219
V_{AA}	additive × additive epistatic variance	127
V_{AD}	additive × dominance epistatic variance	127
V_{DD}	dominance × dominance epistatic variance	127
$V_{\delta x}$	variance of the gene frequency charge in one generation	372
V_k	variance in progeny number	346
v	mutation rate, in reverse direction to that measured by u	442
v_{ij}	viability of genotype A_iA_j	185
$v(x)$	reproductive value at age x	20
w	fitness in discrete generation model	5
w_{ij}	fitness of genotype A_iA_j	179
w_i	fitness of allele A_i	179
x	age	11
x	allele frequency varying stochastically	328
Y	phenotypic value	77
Y_0	phenotypic value at $f=0$	
Y_1	phenotypic value at $f=1$	
Z	p_1p_4/p_2p_3, a measure of linkage disequilibrium	197

2. Greek Symbols

α	proportion of double reduction	52
α	departure from Hardy-Weinberg proportions	356

α_i	average effect of allele A_i	117
Γ	gamma function	391
Δ	a determinant	275
Δ	an increment, change in one generation	6, 179
δ	an increment	230
δ	Dirac delta function	383
∂	partial derivative	
θ_{ij}	P_{ij}/p_ip_j, a measure of departure from Hardy–Weinberg proportions	212
λ	eigenvalue	14, 506
λ	limiting value of H_t/H_{t-1}	88
λ_1	largest eigenvalue, rate of population growth at stable age distribution	15
μ	mean	486
μ_n	nth moment about 0	332
Π	product symbol	480
π	true probability	497
Σ	addition symbol	480
σ	standard deviation	483
ϕ	$F - f^2$	97
$\phi(p,x;t)$	the probability that the gene frequency is x at time t, given that it was p at time 0	372
ϕ_{ij}	$\log_e \theta_{ij}$	
χ^2	$\frac{\Sigma(observed\text{-}expected)^2}{expected}$	493
Ω	probability of coexistence of 2 alleles in a population	386

3. Other Symbols

!	factorial $[n! = n(n-1)(n-2)\ldots 2\cdot 1]$	491
′	used to designate the next generation	184
‾	mean value	480
^	designates an equilibrium value	146
*	designates that the quantity applies to males	45
**	quantity applies to females	45
~	harmonic mean	302

INDEX OF NAMES

INDEX OF SUBJECTS

70 71 72 73 7 6 5 4 3